Transportation Engineering and Traffic Management for Road Safety Improvement

Transportation Engineering and Traffic Management for Road Safety Improvement

By

Sandipan Goswami

Cambridge
Scholars
Publishing

Transportation Engineering and Traffic Management
for Road Safety Improvement

By Sandipan Goswami

This book first published 2026

Cambridge Scholars Publishing

Lady Stephenson Library, Newcastle upon Tyne, NE6 2PA, UK

British Library Cataloguing in Publication Data
A catalogue record for this book is available from the British Library

ISBN: 978-1-0364-7408-9
ISBN (Ebook): 978-1-0364-7409-6

"I dedicate this work to all those people across the world who lost their lives in road accidents....

and also

to all my colleagues and friends in my professional career and authors of technical papers who helped me learn and gather knowledge on the subject and taking a few excerpts from their technical papers".

Contents

PREFACE

The primary considerations for transport planning, design, engineering and management are safety, mobility and economy. This book has been in demand and long-awaited by engineering universities and institutions with civil engineering, transportation engineering, highway engineering and social science departments. There have been many technical papers published by eminent authors, but as such, no book covering all the essential items is available on the international market; therefore, this is the first reference book for all essential needs on the subject. Government transportation departments such as Federal Highway Administration, USA, Transportation Research Board (TRB), USA, 50 departments of transportation (DOTs), the National Highway Traffic Safety Administration (NHTSA with Traffic Safety Marketing), USA, the Ministry of Road Transport and Highways, India, the Central Road Research Institute, India, the Indian Academy of Highway Engineers, the Indian Roads Congress, 35 Indian institutes of technology and thousands of engineering institutions, funding agencies such as The World Bank, the Asian Development Bank, the African Development Bank, UNDP, Institution of Highways and Transportation, 6 Endsleigh Street, London WC1H 0DZ, United Kingdom, and road research centres, consultants, professional and researchers across the world, will benefit from this book.

Road safety and accidents are a burning social issue at present. This book identifies the root causes of the issues and takes countermeasures for resolving them by governments, the private sector and NGOs. It discusses safety training and various measures that each developing and developed country can implement. It will help generate public awareness, various provisions, safety campaigns, funding for implementation, educating children, residents, pedestrians, drivers, police, etc., along with engineering measures to a certain extent.

Road accidents cause a significant loss of lives and assets. The cost to every government in countries for compensating victims is significant and is also discussed in this book. This book is intended to be used by senior government officials at the director general or secretary level in sectors who have responsibility for road safety. The book can also act as a basic source of reference for faculties in institutions, aid agency officials and other decision makers who are often assigned the job of improving road safety; they may be at least familiar with the key issues, elements, and opportunities relating to the improvement of road safety.

The article "Traffic Accidents Take Toll on Social, Economic Progress, General Assembly Agrees, Adopting Resolution Improving Road Safety" issued by the United Nations Department of Public Information states that adopting a resolution to improve road safety as deaths and injuries from road traffic accidents (RTAs) pose a serious threat to global health and have a negative impact on social and economic progress, as well as sustainable development.

This book consists of eight chapters followed by the three appendices A, B and C. The arrangements are described in the following paragraphs: the improvement of road safety through a multidisciplinary and comprehensive approach, measures that need to be undertaken in each sector and the implementation of the three priority actions in each sector to initiate improvements towards and strengthen the ability of each country to begin to address its safety problems.

Appendix A mentions important software products with their websites and various documents of importance to all readers.

Appendix B includes contacts of international organisations dealing with road safety, encouraging the formation of a network of road safety professionals and experts mainly in the Asian and Pacific regions.

Appendix C mentions the chapter-wise references.

An intelligent transport system (ITS) uses advanced technology for traffic management and is the starting point of an overall improvement in the traffic system that has a significant and effective contribution towards improving road safety. The various issues are discussed in detail along with the technology presently available.

A green highway project enhances road safety, which can be achieved by adopting a natural environment through planting vegetation. Electric vehicles can be better monitored by ITS-usable systems and devices, even when stationary. Using a solar panel may keep the cells charged. The upgrade, rehabilitation and new construction of roadway infrastructure projects could have green cover and use reusable or recycled material.

Further reading

"Traffic Accidents Take Toll on Social, Economic Progress, General Assembly Agrees, Adopting Resolution Improving Road Safety" issued by the United Nations Department of Public Information https://www.un.org/press/en/2014/ga11498.doc.htm

World Bank: road crashes are the biggest safety challenge for younger people.

https://blogs.worldbank.org/endpovertyinsouthasia/road-crashes-are-biggest-safety-challenge-youths

ResearchGate on Social Justice Approach to Road Safety in Kenya: Addressing the Uneven Distribution of Road Traffic Injuries and Deaths across Population Groups

https://www.researchgate.net/publication/45389690_Social_Justice_Approach_to_Road_Safety_in_Kenya_Addressing_the_Uneven_Distribution_of_Road_Traffic_Injuries_and_Deaths_across_Population_Groups

1. Springer Open Declaration on Road Safety: challenges and opportunities in Latin America and the Caribbean. An estimated 1.3 million people die in traffic accidents each year worldwide, and millions more are injured, with developing countries disproportionately affected. It is predicted that the number of global traffic deaths will be around 1.8 million annually by 2030, making it the eighth leading cause of death in the world. This paper provides an overview of salient road safety issues in the Latin America and Caribbean (LAC) region. It also reviews existing theoretical and causal empirical evidence of interventions to improve road safety outcomes, concentrating on three core areas: safer roads and mobility, safer vehicles, and safer road users. This paper discusses promising areas for policy experimentation and future research.

https://latinaer.springeropen.com/articles/10.1186/s40503-019-0078-0

2. Clinical Epidemiology and Global Health Despite flows in Transport Policy, it is noticed that many more proven wide social factors distressing the number of road traffic accidents, including injuries and deaths. A wider social determinant among low socioeconomic groups,

including identity crises, largely contributes towards an increased rate of road accidents in India.

https://cegh.net/article/S2213-3984(20)30159-7/fulltext

3. Asian Development Bank (ADB) on Technical Assistance for Socioeconomic Impact of Road Crashes (Financed by the Poverty Reduction Cooperation Fund)

https://www.adb.org/sites/default/files/project-document/69005/tar-stu-38081.pdf

ACKNOWLEDGEMENTS

Diana Marie T. Hernandez Transport Officer
Sector Advisory Service Cluster
Sustainable Development and Climate Change Department, Asian Development Bank
Tel: +6328683-1817
www.adb.org

For permitting me to use figures, tables and text excerpts from the *ADB Guidelines on Road Safety*.

I credit the whole report when using some figures/text from the report.

Techsoft Engineering Services for permitting me to use figures, tables and text excerpts from various documents in their offices.

www.techsoftgloblal.com

PROFILE OF: SANDIPAN GOSWAMI

Sandipan Goswami received his Bachelor's Degree in Science from the University of Calcutta in 1978, Bachelor's Degree in Civil Engineering from the University of Calcutta in 1983 and Master's Degree in Civil Engineering, with specialization in Soil Mechanics and Foundation Engineering (Geotechnical Engineering) with Rock Mechanics and Tunneling from Indian Institute of Technology, IIT - Kharagpur in 1985.

Sandipan Goswami pursued a career in transportation engineering. His areas of work include the reliability of the design for the construction of highway, tunnel and bridges of concrete and steel structures in the perspective of users' safety and cost-effective design. He worked in The World Bank in 2009 and 2012 to review the DPRs prepared by the states of Assam, Kerala and Gujarat and identifying the safety issues as well as the countermeasures for the road users. He reviewed DPRs prepared by various consultants, worked for design audit, safety audit for highways, bridges, junctions and intersections in various State highway projects for 'The World Bank'.

Over the span of 40 years, he became well known for his technical contribution to the IITs, NITs, and other engineering universities, National Institute for training of highway engineers (Presently Indian Academy of Highway Engineers under The ministry of road transport and highways), Project Authorities, engineering consultants and construction companies by lectures, seminars, workshops, presentations for making Feasibility studies, DPRs and construction projects in EPC and BOT modes.

Sandipan Goswami was honored with titles 'Fellow of the Institution of Engineers' in India and Professional Engineer (PE) as Foreign Engineer, by Board of Engineers, Malaysia in 1997. He is life member of Indian Roads Congress (IRC) and is certified as 'Chartered Engineer' by the institution of engineers.

Professional career:

Sandipan has provided leading engineering services for large Government Projects in India, Malaysia, Bangladesh and Nepal.

He worked in the position of "K1 - Team Leader / Senior Tunnel Engineer (International)" of Project Management Consultant, by JV of ERMC (Nepal - Lead Consultant), Yongma (Korea) and Dongsung (Korea), for "Consultancy Service for Project Management Consultant for the Execution of Planning, Design, Engineering, Procurement, Commissioning and Five Years of Service Operation & Maintenance of one Double Lane Standard Highway Tunnel, Rock Shed Structure, Rock Fall Mitigation and Road Works of Siddhababa Tunnel Project at "Butwal, Nepal", on EPC Basis (Contract Identification Number DoR_QRDC/33701168/079-33) for Department of Roads (DoR), Kathmandu, Nepal. Contractor - China State Construction Engineering Co. Ltd., No. 15 Sanlihe Road, Haidian District, Beijing, China. Project Authority: Quality Research and Development Center (QRDC), Department of Roads, Ministry of Physical Infrastructure and Transport, Government of Nepal, Under Swiss Technical Assistance to Government of Nepal for the Implementation of Siddhababa Tunnel Road. The work has to comply with the import of goods and services prohibitions in the Employer's country when by an act of compliance with a decision of the United Nations Security Council taken under Chapter VII of the Charter of the United Nations.

He worked as:

1) Team Leader of Authority's Engineer for the construction of Access Controlled Nagpur - Mumbai Super Communication Expressway (Maharashtra Samruddhi Mahamarg) in the State of Maharashtra on EPC Mode for Package - 14, project of Maharashtra State Road Development Corporation Ltd. (MSRDC). This included NATM Drill and Blast Twin-Tube 6-Lane continuous Tunnel of length 7.74 km, Via-ducts as pre-stressed concrete bridges (60 metres high) at either end and were with Intelligent Transport System (ITS) features.

2) As Technical Advisor for Deloitte-Touché-Tohmatsu for institutional development, capacity building with e-governance and construction management for the road infrastructure development corporation of Govt. of Himachal Pradesh, a project by the World Bank.

3) As Technical Advisor for Deloitte-Touché-Tohmatsu for detail engineering design and financial transaction advisory services for 24 km long (approx.) six-Lane expressway with about 10 km elevated section with at grade service roads in 'Public-Private-Participation (PPP) model' in Bangladesh.

4) As Technical Advisor for Deloitte-Touché-Tohmatsu for Road Network Expert for development of Road Connectivity for promoting MSME Sector, in the State of Bihar, India, a World Bank project.

5) As Team Leader of Independent Engineer for the construction of National Highway NH34 (Old) / NH12 (New) project in the state of West Bengal.

6) Worked in The World Bank (IBRD) in 2009 and 2012 for the engineering audits of Detail Project Report for State Road Improvement Projects in the states of Assam, Kerala and Gujarat.

7) Worked for Technical Assistance Team of Asian Development Bank for Rural Road development project in the state of West Bengal.

8) Worked for different engineering companies as Halcrow, SNC Lavalin, Egis, Lea Associates, CES (Jacobs), EMA-Unihorn, Deloitte Touche Tohmatsu, Adani Road Transport Limited etc.

9) Participated in the lead roles in the design and construction of thousands of kilometers highways, hill roads, low-cost rural roads, tunnels, bridges, Viaducts and structures in India, Malaysia, Bangladesh and Nepal.

10) Studied for various safety issues at road intersections, interchanges, roundabouts, high speed non-urban highways in inter-state highway projects.

11) Designed Dhaula Kuan Multi-level Grade-separated Interchange in New Delhi, Kona Expressway Interchange in West Bengal, Mumbai

Nasik Expressway, Connector to Port Klang by JICA (Japan International Cooperation Agency), etc.

12) Worked for detail engineering design of 256 km long four-lane West Coast Expressway in BOT model in Malaysia.

13) Worked as Senior Design Engineer for design and as project manager for the construction of six cut and cover vehicular tunnels in Kuala Lumpur City Centre, Malaysia.

14) Worked as Team Leader for the preparing of DPR for State Highway Project, Package 19 in the state of Bihar, worked as Team Leader for detail engineering design for DPR of Ropar to Phagwara four lane highway (approx. length of 80 km) in the state of Punjab, by Punjab Infrastructure Development Board (PIDB), including three level Railway Over Bridge (ROB having on-ground level highway-elevated bridge and underground tunnel), junction improvements etc. on behalf of Consulting Engineering Services (I) Pvt. Ltd. (Presently Jacobs).

15) Worked as Team Leader in the Feasibility Studies for six-Laning of NH 5, Vishakhapatnam to Anakapalli to Rajahmundry in the state of Andhra Pradesh, on behalf of Egis India,

16) Worked for design and construction of various tunnel projects in national and state highways, The tunnels were by 'New Austrian Tunneling Method (NATM by Drill & Blast) for Jammu-Udhampur section of NH 1A,

17) Worked for design and construction of 6 km long Railway Tunnel by HCC-Alpine JV for Delhi Metro Airport Express Line (DMAEL) project,

18) Worked for design and construction of Drill & Blast Tunnel for Hydro electric Power project in Himachal Pradesh,

19) Worked for various Airport Planning and detail engineering design projects including NSCB International Airport in Kolkata.

Delivered lectures as visiting faculty in National Institute for Training of Highway Engineers (NITHE, presently Indian Academy of Highway Engineers, IAHE), under Ministry of Road Transport and Highways, Govt.

of India, on Project Preparations for Highway, Bridge and Tunnel projects for engineers who came from various government departments of India, Africa etc.

Authored the widely appreciated technical paper on "Project Development – Development of Rural Road Infrastructure, IRC" and many other technical and safety related articles in ResearchGate (Germany) and Academia (USA).

Sandipan Goswami received the awards namely, i) Rashtriya Gaurav (National Pride) Award, ii) Award from the CMD of NPCC, for proposing ERP implementation for 'Managing Projects – A Systems Approach', iii) Recipient of Millennium Achievers Gold Medal by International Institute of Success Awareness.

Note: Sandipan Goswami is the author of various academic text books as mentioned below.

All these books include various theory descriptions, examples, tutorials and software applications on the real project applications. The softwares have been in use for last 28 years in Highway, Railway, Bridge, Tunnel, Pavement, Metro Railway, LRTs, MRTs, Elevated transport corridors, Airports, Irrigation, Open Pit Mining projects in various countries.

- Text Book '**Computer Aided Highway Engineering**' published by Taylor and Francis, Boca Raton, Florida, USA https://www.taylorfrancis.com/books/mono/10.1201/9781003045830/computer-aided-highway-engineering-sandipan-goswami-pradip-sarkar ISBN 978-0-367-49338-7

- Text Book '**Computer Aided Bridge Engineering**' published by Nova Science Publishers, New York, USA https://novapublishers.com/shop/computer-aided-bridge-engineering-detail-design-of-pre-stressed-concrete-i-girder-box-girder-bridges/ ISBN 978-1-68507-413-5

- Text Book '**Pavement Engineering – Design Construction and Maintenance**' published by PHI (Formerly Prentice-Hall) Learning, ISBN 9789391818104 https://www.phindia.com/Books/BookDetail/9789391818104/pavement-engineering-goswam i https://www.phindia.com/Books/Author/OTc4OTM5MTgxODEwNA

- Text Book '**Tunnel Engineering with Rock Mechanics and Engineering Geology for Drill & Blast and NATM Tunnelling**' being published by Khanna Book Publishing Company Pvt. Ltd., New Delhi, India and New Jersey, USA

- Text Book '**Transportation Engineering and Traffic Management for Road Safety Improvement**' being published by Cambridge Scholar Publishers, United Kingdom,

Articles, Technical Papers and Honorariums:

- Article 'Managing Projects – A Systems Approach' – NPCC CMD's Award winner

- Article 'Project Development – Development of Rural Road Infrastructure', Published by IRC
- Recipient of RASHTRIYA GAURAV (National Pride) Award of 2004
- Recipient of Millennium Achievers Gold Medal by International Institute of Success Awareness
- Faculty – NITHE (Presently IAHE), Ministry of Road Transport & Highways,
- Articles on Design of Bridges, Structures, Tunnels in ResearchGate-Germany, Academia-USA, etc.

The resources mentioned above appropriately suffice the requirements of Engineering Universities, Research Institutions, Faculties, Students, Engineers in the Industry, Professionals, Consultants, Contractors, Project Authorities, Government Departments and agencies.

- Tutorial videos:
  https://www.youtube.com/channel/UCLY751jDWngqMfhKrlRcVwA/playlists
- Academia (San Francisco, California, USA) Profile:
  https://www.academia.edu/search?q=sandipan%20Goswami
- ResearchGate (Germany) Profile:
  https://www.researchgate.net/search.Search.html?query=sandipan+goswami&type=publication
- URL for Book Tutorials / Author's Profile:
  https://www.roadbridgedesign.com/#
- Social Media (Facebook):
  https://www.facebook.com/sandipan.goswami.18/

CHAPTER 1

ISOS AND ROAD SAFETY AUDITS

Abstract: This chapter covers various topics such as a general overview on road safety using the traffic management systems covered by ISO 39001, ISO 39002 and ISO 39003, a road safety audit (RSA) and its objectives, understanding the road safety audit, the primary elements of the RSA process, its focus, use, application, effectiveness and benefits and liabilities, in connection with existing roads and new developments

1.1 General overview of road safety using traffic management systems

Road safety using traffic management systems is well stipulated in ISOs 39001, 39002 and 39003, and each is elaborated on in the following sections.

1.1.1 ISO 39001

This is the best template to use for road safety traffic management. To improve road safety performance, pointers are provided in this document for organisations wishing to use it. To describe the development and implementing of ISO 39001, this guide provides a brief introduction that encourages further enquiry and consideration.

Step 1: Scope and Context

Identify the impact an organisation can have on road traffic safety (RTS) management system, map that impact across interested parties and determine the organisational scope of an RTS management system (refer to clause 4 of ISO 39001).

The standard requires the organisation to commit to a long-term goal of the eventual elimination of death and serious injury, while allowing it to determine the scope of the management system that it puts in place. The scope of the management system is best determined by considering the following activities that have an impact on road user safety:

- Use of the RTS management system by employees to and from work, or on duty.
- Under the terms of the RTS management system, goods and passenger transport are carried out by theorganisation, or this is contracted to other organisations.
- Organisations such as supermarkets, schools and places with many visitors have activities that control or influence the traffic generation to and from these locations.
- The RTS management system covers services and products related to road construction, maintenance, emergency trauma response or enforcement. Organisations can increase or decrease the risk of fatality or serious injury on the road network or their premises. During this process, interested parties need to be engaged, as they are likely to need to play a role in implementation.

Step 2: Leadership

To eliminate death and serious injury and provide resources to establish, implement, maintain and continually improve the RTS management system, commitment from management needs to be established by adopting a long-term vision, and organisational responsibilities (refer to Clause 5 of ISO 39001) assigned by establishing, documenting and communicating the RTS management system policy.

ISO 39001 requires leadership and commitment from the top management of the organisation in a variety of ways, including:

- Setting a long-term RTS objective for the elimination of death and serious injury in road traffic crashes. It is also essential to decide on the RTS results to be achieved in the interim.
- To achieve the established RTS objective(s) by developing a safe RTS; working in partnership and collaboration may be effective with interested parties.
- It is essential to communicate the importance of effective RTS management and conform to the RTS management system requirements.
- To establish, implement, maintain and continually improve the RTS management system, Nominating and allocating the human and financial resources as necessary.

A knowledge-based policy statement that reflects the commitment by top management to the elimination of death and serious injuries, and for continuous improvement of the management system, is an important aspect of leadership. For developing objectives and targets, and mobilising the resources needed for their delivery, the policy becomes a guiding document.

Step 3: Planning

RTS performance factors that are relevant to the road authority are to be identified. The risks and opportunities through assessment of current performance are to be determined. Objectives and targets for each performance factor are to be set, and action plans are to be developed as per Clause 6 of ISO 39001.

Identification of safety performance factors that the organisation must consider when developing its objectives and targets is the defining feature of ISO 39001. As specified in the standard, the following are the three safety performance factors:

- Risk exposure factors, such as for different road user groups, the traffic volume on the road network, including pedestrians, cyclists and motorcyclists, who are vulnerable road users.
- Intermediate safety outcome factors, such as those that have an evidential base linking their improvement with improved safety.
- Final safety outcome factors, such as on the road network, the number of deaths or serious injuries.

Intermediate Safety Outcomes

The intermediate safety outcome factors need the most attention, as these are causally linked tothe final safety outcome factors, while risk exposure factors and final safety outcome factors need continuous monitoring. This is because the intermediate safety outcome provides the best means of assessing the underlying level of safety, rather than relying on relatively rare instances of crashes or injuries. In successful road safety work, the targeted and/or monitored typical intermediate outcomes or safety performance indicators are the following:

- Average mean speed or levels of excessive speed.
- Levels of use of seat belts, child restraints and crash helmets.
- Levels of disciplined driving.
- Safety ratings of the road network using iRAP scores.

- Safety ratings of the vehicle fleet using NCAP ratings.
- Response times of the emergency medical system.

Relevant institutional outputs can also be measured, targeted and monitored. For example, the efficacy of the reminder system for seat belts may improve levels of seat belt use. Within ISO 39001, the general categories of RTS factors that need to be considered are:

- Road design, side areas, junction design, and safe speed by considering separation, especially for oncoming traffic and vulnerable road users.
- How appropriate the road is, in respect of vehicle type, user, type of cargo and equipment.
- The use of safety equipment by road users, especially seat belts, child restraints, bicycle helmets and motorcycle helmets and the means to see and be seen.
- Safe driving speeds factoring in vehicle type, traffic and weather conditions.
- Drivers' fitness and ability, especially considering fatigue, distraction, alcohol and drugs.
- Safe travel planning considering the need to travel, the amount and mode of travel and choice of route, vehicle and driver.
- Safety of vehicles, concerning the protection of passengers, protection of other road users, whether vulnerable or other vehicle occupants, road traffic crash avoidance and mitigation, roadworthiness, vehicle load capacity, and securing loads on the vehicle.
- Appropriate authorisation to drive the class of vehicle.
- Removal of unfit vehicles and drivers from the road network.
- Post-crash response, first aid, emergency incident management and post-crash recovery, along with rehabilitation.

To support the achievement of these targets and goals, objectives need to be established by the authority for intermediate outcomes, along with specific action plans to determine which safety performance factors the authority wants to work with. A systematic improvement can be achieved by focusing, measuring and following up on the intermediate outcomes that will improve RTS performance considerably.

Step 4: Implementation

The RTS management system is to be implemented and operated by ensuring that sufficient capacity is provided for the objectives and by meeting the targets, as referred to in clauses 7 and 8 of ISO 39001.

The first three steps of the development of the management system define the features of ISO 39001 as a management standard. The implementation, operation, evaluation, and improvementof the management system are more generic, reflecting good management processes. The following are the key implementation functions emphasised by ISO 39001:

- For the successful operation of the management system, effective coordination of activities both within the organisation itself and with external stakeholders is essential.
- To implement the safety management system, an ongoing allocation of resources is needed, and to support improvements in the system, additional resources are likely to be needed.
- Staff are required to understand its purpose, the key processes and approaches that are necessary for its implementation, and the processes to ensure new safety practices and techniques are disseminated and adopted over time.
- The SMS and its requirements need to be communicated and regularly promoted both internally and externally to build support for the system forcontinuing reductions in road traffic risk.
- The processes for RTS need to be set, documented and followed. The documentation needs to support the management task and not become a burden.
- To effectively control safety risk and improve safety results over time, the documentation and reporting need to be sufficient.

Step 5: Monitoring and Evaluation

RTS performance is to be monitored and evaluated by conducting internal audits and periodic reviews of the RTS management system to identify opportunities for continued improvement by referring to Clause 9 of ISO 39001.

Performance and achievements of the management system need to be monitored and evaluated over time. The audit and review processes for monitoring and investigation need to be established, with a focus on critical safety issues—that is, the safety performance factors that the organisation has identified. A periodic evaluation for compliance with the standard is important.

Step 6: Continual Improvement

Improve RTS on a continuous basis by reviewing its performance, by referring to Clause 10 of ISO 39001.

Organisations need to look ahead for substantial improvements in their operations and work towards achieving such improvements. To achieve the objectives of the RTS management system and to make the necessary changes to the management system, ISO 39001 requires top management to undertake a critical review.

Number Of Organisations Certified to ISO 39001

In Sweden, there are seven organisations certified to ISO 39001 by an accredited certification body: two in public transport and five in cargo transport. There are around 110 organisations certified to ISO39001 by the road transport company association (non-accredited): transport/logistics organisations.

In Japan, there are 144 organisations certified to ISO 39001 (October 2016) and 17 private certification bodies. These are mainly organisations involved in freight transport.

Some national governmental leadership has resulted in the adoption of ISO 39001 in public procurement (e.g. Sweden), in governmental fleet safety policy (e.g. Colombia, Malaysia (Royal Police) and Mauritius), the availability of financial incentives for the development of ISO 39001 in organisations (e.g. Japan) and the production of a starters' guide for road agencies (e.g., Australia). Research reports, materials for guidance and training in ISO 39001 are used in several countries (e.g., the UK, Malaysia).

1.1.2 ISO 39002

1.1.2.1 Scope

This document provides guidelines for good practices that organisations can adopt to reduce the number of fatalities and serious injuries, the severity of injuries, and minimise damage to property and economic loss. Any authority may protect commuters, including vulnerable road users (VRU), through the adoption of a proactive approach to managing risks by the application of this document. This document is also applicable to and beneficial for commercial transport organisations as well as schools.

1.1.2.2 Normative References

The documents are referred to either in full or in part if their content is suitable for the requirements of this book. For a dated referenced document, only the cited editions are applicable. For undated referenced documents, the latest editions, including any amendments, are applicable as per ISO 39001, RTS management systems requirements with guidance for use.

1.1.2.3 Terms and Definitions

The terms and definitions given in ISO 39001 and the following are applicable to this document. In standardisation, ISO and IEC maintain terminological databases for use at the following addresses:

- ISO Online browsing platform, available at: https://www.iso.org/obp
- IEC Electropedia, available at: http://www.electropedia.org/

1.1.2.4 Commuting Crashes

This is a crash in which a commuter is involved while travelling:

- Between their places of stay and places of work/study.
- Ona journey that is connected to their employment/study.
- Between their place of work and the place where they eat during an authorised break.

1.1.2.5 Organisation

A person (or group of people) that functions with responsibility, authority and [illegible] relationships to achieve its objectives, which includes schools and other education establishments.

1.1.2.6 Factors Affecting Commuter Crashes

Three main factors are involved: road users (drivers, cyclists, pedestrians and passengers), vehicles, the road and the environment.

1.1.2.7 Road Users

Three elements influence the physical and psychological conditions of road users:

- **Competency**
  To adopt a safe driving style, to take prompt and appropriate actions, and to avoid road crashes or minimise the impact of a road crash, drivers should have essential abilities. To ensure appropriate riding/driving, the limitations of the vehicles should be understood by the drivers, within the context of the road facilities and the environment. Pedestrians, whenever possible, should use the segregated pedestrian zones or pavements, stay behind safety barriers/fences or on designated footpaths. While walking along or crossing roads, they must always be fully alert and face the traffic. The pedestrians should be on the side of the roadfacing the oncoming traffic, in the absence of a pavement, unless there is any special circumstance or specific danger.

- **Health and fitness**
  Factors that can affect road safety are health conditions, fatigue, the influence of alcohol and drugs, and visual problems (short- or long-sightedness, astigmatism, etc.) Psychological conditions that can cause crashes are an unstable state of emotion, stress, lack of sleep, etc. Before and throughout the journey, the road users need to be in a stable emotional state.

- **Behaviour**
  Speeding, tailgating, weaving and red-light running are major causes of road crashes, which are unsafe and inappropriate behaviours. While on the road, road users can be distracted if they are eating, drinking, smoking, texting, answering or making calls, etc. Driver behaviour can be negatively influenced by 'organisational factors', such as pressure for staffto be 'reachable' during their commute. On public transport, passengers need to ensure that their behaviour keeps them safe during the journey.

i. To minimise disruption to other road users and traffic, passengers must wait at designated public transport stops.
ii. Alighting and disembarking must be at the designated public transport stops.

iii. On the way to the public transport stops, either as a cyclist, pedestrian or other, rules and regulations must be obeyed.
iv. Before alighting or after disembarking, cross at designated crossings or make use of the provided channelling to cross at safe places.

1.1.2.8 Vehicles

The three factors that can contribute to crashes are safety features and devices, including: engineering, design and structure of vehicles, active and passive safety devices, such as electronic stability controls (ESC), anti-lock braking systems (ABS), side impact protection or seatbelt reminder systems, and materials used for the manufacturing of vehicles.

- **Suitability**
  The vehicle's selection should be relevant to its purpose, such as a four-wheel drive should be used in an agricultural setting or snow-covered areas of work.

  NOTE 1: The possible existence of local legislation should be considered regarding the modification of vehicles required for persons with disabilities.

- **Condition**
  To ensure that the vehicle is functioning correctly and has a long life expectancy, acomprehensive periodical maintenance of vehicles is critical.

- NOTE 2: The possible existence of local legislation may be applicable to vehicle roadworthiness.

1.1.2.9 Road and Environment

The road and environment conditions, besides users and vehicle factors, are also consideredas contributory factors. To be aware of environmental factors such as the topographical conditions and the road type (whether single-or dual carriageway, etc.) throughout the entire commuting route is important for road users. Appropriate actions and extra care can be taken when commuting, upon knowing the potential hazards and risks along the route. Attention should be given to:

a) Road conditions:
   - Road surface has potholes or is slippery or greasy.
   - Geometric features such as sag curves, crest curves and lane width.
   - Road under construction.

b) Road furniture available along the route, such as the type of guardrail used, road signage and road markings.

c) Road environment:
   - Terrain type, whether flat or hilly/mountainous.
   - Signage, whether inadequate, confusing, or obstructed.
   - Lighting.
   - Haze.
   - Thick fog.
   - Snow, ice, heavy rain, flooding, landslide or crosswinds.
   - Animals crossing.
   - Illegally parked vehicles impacting sightlines.
   - Any significant change to the road environment and condition affecting safety, such as fallen
   - objects, landslides, etc.

d) Traffic volume and conditions, such as peak hours or festive seasons.

1.1.2.10 Roles and Responsibilities

1.1.2.10.1 Organisation

Crucial success factors include the organisation's leadership and commitment to safety management programmes and initiatives in the workplace. The health and safety of their employees/students, their attitude, vision and mission, and their willingness to allocate resources to make an impact on commuting safety are directly related to the awareness of safety management programmes and initiatives in the workplace. Understanding the importance of safety is one of the main responsibilities of an organisation. The management of the organisation should be aware of occupational health and safety and be able to monitor related activities that are performed within the organisation, including the assessment of its achievement and effectiveness. Organisations need to identify the relevant road authorities that are responsible for roads connected to their premises so that they can report any damage to the road or unsafe conditions.

- Human factors & interface/training/misuse.
- Priority for the emergency services.
- Passive/aggressive driving approach/behaviours.
- Sustainability issues.
- Normal driving behaviour.

3. Additional Safety Ethical Considerations or Manufacturer's Special Features

1.2 Road Safety Audit (RSA)

Traffic engineers in the UK developed the idea of an RSA as a check for new and existing road projects and schemes in the early 1980s. Subsequently, countries like Australia, Denmark, New Zealand, Canada, France, Greece, Hong Kong, Iceland, Ireland, Italy, Malaysia, the Netherlands, Perm, Singapore and the US have developed RSAs. In several Canadian provinces, there is strong support for audits, and pilot audits were carried out in various states in the USA, which shows that the interest is growing rapidly. Audit guidelines have been developed for developing countries such as Nepal and Bangladesh by consultants working for the World Bank. Vietnam and China have had RSAs carried out by Asian Development Bank and World Bank consultants. Thailand has had a national seminar on RSA plus training courses. Committees such as ITE, PIARC, TRB and Austroads are addressing RSA issues. RSAs are progressing rapidly around the globe. An RSA is a positive process—the earlier, the better, making it safer and cheaper. An RSA may be the only means by which road safety is explicitly considered in a project. RSAs are a small cost, around 1-2% of total design costs, but offer great benefit. In the first year, the rate of return may be up to 120% if audit recommendations are followed (Jordan study 1998).

1.2.1 Scope of the RSA

- To adopt safety-conscious planning and design to minimise the likelihood of crashes occurring.
- To ensure that any injury is minimised in case a crash occurs (with the provision of anti-skid surfacing and crash barriers).
- To ensure that critical sight distances and other safety-related design criteria have been met.
- Managing risks in such a way that the risk of major safety problems is less than the risk of minor problems.

- Adopting safety provisions to reduce the whole-life cycle costs of a design, as poorly designed infrastructure is expensive to correct afterwards.
- Ensuring minimum risk of crashes on the existing road network (particularly at junctions) as well as on any new road scheme.
- Including road safety engineering in highway design work, and enhancing consideration of the safety of all road users in all new and existing schemes.

1.2.2 Not in the scope of the RSA

- To assess or rate a project as good or poor.
- Ranking or justifying one project against others in a works programme.
- Evaluate and rate one option against another.
- Checking for compliance with technical/ design standards.
- Investigation accidents.
- Redesign a project.
- Check the designer's competence.
- Technical audits.
- Relevance of application only for high-cost projects or projects involving safety problems.
- To describe informal checks, inspections or consultations.
- Raise subjective personal concerns.

1.2.3 Specialists for a safety audit

To take a fresh look at the project, focusing on safety, the safety auditors should be independent of the design process. The audit needs to be carried out by specialists to make it effective. Expertise and experience in road safety engineering are essential qualities of the audit team.

1.2.4 Types of projects to be audited

An audit is required at the planning stage to identify the requirements for new motorways, major four-lane projects, reconstruction and realignment projects, junction projects, both with and without signals, pedestrian and bicycle routes, deviated access roads near project roads, local area traffic management schemes and accident reduction schemes.

Austroads 2001 states that "A road safety audit is a formal examination of a road/traffic project in which an independent, qualified team reports on the project's accident potential". The *Manual for Safety in Road Design* (India) states that a "Road Safety Audit is a formal procedure for assessing accident potential and safety performance in the provision of new road schemes, the improvement and rehabilitation of existing roads and in the maintenance of existing roads".

1.2.5 Objectives of an RSA

- On new road projects, high levels of safety must be ensured.
- Whole-life costs of projects are to be reduced.
- On the adjoining road network, accident risk is to be minimised.
- The safety of all road users is to be promoted.
- Road safety is to be promoted by proper engineering.

An RSA combines assessing how road users will use the road with the science of proven road safety engineering principles.

1.2.6 Understanding the RSA

- The audit is complementary to accident reduction and prevention.
- The aim is to ensure a safe road system.
- The use of safety checks or safety audits is the key component of accident prevention.
- The focus should be on the planning and design of new roads and traffic schemes, and occasionally on existing ones.

1.2.7 Definition and objectives of the RSA

- For significant improvements, safety must be systematically designed into each project; it isnot created automatically.
- Highway designers must consider the safety experts specific to each project and apply soundsafety and traffic engineering principles.
- An RSA is the process aimed at preventing accidents using road safetyengineering.
- The RSA is a means for accident prevention by checking the design, implementation.
- and the operation of road projects using a set of safety principles and treatment.

1.2.8 Primary elements of the RSA process

- The safety audit is a formal process and not an informal check.
- The safety audit is an independent process.
- A specialist with appropriate experience and training carries out the safety audit.
- The RSA should be restricted to road safety issues.

1.2.9 Focus of the RSA

- Minimise the risk and severity of road accidents on the project road or the nearby network.
- Minimise the need for remedial or improvement work after construction.
- Reduce the whole-life costs of the project.
- Improve the awareness of safe design practices by all those involved in the planning,design, construction and maintenance of roads.
- The RSA can work in two ways: rectifying preventable crash-producing elements at the planning or design stage, suchas any inappropriate junction layout and resolving all remaining or existing issues.
- Introduce features suitable for reducing accidents, such as anti-skid surfacing and the provision of guard rails.
- Provision of adequate traffic control devices, and provide delineation, etc.

Highway designers and traffic engineers have always practised a form of safety audit. What isimportant about the recent emergence of the practice is its specific incorporation as:

- A discrete phase.
- Independent from the designer.
- Development of defined auditing procedures with a checklist on various aspects, followed by the road or traffic authority.
- May be incorporated within an overall quality management or quality assurance process by the road or traffic authority.

1.2.10 Use of the RSA

The UK introduced the concept of RSAs in the 1980s. The government's target was to reduce road fatalities by one-third by the year 2000. It was given impetus by the preparation of key publications, the *Road Safety Code of Good Practice*, Local Authorities Association, 1989, and *Guidelines for the Safety Audits of Highways*, Institution of Highways andTransportation, 1990.

- For all national trunk roads and motorways, an RSA was made mandatory in 1991.
- Following the success in the UK, an audit process was also adopted in New Zealand in 1992.
- Since 1993, a safety audit has been mandatory in many state highway projects.
- In 1992, the World Bank began to show an interest in RSAs.

1.2.11 Application of the road safety audit

An RSA can be carried out at any or all following five stages.

Stage 1 Feasibility Study

- A safety audit may influence the selection of a project route
- To apply appropriate design standards and design criteria
- To assess the impact on the existing road network
- To maintain route continuity
- To select types of junctions
- To control access
- To decide the number of lanes required
- To consider the relevance of terminals and development stages, etc

Stage 2 Draft Design

This audit is to be undertaken on completion of a draft plan or a preliminary design.Typical considerations include:

- Speed, sight distances in the design of horizontal and vertical alignments, super elevation lanes and the width of the hard shoulder.
- Overtaking lanes.
- Junction layouts, with or without signals, and median openings.
- Provision for parked and stationary vehicles.
- Provision for cyclists and pedestrians.
- Underpasses, foot bridges, etc, at petrol stations, schools and hospitals.

- The effects of deviation from the standards and guidelines as per the site situation.
- The safety of traffic and workers during construction, etc.

Stage 3 Detailed Design

This stage is on completion of a detailed design, normally before preparation of contractdocuments. Considerations include

- Lane and other road markings.
- Road signage.
- Delineation.
- Lighting.
- Junction details.
- Clearance of roadside objects.
- Provision for road user groups with special requirements.
- Temporary traffic management and control during construction.
- Storm water drainage and snow clearing.
- Roadside objects.
- Landscaping, avenue and median plantation.
- Batters (inclined, raised areas)
- Guard rails, fencing, etc.

Stage 4 pre-opening

Prior to the opening of the road to traffic, the audit would involve inspection by driving and walking around the project tocheck the fulfilment of the needs of all road users and would include inspecting visibility at night-time, during rain and fog, in both wet and dry conditions, and to consider similar issues raised in Stages 2 and 3, but with the view of assessing theiradequacy as constructed. Any specific variations that might have occurred from the plans in the process ofconstruction to be noted and studied.

Stage 5 In-service

Sections of the existing road network need to be examined systematically to assess the adequacy of the roads, junctions, verges, etc, from a safety viewpoint. This can have two applications: For some considerable time, after the Stage 4 audit, any new scheme is to be monitored once the road is opened to traffic. A safety audit of an existing road network should be done to identify any safety-related deficiencies.

1.2.12 Process of an RSA

- Commitment from the management is a key requirement
- Auditors are to review and find things wrong with the planning and design work
- Specialists' advice to the design team is required during the audit process
- In implementing the SMS in the US, Hall (1993) was concerned with:

    - Interdepartmental conflict—other units may think that the accomplishmentof their goals is threatened by the infringement of safety management initiatives into their territory
    - Care should be taken to maintain an atmosphere of cooperation and mutual respect among affected functional areas throughout system implementation
    - An agreed process for the RSA

The process aims to reduce the total cost of a project, minimising remedial measuresduring the operation and maintenance period. There will be a cost for an audit process, which is worthwhile to offset against the potential savings elsewhere. The savings may be achieved by taking measures at the planning stage by avoiding alterations in future. It is much cheaper to change a detail in the planning stage than to replace a feature once implemented. The aim is accident prevention. The costs resulting from litigation shall be reduced. The auditor or audit team should be independent and not be driven by management. To act upon the audit report, there must be a designated procedure. When a specialist team is deployed, one of three procedures is to be followed

- Prior agreement to accept the recommendations of the safety audit
- Assessment of the audit report by the client
- Assessment of the audit report by the designer

An agency conducting an RSA process must determine which of these procedures to follow, depending upon its expertise and the role of safety auditing within a wider institutionalframework such as quality management. No matter which procedure is adopted, the key factors are as follows:

The audit team must possess specialist knowledge of road safety engineering. Safety audit findings should be formally documented and reported at each step of the auditprocess. A set of checklists on different aspects of the project, checks are different for features. Use of checklists on issues that can potentially arise at the relevant stage of theproject acts as a memory prompt and ensures problems are less likely to be overlooked. The main benefits of checklists are that designers can follow them before the audit.

Audit team members are to keep abreast of the latest standards and specifications. Depending on the size and complexity of the project, the audit team should be adequate. British experience says that at the feasibility or layout design stage, a three-person team would be suitable, comprising a road safety specialist with experience in crash investigation and safety engineering, principles and practice, a highway design engineer and someone with experience in safety auditing, who can generate discussion and assist in the procedure. There should not be a substitute for expertise.

Monitoring and evaluation: the process involves three aspects: procedures, problemsencountered, and countermeasures effectiveness to the system and an evaluation of costs and resources by scheme type

1.2.13 Liabilities

- The Australian guidelines mentioned legal issues, with the following conclusion
    - No case involving an RSA has yet come before a court.
    - The legal implications must be considered.
    - The predictions are not to be guesswork; they are to be based upon well-establishedprinciples of tort law.
- The safety audit is to create a safer road environment.
- The safety audit is to be encouraged by the legal system; a major objective of litigation is toencourage safety.
- By using various methods, black spot roads can be made safe.
- The safety audit is to include periodic inspection.
- The safety audit is to check for the adoption of the highest standards of engineering practice.
- There should be an allocation of funds to support RSAs.
- Highway authorities are to decide the best selection for a given project.

1.2.14 Safety audits of existing roads

Inclusion of a safety audit of existing roads can be an important component of an overallaudit process to identify any existing safety deficiency in the design, layout and street furniture which are not consistent with the functioning of the road.

- There should be consistency in the standards.
- The safety audit process is to ensure that many items can be placed on the maintenance
- programmes, which are related to the maintenance of the road.

1.2.15 Safety audit for development projects

An extension of the RSA in some local authorities is to require that development proposals be audited, which may include

- A new commercial development that will generate traffic on an existing road.
- A residential development that involves street construction.

These would need to be independently audited, and a report submitted as part of the development application.

1.2.16 Audit reports of development projects may address

- The safety impact during peak congestion.
- Pedestrians' and cyclists' movements across existing roads.
- The safe provision of public transportation.
- Vehicular and pedestrian access to the site.
- Safety of the parking provision.
- Conflicts between pedestrians and vehicles on and adjacent to the site.
- Suitability of new road alignments with the type and layout of junctions.
- Posted speed of vehicles within the site.
- Visibility in the worst situations.

1.2.17 Effectiveness of road safety audits

Evidence is emerging that a safety audit is a cost-effective safety measure, although a new technique. If a project is subjected to a safety audit, it will very likely lead to improved safety. The resources needed for safety audits are, in fact, quite small.

- As suggested by the UK's experience, individual safety audits have the potential to remove up to one-third of crashes, and one safety auditor can cover and improve an area experiencing 1000 crashes per year.
- As suggested by the Australian and New Zealand experiences, the safety audit adds to road design costs is negligible compared to the saving of lives.
- Attempts have been made to quantify the benefits of a safety audit. One highway authority in Scotland, in 1991, estimated that one-third of accidents could be preventable with a road safety improvements audit. A one per cent saving per year is worth 1.5 million (benefit) with a resource cost of 100,000, and therefore has a benefit-cost ratio of 15:1.

1.2.18 Benefits of an RSA

The benefits of an RSA were summarised by Austroads (1994) as below:

- The possibility of crashes can be reduced on the road network.
- The severity of crashes can be reduced on the road network.
- In the minds of road designers and traffic engineers, road safety is given greater prominence.
- The need for costly remedial work is considerably reduced.
- The total cost of the project to the community is reduced.

References

International Organisation for Standardization: https://www.iso.org › developing_standards › docs

Startup guide to ISO 39001: road traffic safety management. https://www.iso.org/files/live/sites/isoorg/files/developing_standards/docs/en/ISO_39001_Startup_Guide_2017-06.pdf

CHAPTER 2

ROAD SAFETY AND ADMINISTRATIVE SETUP

Abstract: this chapter covers various topics such as a general introduction and overview, guidelines to establish an effective system, trends in road safety, road safety improvement plans, road safety management, road accident data maintenance, road safety funding by the insurance industry, planning and design of safe roads, improvement of hazardous locations, road safety education for children, training and testing of drivers, publicity and campaigns for road safety, traffic legislation, traffic police and law enforcement, emergency assistance for road accident victims, safety road safety research and costings of road accidents.

2.1 General Introduction and Overview

This chapter is designed for ease of use by different interest groups. It consists of 16 sections as mentioned below, and the arrangements are described in the following paragraphs. The improvement of road safety is possible through a multidisciplinary and comprehensive approach, such as the measures that need to be undertaken in each sector that affects road safety. Implementation of the three priority actions in each sector would at least initiate improvements in road safety and strengthen the ability of each country to begin to address its safety problems.

2.1.1 Guidelines to Establish an Effective System

In this chapter, each section summarises the key points and priority actions relating to a specific sector, followed by descriptions in more detail of what needs to be done to develop effective road safety activity in that sector. The necessary coordination and management of road safety is elaborated in each of the above sections.

Improvement of road safety never finishes; it is a continuous process. This chapter focuses on establishing an effective system for preparing improvement guidelines, carrying out safety research studies, tentative costing and implementing well-planned measures for achieving safe travel by road users in the developed and developing countries, including the Asia Pacific regions.

This chapter provides guidelines to establish an effective system by the road authority of the country for the improvement, upgrading, maintenance and management of the system.

In addition to the guidelines, road safety for developing countries is presented with several technical notes and reports documenting various aspects of road safety in the region. The various sections of this chapter have focused on the problems faced by and facilities for the most vulnerable road users. The guidelines are found to have been most effective in developed and developing countries and have been prepared based on experience studying road safety issues in various countries.

2.1.2 Scope

The chapters are devised to help countries overcome some of the deficiencies and impediments preventing the effectiveness of road safety. Guidelines are provided on the types of interventions and countermeasures that can be implemented in various major sectors affecting road safety, and how these different interventions can be coordinated within national road safety action plans and programmes to bring about improvements. It is intended to promote comprehensive coordinated approaches for road safety improvement and to encourage networking and sharing of information.

2.1.3 Aim

The various guidelines are aimed at encouraging the development and implementation of initiatives in all the major sectors affecting road safety. The specific objectives are as follows:

- Provide a single comprehensive source of reference and guide for senior government officials in developing countries on various actions and interventions that need to be taken in each of the different sectors affecting road safety.
- Provide an easily accessible format that can be updated easily, containing information for policymakers so that senior government officials can discuss these issues more easily with the ministers and officials of the planning commissions.

- Bring together the practical experience of specialists from developing and developed countries in a single document.
- Improve the capability of developing countries significantly to respond to their growing road safety demands effectively.

2.1.4 Guidance and Advice

It is a fact that almost all developing countries in the world suffer from a lack of financial resources; it is therefore essential that the limited available resources should not be wasted. Road accidents result in high economic losses annually, and action is necessary to reduce these losses. The suffering of many countries is common due to the absence of adequate technical resources and expertise to ensure the best possible road safety. The availability of trained professionals is much less than required. Their knowledge is also not complete with respect to various recent developments and techniques that can improve road safety. It is hence felt to be an urgent need to combine the collective knowledge and experience of developed and developing countries that have faced and resolved such issues. This book attempts to fulfil such requirements. This chapter provides a set of guidelines that should assist developing countries in addressing their road safety problems more effectively.

2.1.5 Responsibilities are Addressed

The senior government officials at the director general or secretary level of the road transport ministry, faculties in the institution, aid agency officials and other decision makers who are often assigned the job of improving road safety, may be at least familiar with the key issues, elements, and opportunities relating to the improvement of road safety as well as their involvement. The various sectors that have the responsibility for road safety in a country need a basic source of reference for working on maintaining various records, studies, analyses and safety improvement activities.

2.1.6 Structured Approach

Various sections of this chapter provide information about the change in traffic trends and accidents in detail, along with the need to attend to road safety by developing and developed countries by following guidance in administrative settings, discuss the motorisation and growth in vehicle fleets, and illustrate recent trends in road accidents in those countries. It also discusses the socioeconomic cost of road accidents, showing that the problem has now become more acute, and argues for urgent action to be taken.

This chapter discusses how road safety action plans and interventions can be designed for maximum effect. It discusses the stages of road safety development and illustrates the kind of coordinated action and intervention that is needed. This proposes raising awareness to develop prioritised road safety action plans and eventually developing three- or five-year road safety programmes. This chapter also discusses road safety goals and targets that can bring desired changes in road users' behaviour, such as wearing seat belts, helmets, avoiding mobile phones and alcoholic drinks, etc., thus resulting in casualty reduction. Advice is also provided on how road safety initiatives can be financed.

2.1.7 Sources of Information and Networking

The appendices provide a list of the key publications considered to be the most important for road safety professionals. In addition, sector-specific references are provided within each technical sector in this chapter. Appendices also provide telephone/fax numbers, email and postal addresses for many organisations from whom additional information can be acquired.

2.1.8 Guidance on the Use of the Sections

This chapter has been prepared as freestanding sections that are internally cross-referenced and has been designed with separate sections for various essential facets of the implementation, so that senior government officials can extract relevant sub-sections for discussion with officials of planning departments as appropriate.

2.1.9 Limitations of the Sections

It is quite difficult to provide a general road safety reference covering all sectors of safety in developing countries. It may not meet the comprehensive needs of individual countries because of their diversity. It, therefore, may hopefully address common issues and problems. The attempt is genuinely difficult to cover the wide mix of countries at different stages of traffic development and with different road safety problems. The countries in the Asia and Pacific regions contain some of the largest and smallest populated countries, some countries with fast economic growth, and others are earlier in the development stage. It contains highly industrialised city-states such as Singapore on the one hand, and on the other hand, countries like Nepal, where more than half of its districts do not have roads suitable for motorised vehicles.

It is, therefore, not feasible to provide detailed guidance for all situations. However, an attempt has been made to provide a basic overview of what needs to be done in each sector. Guidance, references and contacts are provided for readers interested in exploring topics in more detail. The knowledge, views and opinions of safety experts were gathered to ensure that the guidelines are as practical and relevant as possible to the needs of the region. It is the intention that the informative sections and guidelines reach a greater number of readers across the world as a resource for various people involved in the improvement of road safety to save lives. Road safety is one of the important social issues, as nothing is more valuable than people's lives.

2.1.10 Safety of Vulnerable Road Users (VRUs)

In some developing countries, data analyses show that VRUs are involved in a high proportion of traffic accidents. To quantify the nature and scale of the road safety problems faced by these road users, some specific efforts were made. The definition of VRU varies depending on the context of a particular study. In response to the needs of the developing countries to provide road safety to VRUs, the focus is on:

1. Pedestrians.
2. Pedal cyclists and other non-motorised vehicles.
3. Motorcyclists.

Although the focus of this section of the book is on current engineering practices, these measures in isolation do not necessarily solve all VRU safety problems. Consequently, complementary enforcement and educational measures are also considered. Because of the limited availability of detailed data, the study is not always able to explore specific casualty types within each VRU group, such as the very young or very old, who may have a higher risk of being involved in accidents.

Data Overview: pedal cycle and cycle-rickshaw data cannot generally be separated. However, from alternative sources, the proportional split of the non-motorised vehicles (NMV) types has been shown wherever possible. Examples of accident data for the region are summarised in Tables 2.1 and 2.2, which show that pedestrians and motorcyclists are involved in the most fatal and injurious traffic accidents. In Malaysia, Singapore, and Taipei

(China), more than half of all fatal accidents involve motorcyclists, while in Hong Kong (China), more than two-thirds of all fatal accidents involve pedestrians.

2.1.10.1 Pedestrians

Research conducted by the Transport Research Laboratory (TRL) in the UK identified pedestrian safety as a problem area for developing countries. Pedestrians, in general, are involved in more accidents in developing countries than in developed countries. This can be partially explained by the low motorisation levels in developing countries, resulting in more people walking. Traditional transport planning and traffic management have generally tended to focus on the safety of vehicular traffic and have largely overlooked pedestrians. In addition, due to the unaffordability of private transport and inadequacy of public transport, there tends to be a higher percentage of pedestrian trips in the cities of the developing world.

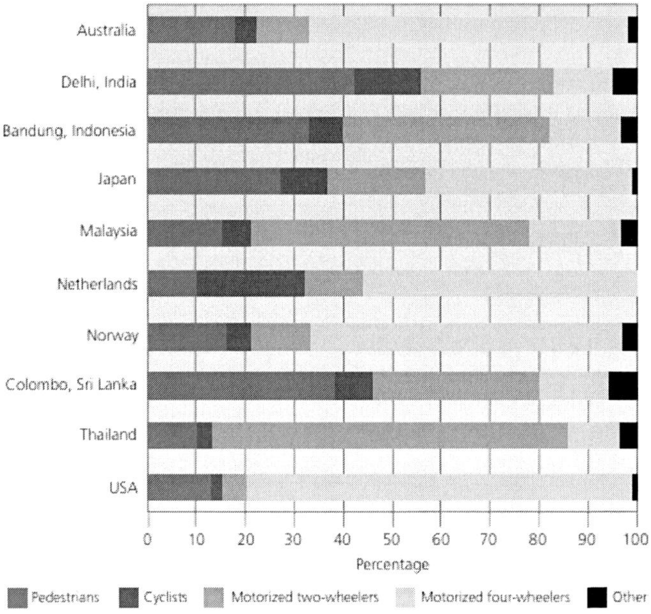

Figure 2.1 Road users killed by various modes of transport as a proportion of all road traffic deaths
Source: D. Mohan, Journal of Transport and Infrastructure, 2002, 9: 79–94.
Ref: https://www.who.int/violence_injury_prevention/publications/road_traffic/world_report/chapter2.pdf

Table 2.1 Percentages of Fatal Accidents Involving Vulnerable Road Users (1993)

Sub-region	Country	Pedestrian (per cent)	Cycle-based vehicle (1) (per cent)	Motorcycle (per cent)
NIEs	Hong Kong, China Republic of Korea, Singapore, Taipei, China	67 48 25 19	4 0 8 (2) 6	10 7 51 64
Central Asia	People's Republic of China (1994)	(2) 11	(2) 9	10(2)
Southeast Asia	Malaysia (1994) Thailand	(2) 15 9	(2) 6 4	(2) 57 NA
South Asia	Sri Lanka	45	17	16
PDMCs	Fiji	43	3	1
Developed Countries	Australia, New Zealand	19 15	3 3	NA 16

Table 2.2 Percentages of Injuries Involving Vulnerable Road Users (1993)

Sub-region	Cou ntry	Pedestrian (per cent)	Cycle-based vehicle (1) (per cent)	Motorcycl e (per cent)
NIEs	Hong Kong, China Republic of Korea, Singapore, Taipei, China	35 na 17 13	4 (2) 0 5 5	17 6(2) 69 62
Central Asia	People's Republic of China (1994)	(2) 9	na	74(2)
		(2)	(2)	(2)

Sub-region	**Cou ntry**	**Pedestrian (per cent)**	**Cycle-based vehicle (1) (per cent)**	**Motorcycl e (per cent)**
Southeast Asia	**Malaysia** (1994) and **Thailand**	6	5	57
South Asia	**Sri Lanka**	48	19	26
PDMCs	**Fiji**	37	6	9
Developed Countries	**Australia New Zealand**	15 9	7 9	Na 14

'na' Data not available.
Includes two- or three-wheeled cycle-based vehicles.
Data relates to deaths.
Source: RETA project data.

Table 2.3 Percentages of Total Vehicles (Excluding Handcarts) By Vehicle Type in Selected Asian Countries

City (Country)	Bicycle	Cycle-rickshaw	Animal cart	Motor-cycle	Bus	Other motor vehicle
* Dhaka (Bangladesh) (1990)	10.7	53.3	0.0	16.0	1.3	18.8
Phnom Penh (Cambodia)*	47.1	4.2	0.2	43.6	0.0	4.9
U Shanghai (People's Republic of China) (1990)	96.9	1.4	0.0	0.5	0.3	1.9
n Kanpur (India)(1992)	54.7	3.5	0.6	18.6	0.1	22.4
w Surabaya (Indonesia) (1992)	40.1	4.6	0.0	38.7	0.3	16.3
Tokyo (Japan) (1992)	59.9	0.0	0.0	13.8	0.1	26.2
* Penang (Malaysia) (1992)	49.6	0.6	0.0	29.0	0.0	20.8
Metro Manila (Philippines)*	12.6	0.7	0.0	8.4	0.7	77.6
U Chiang Mai (Thailand) (1992)	4.2	0.6	0.0	72.3	0.0	22.9
n Hanoi (Viet Nam) (1992)	84.6	0.4	0.0	10.8	0.0	4.12

(*) Unknown year of source.
Source: PADECO Co., Ltd., 1996.

Pedestrian Facilities

Most countries have provided a range of pedestrian facilities, but in most cases, the overall provision was inadequate. There are often good examples worth considering for use by other countries. The more developed countries typically have had a longer history of pedestrian facilities. Some good and bad examples of measures and facilities are discussed next. Pedestrian facilities attempt to take care of pedestrian safety, convenience and vehicle movements. There will always be a trade-off between these needs, and evaluation of the effectiveness of the various levels of provision is necessary. The kinds of measures that have been implemented to assist pedestrian safety are described below. However, given the limited transferability of other countries' findings and insufficient accident data, the effectiveness of pedestrian facilities and remedial measures should be closely monitored.

Roadside Facilities

i. Footpaths On Urban Roads

As modal segregation offers the best protection for pedestrians, footpaths are necessary to provide them with an alternative to sharing the carriageway with vehicles. Yet despite the heavy reliance on walking in developing cities, footpaths often do not receive priority and are constructed only if adequate space is available after the necessary road width has been provided. As pedestrian counts are rarely conducted, the justification for additional or widened footpaths is rarely documented. To meet the demands on prime urban space for economic and social activities in developing countries, footpaths also need to be maintained for pedestrian usage, with a view to minimise conflict with vehicles on the carriageway. In the case of maintaining footpaths by removing obstructions and encroachment, such as roadside sellers, it will be difficult; priority locations, such as the busiest junctions or known accident black spots, should be kept clear of obstructions so that footpaths are available for pedestrians. Parking on footpaths should not be allowed.

Figure 2.2: Absence of footpath provision on a road bridge in New Delhi, India.

ii. Hard Shoulders

Where pedestrian traffic is significant but insufficient to justify a footpath, the provision of sealed hard shoulders may provide a smooth, compacted surface for pedestrians. Edge lines should be marked to provide pedestrians with a visual guide of where the shoulder ends. A coloured shoulder surface may not be affordable, but some barriers may ensure better safety for pedestrians.

iii. Pedestrian Precincts

Pedestrian zones are found in some urban streets but are rarely found in non-urban highways in developing countries, despite high pedestrian density and the predominant share of walk trips. Rare examples include the redesigning of road space and service areas for the convenience and safety of pedestrians at two railway stations in Mumbai, India, in 1982. In the People's Republic of China (PRC), some pedestrian-facilitated streets allow pedal cycle access at certain times of the day. As a compromise, cycle sharing with pedestrians is preferable and but can force the NMV to use dangerous alternative routes where they may conflict with motorised vehicles. Provisions for pedestrians have limited potential for improvement in the developing world.

iv. Crossing Facilities

Efficient pedestrian crossing facilities could arguably be the most important pedestrian safety point, as the majority of accidents involving pedestrians occur while crossing a road.

v. Pedestrian Refuges

Pedestrian refuges are a common crossing feature in developed countries, yet they are rarely used in developing countries. In South Africa, a properly designed refuge island is expected to reduce pedestrian risk by 50 per cent (ref. Centre for Scientific and Industrial Research [CSIR], 1992). Where pedestrian and vehicle flows are insufficient to justify controlled crossings, pedestrian refuges allow pedestrians to cross the road in two stages. Pedestrians should not have to cross more than two lanes of traffic at a time with a central median provided for refuge. Many of the capital cities in the Asia Pacific region have wide roads and would benefit greatly from the introduction of pedestrian refuges to ease the crossing problems. Pedestrian refuge islands should be at least 2 m wide and preferably painted with reflective markings to maximise attention. Where space does not allow a 2 m refuge, a 1.5 m refuge can be used. In Beijing, PRC, the standards are more restrictive with a preferred minimum width of 1.8 m, although the absolute minimum width is no less than 1.2 m.

vi. Medians

For pedestrians, the benefits of medians are rarely appreciated. In developing countries, on most roads, crossing movements are uncontrolled, and where kerbed medians are used, the design should be pedestrian-friendly and avoid introducing unnecessary obstacles or risks to pedestrians. Along some medians, trees are also planted, causing visibility problems, mainly on bends. The central medians are often built with barriers that provide no ingress or egress for vehicles. Medians constructed from low-concrete blocks provide little protection from traffic. The design principles recommend a minimum central reserve width of 2 m; the absolute minimum allowed at restricted sites is 1.2 m.

Research was conducted in India by the Central Road Research Institute on the types of medians being constructed in New Delhi and their relative effectiveness. Dual carriageways may cause negative impacts on speeds, but are appreciated for improving road safety by separating traffic flows, preventing head-on collisions, and allowing pedestrians to use the median to cross in two stages. However, in some cases, increased speeds may outweigh the safety benefits. For instance, a 300 m section of a highway in Papua New Guinea was upgraded to a dual carriageway and thus, while head-on, rear-end, and sideswipe accidents were reduced significantly, accidents involving pedestrians and off-road accidents doubled. Additionally, while the multivehicle accident types were minor, the sideswipe accidents, although smaller in number, involved mainly serious and fatal accidents (see Hills et al., 1991).

vii. Zebra Crossings

In some developing countries, pedestrians are provided with the 'zebra' crossings, but in most cases, the actual benefits are too little. Studies conducted over the years in Dhaka (Bangladesh), Surabaya (Indonesia), Karachi (Pakistan), Colombo (Sri Lanka) and Bangkok (Thailand) have confirmed that drivers mostly do not stop for pedestrians using crossings (ref. Sayer, 1994). Zebra crossings are the most affordable formalised pedestrian crossings, but the disrespect shown to them by drivers renders them of little practical use in any event. Enforcement by surveillance cameras and electronic 'tags' (FASTag) may be implemented as part of an intelligent transport system.

Figure 2.3: Poor planning of a pedestrian crossing in New Delhi, India, leading to a heavily planted verge.
Figure 2.4: Barriers erected to prevent motor vehicles parking on the footpath obstructs pedestrian access to the road

Figure 2.5: A traffic signal-controlled junction with a pedestrian phase. Note the lack of a formal paved. footpath leading to the crossing facility.
Figure 2.6: A confusing pedestrian crossing arrangement at a major junction in New Delhi, India.

Figure 2.7: Visibility of pedestrians, especially children, has been markedly improved where special railing, such as Visirail, is used.
Figure 2.8: An example of a pedestrian crossing in New Delhi, India, with central guard railing, Note, however, the narrow central median.

Figure 2.9: School children using a raised at a pedestrian crossing in Fiji.
Figure 2.10: School children are crossing the road media under the direction of a crossing patrol. Note the barriers at either side of the crossing.

Traffic Signal-controlled Pedestrian Crossings

Signalised pedestrian crossings, unlike zebra crossings, offer priority to vehicles, with pedestrians allowed to cross only on signal. Unfortunately, like zebra crossings, they are not self-enforcing and rely on driver compliance with crossing regulations. In Bangkok, Thailand, surveys found signalised crossings to have a higher compliance rate (39 to 74 per cent) than did police-controlled (20 per cent) crossings. Where pedestrian volumes are high and pedestrians need to be platooned, signalised pedestrian crossings are recommended to avoid continual disruption of vehicle streams. Pedestrian-operated signal-controlled crossings (pelican crossings) have been proposed at several locations due to this factor. Phase length is determined by walk speed, and 1.2 metres per second (m/s) is the common standard used, although slightly higher values have been used in Beijing (PRC) (1.3 m/s) and Bangkok (Thailand) (1.6 m/s to 1.7 m/s).

i. Raised Pedestrian Crossings

Raised pedestrian crossings were introduced in Karachi (Pakistan) in 1987 at three locations. In addition to forcing drivers to stop for pedestrians, raised pedestrian crossings had the potential benefits of (1) reducing approach speeds, reducing the likelihood of an accident and its relative severity. Pedestrian judgement perception would benefit as the safe gap misperception increases sharply with vehicle speed. (2) Increased pedestrian usage as the crossing, unlike other crossing points on the road, is at the same level and is an extension of the footpath and (3) increased visibility and conspicuity is achieved if the height of the pedestrian crossing is raised so they are more easily visible to drivers.

The recommended layout of the crossing must include a triangular warning sign preceded by two sets of road humps of increasing height, followed by a raised zebra crossing of further increased height. In implementing the measures, care should be taken that the preceding alerting and warning humps are not constructed higher than recommended, while the hump at the crossing is not lower.

ii. Grade-separated Pedestrian Crossings

There is an increase in traffic levels both by vehicles and pedestrians, and a lack of compliance with pedestrian crossings. This has resulted in the use of pedestrian overpasses more often, as they can accommodate pedestrians without interference from vehicle flows. The 'Sixth Malaysia Plan' funded the construction of 29 pedestrian overpasses, and in Dhaka (Bangladesh) and Kathmandu (Nepal), permanent and temporary steel footbridges were constructed. Overpasses will not be used if they are perceived to require longer travelling time, so railings are often required to prevent pedestrians from crossing, despite it being more dangerous. Usage of pedestrian overpasses in Malaysia ranged from under three per cent to more than 96 per cent, but no information was given on the presence of railings. A 6.5 m high pedestrian overpass was constructed outside a boys' school in Karachi (Pakistan), and to promote usage, 1.5 m railings were put up on the central reservation and for 150 m along each side of the road. Despite these measures, surveys found more than two-thirds of pedestrians crossed at-grade-level by either climbing over or finding gaps in the railings. It was estimated that crossing on the overpass took 12 seconds longer (a total of 45 seconds) than crossing at-grade.

iii. Junctions

It appears that most accidents involving pedestrians, as seen in the data, occurred away from junctions. This might be due to dispersed crossing demand. The lack of pedestrian consideration generally shown at junctions gives pedestrians little incentive to go some distance away to cross at a junction. The locations for pedestrian crossings must be carefully considered, as they are often placed awkwardly, obstructed by trees or street furniture. As pedestrians do not always require the full width, driver violations often occur, and a revised value of between 4,000 ped/h/m and 5,000 ped/h/m has been recommended, with crossing widths of 2.5 m instead of 4 m, allowing more crossing time. Pedestrian refuges have also been recommended for junctions to allow a two-stage crossing. If the refuge is situated further back, where the road is narrower and vehicle movements are more uniform, the pedestrians will still choose to use the refuge.

iv. Traffic Signal-controlled Junction Pedestrian Crossings

The all-red pedestrian phases are extremely rare, and pedestrians often must compete with the flow of vehicles, especially those turning right. The Calcutta Traffic Engineering Project (CTEP, India), in the late 1970s, recommended banning left turns when there were more than 200 vehicles per hour. A recent TRL Overseas Road Note, "The Use of Traffic Signals in Developing Countries", summarised the range of pedestrian facilities at signalised junctions, as shown in Table 2.4. An all-red phase that is built into the cycle would still be red if no pedestrians were crossing and would be likely to incite noncompliance with traffic signals by drivers; it is therefore discouraged. The pedestrian crossing stage should follow the vehicle stage on the critical approach.

Table 2.4 Summary of Pedestrian Facilities at Signalised Junctions

Type of facility	Characteristic
No pedestrian signal	Traffic signals, even without signals for pedestrians, can help pedestrians to cross by creating gaps in traffic streams. Especially applicable where there are refuges and on one-way streets.

Full pedestrian stage	All traffic is stopped, as demanded by push buttons. More delays to vehicles than the combined vehicle/pedestrian stages.
Parallel pedestrian stage	The combined vehicle/pedestrian stage is often accompanied by banning vehicle movements. Useful across one-way streets.
Staggered pedestrian facility	Pedestrians cross one half of the carriageway at a time. A large storage area in the centre of the carriageway is required. Staggered preferably to face oncoming traffic.
Displaced pedestrian facility	For junctions close to capacity. The crossing point is situated away from the junction but within 50 m. Normal staging arrangements as above.

v. Roundabouts

On properly designed roundabouts, the overall crossing area is usually increased; here, adequate deflection will result in the reduction of vehicle approach speeds. Vehicle channelling is created when pedestrian refuges are incorporated within the traffic islands. In case no safe and convenient crossing exists, the pedestrians are then likely to choose the most direct route and cross onto the centre of the roundabout. This is common behaviour in most countries in the developing world.

(a) General Design Considerations

i. Visibility

Extreme low levels of driver compliance cannot be fully excused by visibility obstruction. In Bangkok (Thailand), pedestrian delay was reduced where pedestrian crossings had flashing beacons installed. In developed countries, zigzag markings on the approach and exit of zebra and signalised crossings are used. In most developing countries, motorists are not allowed to wait, park, or overtake as these could hide a pedestrian while crossing, but they do not have any zigzag markings on the approach and exit of zebra and signalised crossings, additional lighting, or warning signs. Road signs are also outdated, small and almost never reflective.

ii. Lighting

Visibility is a basic requirement of all pedestrian crossings, and lighting of even a modest standard can reduce nighttime accidents by about 30 per cent on main roads. Accidents involving pedestrians appear to be low at night, and that could be due to reduced pedestrian movements. Furthermore, nighttime accidents are not accounted for due to a lack of count data, but may be over-represented in terms of trips. All at-grade-level pedestrian crossings (mid-block and junction) should be checked for adequate lighting. The whole crossing should be properly lit, and the pedestrians should not walk in the shadows; to ensure this, supplemental lighting may be provided. The design of street lighting is discussed in detail in Chapter 3.

iii. Railings

To direct pedestrians towards safe crossings and away from high-risk locations, railings can be installed. Research conducted in the UK found that the presence of railings on roads with zebra crossings located every 92 m reduced the percentage of crossings in between the zebra crossings. At accident black spots, on primary or higher-class roads, where vehicle flows have priority over pedestrians, railings should be used to restrict pedestrians from crossing. On local roads, provisions of railings should be restricted at accident black spots or vulnerable areas, such as at schools or parks. At present, railings seem to be used primarily at pedestrian overpasses. Railings in developing countries will need to be higher than those used in developed countries, as they should prevent pedestrians from climbing over, and should be see-through so they do not hamper visibility.

2.1.10.2 Non-motorised Vehicles (NMVs)

In cities and towns throughout the developing world, particularly in Asia, the NMVs such as pedal cycles, cycle rickshaws, animal carts and handcarts have traditionally been an essential means of transporting people and goods. However, in recent years, they have come under an increasing threat, and that has resulted in them disappearing in some major cities in Asia, for example, Kolkata, India and Jakarta, Indonesia. A study commissioned by the World Bank and carried out by PADECO Co., Ltd. in 1995 reported that the following factors threaten the future of NMVs in Asian cities:

(a) Increased motorisation and an increased use of motorcycles have created more traffic congestion and a consequent reduction in street space available, which has reduced the speed of vehicles and made roads safe for NMV use.
(b) Changes in metropolitan spatial structure caused increased trip lengths.
(c) Inadequate facilities for NMVs have resulted in the exclusion of NMV needs in urban transport planning and investment programmes.

(d) The modernisation of cities in the developing world has promoted the attitudes that 'NMVs are backward',
(e) A tendency has generated to believe that NMVs are the cause of urban congestion.
(f) Restrictions, limitations on NMV ownership, excessive and often inappropriate regulation of NMV operation, and total or partial banning of their use.

However, in several countries, government road and planning departments were observed to be providing facilities specifically for NMVs. These facilities provide significant potential for solving safety and capacity problems. All NMVs are also VRUs. However, in many countries, reported accident rates were observed to be relatively low, particularly when compared to the statistics for motorcycles and pedestrians. A high level of accidents involving cyclists occurs in certain cities in Beijing, the PRC, where cyclists are involved in about 70 per cent of the total traffic accidents. In the PRC, in most cities, cyclists generally constitute the largest proportion of the traffic flow. It is recognised that pedal cycles, cycle rickshaws and other NMVs all have different operating capabilities and thus have different associated problems, although the data for the separate categories of NMVs may be included in the report. It should be noted that any facility should consider the expected NMV traffic composition during the design stage.

NMV Facilities

It was observed in the previous section that large numbers of NMVs are in daily use in many developing countries. These have obvious importance as a mode of transport, and a few cities in the region specifically provide measures to assist NMVs. However, many roads and networks in some cities are still managing the traffic without any consideration for the needs of NMVs. Indeed, not one city was observed to provide a linked network for popular journey patterns of NMV users. Instead, some interesting and useful solutions can be observed operating successfully throughout the region to help with high NMV levels.

This section is intended to illustrate and highlight successful NMV facilities from around the region. However, these discussions only indicate the general form that these measures can take. The facilities discussed are suitable for pedal cyclists and other NMVs, but it should be noted that the design of any facility should consider the expected NMV traffic composition.

a. Road links

Road links are sections of roads between junctions. To assist NMVs, various facilities can be provided along the road links. These facilities can be categorised into two segregated measures, either by physical or visual means, or measures that help NMVs within a mixed traffic environment. Segregating measures usually separate motorised and non-motorised traffic, but sometimes they consider four-wheeled motorised vehicles separately from the rest of the traffic. The most suitable type of segregation depends on the factors associated with the road hierarchy and land use, such as traffic speed and volume, traffic composition, and vehicle access. High cycle flows may warrant the provision of a segregated facility and will help govern the actual physical dimensions of the facility provided. Total segregation can also be provided where NMVs run on a separate network of paths to motor vehicles, but no examples of this type have been identified in the region. Where NMV traffic is prevented from using certain roads, it is important that an alternative facility exists for NMV users and that these alternative routes are attractive in terms of safety, comfort and journey time.

i. Physical Segregation

Physical segregation between non-motorised and motorised vehicles using a barrier or raised track is intended to maximise safety along road links that have high vehicle speeds, flows or a mixture of both. The actual barrier or height difference between the surfaces prevents the interaction of the two types of vehicles. The infrastructure involved in physical segregation requires high initial implementation costs compared to visual segregation. Delineating dedicated cycle lanes alongside major roads is helped by using raised kerbs. With the provision of adjacent verges, safety can be increased by further separating the two modes of vehicles (Figures 2.11 and 2.12). The height of a raised track or verge edge should be sufficient to stop motor vehicle encroachment onto the segregated track. Furthermore, it is also often necessary to prevent illegal use of these paths by cars.

Figure 2.11 Physical segregation by a tree-lined verge Continuous physical segregation in Ho Chi Minh City, Viet Nam.

Figure 2.12 Continuous physical segregation in Ho Chi Minh City, Viet Nam. kerbing along an arterial road in Yogyakarta, Indonesia

Alternative segregation methods can include periodic barriers, spaced at intervals along the route that may be in the form of upstanding block work, bollards, posts, or even old oil drums, as well as actual continuous barriers such as low railings and walls (Figure 2.13).

Figure 2.13: Railings used to segregate cyclists and motorised vehicles in the PRC

Physically segregated paths can be used for two-way cycle flows. However, their design must ensure that a satisfactory width has been provided to accommodate two- and three-wheeled NMVs, along with any overtaking manoeuvres (Figure 2.14). Care also needs to be taken in the design of merging and diverging areas where non-motorised and motorised vehicles mix. Segregated facilities for NMVs at pinch points, such as road bridges throughout the region, are rare. However, a few good examples do exist, as in Dhaka (Bangladesh) (Figure 2.15).

Figure 2.14: Slow-moving two-way vehicle lane in Yogyakarta, Indonesia.
Figure 2.15: Physical segregation for NMVs traffic over a road bridge in Dhaka, Bangladesh.

Good examples of carriageway separators can be found for slow-moving vehicles (SMV), which are intended for one- and two-way flows, according to the road hierarchy. These lanes allow safe and direct travel for NMVs in the town centre and on the arterial routes into and out of the city. Lane widths tended to be about 5 m wide, allowing three cycle rickshaws to pass side by side. Signs limit the lanes for NMVs and motorcycles, and cars and lorries are explicitly banned. The provision of further SMV lanes is also recommended (Figure 2.46). Some geometric designs include NMV lanes on national, regional, and feeder roads. The provision of cart or bull tracks is also discussed for feeder roads (Figure 2.46 (c)).

ii. Visual Segregation

White line markings on the carriageway provide a visual segregation of non-motorised and motorised vehicle traffic. In developed countries, these nearside carriageway markings can take the form of either broken white lines, termed advisory cycle lanes (Figure 2.15), or continuous solid white lines, termed mandatory cycle lanes (Figure 2.16). These lanes, where they exist, are recommended for cyclists to use. To enter these restricted lanes only when necessary ensures that the cyclists are not impeded. These facilities are dependent on driver compliance and effective enforcement of the traffic regulations. The type of lane to be implemented depends on road hierarchy and land use factors such as carriageway width and parking requirements in the area. For instance, parallel parking alongside the pavement will discourage cycle use in the lane and render it ineffective. Flows along these lanes should be in the same direction as the traffic.

Lane widths should not be so wide as to encourage two-way cycle flows or to encourage illegal use by motor vehicles. The integrity of the actual lane can be enhanced by providing wide markings, double-line markings or even hatched areas to denote the separation. Alternatively, yellow paint has been used to help denote the lane. Lanes observed in some developing countries tended to be 1.5 m to 2 m wide, making them suitable for bicycles, but wide enough for only a single rickshaw, which forces overtaking NMVs to stay outside the lane.

Although standards in developed countries recommend that physical segregation also be provided at the start and end of contra-flow cycle lanes, which are usually denoted by a continuous white line, contra-flow cycle lanes can help provide an area-wide network for NMV users by creating direct and cohesive routes. Safe design requires crucial consideration of the lane widths to reduce the tendency to travel out of the lanes onto the main carriageway.

Figure 2.16: Visual segregation for SMVs using broken white lines in Yogyakarta, Indonesia.
Figure 2.17: Visual segregation lines in Dhaka, Bangladesh

Figure 2.18: Contra-flow cycle lanes in (PRC). Shanghai.
Figure 2.19: One-way motor vehicle flow and one-way cycle flow system in Shanghai (PRC).

iii. Mixed Traffic

Under certain conditions, the planned use of mixed traffic can be relatively safe, such as low speeds and flows of motor vehicles, along with an appropriate traffic stream composition. Safety aspects relating to VRUs in terms of speed and traffic reduction measures are to be monitored. Having improved the surface condition by road rehabilitation projects, there is an increasing number of accidents occurring on rural roads. The improved road surface encourages higher vehicle speeds, which accentuate even further the differences between motor vehicles and NMVs. When such rehabilitations are undertaken, additional safety measures need to be provided to avoid problems. This is particularly to be considered where large numbers of pedestrians, NMVs or SMVs can be expected.

Where high pedestrian flows exist, the provision of facilities for joint use by pedestrians and NMVs without segregation is not recommended and should not be the preferred solution if there are alternatives. On busy radial routes, this approach was observed.

The widening of nearside road lanes by altering the position of lane markings is being carried out in Australia, Europe and the US, where mixed traffic does exist. A greater clearance distance between cyclists and overtaking motor vehicles is allowed by this practice. Care must be taken, however, to ensure this inside lane width is of a size to accommodate two rows of motor vehicles during periods of traffic congestion.

b. Junctions and Crossings

At junctions, segregation of traffic is not always possible or suitable because the potential for conflicts between non-motorised and motor vehicles is greatest. Conflicts at junctions account for a large proportion of all cycle accidents. Examples of conflicts between the pedal cycle and motor vehicle at junctions are shown in Figures 2.20 and 2.21.

Considering their specific vehicle characteristics, NMVs must be given full and proper consideration at junctions as part of the network design process. An example of poor planning is causing a potential conflict point due to the right/left-turning traffic having to cut across the cycle lane with no priority markings or signs provided, as shown in Figure 2.22

Figure 2.20: A potential conflict at a junction between a cyclist and a lorry in Hanoi, Viet Nam
Figure 2.21: Example of poor planning at a junction in Shanghai, PRC. Note that traffic turning right must cut across the adjacent cycle lane.

Figure 2.22: Conflicts between cyclists, motorcyclists and motor vehicles at a junction in Shanghai, PRC.

Next, the measures to assist cyclists and other NMVs at priority junctions, signal-controlled junctions and roundabouts are discussed below, along with grade-separated facilities such as footbridges and underpasses. To provide clear instructions and directions to motor vehicle drivers and NMV users, all junctions should be appropriately marked and signposted. The design of any junction with specific NMV provision should include all the normal geometric factors, including sightlines and visibility requirements that would be considered for any other vehicle.

i. Priority junctions

NMV users should not be prevented from carrying out a full range of legal turning manoeuvres at junctions and should expect the same privileges as drivers of motorised vehicles whenever possible. The junctions need careful design for manoeuvres like weaving by cyclists and motor vehicle drivers, or to be in the appropriate turning lane, as a common practice. For example, turning NMVs are often extremely exposed when they are waiting in the centre of the carriageway for a gap in the traffic. It is suggested that priority junctions be designed to minimise the interaction time by reducing the distances to be crossed, say, by reducing turning radii, which helps reduce conflicts. Short turning movements can be helped by leading cycle lanes away from the junction and onto the adjacent junction arm. There are no specific measures that have been identified to assist NMVs at priority junctions in the region until the study was done.

Turning restrictions via physical measures on motor vehicles can have beneficial effects on safety, as the conflicting interaction between vehicles is reduced. There are beneficial safety consequences for NMVs with reduced vehicle speeds and reductions in motor vehicle priority. In some developing countries, the traffic police were commonly observed controlling flows and access at junctions. This can be an effective method of traffic control with benefits to NMVs when the police are properly trained.

ii. Signal junctions

In some countries, some traffic signals for bicycles exist on roads with heavy bike use (Figure 2.23), which respond to bicycle-sensitive inductive loops in the road surface.

It is to be noted that a complex signal-controlled junction is an expensive solution. For existing widespread NMV usage, other less expensive traffic management alternatives are to be considered first at signal-controlled junctions where traffic levels are high.

Figure 2.23: Separate traffic signals for cyclists at a junction in Beijing, PRC.
Figure 2.24: An advance stop line for cyclists at a signal-controlled junction in Market Harborough, UK.

To reduce the number of potential conflicts, all opposing traffic movements, regardless of vehicle type, are to be prohibited, where opposing turning manoeuvres have been banned over a series of junctions before a roundabout. Cyclists are particularly vulnerable at roundabouts, and a variation in the above method may be considered. The variation prohibits opposing vehicle turning movement but allows NMVs to cross the major road from the minor road using a straight crossing manoeuvre.

iii. Roundabouts

At roundabouts, cyclists are particularly vulnerable due to the high levels of conflicting movements and speeds; accidents can occur there at any time. The design of roundabouts is required to be done carefully to help reduce approach speeds. In developed countries, research has led to some special measures to help cyclists at the roundabouts. There are three types and involve either mixed flows on the roundabout itself, or some physical segregation on parts of the roundabout, or a separate track for cyclists around the outside of the roundabout. Mixed flow roundabouts are featured with small central islands and single-lane approach roads with single-lane circulatory movement, while segregation measures reduce the risk of NMVs being hit on the roundabout by vehicles entering the system. In some cities, movable barriers have been provided at the roundabout, although their effectiveness is yet to be known.

iv. Grade-separated crossings

Sometimes pedestrian foot bridges (FOBs) have a central wheel ramp. These ramps were initially intended to help cyclists push their cycles along, but eventually, motorcyclists were observed riding up and down the ramps (Figure 2.34). Compared to underpasses, FOBs have a lower construction cost but are less ideal for joint cycle and pedestrian usage. These are generally narrower than underpasses, thus increasing the potential for conflict. Also, the rider must dismount the bicycle and wheel it along the ramp, which is not convenient.

Underpasses provide safe grade-separated crossings of carriageways that would otherwise be too dangerous or inappropriate to cross. To be well-used, underpasses, as with FOBs, need to be conveniently situated. That may attract a high volume of pedestrians and cyclists. Dimensions of underpasses are important in allowing access to them for cyclists. Wherever possible, pedestrians and NMVs should be segregated to provide safe access for both. Approach ramps should not be so steep that they invite high speeds through the underpasses or make the exit too strenuous. Underpasses with specific NMV provisions are rare.

c. Parking facilities for NMVs

The provision of parking facilities to provide safe storage of the cycle and reduce the threat of theft or vandalism is an essential part of the cycle network. Misuse or illegal provision of cycle stands located in areas with high-volume traffic causes congestion. However, the threat of theft is reduced by the presence of an attendant. Such stands that fulfil these requirements are increasingly being provided with shelter from the weather.

Official waiting areas for cycle rickshaws are rare, and many clusters of cycle rickshaws can be seen operating from various street corners and major junctions in Asian countries. Rickshaws waiting at the edge of the carriageway, reducing the effective carriageway width and hence the road capacity, are shown in Figure 2.24.

Pedal cycles parked on the footpath block pedestrian movement and force pedestrians to walk in the road. This is a problem associated with the lack of suitable parking arrangements and is shown in Figure 2.22.

Figure 2.25: Rickshaws parked on the side of the street due to a lack of suitable waiting bays in Yogyakarta, Indonesia
Figure 2.26: Pedal cycles parked on the footpath blocking pedestrian access in Hanoi, Viet Nam.

2.1.10.3 Motorcyclists

Motorcycles account for a large proportion of all vehicles in various developed and developing countries. Motorcycles, due to their comparatively low cost, are purchased and used by young drivers, and these tend to be their first affordable motor vehicles. Unfortunately, they all contribute towards making young riders high-risk road users, for having high-risk thresholds, limited training and testing, and, obviously, a lack of experience.

As large motorcycle fleets exist relative to the number of motor vehicles, then unsurprisingly, accidents involving motorcycles tend to be higher. This vulnerable mode of transport has continued growth, which may accelerate motorisation-related death rates in some countries. In several countries, motorcycles are involved in a substantial share of fatal and injurious accidents, as given in Tables 2.1 and 2.2. Table 2.5 shows the extent of the involvement of motorcycles in fatal and injurious accidents.

Table 2.5 Percentage of Fatal and Injurious Accidents Involving Motorcycles

Sub-region	Country	Fatal accidents (per cent)	Injury accidents (per cent)
NIEs	**Hong Kong,**	**10**	**17**
	China Republic	**7**	**6**
	of Korea	**51**	**69**
	Singapore	**54**	**62**
	Taipei, China		

Central Asia	**People's Republic of China (1994)**	**10**	**NA**
Southeast Asia	**Malaysia (1994)**	**57**	**74**
South Asia	**Sri Lanka**	**16**	**26**
PDMCs	**Fiji**	**NA**	
Developed	**New Zealand**	**16**	**14**

a. Motorcycle Facilities and Accidents

Although few physical facilities may improve motorcycle safety, some measures have been identified and are considered important. Motorcycle safety can be increased with the separation of two- and three-wheeled motorcycles from large, high-speed vehicles. Furthermore, the safety of motorcyclists will benefit from speed reduction measures where there is mixed traffic. This can be seen in the section on traffic calming and speed reduction measures.

Table 2.6 Percentage of Registered Motor Vehicles that are Motorcycles and Their Rate of Growth

Sub-region	**Country**	**Percentage of motorcycles out of all Motor Vehicles (1993)**	**Percentage change in registered motorcycles (1989-1993)**
NIEs	**Hong Kong, China**	**6.3**	**+41**
	Republic of Korea	**23.6**	**+63**
	Singapore	**20.5**	**-1**
	Taipei, China	**73.7**	**+44**
Central Asia	**People's Republic of China (1994)**	**36.8**	**+139**
Southeast Asia	**Indonesia**	**68.4**	**+22**
	Lao PDR	**NA**	**+80**
	Malaysia	**56.8**	**+30**
	Philippines	**26.8**	**+65**
	Thailand	**66.6**	**+75**
	Viet Nam	**NA**	**+175**
South Asia	**Bangladesh**	**NA**	**+48**
	India	**67.3**	**+60**
	Pakistan	**48.1**	**+32**

Sub-region	Country	Percentage of motorcycles out of all Motor Vehicles (1993)	Percentage change in registered motorcycles (1989-1993)
	Sri Lanka	NA	+68
Pacific	Tonga	NA	+45
Developed Countries	Australia Japan New Zealand	2.8 21.0 NA	-8 -10 -38

b. Road links

By separating motorcycle traffic from other motor vehicles, increased safety can be achieved. This segregation can take any one of two options. Either exclusive motorcycle lanes can be created with the provision of a raised median, or joint lanes can be provided for motorcycles and SMVs. Pedal cyclists and other NMV can also use these joint lanes.

Table 2.7 Registered Motorcycles in the PRC

Year (10,000)	Motorcycles (10,000)	Motor vehicles	Motorcycles as a percentage of total motor vehicles
1990	359.33	1318.53	27
1991	421.28	1,476.26	29
1992	506.15	1,657.66	30
1993	647.49	1,946.03	33
1994	858.79	2,331.64	37
1995	1,093.81	2,736.60	40

Figures 2.26 and 2.27 show examples of shared motorcycle and pedal cycle lanes in Kuala Lumpur, Malaysia, and Viet Nam. Differences in lane width are clear from the photographs; sufficient lane widths can be determined from traffic counts and forecasted flows of vehicle modes.

Figure 2.27 Segregated motorcycle lane and shared pedestrian and pedal cycle lane in the PRC.
Figure 2.28 Motorcycle and pedal cycle SMV Lane in Ho ChiMinh City, VietNam

Figure 2.29 SMV Lane at a junction in Ho Chi Minh City.
Figure 2. 30 Shared motorcycle and pedal cycle lanes In Kuala Lumpur, Malaysia.

Figure 2.31 Shared motorcycle and pedal cycle lanes in Kuala Lumpur. Notice the explicit signs.
Figure 2.32 Motorcyclist using the main carriageway in the PRC. Note the segregated cycle lane.

Figure 2.33: Motorcyclists forced to use the carriageway in Viet Nam.

Figure 2.34: Posts placed across a road prevent four-wheeled vehicle access in Bandung, Indonesia

Figure 2.35: Motorcyclist using a central ramp intended for pedal cyclists on a footbridge in Jakarta, Indonesia
Figure 2.36: Motorcycle parking signs in Yogyakarta, Indonesia.

Figure 2.37: Motorcycle parking on the footpath blocking pedestrian access.
Figure 2.38: Motorcycle parking blocking a lane on the carriageway.

Clear road signs are also evident for the joint use of the lane, as shown in Figure 2.31. The provision of slow-moving lanes to accommodate motorcycles and pedal cycles was also proposed, with separate lanes of about 3.5 m to 4.5 m wide on primary collector roads. In certain circumstances, the shared use by NMVs and motorcycles of separate SMV lanes is generally not allowed, and motorcycles are directed to use the main carriageway, as shown in Figures 2.32 and 2.33.

Sometimes, motorcycle and motor vehicle separation can be achieved by allowing the shared use of bus lanes. However, full consideration of the traffic flows of both types of vehicles is important, because the shared use at specified times of the day could only be a possible acceptable measure. Care must be taken not to encourage the sharing of all facilities, such as pedal cycle measures at junctions or even on footbridges, due to the differences in respective vehicle speeds. Figure 2.35 illustrates the misuse of a central cycle ramp by a motorcycle.
Alternative road link measures include posts at the entry/exit points to prevent access of four-wheeled vehicles, as shown in Figure 2.34.

c. Road surface and traffic engineering

Several Australian safety specialists have highlighted the need to consider street maintenance and traffic engineering, as these can have a serious effect on motorcycle safety. The specialists suggested the areas for consideration regarding traffic engineering and motorcycle safety as follows:

i. Highway geometry:

For motorcyclists, an adverse camber is especially hazardous because of the need to lean even further for balance. Combined with uneven road surfaces or localised low-friction areas can be dangerous for motorcyclists.

ii. Road surface:

1. Drainage covers and steel plates to cover road works may often become slippery. The application of an abrasive coating reduces the danger.
2. Poorly designed speed breakers and traffic calming measures can be dangerous to motorcycles. Designs such as those advocated by TRL and the Australian Road Research Board should be used.
3. On corners of sealed roads, loose gravel exists on unsealed shoulders and can be dangerous to motorcyclists. Such problems can be resolved by sealing a curved section of the kerb and gutter at junctions on roads.

iii. Junctions:

Motorcycles are frequently not noticed as they can be hidden by other vehicles, so at signalised junctions, wherever possible, fully controlled turns should be used to avoid accidents. For motorcycles, sharply flared junctions and small roundabouts with painted central islands can be dangerous. Road surface loop detectors are sometimes too insensitive (due to the lightweight of motorcycles) and can fail to detect motorcycles or allow them to clear the junction before the lights change.

iv. Road markings:

Different types of road markings (for example, painted arrows, lane and edge lines or zebra crossings) may cause variations in skid resistance, during wet weather or on bends. Raised markers that are used to delineate lane and edge lines can also cause problems, especially on bends.

v. Safety barriers:

A collision with a normal W-beam (Armco) steel barrier can be dangerous for motorcyclists, who can be catapulted over or under the barrier and reflectors on the barriers can cause injury to a rider sliding along the barrier. Wire rope barriers and their support posts, along with end treatments, can cause severe injury to motorcyclists. From a motorcyclist's viewpoint, the concrete tapered barrier, similar to the types in New Jersey, is probably the least damaging in the event of an accident.

vi. Roadside furniture and objects:

Roadside vertical kerbing can cause additional dangers to motorcyclists sliding along the road after an accident. It should be of a rounded semimountable design. Roadside poles, if placed too close to the kerb at corners, can be dangerous as motorcyclists must lean over on super-elevated carriageways as they go round a bend. It is now clear that particular care needs to be taken in the design of road and traffic engineering facilities where many motorcyclists can be expected in the traffic stream. Although such measures will not eliminate motorcycle accidents, they will minimise their occurrence and reduce their severity when they do occur.

d. Parking facilities

Parking facilities are an essential part of overall motorcycle requirements. Facilities should be well signposted and be able to cope with the expected number of motorcycles. Parking exclusively for motorcycles could be considered, where motorcycles form the largest share of all modes in high-volume traffic. Signs are an integral part of motorcycle parking facilities and are essential; they can be seen in Figure 2.36. Figures 2.36 and 2.37 clearly illustrate that the absence of adequate parking provision can block footpaths and roads. To reduce the threat of theft, the parking should have attendants.

Integrated Programmes

Many motorcycle road safety initiatives rely on associated measures such as road safety education and suitable traffic regulations, rather than physical engineering measures. These measures in the past were typically based on rider training and helmet legislation.

a. Enforcement

If correctly legislated and enforced, motorcycle helmet wearing to a specified standard can reduce the severity of motorcycle accidents. Legislation must provide for the compulsory wearing of helmets while riding a motorcycle of any size and speed; this must be backed by enforcement. Traffic police using motorcycles must be seen to wear the correct helmet type, have the helmet fastened, and enforce their correct usage. Pillion passengers should also wear helmets. Regulations concerning the size and speed of motorcycles should also be considered, along with the use of graded licences based upon age and experience.

b. Education, training, and publicity

Training motorcyclists, as with any motor vehicle driver, is an important safety requirement and should be carried out by trained and qualified instructors. Training should cover information not only based on the development of driving skills and the driver's responsibilities, but also on safety factors such as the risk of not wearing an approved safety helmet. Figure 2.44 shows motorcycle training and testing in Singapore.

A Malaysian road accident report (1994) mentioned that motorcycle riders were at fault in 50 per cent of the accidents they were involved in and that more than 50 per cent of the motorcycle riders involved in accidents were aged 16 to 26. This highlights the need for suitable training for young riders. Potential dangers associated with motorcycle riding should be publicised (Figure 2.46) along with enforcement and legislation measures. Safety publicity should be adequately strong to overpower other influences presented in the mass media and draw the people's attention. Images that promote speed as glamorous when advertising powerful motorcycles should be banned.

Figure 2.39: Motorcycle driver training and testing

Figure 2.40: Signs on the side of the road in Yogyakarta, Singapore

One of the most effective areas for education and publicity is to ensure the visibility of motorcyclists. Many accidents occur because other vehicle drivers do not see motorcycles. Consequently, efforts need to be made to encourage motorcyclists to increase their conspicuity. This can be done in the following ways:

i. Keeping headlights on (or always dipped) as this makes the motorcycle much more conspicuous to other road users.
ii. Ride wearing light coloured clothing and preferably reflective jackets.
iii. Use reflective sticky strips on the front and rear mudguards and along the sides of the motorcycle to make it more conspicuous at night.
iv. No overtaking if at least half a kilometre in front is not visible, and if visible, you must see a clear path. Avoid overtaking at black spots.
v. Indicate when overtaking on a highway or street, for example, a large bus or container lorry may shift laterally, which doesn't always mean it is giving way for overtaking; instead, such a shift may be for taking a U-turn to the other carriageway through the median cut. Any overtaking motorcycle may collide with the large vehicle taking the U-turn, as the motorcycle might not be seen in the mirror.

2.1.11 Recent Developments and Planning

The planning process may improve safety through consideration of various components, such as land use, network design, and road hierarchy.

The number and severity of conflicts that occur between motor vehicles and VRUs, or high-speed traffic and low-speed traffic, or heavy goods vehicles and other road users can be reduced by separating vehicles and preventing the interaction of various road users with careful planning. The design of safe roads using standards and specifications is important, although the planning of safe roads is also essential for the prevention of conflict. The design and suitable provision of the facilities relevant to safety for each group are to be considered as the correct provisions.

2.1.11.1 Land Use Planning

The control of land use and the traffic associated with the different types of use is essential to successful planning. Incompatible traffic patterns and vehicle compositions from various land use sources should be separated to minimise potential conflict. Hence, zonal planning should be considered to reduce trip lengths and to prevent VRU interactions with high-speed and heavy volumes of traffic. Access control is also important in preventing inappropriate traffic mixing with VRUs.

Addressing VRU issues of particular importance is the adoption of the principle of locating traffic-generating facilities adjacent to a road suited to the type of traffic expected, such as the provision of coherent and linked cycle paths and footpaths close to a school. Furthermore, by preventing the development of linked land use on either side of main roads, conflicts can be avoided; for example, schools and housing estates on opposite sides of a primary distributor road, which is often crossed by the people there.

Table 2.8 Pedestrian and Pedal Cycle Activity with Respect to Road Category

Road category	Main activity	Pedestrian activity	Pedal cycle activity
Pedestrian street	Shopping, walking	Predominant	Considerable, speed is limited by pedestrians
Access road	Walking, vehicle access, and deliveries	Considerable freedom	Considerable, crossing, weaving, stopping
Local distributor	Start/finish of vehicle trips, bus stops	Controlled	Considerable, bicycle lane, not physically segregated, no crossing facilities
District distributor	Through traffic, public transport	Minimum positive safety measures	Considerable, segregated cycle lane, crossing facilities
Primary distributor	Fast moving, through traffic	No safety measures — segregation	Controlled, segregated cycle lane

2.1.11.2 Road Hierarchy

The roads within a network should be defined and classified according to their function. In a network the roads are traditionally split into levels according to the primary use of the road and thus the associated traffic. Within the planning and design of the network and road hierarchy, in order that their own network of facilities be provided, VRUs should be considered. Hence, coherent, direct and safe routes for VRUs can be provided by measures such as linked footpaths, cycle lanes and road crossing facilities. The VRU networks are to be identified, which can allow for the planning and provision of appropriate facilities where the motor vehicle and VRU networks interact. For example, local access roads that don't need underpasses and footbridges. Similarly, on low-flow and low-speed roads, no special provisions are needed for pedal cycles and other NMV. Table 2.8 shows pedestrian and pedal cycle movements with respect to the typical traffic movements on various categories of roads.

Traffic management can have positive effects on safety through the creation of a suitable road hierarchy and network design. On certain types of roads, traffic management can limit traffic levels and speeds, thus benefiting VRUs. Traffic management should help pedestrians and road vehicles remain on their respective facilities to minimise their interaction with each other. The by-products of good traffic management using suitable road hierarchy planning and design improve traffic efficiency, traffic safety, the environment and reduce congestion levels. Appropriate planning of the road hierarchy can result in maximum priority for groups such as pedestrians and cyclists over motor vehicles. This objective is an environmental concern that promotes easier pedestrian and cycle use as well as general safety by restricting access to facilities for motor vehicles (ref. Table 2.8).

2.1.11.3 Design Standards

Design standards for VRUs should reflect the needs of each group in providing safe facilities for the movement of people and goods to reduce potential conflict. However, for a facility to be fully utilised, such measures should be attractive to potential users in terms of the direct route. Facilities need to be coordinated and cohesive to maximise their potential benefit to all users while balancing the safety requirements. The volume of VRUs within the overall traffic is an important consideration for design guidelines. In the case of NMV facilities, a separate lane for the sole use of NMVs requires a width related to the volume as well as composition of NMV traffic.

However, to define the need and type of cycle lane, it is suggested that the speed, flow and composition of motor vehicle traffic on the road are to be considered as the main factors. Pedestrian volumes are important to the level of facilities provided. Pavements should cater for the expected volume of pedestrians along a route and be of sufficient width to prevent pedestrians from walking on the carriageway, increasing the potential for conflict. A wide pavement provision may have issues with the encroachment by squatter areas, hawkers and roadside stalls, which can create hazardous conditions for all road users. To provide suitable facilities for VRUs, it is necessary to determine potential usage by carrying out pedestrian and traffic counts. This book is not intended as a design manual, and although previous sections provide general information on standards and capacity of different facilities, no attempt has been made to recommend specific guidelines relating to traffic volumes or other such design requirements.

This book only seeks to identify facilities in use in developing countries, to indicate the forms that they can take and, where possible, to comment on their perceived effectiveness.

2.1.11.4 Traffic Counts

The basis for designing traffic systems that cater to all requirements of road users is provided by the traffic counts, along with accident recording. Identifying the needs of pedal cyclists and other NMVs depends not only on the number of pedal cycles/NMVs, but also upon the number of motor vehicles in the traffic and how the two categories interact. Defining the road hierarchy and traffic flows within various land use zones relates to these two matters.

At or near major traffic-generating roads, the traffic volumes will be highest, and this should be considered during the planning of traffic counts. The details of existing traffic movement, how this flow has changed over time, and how it is likely to change in the future, can be provided by data in the form of vehicle and pedestrian counts. These are all used in the design of adequate facilities to assist road users and improve road safety. It is to be noted, as has been shown, that only limited data exists for NMV traffic, while pedestrian traffic flow is almost never counted. Data is available for motorcycles because of their place in mainstream traffic, as well as the requirement for registering vehicles, which provides some details of ownership levels.

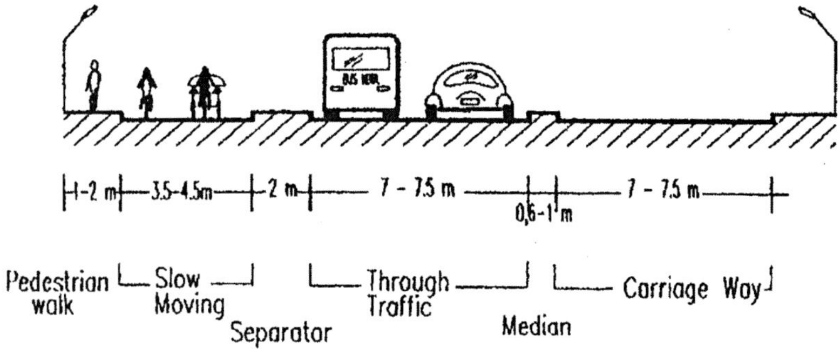

Figure 2.41: Primary collector road through an urban area.

2.1.11.5 Accident Data Systems

Data on accidents determines not only the extent and type of safety problems and the kinds of remedial measures most likely to be effective, but also provides the objective and scientific basis. The safety of a nation's road network cannot be assessed and monitored without the study of accident details. A database analysis allows an understanding of the circumstances and patterns of accidents. To complete the process and expect positive actions towards road safety, information from the analysis system must be disseminated to relevant organisations. Accident data should include general accident details such as road type, environmental features and location, as well as the vehicle/driver details and casualty details. To record and manage the accident data, the traffic police are ideally placed. Within this data collection, the pedestrian action and location should also be included. A good accident reporting system also enables investigation of hazardous locations and identifies any patterns or similarities between accidents. After identifying what could, with justification, be called the world's worst black spot, the accident analysis computer software Microcomputer Accident Analysis Package (MAAP) can also check for similarities in accident timings, vehicle manoeuvres and other factors such as the vehicle types involved.

It appears that VRU accidents are not monitored in many countries, or at least they do not publicise data findings. It is found in the UN/ESCAP survey; while accident data are centrally stored in 13 of the 15 respondent countries, only seven countries use microcomputers for accident data storage, and four countries use mainframes.

2.1.11.6 Foreign Aid Projects

In the past, foreign-funded road and transport improvement projects traditionally had a narrow focus, being concerned primarily with maximising traffic efficiency. However, in recent years, the safety problem has been receiving growing attention, along with environmental issues and the social impact created during the design of new transport infrastructure. When addressing VRUs' requirements and their true impact on recent transport developments, it has been observed that due to the lack of information on accident involvement rates and the volumes within the traffic flow, VRUs' safety and activities are still largely unknown. In the past, this lack of knowledge has contributed to the lack of consideration of VRUs in the design and construction of many projects.

2.1.11.7 Traffic Calming and Speed Reduction Measures

In Europe and on other advanced continents, two key road safety engineering ideas have been developed and implemented: traffic calming and safety audits. These have been credited with achieving accident reductions, improving the environment and avoiding expensive construction modification costs.

To calm traffic, engineering measures are used to reduce the speed and volume of motor vehicle traffic and have proven successful in developed countries. These not only reduced the risk of accidents but also the accident's severity. These have benefited all road user groups with the creation of a better environment, but are particularly beneficial to the safety of pedestrians, pedal cycles and other NMVs. Traffic calming methods involve road geometrics with respect to horizontal and vertical realignment of roads at selected locations using various methods, such as road humps and speed tables, chicanes, pinch points and footpath widening. Junctions can also be altered by raising the whole junction layout, providing entry across junctions and the use of mini roundabouts to help control speeds. Such measures should be implemented in conjunction with an area-wide set of all measures.

The benefits of speed reduction have received much attention in recent years. An international review of speed and accidents concluded that, on average, each 1.6 km/h reduction in speed would result in a seven per cent reduction in fatal accidents. Another finding is the effect of speed on pedestrian casualty severity. At 32 km/h, five per cent of pedestrians hit will die, compared to 64 km/h, where 85 per cent of adults and 95 per cent of children hit will die.

Traffic calming is meant to benefit a wide-ranging group by reducing vehicle speeds and volumes, and no specific measure for NMVs is necessary, provided that their needs are fully considered during the planning stage.

Figure 2.42: Traffic calming in Tokyo, Japan
Figure 2.43: Town centre traffic calming in the UK

Figure 2.44: Raised pedestrian crossing in Fiji.
Figure 2.45: Gateway in Fiji.

2.1.12 Safety Audit

Safety audits are primarily two-fold; first, a study of the site is carried out, and accident data is collected, and various measures for road safety improvement are proposed at the planning stage before the design and construction of the road/highway, and that is discussed in this section. The second safety audit is carried out during the construction of the road/highway to inspect two further aspects: one is whether the construction workers at the site follow the standard safety norms during construction, and the other is whether the proposed safety measures for the road users, as contractual obligations, are adopted in the design and construction of the road/highway. Any additional requirements may also be addressed at this stage, for example, additional median cuts at a hospital, or a FOB at schools, etc.

Safety audits at the planning stage refer to the formal process of reviewing road projects specifically to identify potential or existing safety hazards to allow for timely proactive correction, especially for pedestrians and other VRUs whose safety needs are often neglected in traditional transport planning and design. This is valid in most developing countries. Project authorities have a very nominal budget for road safety measures, which leads to the removal of various items from the construction schedule to meet the budget. Such items are underpasses, overpasses, railings, pavements, covered drains, service roads, footbridges and the improvement of black spots, etc. Even road markings are considered 'ornamental' at the time of new construction or any road resurfacing project.

Without road marking and CCTV cameras, the enforcement of traffic rules becomes a farce. The surveillance camera, in combination with the FASTag system, helps obtain the details of the vehicle and driver for undisciplined driving. Highway authorities should not hesitate to establish an intelligent transport system (ITS) for road safety improvement, and the safety audits should have a focus on provisions of such advanced technologies. Safety checking or safety auditing, which is a more formalised process, offers opportunities to create safer roads for VRUs.

Before a formal safety audit, the audits on road schemes can be carried out with specific reference to the VRUs' needs. In developed countries, audits of cycle routes have become the norm. The elements that can be checked at various stages for each VRU are shown in Table 2.9. It should be noted that these considerations are factors within the overall formal safety audit. A safety audit is described in detail in Chapter 1.

Many countries, in response to the Road Engineering/Safety Infrastructure section of the UN/ESCAP survey, claimed to have conducted safety audits or safety checks before construction. The extent to which these safety audits exist in practice is not known.

The key principles of the road safety audit process in Nepal include:

- Designing for all road users.
- Encouraging appropriate speed and behaviour with appropriate design (Transport Engineering and Safety Unit [TESU], 1996).

The first principle refers to ensuring that all road users' needs are considered in the design process. The traditional road engineer's focus on surface design and four-wheel motorised vehicles is not the only thing to consider. The second VRU-related principle given is that good design must enforce the appropriate driving speed. For example, when major national highways pass through small communities or towns, speeds that were perfectly acceptable on the rural/non-urban sections of the roads will no longer be appropriate. Thus, the design makes drivers understand that they are entering a different 'zone' where high speed is no longer acceptable. Use of such features as rumble strips and gateway features as motorists approach communities gives a clear indication of a change in road character, and this should result in reduced speed as motorists pass through the community.

Table 2.9 Safety Audit Considerations for VRUs

Audit stage	**Vulnerable Road User Design Considerations**		
	Pedestrian	**Non-motorised vehicle**	**Motorcycle**
a **Feasibility** .	Impact of road improvements in areas of high activity. Severance of the existing route. Expected flows cross the road. Pedestrian provisions —footpath and crossings. Accident data for similar schemes.	Impact of road improvements in areas of high activity. Severance of the existing route. Expected flow on/across the road. Cycle lanes/crossings. Accident data for similar schemes. Junction numbers are minimised and of the correct type.	Scheme consistent with adjacent road standards. Severance of the existing two-wheel vehicle route.
Preliminary design	Establish sightline design standards. Adequate footpath provision. Crossing provisions at major/minor junctions and traffic signals. Suitable crossing locations and types.	Establish sightline design standards. Merging/diverging areas with motor vehicles. Crossing facilities. Suitable junction types and size considered re: movements. Adequate cycle path provision.	Suitability of separate lanes merging/diverging locations, sightlines, and speeds.
Final design	Signs/markings/street furniture correct position, conspicuity and visibility. Adequate footpath provision with cohesive/direct routes. Pedestrian railing provision and location. Landscaping effect on visibility. Correct materials specified. Lighting requirements. Approach speeds of vehicles at crossing locations. Dropped	Signs/markings/street furniture correct position, conspicuity and visibility. Landscaping effect on visibility. Correct materials specified. Lighting requirements. Approach speeds of vehicles at crossing locations. Dropped kerbing. Turning conflicts at junctions are minimised. Roundabout approaches/circulatory lanes. Overtaking at refuges is	Signs/markings/street furniture correct position, conspicuity and visibility. Landscaping effect on visibility. Advancing warning of crossings. Turning conflicts at junctions are minimised. Motorcycle paths of suitable design standard. Traffic regulations to allow specified access.

Audit stage	Vulnerable Road User Design Considerations		
	Pedestrian	Non-motorised vehicle	Motorcycle
	kerbing suitability Zebra, pelican crossings and pedestrian phase at signal junctions—signal heads visible, central refuge, signal timings, and banned turns, pedestrian visibility to drivers. Refuges— size, visibility, location. Overtaking prevented.	prevented. Cycle crossings provided. Adequate and suitable cycle path provisions and dimensions. Traffic regulations ban waiting vehicles on lanes.	
Pre-opening	Adequate footpath provisions. Footpaths connected—direct, cohesive, safe routes. Signs and railings checked. The route should be walked, cycled, and driven along during the day and night.	Cycle paths of adequate dimensions and linked. Smooth transitions between riding surfaces. The route should be walked, cycled, and driven along during the day and night. Publicise that the path exists.	Motorcycle paths of adequate dimensions and fulfilling design standards. Smooth transitions between riding surfaces. The route should be walked, cycled, and driven along during the day and night.

2.2 Trends in Road Safety

2.2.1 General Trends

The rapid increase in vehicle sales in developing countries in recent years has placed considerable pressure on road networks and on their traffic control devices. Many of these control devices were never designed for the traffic flows presently using them; drivers and pedestrians are still provided with inadequate facilities. Over a span of ten years, the number of people killed annually has increased by almost 60 per cent. It is estimated that, in addition, at least three to four million people are injured in road accidents each year. Official statistics show that many of these casualties received crippling injuries that may handicap them for the rest of their lives and will impose a financial burden on their families and on their communities. The road accident situation is slowly improving in the high-income industrialised countries, e.g., Australia, Japan and New Zealand. But most of the developing countries face a worsening situation, and traffic accidents account for more deaths in the Asia Pacific region than common diseases. As infectious diseases are increasingly brought under control, road deaths and injuries have, in recent years, risen. In Thailand, for example, more years of potential life are now lost through road accidents than through tuberculosis and malaria combined. Road accidents are the second-highest cause of premature death among core segments of the population (5–44 years of age) in many developing countries.

What is even more worrying is that road accidents are underreported, so the official statistics underestimate the situation. The problem is certainly much more serious than even the analyses of trends would suggest, as these are based on official police statistics. The following sections present comparative analyses of current trends and statistics for the Asia region.

If these trends in the growth of the number of vehicles continue, there will be severe consequences for the region. In the following paragraphs, it is indicated that urgent action is needed if the present systems are not to be swamped by a deluge of vehicles in the future. Recent changes in motorised vehicle fleets are shown in Figure 2.54. The PRC, India, the Republic of Korea and Thailand have all substantially increased the size of their motorised vehicle fleets in recent years. In the Republic of Korea, for example, the vehicle fleet has increased eightfold in only 12 years, and most of this growth has arisen out of increased demand for private vehicles.

2.2.2 Motorisation and Vehicle Fleet Growth

During the last decade, the rate of increase of gross domestic product (GDP) of developing Asian countries has consistently been double the global average. Tourism into and within the region has grown rapidly in recent decades, and the urban populations of the Asia Pacific region have almost tripled since 1960.

This economic growth, increasing disposable income, and increasing urbanisation are creating greater demand for transport, and the number of vehicles on the region's road system is growing rapidly. Many countries have experienced annual growth rates of about 16 or 17 per cent, e.g., the PRC (18 per cent), India (17 per cent), Malaysia (15 per cent) and Viet Nam (18 per cent). This is equivalent to doubling vehicle fleets in five years and tripling them in eight years. This factor, allied to the high proportion of two- and three-wheeled motorised vehicles and the relatively young age of most of the population, all contribute to the region's serious road safety problems.

The freedom, benefits and convenience enjoyed by owners of a private vehicle are such that most individuals aspire to own one as soon as they can afford it. In developed countries, this has resulted in large increases in the ownership of cars as the economic conditions of the people have improved. In developing countries, private motorisation includes motorcycles as well as cars, as these are commonly the first affordable motorised vehicles that people in the developing world can possess.

Therefore, in the developed countries, private vehicles are mostly cars, while in developing countries, they encompass a much wider range of motorised vehicles. The consequences of these two scenarios are discussed in the following paragraphs.

The private vehicle ownership level in many developing countries is still low in comparison to that of industrialised countries. In these more motorised countries, it has been found that the shape of the private vehicle ownership curve over time approximates an S-curve (Figure 2.55).

In the early years (at the tail of the curve), the vehicle ownership growth is slow and steady but this is followed by a period of rapid expansion or 'explosive' growth (i.e., the steep part of the curve), before growth slows down again towards the top of the curve as it approaches saturation level, which happens when it is assumed that all who want vehicles will have acquired them. At that stage, the rate of increase is expected to drop until it just keeps pace with the number of new drivers, since all existing demand will already have been met. In more motorised countries, it has been found that vehicle ownership growth rates do not begin to level out until ownership levels of about 500 vehicles per 1,000 population or higher are achieved. Consequently, this observation suggests that, despite the rapid increases already being experienced in the region, many countries are still at the bottom of the growth curve and have yet to enter the very steep 'explosive' growth part of the vehicle ownership curve.

If motorisation continues as it has in all other industrialised countries, then growth in ownership is certain to continue to reach the level of at least 500 vehicles per 1000 people. Australia, the United States and many countries already have ownership levels of well above this theoretical limit.

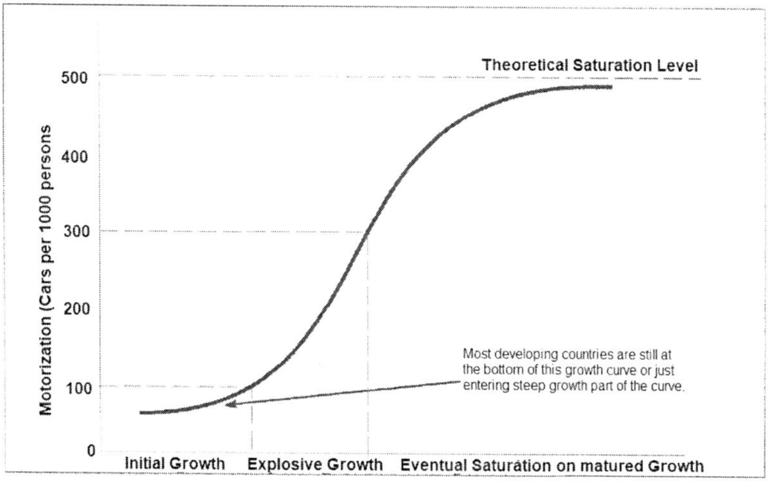

Figure 2.46 Generalised Vehicle Ownership Growth Curve

While the number of motor vehicles per 1,000 persons has surpassed 600 in many industrialised countries, such as Taipei (779), the next highest motorisation level is 344 in Malaysia. The other countries that have relatively high motorisation levels are Thailand (225), Singapore (207), the Republic of Korea (184) and Fiji (121), and all other countries have motorisation levels lower than 86.

Compared with the other sub-regions of the Asia Pacific region, Southeast Asia shows a modest rise in the number of motor vehicles. The growth varied from 1.3 times (Singapore) to 6.3 times (Thailand) from 1981 to 1996. In several countries, vehicle fleets have doubled since the 1980s. The only country with extraordinary increases in the number of motor vehicles is Thailand. It could be that in Southeast Asia, the relatively modest rise in vehicle ownership in the 1980s follows an earlier explosion in the number of vehicles (especially motorcycles).

Discouraging the use of private vehicles may be possible in four ways,

- Legislation followed by enforcement of the use of shared vehicles to and from workplaces; this is possible using carpools or workplaces picking up and dropping off their staff using company vehicles.
- Government action to develop an integrated and efficient public transport system. The public transport system must be comfortable, affordable and faster compared to using private vehicles. A metro

rail system, or a light/mass transit system, may be effective in the reduction of vehicle usage.

- Apply a weekly limit on fuel sales to owners of private vehicles. This may force them to use the two options mentioned above.
- Divert motorcycles and car traffic to an alternative, longer route during strategic hours of the day.Much of the data from the Pacific region is erratic and may be highly unreliable. It is highly improbable that vehicle numbers would fluctuate to the degree they appear to be doing in some Pacific countries. It is more likely that the political upheavals in the sub-region during the late 1980s led to inadequate recording of the number of motor vehicles, causing erratic fluctuations in what might otherwise be a straight line.

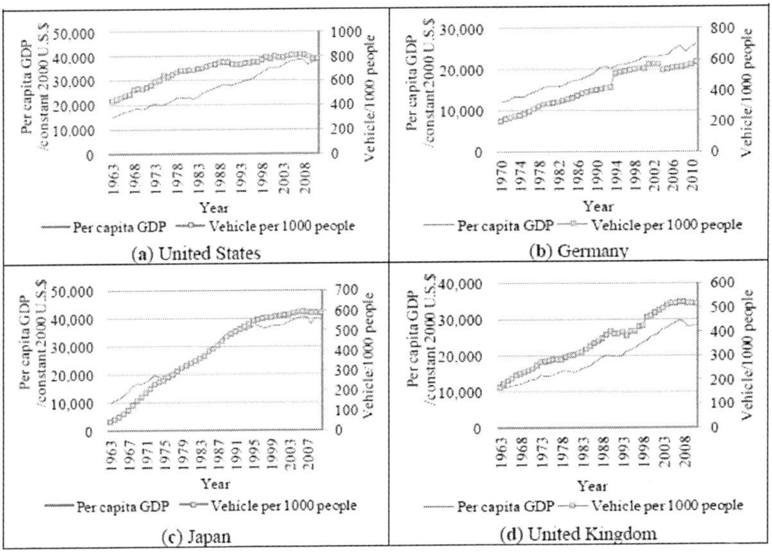

Figure 2.47 Annual Vehicle Ownership and Economic Development
Data Source: www.mdpi.com/journal/sustainability Sustainability 2014, 6, 4877–4899; doi: 10.3390 / su6084877

As mentioned above in relation to vehicles per capita, trends in the developed countries of the region are heavily influenced by the large number of vehicles in Japan. In terms of vehicle fleet, Japan experienced a 44 per cent increase between 1982 and 1994, whereas Australia and New Zealand had increases of 25 and 10 per cent, respectively. The S-shaped motorisation curve described earlier is based on developed countries' experience, where two- and three-wheeled motor vehicles did not play an important role. In many parts of the Asia Pacific region, two and three-wheeled motor vehicles dominate the motor vehicle populations and the vehicle growth rate. With motorcycles being a much more affordable vehicle type, motorisation levels may well increase at even higher rates than found in the developed world.

When a boom in motorisation occurs, road traffic accident deaths are expected to rise dramatically. But while deaths have been shown to have increased significantly, the frequent fluctuations in the data from many developing countries indicate data weaknesses that hide the actual road accident trends. Existing vehicle fleet growth at 15 to 18 per cent per year is already causing considerable traffic congestion and road safety problems, stretching the resources of the transport authorities with responsibilities for maintaining the road networks operating efficiently and safely. Future increases in the explosive stage of development will place even more strain upon road networks and organisations unless procedures and systems, such as access control, improved traffic management using an ITS and effective public transport facilities, are established before the problem becomes unmanageable. This requires urgent action to be taken.

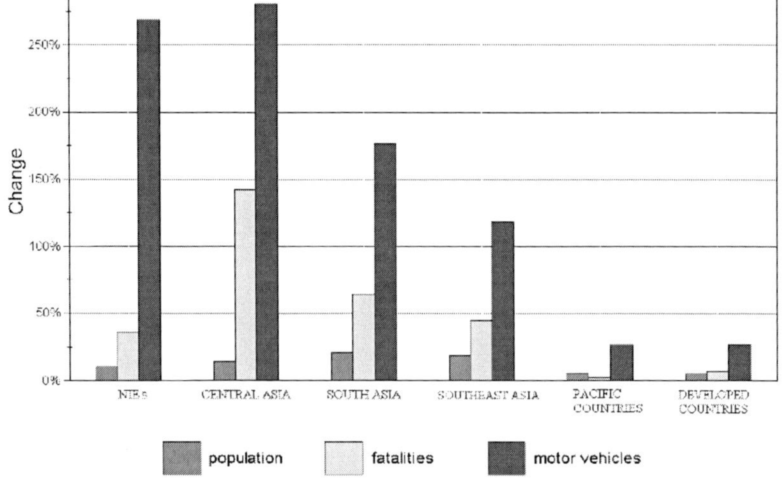

Figure 2.48 Increase in Population, Road Casualties and Motor Vehicles in the Asia Pacific Region during 1984 to 1993.

2.2.3 Observed Trends in Traffic Accidents

a. Trends in Road Accident Fatalities and Population

Figure 2.48 shows the gross number of road deaths over time for Asia Pacific countries, African countries, and industrialised countries. The statistics may be compared with the changes in population in these countries. The number of deaths in the Asia Pacific region has been factored to take account of differences in the definition of death due to a road accident. European Union standardised 30-day death factors were used to create a common base.

Figure 2.48 also shows that by far the largest increase in the number of accident deaths (89 per cent) is in the Asia Pacific region, with African countries and industrialised countries exhibiting a figure of 15 per cent and minus 5 per cent, respectively. In terms of population growth, however, the increase for the Asia Pacific region (24 per cent) lies between that for African countries (41 per cent) and developed countries (5 per cent). Given these figures, the most likely explanation for the rise in the number of deaths and injuries from road accidents in the Asia Pacific region is due to the massive increase in the numbers and usage of motorised vehicles.

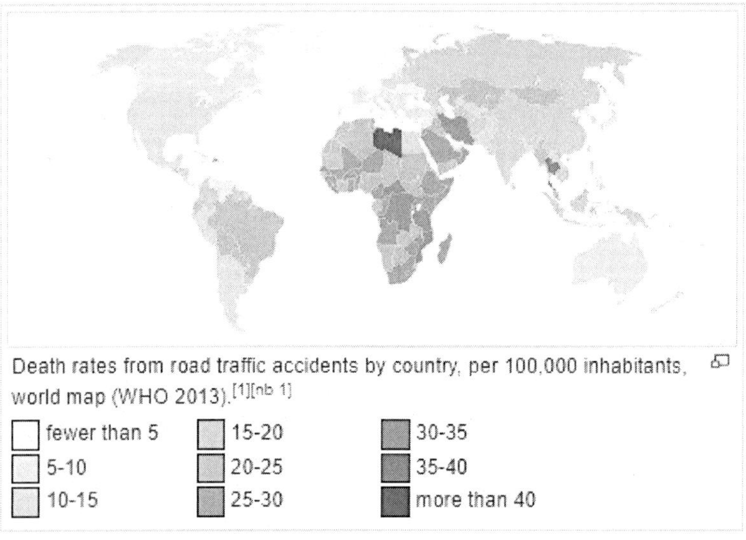

Figure 2.49 Countries with Traffic-Related Deaths
Ref. https://en.wikipedia.org/wiki/List_of_countries_by_traffic-related_death_rate

According to the World Health Organisation (WHO), road traffic injuries caused an estimated 4.35 million deaths worldwide in 2016, that is, one person killed every 25 seconds. Only 28 countries, representing 449 million people (seven per cent of the world's population), have adequate laws that address all five risk factors (speed, drink driving, helmets, seat belts, child restraints, and use of phones). Over one-third of road traffic deaths in low- and middle-income countries are among pedestrians and cyclists. However, less than 35 per cent of low- and middle-income countries have policies in place to protect these road users. The average rate was 18.4 per 100,000 people. Low-income countries now have the highest annual road traffic fatality rates, at 24.1 per 100,000, while the rate in high-income countries is lowest, at 9.2 per 100,000. Seventy-four per cent of road traffic deaths occur in middle-income countries, which account for only 53 per cent of the world's registered vehicles. In low-income countries, it is even worse. Only one per cent of the world's registered cars produce 16 per cent of the world's road traffic deaths. This indicates that these countries bear a disproportionately high burden of road traffic deaths relative to their level of motorisation.

There are large disparities in road traffic death rates between regions. The risk of dying because of a road traffic injury is highest in the African Region (26.6 per 100,000 population), and lowest in the European Region (9.3 per 100,000). Adults aged between 15 and 44 years account for 59 per cent of global road traffic deaths. Seventy-seven per cent of road deaths are males.

The total fatalities figures come from the WHO report and are often an adjusted number of road traffic fatalities to reflect the different reporting and counting methods among the many countries (for example, "A death after how many days since the accident event is still counted as a road fatality?") by international standard adjustments to 30 days, or "to compensate for under-reporting in some countries".

b. Road Accident Fatalities and Vehicles

Figure 2.60 shows the number of deaths per 10,000 vehicles for 1992/1993 for many countries in the UN/ESCAP region. Australia, Japan and Taipei have death rates that are characteristic of those generally found in the industrialised countries. This is usually about two road accident deaths per 10,000 licensed vehicles. In contrast, many developing countries in the Asia Pacific region have high death rates, typically between 20 and 70 deaths per 10,000 motor vehicles. The death rate for Tajikistan appears to be so high as to be implausible, probably resulting from data deficiencies. For this country, it is likely that the vehicle licensing records are incomplete and that there are, in fact, many more vehicles in this country than official statistics suggest.

This single indicator of annual deaths per 10,000 vehicles highlights the severity of the road accident problem in many countries in the Asia Pacific region. Even in emerging, newly industrialised countries, such as the Republic of Korea and Thailand, there are between 11 and 15 deaths per 10,000 vehicles, i.e., about eight times higher than in the developed countries of the region. If Japan had a comparable death rate to, say, the Republic of Korea, the number of people killed on Japan's roads per year would be 116,000, as opposed to the 13,800 "standard 30-day deaths" that occurred in 1994.

c. Fatality Index

The fatality index is the deaths divided by total casualties, including deaths, expressed as a percentage, in industrialised countries, which are typically characterised by low indices of one to two per cent. In Figure 2.50, it can be seen that Japan has an index of 1.6 whereas developing countries in the Asia region have indices between 20 and 42, indicating that the relative risk of death if involved in a road accident is 13 to 27 times higher in these industrialised countries (data for later years is not available).

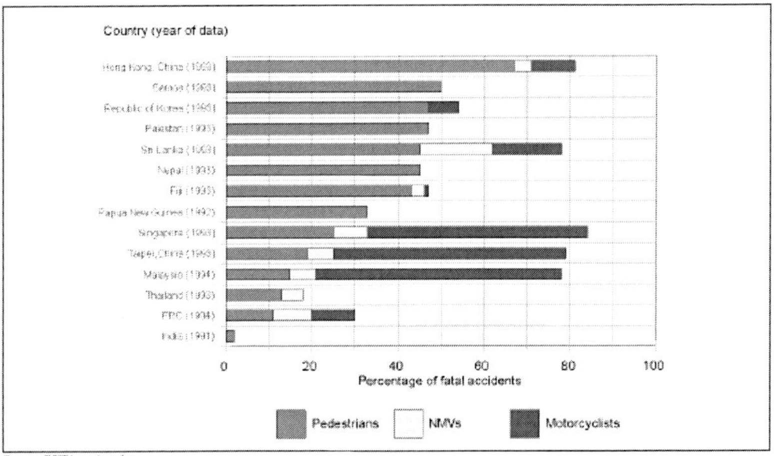

Source: RETA project data.

Figure 2.50 Percentage of fatal accidents involving road users

The reliability of the fatality index depends crucially on accurate reporting of deaths and injuries, but can also be influenced by medical facilities. Unless prompt medical attention is given to injured people at the roadside, their condition will deteriorate. Thus, adequate management of ambulances, hospitals, doctors and hospital beds is required to keep the figures low. In an earlier regression analysis, the TRL of the United Kingdom found a significant relationship between fatality index and the number of doctors per capita.

Analyses were carried out by using available data from Asia Pacific countries for the fatality index that was regressed against three medical indicators: doctors per 1,000 population, nurses per 1,000 population and hospital beds per 1,000 population. The results of these analyses clearly show that the quality of medical facilities and attention available to victims of road accidents makes a big difference to their survival once injured in a road accident. There are qualitative differences also in the nature of road accidents between countries. For example, pickups and open-topped lorries are used as public service vehicles in many Asia Pacific countries. Accidents involving such vehicles are likely to cause serious injury to greater numbers of people. Road traffic accidents involving multiple casualties are additive to the problem of limited medical resources.

It is important to note that the underreporting of road accidents artificially inflates a country's fatality index. Most fatalities are reported to the relevant authorities because of their serious nature. Underreporting tends to occur at the level of less serious accidents where injuries are slight. In such accidents, there is less need to involve the police or other official agencies, since any compensation can often be negotiated privately.

If all records of deaths and fewer records of total accidents are made, then the ratio value is higher than actual, which artificially increases the proportion of casualties who die and thus the fatality index. In some countries where underreporting is particularly high, a high fatality index is more due to underreporting than lack of medical facilities. However, it is well understood that emergency medical services and other incidental management would have been much more effective in reducing the very high fatality indices in the region.

2.2.4 Observed Vulnerable Road Users

Pedestrians, cyclists, NMVs and motorcyclists, because of their much greater numbers and the absence of adequate facilities, often appear much more frequently among casualties in developing countries than in developed ones. Figure 2.50 shows that more than two-thirds of all fatal accidents in Hong Kong, China, involve pedestrians, while in Malaysia, Singapore and Taipei, more than half of all fatal accidents involve motorcyclists. The reported national accident statistics involving NMV users were generally low. However, in the PRC, the scale of the problem is such that about 70 per cent of the total traffic accidents involve cyclists in a city like Beijing.

The mass of NMV users' share in total traffic is obviously the main reason. In Shanghai, NMVs comprise about 87 per cent of all vehicular traffic (excluding pedestrians). Figure 2.51 shows the motorcycle share in the motorised vehicle fleet in relation to the percentage of fatal accidents involving motorcyclists in selected countries. Those countries where the lower bar (percentage of fatal accidents) is at a higher level than the upper one (percentage share of the vehicle fleet) have a proportionately more dangerous situation for motorcyclists, notably Hong Kong (China), Malaysia and Singapore—three of the newly booming economies. Hong Kong, however, has a low proportion of motorcycle users.

Developing countries form two categories: more developed and less developed countries. Motorcyclists are the most VRU in some of the more developed countries. In less developed countries, pedestrians and NMV users generally appear to be at high risk of accident involvement and particularly at high risk of being killed.

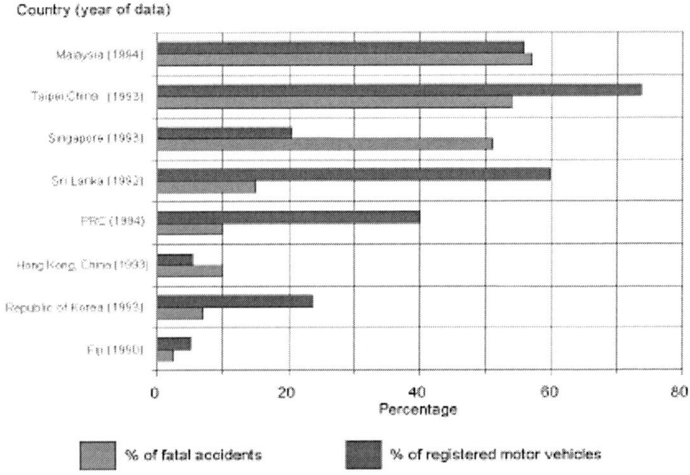

Figure 2.51 Motorcycle share in vehicle fleets and involvement in fatal accidents

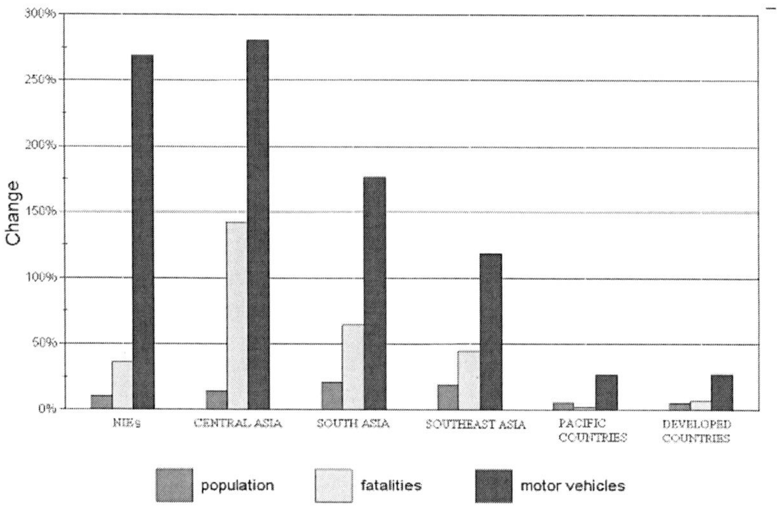

Figure 2.52 Changes in population, road fatalities and motor vehicles in the Asia Pacific region, 1984-93.

2.2.5 Socioeconomic Costs and Impacts of Road Accidents

Traffic accidents result in deaths and serious injuries that represent a considerable waste of a nation's resources and cause anguish and grief for family and friends. Aside from the emotional consequences of accidents, the loss to the community in terms of pure cost is high. Each year, apart from more than 235,000 people killed in the Asia Pacific region, about one million people are reported injured in road accidents. Despite underreporting, it is estimated that at least three to four million people are injured each year in road accidents. It is, therefore, necessary to include an estimate of the productive loss to the country of someone killed or injured. The visible costs of accidents are:

- Seriously injured people can remain in hospital for days, months or for the rest of their lives
- The use of hospital beds and highly skilled medical staff
- Costs incurred by police at accident sites
- Insurance and its processing costs
- Vehicle repair costs

TRL research has shown that the number of working years lost because of road accidents is higher than that lost by other causes of premature death. About 70 per cent of the 'years of life' lost due to road accidents are working years; hence, developing countries lose the most economically active and productive years from such persons. It is evident from the comparison with the other costs of premature death in developing countries, particularly malaria and infectious diseases, those deaths from road accidents seem to be increasing. The economic loss is a major drain upon the economy of a developing country that needs to import medical supplies and vehicle parts. Therefore, many of the losses are often sustained as foreign exchange losses.

Research in several countries indicates that such losses are normally in the range of one to three per cent of a country's GDP. Even if it is the minimum of one per cent of GDP, this still results in high losses annually.

Applying this percentage to the developing countries' cost of road accidents is more than US$20 billion per year. Recurring losses of this magnitude, which are greater annually than the total lending to these countries by the World Bank and Asian Development Bank (ADB) combined, should be of urgent concern to senior government officials in the region as they inhibit economic and social development.

2.2.6 Action Needed

Recently, there has been a steady rise in road accident deaths. Experience from developing countries indicates that in the early years of motorisation, road accidents typically rise in line with the increase in the number of vehicles.

It should also be recognised that there are factors in developing countries that are conducive to the improvement of safety and should not be seen as insurmountable obstacles in the development of effective plans, countermeasures, and interventions. Even where the problem itself cannot be eliminated, mechanisms can be devised to improve such deficiencies. In some respects, despite the problems cited above, the timing for investment in road safety in the region will be right, as there are several helpful factors at play that are now conducive to success in this field. The main reasons why the time is right to stimulate activity are mostly the following:

- There is increasing recognition that road accidents are already a serious problem. The annual toll of more than 235,000 deaths and an estimated three to four million injuries per year is causing serious social and financial problems for the region.

- Based on present trends, the annual deaths from road accidents will rise to more than 450,000 per year within a decade, with millions more being injured or crippled each year. It is now recognised that urgent action is needed to avoid this.
- There is increasing recognition by aid agencies that there is a serious problem confronting developing countries, that action needs to be taken, and financial and technical assistance needs to be provided for such action.
- There is increasing recognition of the economic losses being sustained. Annual losses more than US$20 billion per year (which exceed the total World Bank and ADB annual lending in the region) substantially inhibit the economic and social development of the region.
- The current early stage of development of many of the road and transport sectors in many countries offers significant opportunities to avoid many problems and difficulties that have had to be faced and overcome in other, more motorised countries. Adoption of safety-conscious planning and design approaches in roads, and introduction of effective countermeasures will add only marginally to the cost at the planning and design stage, but would be prohibitively expensive if they were added later as remedial measures; and
- Opportunities and scope exist for avoiding future problems by incorporating procedures, controls, and improvements now in many sectors while the systems are still in a state of incomplete development.

All the preceding points are encouraging factors indicating that, if the road safety sector can be stimulated in the right way and the right institutional environment created for it, countries can gain the benefits of improved road safety on their networks.

Section 2.3 shows how to develop coordinated action plans for road safety to allow a more effective and coordinated application of the various interventions in different sectors.

Section 2.4 discusses road safety management and provides guidance and information on the types of policies and interventions that can be implemented in each of the different sectors that affect road safety. This guidance is intended to provide a source of knowledge and experience gained from around the world so that all developing countries can apply the methods and techniques that have been found to be successful in developed and other countries.

The necessary information should be obtained to take action. Failure to take action will condemn millions of fellow citizens to avoidable death or crippling injuries and greatly support the social and economic development of these countries. Many of these deaths and injuries are avoidable simply by implementing the practical measures and policies advocated in this book.

References

For more information on ADB, visit: http://www.adb.org.

2.3 Road Safety Improvement Plans

2.3.1 General

For managing and implementing national road safety action plans and interventions, an effective action plan, coordination and decision-making are instrumental. An appropriate management structure and functional system are essential to achieve this. The various options available and the functions needed at the different administrative levels are discussed in detail in Section 2.4.

Appropriate conditions must exist or be induced within a country before introducing a road safety action plan for its effective implementation. Various developing countries are at different stages of development with respect to safety activities and interventions, and have different strengths and weaknesses. Most countries go through several stages before establishing the relevant system in all sectors operated by relevant organisations. The following sections discuss the different stages and activities involved in most developing countries' efforts to tackle road safety problems. These stages are as follows:

1) Creating awareness,
2) Prioritised action plans,
3) Five-year programmes for road safety,
4) Goals and Targets for road safety,
5) Financing Initiatives for road safety,
6) Monitoring and Evaluation of road safety.

Each of these is briefly discussed below.

2.3.2 Creating Awareness

a) Typical Situation at Stage 1

The minimum or no safety awareness may exist at the initial stages of safety development; therefore, efforts will first need to be made to enhance the awareness of key decision makers as to the scale and nature of the problem and the activity that is necessary to alleviate it. Typically, in such a country, accident data may not be recorded regularly; even if it is collected, it is likely that the recording system will need extensive improvement. It is usually difficult to clearly identify the trend of road users who are at greatest risk and the main organisations with road safety responsibilities. General interest by the government will usually be low, and it will be only a few individuals (commonly medical personnel) who are interested in bringing the problem to the public's attention.

Normally, there will be no special unit in the police department to analyse or disseminate accident data, and even if it exists, the data will be little used other than to provide overall administrative statistics for use within the police force. Accident data will often be incomplete or grossly inaccurate. Also, there will only be a few personnel actively working to improve road safety through remedial measures.

b) Action Required at Stage 1

In such circumstances, the primary focus of activity is to raise awareness among the decision makers and to demonstrate to them that road safety is a growing and urgent problem. The definition of the scale, characteristics of the problem, brings together key parties that may be able to contribute towards solving the problem, and convinces those responsible for resource allocation that sufficient funds need to be made available to initiate the activities to tackle road safety issues. The two main tasks that need to be undertaken at this stage are as follows: an independent review of the required road safety activities and to quantify the scale, nature and characteristics of the problem to identify deficiencies and weaknesses in terms of road safety in the sectors and organisations related to road safety; and a national road safety seminar needs to be organised involving all senior personnel from key government and non-government agencies with responsibilities or interest in road safety. Six or seven selected agencies should present a short paper in ten minutes on road safety problems, what needs to be done in their particular areas of responsibility to tackle the problem, and what prevents them from taking the necessary action. This seminar should result in an agreement among the key agencies on the actions needed to tackle the problem. An interim working group should also be established by using the senior staff in six or seven of the key agencies represented at the seminar. This should function to keep the safety coordinating mechanism operational until a more formal national road safety council (NRSC) can be established.

These activities may need a period of approximately two to three months. The technical assistance of an independent road safety expert may be needed to carry out an independent assessment of the road safety situation and the actions required. The national road safety seminar should be organised to take place shortly after the completion of the review of road safety, so that the results can be presented at the seminar to all participants.

The ten to 15 agencies with road safety responsibilities or interests should be invited to send a senior representative to the one-day seminar. Participants at the seminar should be drawn both from the public sector, that is, from ministries and departments of central and state governments, and the private sector, which includes large commercial companies, the insurance industry, transport operator organisations, and motor traders' associations, etc. Representative organisations such as the Chamber of Commerce, the Lions Club and the Rotary Club, and media representatives may also be invited to participate in the seminar.

c) Financing Activities at Stage 1

There are several sources of funding to cover the costs involved, and these costs are of two different types.

1. Technical assistance costs: funds for this are commonly available from development banks such as the World Bank or ADB. They often finance short-term technical assistance from their own funds or from trust funds. The purpose of such technical assistance is to review road safety activities and to identify areas that need improvement. Technical assistance is also often available from bilateral agencies such as the United Kingdom's Overseas Development Administration (ODA), the Japan International Cooperation Agency (JICA) and other bilateral aid agencies that sometimes provide a road safety expert for a short period to review ongoing road safety activities. All these types of assistance are on a grant basis.
2. The various central or state governments may also use funds from existing development project loans to carry out such reviews of

safety. Both the major development banks (ADB and World Bank) actively encourage reviews for road safety so that a better understanding of the problems and needs can be identified. Countries should seek such assistance during discussions with ADB and World Bank project officers visiting the country, or by following up with relevant specialists at the institutions (ref. Appendix B).

3. Implementation costs: the only real implementation costs involved are in carrying out the one-day national road safety seminar. Costs involved are minor and can often be financed by sponsorship from the large motor manufacturing companies, the fuel companies or other commercial concerns such as banks and insurance companies, who see involvement in sponsorship of road safety as part of their social obligations and good public relations. The main purpose of these activities is to raise overall public awareness of road safety as an issue through press and media coverage of the national seminar, awakening interest among major players who can influence road safety and raise commitment among government agencies, which can take action to improve the situation.

2.3.3 Prioritised Action Plans for Road Safety

a) Typical Situation at Stage 2

When a country is ready for the next stage of implementation in road safety, the typical conditions that exist are that the government will probably have received some technical assistance to review road safety activities and will have organised a national road safety seminar. The interim road safety working group may already be coordinating activities from time to time and obtaining periodic sponsorship for road safety initiatives, publicity campaigns and materials. Individual ministries responsible for road safety may be starting to address the problem and beginning to take action that will improve road safety.

In particular, the ministries that deal with external funding agencies should have taken the opportunity to request assistance from bilateral aid agencies and multilateral development banks to finance some or all of the proposed road safety improvements to rectify the deficiencies and weaknesses identified by the road safety review undertaken during stage one.

Measures: installing road signs, road markings, lectures while issuing driving licences, educating road users about traffic rules, enforcing traffic rules, installing surveillance, identifying accident black spots and taking protective measures, repairing damaged pavements, improving visibility, speed control measures, underpasses, FOBs, pavements, covered drains, railings, separating slow and NMV traffic, crash barriers, attractive hoardings, awareness campaigns about the dangers of drinking and drug driving, using mobile phones, not wearing seat belts, video screens in the front panel of vehicles and the associated high financial penalties, ITS, incidental management and the medical system.

b) Action required at Stage 2

The main purpose of this stage is to:

1) Develop an overall strategy with a priority action plan for the improvement of road safety.
2) Identify the most urgent improvements for inclusion within a priority action plan.
3) Establish basic systems and procedures to enable more effective activities in road safety; commence implementing the priority action plan.
4) Identify and strengthen the key organisations and individuals who need to work to improve road safety in the longer term.

The primary attention during this stage should be towards strengthening the key organisations, systems and procedures so that once the priority action plan is implemented, local organisations will be able to tackle road safety problems. During this phase, the development of a strategy and implementation of a prioritised action plan are to be undertaken. This usually includes improvement of the accident data registering system, the establishment of a national road safety council or some other coordinating mechanism and implementation of demonstration projects in other important sectors. This is done to provide training for individuals involved in the key agencies and to establish procedures and practices. These shall be conducive to the improvement of road safety in the country.

A small multidisciplinary group of four or five senior individuals deputed from government agencies with the greatest responsibilities for road safety is to participate in training through demonstration projects, and it should make a two to three-week overseas study tour, visiting countries that have demonstrated success in resolving road safety problems.

The prioritised action plan must ensure that basic systems are operating at a minimum level in all the most important areas by the end of the action plan period. Implementation of the plans can be done from 12 months to three years, depending on the size and scope of the action plan, the size of the country and the nature of the problems to be addressed. The general intention is to develop a downstream five-year programme while implementing the most urgent improvements needed to strengthen the key institutions and agencies.

Wherever possible, a timescale of around three years should be allowed to develop and implement all the main improvements needed for the action plan. The actual resources required to undertake the action plan will again depend upon the plan's nature, the scope of work, and the size of the country. External road safety expert advisory input of anything from 25 (five people for five months) to about 50 (five people for ten months) months is typically required in implementing action plans.

For the actual physical cost of the individual interventions implemented as part of the action plan, financial provision has also to be made in addition to the technical input.

An indicative amount of US$3 to US$5 million should be available for carrying out the demonstration and pilot projects. The government itself will finance some of this, and other parts may be eligible for financing from road safety components within aid-funded loans.

In addition to the resources, which may be available from external funding, the government will have to cover the cost of its own staff and specialists who will work as counterparts with the specialist external consultants and will participate in the implementation.

c) Financing Activities of Stage 2

Two different types of activity need to be financed under this phase:

- Technical assistance costs that might be required to provide specialist road safety advice and assistance at key stages.
- Implementation cost of the priority action plan, including funding of the training costs for local specialists.

The above costs are mostly financed through a road safety component within an aid-funded project. Such components typically include funds for technical assistance, equipment, infrastructure improvements (such as improvement of accident black spots) and training (both domestically and overseas) of key personnel. Governments interested in implementing prioritised action plans for developing road safety should be aware that international development banks are keen to provide loan funding for the improvement of road safety, as they have recognised the importance and urgency of the problem. In addition, bilateral agencies assisting a country can also often be keen to assist other countries in resolving road safety issues. Thus, the international funding community is aware as well as willing to provide loans or other funding for road safety interventions. However, a request must be made by the borrowing country or recipient country for road safety assistance to the World Bank or ADB, so that these banks can consider providing funding within existing or proposed projects.

Due to the bureaucratic procedure involved, the allocation of funds for prioritised action plans can take some time. Therefore, requests should be made as quickly as possible so that ongoing or future aid-funded projects can provide funding for safety measures. Sometimes, where funds have been underspent on previous or existing projects, aid agencies are often willing to allow some of the funds to be reallocated for use on road safety. For making such funds available, the governments should seek approval from the aid agency for the reallocation of these funds for safety initiatives.

By the end of the priority action plan, local professionals should have the capability to tackle the road safety problems that the country is facing, and the government ministries should be investing in road safety as part of their normal cycle of budgets in each sector.

2.3.4 Five-year Programmes for Road Safety

a) Typical situation at Stage 3

The government should have received technical assistance and implemented a priority action plan, and an improved accident data system should be in place with suitably trained local staff operating the data system and generating priority tasks. Analysis of black spots, locations and characteristics of road accident groups most at risk should be conducted regularly. An NRSC with several subcommittees should be providing policy guidance and some financial support to safety committees in municipalities and provinces.

Highway authorities should be skilled in basic improvement work of accident black spots, and undertake black spot improvement programmes on national and provincial roads. Each effort is undertaken through the

NRSC to, for example, improve driving tests, vehicle inspection and develop children's traffic education programmes. There may be a core of road engineers and professionals specialising in various aspects of safety and interested in tackling the problems, but they don't have the resources to do so. The priority action plan that is already implemented will have several government and non-government organisations involved and active in improving safety, creating a conducive atmosphere to formulate and implement a set of five-year road safety programmes.

b) Action Required at Stage 3

The activities initiated during the priority action plan are to be consolidated, to develop new activities and interventions that will help reduce the annual toll of road-related deaths and injuries. The first five-year plan should be an extension and consolidation of the previous priority action plan. But subsequent five-year programmes will be building on each previous five-year programme. The real implementation is normally undertaken by way of annual national road safety plans, and these are prepared each year to guide and focus the activities of the various agencies towards achievable targets within the following year, but operating within the context and framework of the overall five-year programme.

c) Financing Stage 3 Activities

Now, significant investment in road safety is becoming necessary, and most of it is to be made by the government of the country and not through external aid funding. In the case of infrastructure improvements for a one-off safety programme of road marking and sign improvements on the national road network, they can be incorporated into aid-funded projects, but in most cases, actual funding will have to be included in the annual budgeting and five-year national programmes of government.

The government gradually becomes aware of the annual losses to the economy and is willing to see expenditure on safety as an investment. That will, in the longer term, reduce such economic and social losses from road accidents. Some countries even allot a fixed proportion of annual gross national product (GNP) towards the reduction of road accidents.

Within the five years, safety plans are usually produced every year in the form of a document that is published and widely circulated so that all agencies have a clear focus on what needs to be done and the targets to be achieved during the upcoming year. The resources required to establish the necessary procedures, systems and practices to reduce the road safety problem can vary from US$15 to US$20 million in a small country and can vary from US$50 to several hundred million US dollars in a large country. Most of the investment in a five-year action programme that is to be undertaken by the country should recognise that expenditure on road safety is an investment, not a cost.

Completion of the first five-year programme on road safety does not mean the end of the problem or that the problem has been solved. It will be necessary to develop a second and probably even a third five-year programme to continue reducing the number and severity of road accidents. This may cost millions of US dollars, but conversely will 'save' many more millions of dollars, through increased productivity and less use of medical facilities.

2.3.5 Goals and Targets for Road Safety

Strategies and plans to improve road safety should set realistic targets. In most industrialised countries, these have been set in terms of a reduction in the number of accidents and casualties expected. These targets must be understood and accepted by those responsible for improving road safety. The outcome of the plan must be regularly observed and reported upon. A national road safety plan should, therefore, be based on measures based on a careful analysis of the problem.

This requires the collection and assembly of factual data on the road safety situation, the identification of the problem areas, the investigation of the true causes of accidents and the development of appropriate countermeasures. Ranking such measures should be based on the estimation of their cost-effectiveness in saving lives or serious injuries.

Implementation of safety plans requires a clear identification of responsibilities between the different departments of the central government. One way of achieving this is to publish challenging but achievable casualty reduction targets, first at the national level and later disaggregated to the states. It is important to ensure that appropriate targets are set in developing countries, as is explained in the following four paragraphs.

In addition to action at the central level, decentralised action is also essential. This must involve government organisations and activities at the local level. Entities involved will not only be those with statutory responsibilities, such as the police and local authorities, but a wide range of organisations, such as automobile clubs, private sector companies, the insurance industry and others that can also be instrumental in explaining and promoting road safety plans and providing feedback.

Police equipment and infrastructure improvements are expensive and must be clearly targeted and tightly controlled to make sure that the country is getting the best value for money. Continuous monitoring and periodic evaluation are also important to verify that the projected benefits are occurring. Training, study tours, twinning with other countries or organisations and improved data and research need to be closely linked to increased road safety programmes and safety funding so that data and appropriate skills are applied to the problem.

Developing countries should not blindly set casualty reduction targets of the type typically being set in industrialised countries. These tend to be statements such as, “Reduction in the number of deaths in 2014 by 30 per cent by the year 2020.” Most industrialised countries are now at the mature stage of motorisation and, in many cases, approach or even exceed theoretical saturation levels (see the S-curve in Figure 2.46 in Section 2.2). Hence, their vehicle fleets are now stabilised, or at worst, are growing slowly, and accident levels too have stabilised. Therefore, they can, with reasonable confidence, set targets for casualty reductions of 30 per cent or more within six to 15 years and many countries, e.g., Australia, Denmark, the Netherlands, New Zealand and the UK, have now done so.

2.3.6 Financing Initiatives for Road Safety

Adequate funding is essential for implementing a comprehensive and effective road safety policy and action plan. Road safety is a highly cost-effective activity involving the saving of human lives and the reduction of economic losses, but public budgets are often limited for such an important need. In many cases, it requires proper upgrading and rehabilitation of existing roads equipped with various road furnishers, surveillance, enforcement and enhanced medical facilities. Therefore, a sharp increase in funding is often justified to finance a well-prioritised action plan based on cost-benefit analyses. Each country needs to decide how best to finance its safety action plans at a level in accordance with its needs. Comprehensive road safety programmes that require substantial funding and potential sources of funding, which are not mutually exclusive, could include the following:

1) Budgets of government departments and agencies are normally the main source of funding for all actions under their control, including safety actions. This applies to central and local governments.

2) Road user charges are directly allocated to road departments for road management and investment. In comparison, charges for vehicle import, purchase, fuel, and road tolls are often considered as general government revenue. Charges for vehicle registration, driving licences and vehicle testing also raise additional revenue.
3) Road funds are one type of mechanism for allocating road user charges to road spending. These are established in several countries, often with the initial support of the World Bank or ADB, to help stabilise the flow of funds to the road sector. They generally rely on user charges such as levies on fuel, tolls, vehicle taxes and sources related to the service provided and incorporate increased accountability regarding priorities in road spending.
4) Road safety funds have a similar function, but are specifically for road safety actions by several different agencies. Revenue is raised from levies on traffic sources (e.g., vehicles or fuel) and expenditure, and priorities regarding safety are agreed with safety interest groups inside and outside government. This ensures priority to safety, which is independent of the concerns of individual ministries, and gives greater impetus to inter-ministerial coordination of safety programmes. Some countries, e.g., Botswana, use a levy on vehicle registrations to provide additional financing for road safety within several departments.
5) Private sector contributions to road safety include sponsorship of media articles or road safety campaigns by motor vehicle dealers, oil companies, commercial firms, and transport fleet operators as part of their public relations.
6) Motor vehicle insurance for road safety has been found in many countries and is particularly appropriate. Involving it in the decision-making process can also be beneficial. Prevention of accidents rather than repairing vehicles is cost-effective for the insurers and also for the rest of society. Both parties benefit significantly through the investment of insurance monies to reduce the number of accidents. In the State of Victoria in Australia, and in Finland in Europe, a considerable amount of revenue is being achieved from small levies on motor insurance. Fees corresponding to US$21 per vehicle in Victoria and US$3.6 per vehicle in Finland used to yield US$56.65 million and US$8 million, respectively, in these two countries. The levy was ten per cent in Victoria and one per cent in Finland. Developing countries should consider imposing a levy of about ten per cent on third-party insurance. This is discussed further in Section 2.6.

7) Enforcement fines for increased penalties on offenders for speeding, drink and drug driving, not wearing seat belts and using mobile phones, etc., generate extra revenue in several countries to be allocated to road safety. This can be used to purchase additional enforcement equipment and to finance other road safety initiatives that will help in the overall reduction of road accidents.
8) Development banks are willing to assist the developing countries in tackling road safety problems. Opportunities should be taken to take advantage of these sources of funds by including road safety components when such agencies are carrying out aid-funded projects in the developing countries. For example, development banks can often assist in all three stated stages of road safety development. In Stage 1,

2.3.7 Monitoring and Evaluation of Road Safety

The activities on road safety should be monitored to ensure that money is being spent effectively. As the funds are always limited for road safety activities, it is imperative that the money available is spent most wisely and effectively to tackle the problem. As each initiative is implemented, the effectiveness of that measure should be monitored. One ideal way of doing it is by looking at accident data from the before and after situation for equivalent periods of time.

Monitoring an overall action plan is aimed at the institution building and strengthening the key project implementation units with road safety responsibilities. The focus must be on whether the objectives of the action plan are being achieved. It is necessary to evaluate and ensure that the activities of the consultants and expert advisers have been effective. The impact of such activities must have the desired effect in terms of strengthening the capability of the country to tackle the problem. The recommended monitoring frameworks generally result in greater success in terms of implementation of action plans.

This framework format identifies what activities are to be carried out as part of the action plan implementation. The performance indicators are to be assessed to see whether the desired impact has been achieved. During implementation, the performance indicators can be used to identify whether the project is progressing as desired in terms of development and institutional impact.

2.4 Road Safety Management

2.4.1 General

As road safety is a multidimensional social problem that must be taken care of by many government agencies, the state has to play a leading role in initiating, organising and coordinating the national assault on road safety. The practical development of road safety capability at a national level must consider the following five important aspects:

- Designating either the prime minister's office [PMO] or the ministerial task force to be responsible for the overall road safety policy, as a member of the government.
- Establish a multidisciplinary team led by a high-level government official or a very prominent executive director from the private sector, who shall be assigned the task to initiate and coordinate the state's actions.
- Setting up an NRSC or an official committee as a permanent group to steer the actions.
- To carry out the work, the NRSC is to be assigned planning and adequate technical and financial resources.
- The action outcomes are to be evaluated.

All activities on health, transport, police and education of the ministries involved in road safety must be complementary, and coordination must also be developed with regional and local organisations so that road safety work is undertaken at all levels.

In optimising road safety efforts and their effectiveness within any country, the central coordination and management of road safety is one of the most significant factors. Advice on establishing national and local road safety councils and examples of their successful operation in various developing countries are given by the sector guidelines.

In the next sections, other related matters, such as the setting of specific targets for accident reduction and coordinated road safety plans, are discussed.

2.4.2 Coordination

Road safety activities tend to be fragmented and underfunded, and thus ineffective when a country lacks an effective coordinating body. Typically, in such countries, a few organisations carry out intermittent and fragmented efforts with inadequate budgets. These scattered efforts are undertaken in an uncoordinated way, covering only a few of the various sectors. The activities are done with little technical support and less understanding of the underlying problems to be solved in the most effective way of dealing with them.

1) To tackle road safety issues efficiently and effectively, the following are needed: adequate and effective political support.
2) Adequate funds.
3) Adequate technical and administrative resources.
4) An understanding of the problems.
5) Coordinated efforts among the organisations working in the fields of:

i. Traffic and highway engineering.
ii. Education, lectures at the time of issuing licences, registrations, etc.
iii. Traffic enforcement, surveillance and round-the-clock monitoring.
iv. Health services with adequate emergency facilities.
v. Publicity for communication and awareness.
vi. Driver testing for aptitude, reflex and physical ability.
vii. Vehicle testing for fitness, etc.

The central government must create an organisation dedicated to initiating and coordinating road safety activities. This will ensure that budgets are available and protected and that trained staff are assigned the activities. This is normally best done by establishing an NRSC or a road safety directorate. Such a body can coordinate all the agencies working in the various disciplines, ensuring that duplication is avoided and that the activities of different agencies complement each other. To carry out similar activities at a state or local level in a large country, safety committees at regional or provincial or, in some cases, municipal level have been established. In developing countries, many road safety sectors typically need improvement, and the NRSC and its local sub-committees can coordinate such activities.

2.4.3 Key Elements

Numerous countries have demonstrated that there are three fundamental requirements of critical importance when setting up an organisation to combat road safety problems in developing countries. The implementation will vary according to circumstances and national contexts, but the essential three elements are as follows:

- Political support.
- Technical secretariat.
- Effective coordinating mechanisms.

Political Support

A senior member of the government is to be assigned responsibility for road safety policy and to oversee and coordinate work by their colleagues. This can be the PMO or the head of the government ministry of road transport, most involved in road safety, and has been acting with the authority of the prime minister, where the PMO reserves the right to arbitrate when necessary. The PMO is responsible for road safety in various countries like France, Japan and the Republic of Korea. Committees need to be established at lower operational levels to implement the decisions of the high-level committee.

Technical Secretariat

A high-level official, such as the director general, must be designated to act as coordinator. A small multidisciplinary team of officials will be needed to assist in coordinating activities in the various government ministries. Where an official committee is to be established, this approach shall be appropriate. The first task is typically to establish such a committee as the NRSC or to prepare the documents to establish an NRSC. Where an NRSC is established, a multidisciplinary NRSC secretariat is to be established, appointing a high-calibre executive director for the secretariat, who will operate as the executive arm of the NRSC in coordinating, implementing and following up on NRSC decisions.

Effective Coordinating Mechanism

There is a need to set up either an independent commission, an official committee or a wider-ranging NRSC, including the principal civil servants responsible for issues directly or indirectly related to road safety. Its structure may vary from country to country. The key advantages or disadvantages of each option are outlined in Table 2.10.

Table 2.10 Advantages and Disadvantages of Alternative Options.

Type of coordinating mechanism	Advantages	Disadvantages
1. Independent commission	Independence Ability to bring in non-government interests Nonpartisan	Advisory only, so no power Outside the government system No access to government support staff/resources
2. Official committee	Key decision makers Power to act Can use government Support staff/resources	Inter-ministry rivalries Private sector excluded Less access to private sector funding Wider interests not represented
3. National Road Safety Council	Key government decision makers Private sector skills can be included Access to private sector funding Access to government staff / support via members	Needs secure funding Needs secretariat Needs legislative support

The private sector also has an important role to play within the safety council and can contribute fundraising skills, sponsorship, commercial experience and contacts. Representatives of motor insurance, fuel and other transport-related industries should be invited to come forward and participate as members. Ideally, they should be from industry or trade organisations rather than from individual firms. The work must cover different areas of activity and will be carried out through several subcommittees of the council.

NRSCs provide the best mechanism for achieving the benefits of both public and private sector organisations. The following sections of this chapter will assume that an NRSC is the mechanism in use. For countries using the alternatives of NRSC, most of the comments or advice are considered equally relevant.

2.4.4 Development Stages

To achieve a coordinated road safety effort using national and local level coordinating mechanisms, a developing country usually needs certain typical steps to proceed through, as described below:

- A national road safety seminar at which an interim inter-agency working group is established and a joint statement that key participants sign, is to be published on the need to improve road safety within the country.
- An agreement on the defined role of the NRSC and potential mechanisms for its funding.
- For the establishment of the NRSC and its technical subcommittees, the necessary legislation has been developed.
- Establishment of the NRSC.
- The provision of technical and financial support for NRSC activities.
- Establishment of the local coordinating bodies (PRSCs).
- A programme for short-term activities is to be developed, and
- Implementation of a priority action plan and development of a five-year programme.

Interim Working Group

During a national road safety seminar, an interim working group is to be established comprising senior officials from some most important organisations with responsibilities or potential to assist in improving road safety. Typical membership should include individuals and agencies such as:

1) Director general, traffic police.
2) Director general, roads and highways.
3) Principal secretary, transport department (vehicle and driver testing).
4) Principal secretary, public health.
5) Principal secretary, primary school education.
6) Insurers' representative, motor insurance industry.
7) Secretary General, Chamber of Commerce.
8) Representative from the PMO.

1. Role And Potential Funding Mechanisms for NRSC (Or Equivalent Organisations)

An NRSC is to promote, supervise, and coordinate road safety activities. The NRSC, therefore, must include among its membership all key organisations. Conversely, if there are too many organisations involved, it often becomes difficult to do anything constructive. It is recommended that the number of member organisations should be limited to ten to 20.

Smaller five to six-person subcommittees to be responsible for specific areas, such as fundraising and finance, safety education in schools and publicity. Other external agencies or specialists can be co-opted onto such committees as and when necessary. Most of the day-to-day work of the NRSC will be done by the subcommittees and the NRSC secretariat, organising the whole NRSC meeting only periodically.

For funding, the common options are either by an annual grant from the government or via a levy on motor insurance policies, driver licences, fuel or other sources, which will increase in line with the growth of the vehicles. To avoid undue reliance on government funding, which may often be subject to severe fluctuation or cutbacks, the following sources may be considered as funding provisions:

- About 60 to 70 per cent should come from such a guaranteed levy.
- About 20 to 30 per cent should come from commercial sponsorship and NRSC fundraising activities.
- About ten per cent should come from the government.

2. NRSC Legislation

For effective action, the NRSC should be given statutory responsibility to coordinate and supervise the development of activities for road safety. There should be an obligation to report back on the annual progress to legislators.

The legislation defines:

- The role, responsibilities, and function of the NRSC.
- The sources of funding.
- Empowerment of the NRSC to raise additional funds through sponsorship.
- Permit the NRSC to establish regional, provincial, or municipal committees.

3. Establishment of an NRSC

This event is important and should be launched with attendance by the media. It may be highly useful to involve the prime minister or a senior politician and at least a few ministers at the formal launch of the safety council. There will be an opportunity to feed the media with statistics and facts on the scale and nature of the road safety problems and their economic and social impact on the nation. Various safety-related articles, for example, on seat belts, use of mobile phones while driving and driving in a drunken state, can be placed in the media to raise awareness.

4. Technical and Financial Support

Establishing a multidisciplinary team to work under the high-level director of the safety directorate or by establishing an NRSC secretariat with its own high-calibre executive director could be helpful in achieving the required technical support. In the latter case, some staff could be available to the secretariat from member organisations of the NRSC who can hire additional staff of their own. Funding should be organised via the legislation that established the NRSC, and the NRSC secretariat should raise additional revenue through private sponsorship.

5. Establish Local Coordinating Bodies

Road safety problems must be tackled in various sectors as well as at different administrative levels. Establishing local provincial road safety committees, or PRSCs, is advantageous, modelled on the NRSC. These committees, including local branches of the same organisations, are to operate under the NRSC. This enables carrying out road safety activities at a grassroots level.

PRSCs provide the same type of coordination function at the provincial/state level as that provided by NRSC at the national level. They should receive general publicity and education materials developed by the NRSC and receive some funds from the NRSC. This helps meet some of the local administrative costs. This also allows the PRSC to develop and print local publicity material relevant to local needs and problems. This local publicity material supplements the material, posters and safety pamphlets received from the NRSC.

The PRSCs, like the NRSC, can raise an additional 30 to 40 per cent of their income from local sponsorship and fundraising activities in their geographic regions. This is helpful in raising awareness of safety issues as well as generating revenues.

6. Programmes of Activities at the National and Local Levels

Activities at the local level are to be harmonised with initiatives being taken at the national level. The NRSC should prepare a programme of proposed activities and circulate this to the PRSCs so that their local activities can be integrated with national campaigns and initiatives. In some countries, the NRSC provides exhibition vehicles and specialist presenters to support local road safety committees.

7. Implement Action Plan and Devise Five-Year Road Safety Programmes

The NRSC must supervise the actions of urgent improvements and coordinate the efforts of the different agencies implemented under a five-year safety programme. Each agency and government ministry that is a member of the NRSC, should prepare for its individual activity and influence the activities of a five-year programme to enhance or improve road safety, for example, the Ministry of Health may try to enhance their emergency ambulance system; or the public works department may try to identify and improve about a hundred of the worst accident black spots on the road networks.

Combining these various individual sector-specific initiatives into a single platform constitutes the national road safety plan. The role of NRSC is to promote and support road safety efforts in each of the individual departments with road safety responsibilities. Thus, raising public and political awareness may ensure adequate funds and resources are made available to allow the problem to be tackled effectively.

2.4.5 Receiving Benefits and Effects

The formation of road safety committees at the national or central level, like an NRSC, and at the local provincial levels, as PRSCs, typically leads to the following:

- An increased awareness and understanding of the problem, the political will to address it, and serious coordinated efforts to improve road safety.

- Well-planned road safety campaigns and remedial measures within a coordinated road safety action plan.
- A source of revenue to finance some publicity and education activities on road safety. This may come either from the central government, or perhaps through a fuel or insurance levy or other funding source from the private sector.
- The need for a good accident data system to define the problems has achieved recognition.
- For safety interventions in all relevant sectors, increased technical support is achieved.
- A synergy develops between the various disciplines through shared experiences.
- Updated traffic legislation, traffic system, and highway design standards.
- An initial pause to the rising accident rate eventually lowers the accident rate and accident numbers.

2.4.6 Implementation

So far, many countries have demonstrated effective coordination of their road safety plans and programmes. As there is no single model applicable for different countries, they have devised different solutions depending on their needs and circumstances. Some successful coordination mechanisms are as follows:

The state of Victoria, Australia, set up a two-member task force to take up the road safety issues. Two ministers from the police and transport departments represented the ministries with the highest possible responsibility. An official committee of technical specialists drawn from the main agencies and organisations with safety responsibilities provided necessary assistance. The two ministers were able to make significant improvements in road safety by reducing the number of people killed in road accidents.

Under one ADB-funded road safety action plan, Fiji has established an NRSC. This has legislative responsibility to coordinate activities and improve road safety. It has established a network of three divisional and about fifteen district road safety committees to carry out activities at the local level.

There has been an NRSC in Japan for many years, chaired by the prime minister with all key ministers as members. Several other safety councils also exist at the lower level, chaired by ministers and senior government officials. In addition to these councils, there have been committees. Under the management and coordination agency of the PMO, Japan has established a Road Safety Policy Office staffed by a multidisciplinary team under a senior government official. Their role is to coordinate, promote and supervise road safety activities and the five-year road safety programmes. The official committee was then replaced by an NRSC comprising chief executives of key agencies supported by several road safety advisory groups drawn from interested government and non-government organisations. Although these examples of good practice vary from country to country, those that are more successful have several features in common, as follows:

- High degree of political support, either via political leadership or willingness to establish a legislative framework for an NRSC.
- Multidisciplinary council with all key agencies and senior officials involved.
- Involvement of the private sector.
- Multidisciplinary secretariat or support team led by a high-calibre, highly motivated executive director or senior government official, and
- Adequate funding from government and, especially, non-government sources to enable the work to be carried out effectively.

Irrespective of which coordination mechanisms are adopted, efforts should be made to incorporate the above elements to ensure the likelihood of success.

2.5 Road Accident Data Maintenance

2.5.1 General

For an accurate assessment of the road safety situation, a properly maintained accident database is needed. The data shall be useful if its coverage includes data on casualties and the circumstances of the accident. These databases help to devise and implement appropriate measures designed to combat specific problems. The main processes involved in producing an accident database include the following:

- An accident reporting and recording system.
- A storage and retrieval system.
- An analysis system, and

- An effective dissemination system.

Traffic police are the most ideally placed to record and maintain accident data. The police need to be convinced of the usefulness of accident data and devote time to collecting this data. They need to have adequate resources in terms of staffing, training and computer systems and be motivated by periodic rewards for their efforts. The data collected for all recorded accidents needs to answer the following questions:

- Where accidents occur.
- When accidents occur.
- Who (vehicles, drivers, helpers, passengers) were involved.
- The result of the collision.
- The road, environmental and visibility conditions.
- The reason for the collision.

Having formulated an effective database system is to be introduced, it is important to ensure that the data is utilised as effectively and widely as possible. The annual accident reports from the police could be circulated widely, and national decision makers can use the data. The data should also be made readily accessible to relevant organisations for designing appropriate countermeasures, producing plans, monitoring effectiveness and carrying out research.

2.5.2 Need for Priority Actions

1. Review formats of police accident reports to ensure that they are easy to fill in, use nationwide and cover sufficient information to meet the needs of all potential accident data users.
2. Establish an easy-to-use computerised data storage and analysis system that provides an understanding of the scale and characteristics of the problem and enables the development of appropriate countermeasures devised for high-risk target groups.
3. Ensure accident data statistics and analyses are distributed to those working for the improvement of road safety for designing and monitoring countermeasures.

This section includes guidance on the format that a police accident database should contain, the software requirements and indicates how such a database is best utilised. It also describes the applications of the MAAP system from the TRL. That is the only microcomputer-based accident data system specifically designed for use in developing countries. TransPlan for traffic analysis is recommended, which is available for download from www.techsoftglobal.com.

2.5.3 The Accident Data System

A road accident is best described as 'A rare, random, multifactor event that is always preceded by a situation in which one or more road users have failed to cope with the situation, resulting in a vehicle collision.' Although events are relatively rare in terms of the passage of time and traffic movements at any particular site, they add up to an increasingly worrying problem for governments and the mobility of a country's population. If these events were purely random, then it is unlikely that anything practical could be done to prevent their occurrence.

However, research from all over the world has demonstrated that accidents tend to cluster at points or areas on a road network, e.g., within 50 metres of a particular junction or among groups of road users, e.g., drivers aged 18–21, schoolchildren below age 18, and pedestrians over the age of 70. Although the causes of accidents are multi-factorial, there are likely to be common reasons for the clustering; i.e., why different levels of risk exist (e.g., due to poor road geometry that often results in a driver losing control), poor visibility due to fog, darkness or obstruction, broken road surface forcing drivers to steer vehicles unexpectedly, use of phones, drink and drug driving, attractive hoardings, absence of railings, sudden median opening, road side parking and stoppages, tiredness and lack of or deterioration in skills of a road user group.

High-speed driving with frequent overtaking on rural highways is often observed, and in the absence of a lane-marking FASTag camera system, the offenders cannot be penalised.

To identify particular problems that are treatable and for which specific appropriate action can be designed, a reliable accident database, as comprehensive as possible, is essential. There are also several other groups of people at the local level with road safety interests who require accident data. These include government officers in the transport department, highway engineers, police, lawyers, research groups, politicians, teachers, statisticians, insurance companies and members of the public. They all tend to have slightly differing needs and reasons for wanting the data. These include:

- Investigation of sites or road user groups.

- Designing safety schemes and devices.
- Justification for highway planning.
- Enforcement planning or prosecutions.
- Education and training, and
- Insurance claims.

Whatever the above purpose, the underlying aim should be to acquire as much relevant knowledge as possible from the data to help prevent accidents.

2.5.4 Key Elements

There are four basic components to the way an accident data system operates, as follows:

- accident data for the reporting and recording system,
- accident data storage and retrieval system,
- accident data analysis system, and
- accident data dissemination.

2.5.4.1 Accident Reporting and Data Entry

Most countries have found that there is a need for a legal requirement for road accidents to be reported to the police, for severities of accidents, e.g., involving all personal injuries. It should be reinforced by the insurance company's rules. This will require the claimants to follow this law and encourage reporting of accidents to the police. The best source of validated accident data will generally, therefore, be the traffic police, either the traffic police attending the accident or when the parties involved or witnesses reported the accident at a police station.

In some countries, different types of accidents or accidents on different types of roads are reported to different sections of the police or even by separate organisations because of jurisdiction reasons. This leads to serious problems, causing harassment to the parties informing of an accident, and it is strongly recommended that there be only one organisation responsible for collating and compiling the national accident database, and that the police are the most appropriate group to carry out this important task. Without redirecting the people to another police station, any police station should take the report and send it to the relevant police station.

It must be noted that there will inevitably be a substantial number of road accidents that are not reported to the police. The levels of this underreporting, although significant, vary considerably from country to country. Despite having a legal requirement to report those accidents only involving personal injury, the same is not done, and the studies of hospital data have demonstrated considerable under-reporting. This tends to increase as accident severity increases. Underreporting road accidents is another serious problem in many developing countries, and wherever possible, efforts should be made to cross-check the completeness of police data by comparing it with hospital data.

Ideally, the information required for each accident should be completed at the scene on an easy-to-complete form or booklet. It is also recommended that, if possible, a single form be designed for all purposes (i.e., used for court procedures, filing and computer data entry) to eliminate the need to transcribe data further. The form or booklet will also need to provide space for the driver, pedestrian and witness statements, a written summary of the accident and sketches or photographs.

The forms (or copies) shall be sent to the police headquarters for logging onto the computer. Eventually, the data will need to be collated into a national database, and this can be done electronically via a network or other means of computer linkage.

2.5.4.2 Accident Database

A single accident report form or booklet should contain information about a single accident. Ideally, this same form should be useful for computer data entry. Various governments, or road and police authorities, have their own and different views on what accident information should be recorded. So, it is unlikely that a single common form would be applicable internationally. It isn't easy to list the factors required in all cases. However, the following basic questions the database should be able to answer:

- Where accidents occur, location by map coordinates, road name, and class.
- When accidents occur by year, month, day of week, and time of day.
- Who was involved: people, vehicles, animals and roadside objects.
- What was the result of the collision: worst severity of injury or property damage?
- What environmental conditions were present—poor light, weather, road surface conditions?

- Why or how did the collision occur: collision type, driver fault type?

Table 2.11 contains a suggested list of factors appearing on accident report forms in many countries of the world that have generally been found to be useful. Some of the more important factors are discussed below.

Table 2.11 Recommended Factors for Inclusion in Road Accident Database

General Details /Attendant Circumstances Police Reference	**Road Type**	**Environmental Features**	**Precise Location**
Year Month Date Time Region/state Police station reference Severity Collision type Number of vehicles involved Number of casualties Contributory factors code	Class of road/road no. Carriageway type / no. of lanes Speed limit Junction type Road width Road shoulder width	Light condition Road lighting Road surface condition (dry, wet, etc.) Road surface quality (potholed, rutted, etc.) Weather Junction control Geometry (curvature, incline) Hit and run Road works	Map reference X-coordinate Y-coordinate Node1} optional Node2} Kilometre post To nearest 100 m (e.g., 8 = 0.8 km) Plain language Location description (free text abbreviated) Accident description (free text abbreviated)
Vehicle / driver details			
Vehicle type Vehicle manoeuvre Vehicle damage Length of skid marks		Driver age Driver sex Licence number Seatbelt/helmet Alcohol/drugs suspected	
Casualty details			
Type of road user Age Sex Severity of injury		Pedestrian location Pedestrian movement Passenger location School student	

2.5.4.3 Accident reference—unique identifier

Once an accident record has been entered on a computer, there is always the possibility that the record may be mistakenly entered a second time, possibly by a second computer operator, or even be copied in error electronically. It is therefore important that each accident can be identified uniquely by a particular number or combination of recorded fields, for example:

i. Police station incident number.
ii. Year.
iii. Police station identifier, which may require a combination of the station and region codes.

This provides a computerised check that no two accidents can have the same combination of values for these specified key fields, thus avoiding the possibility of duplication of records.

2.5.4.4 Accident location

An important detail, which is unfortunately often neglected by many police authorities, is a precise and easily computerised accident location system.

When a road authority is considering how to tackle accident problems locally, it is not possible for it to focus initially on the worst sites (to obtain the greatest potential saving in accidents) or ultimately to evaluate the effect of its action unless it can be certain that all accident locations have been correctly pinpointed over a network. Several possible location coding systems are discussed in the following pages.

i. X-Y coordinates

It has become important to have location data plotted on maps. The national grid coordinates are recorded for each accident location to produce accident maps or possibly incorporate the data within other geographic information systems (GIS). For the plots to be useful, a 10 m grid accuracy is to be specified. However, the level of detail possible will depend on the maps available.

ii. Kilometre posts

Unfortunately, it is often difficult for the police or coders to mention the locations of rural accidents on a map. They can be identified with X-Y coordinates using GPS, along with satellite imagery from Google Earth. On intercity roads where a good system of kilometre posts has been installed, preferably with intermediate 100 m marker posts (see Figure 2.53), this can be used easily by the police as a referencing system, i.e., they note down the nearest post to the accident site and the distance to it.

Figure 2.53 100 m marker posts on an expressway in Malaysia

Where 100 m posts are not available, but the culverts are numbered, then strip maps (Figure 2.54) can be prepared that show each culvert in relation to its distance along the road. The police officer has only to indicate the distance of the accident site from the nearest culvert marker or other physical location. The exact location can then be pinpointed in the office by referring to the strip map (see Figure 2.54).

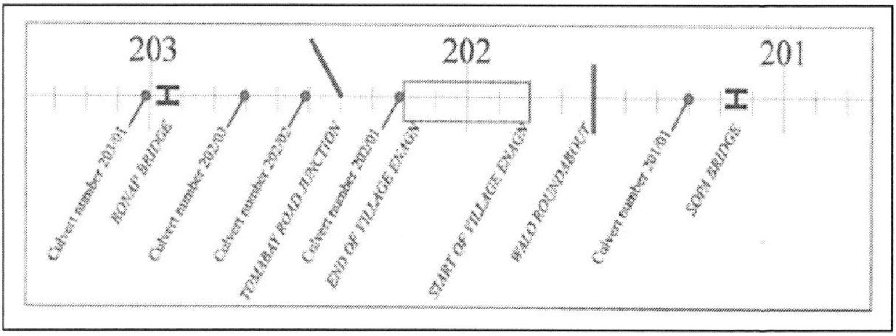

Figure 2.54 - Simplified example of a strip map showing physical locations along the road.

iii. Node-link-cell

In general, there is a great density of junctions in the towns and cities, and a node-link-cell system is recommended, where each major junction on a map is assigned a unique node number. Links are roads between two adjacent nodes. Cells are the least accurate as they are squares covering, for example, a housing estate having many minor junctions. These numbered locations are much easier to find on a map with less chance of error.

1. Plain Language Location Description

Another recommended referencing system is to use free text for an abbreviated description of the location where the accident occurred, by using street names and other landmark features. These serve as a useful check on the coded location and can give a more precise description of the location of an accident.

2. Location Sketch

There should be a collision sketch in all police accident reports, and an accident location sketch, for which a provision of a specified space on the report form is strongly recommended. This sketch should be a simple line drawing of the roads marking the accident site precisely in relation to prominent landmarks, such as bridges, junctions or kilometre posts, and their distance from these, with the direction of the compass point north. The details must be sufficient in the sketches to guide an investigator to the exact location of where the accident occurred, even if the investigator has no prior knowledge of the spot.

2.5.5 Storage/Retrieval System of Accident Data

Irrespective of whether the accident data are stored in a manual or local server-based, or global Internet-based storage system, the data must be easy to store and retrieve as needed. The true benefits of the data shall be harnessed when it can be retrieved easily for analysis.

The data structure has to be such that information stored on individual accidents can be retrieved with filter options, either as a single record or in combination with other records. This enables cross-tabulations and other analyses to be carried out more easily and permits more useful annual reports to be produced.

By using Relational Database Management Software (RDBMS), for example, Microsoft ACCESS, various configurations may be applicable to data systems and analysis reports as required by local authorities in individual countries and by funding agencies as well. The software, irrespective of the system, should facilitate standard analysis techniques such as the listing of worst accident sites and stick diagram analyses.

2.5.6 Development Stages

Five development stages can be identified in the development of road accident data systems.

Stage 1: Accidents Included in General Incident Reports
The police collect accident data using standard police incident reports at the earliest stages, treating it in the same way as all other crime records, with the focus on prosecution. Only the most basic details are collected, and the standard incident form is used the same as for aburglary or any other incident that might occur. There is little information relevant to the circumstances leading to the accident. The concern at this stage is more with identifying the parties involved and reporting the incident to the authorities.

Stage 2: Basic Accident Data Form Introduced
The need for additional information to be put on record, specifically on road accidents, requires a separate accident data form. Usually, the storage, retrieval and analysis are manual; hence, using the data, little real analysis is done. The work is produced for basic statistic tables with little interpretation or analysis of them. The data is used mainly for administrative purposes within the police to be able to report how many accidents happened and in which jurisdiction they happened.

Stage 3: Uniform Accident Form Introduced After Consultation with Other Parties
Eventually, there is a need for better accident data as the number of accidents increases. To cooperate with the agencies, such as the engineers and others, it needs to be a form that meets the data requirements of all key parties. This is to be agreed by the police that their willingness is essential to improve the system. A uniform accident form needs to be designed in consultation with interested parties and introduced nationwide after pilot testing. Often, an Excel worksheet-based accident data system is introduced at police headquarters. This must result in an annual accident report that is produced at the end of each year and circulated to all the key agencies involved.

Stage 4: Accident Data in Use in Other Agencies
At this stage, the various other road safety-related agencies should have started to use the data. Detailed annual reports and the database would be

published and distributed widely to all the agencies. The NRSC could use the data for better publicity and education. Highway authorities would be identifying and improving the accident black spots on various roads. The police would be carrying out more effective enforcement at locations and times known to involve speeding or drink driving accidents. The database would also be used by researchers in universities and research institutes to increase understanding and knowledge about the causes of road accidents required for safety audits and the effectiveness of countermeasures.

Stage 5: Supplementary Data Collected
Once a police data system begins to be used by the key agencies, it will be recognised that a considerable number of accidents are still unreported to the police. Other data sources are to be explored to identify the scale of underreporting. This normally implies auditing hospital data records and insurance industry records by researchers to try to assess the degree of underreporting. For this purpose, the police accident database is to be made widely accessible for use by several organisations. Researchers will be active in comparing police, hospital and insurance records to try to reconcile the underreporting problem.

2.5.7 Receiving Benefits and Effects

As the base measure for safety, accident data is essential to make politicians, planners, engineers, police, educators, publicity specialists and researchers aware of the scale and nature of safety problems over a road network. The database should be used to help decision makers formulate policy at the national level, such as compulsory wearing of front and rear seat belts, helmets, no use of mobile phones while driving, no drinking and driving age restrictions, and other legislation, and to produce a national action plan for improving safety. Even within districts or states, the data can be used to target road user groups at risk, e.g., drunk-driving and no use of mobile while driving publicity campaigns, school education programmes, motorcycle rider training, and police enforcement campaigns. The accident database should be used to draw up local action plans where the worst accident occurrence sites on the road network are identified. In consultation with the police, appropriate accident countermeasure schemes should then be planned and implemented by road engineers to bring about the most cost-effective accident prevention.

2.6 Road Safety Funding by the Insurance Industry

2.6.1 General

Priority Actions Needed

Although funding for road safety is primarily the responsibility of the central government, and sometimes, as necessary, it is supported by regional or provincial governments, the private sector and especially the insurance industry should play an important role in funding road safety. However, it does not yet do so in developing countries overall. While vast sums are spent on accident claim compensation, little thought or financing is directed at road accident prevention, thus leaving a very limited role for the insurance industry at the post-accident stage. Increased motorisation and the associated rise in accident claims require the active involvement of the insurance industry, as it bears most of the costs of road accidents, and should assume greater responsibility for financing and directly promoting road safety. The insurance industry can participate by providing funds and by collaborating with the government in improving a country's road safety problems.

Priority Actions

1. The government must invite and encourage the private sector, e.g., the motor insurance industry, fuel companies and transport operators, for their active involvement in funding relevant road safety activities.
2. Legislate a mandatory requirement for third-party motor insurance on all vehicles, and share five to 10 per cent of the premium as a levy for road safety activities.
3. Motor vehicle insurance regulation should be enforced with a high rate of coverage to maximise the contribution from insurance to road safety.

The various guidelines form a set for policymakers. The role of the motor insurance industry is interconnected with the road safety process, and in many countries, the result has been beneficial and significant. Although many private sector companies have contributed financially to improving road safety measures, this section has a particular focus on the insurance industry, who are key agents in the private sector for improving road safety.

2.6.2 Funding from the Insurance Industry

Although the central government is responsible for developing and implementing a coordinated road safety programme, most of the adverse effects of road accidents, such as cost, are faced by the private sector industry. Hence, the funding for road safety improvements should be shared with those who stand to gain from accident prevention measures.

In most countries, the insurance industry has traditionally limited its involvement to post-accident compensation payments. A no-claims discount was most probably the only incentive to promote accident prevention. In developing countries, this plan of business has proven unsustainable for many insurance companies, incurring excessive losses due to ratios of premium revenue to high claim costs. The situation becomes more aggravated, as accident claims increase rapidly in the countries undergoing rapid motorisation. Traditionally, motor insurance companies in many industrialised countries have been involved in supporting road safety improvements through the provision of money, either voluntarily or through legislation, and through other means, including research, lobbying for change and campaigns. While the government is faced with multiple serious economic and social needs, the insurance industry has a vested interest in seeing road safety receive adequate attention and in efforts being made to reduce accidents.

2.6.3 Key Elements

2.6.3.1 Legislation

To receive significant support from the motor insurance industry, it must be well established and of adequate size to provide funds that can have an impact on road safety. Such an industry emerges with legislation that requires all vehicle owners to have at least third-party insurance and then grows as the vehicle fleet expands.

Table 2.12 Potential annual income from a US$10 levy on motorised vehicles

Country	Annual Income (US$ million)
People's Republic of China	273.0
India	252.0
Indonesia	111.0
Republic of Korea	80.0
Lao PDR	1.3
Malaysia	66.6
Philippines	21.2
Sri Lanka	8.6
Taipei (China)	148.0
Thailand	110.6
Viet Nam	33.9

Third-party motor vehicle insurance requirements for all motor vehicles have been introduced in many countries. Two exceptions are Bangladesh and Nepal. In Bangladesh, government vehicles are exempt, and in Nepal, the vague wording of the legislation makes the requirements of motor vehicle insurance unclear for private vehicles.

2.6.3.2 Enforcement

The enforcement of motor vehicle insurance regulations is required in a developing country to realise its full potential earnings, where the road safety fund is primarily based on insurance premiums. Enforcement of regulations has not strictly followed the compliance with motor vehicle insurance legislation, and many vehicles are driven without insurance. To ensure compliance with motor vehicle insurance regulations, each vehicle is required to have proof of insurance as a requirement of the periodic vehicle inspection for roadworthiness. Vehicle fitness certificates should not be given without proof of insurance. In a computerised system, it should work automatically on the vehicle registration number. It is to be remembered that in a manual system, the extent to which the inspection authorities enforce this stipulation may vary widely. This inspection may be supported best if there is a mandatory display of a valid roadworthiness sticker or certificate as a legal requirement, and that must be clearly visible on the windscreen of the vehicle. In Kazakhstan, motor vehicle insurance can be purchased at the vehicle inspection centres to promote compliance with the new insurance regulations. Proof of insurance coverage can also be confirmed at the stationary checkpoints where traffic police inspect other required documentation, such as ownership, fitness and pollution control certificates, and route permits. Proof of insurance can also be checked, as the driving licences are checked when traffic warnings and citations are being issued. The overdue fines for insurance coverage should be set as a penalty several times the cost of the insurance policy premium value.

2.6.3.3 Promoting Road Safety

If the insurance industry is involved in the road safety organisation and takes an active part in determining the use of their funds, that will make their support and commitment greater to road safety. This will create a more active partnership between the insurance industry and the government, and the role of the insurance industry should not be limited to a passive funding source for road safety. If not in the main body of the NRSC, the insurance industry should be represented on the finance subcommittee. The insurance industry can do much to promote road safety in its insurance policies, as well as by sharing data through the direct organisation of road safety campaigns.

The insurance industry is a good source of accident data, and in many countries, it can be more comprehensive than police data. The total cost of the consequences of road accidents often needs to be calculated before governments begin to invest in road safety.

2.6.3.4 Non-insurance-related Private Sector Support

Various private sector companies that invest in road safety have benefited from an improved corporate image; some private sector organisations outside the insurance industry have a direct incentive to promote road safety. Businesses with large fleets or expensive vehicles are already interested in reducing road accidents among their own vehicles. Safety-conscious policies should include strict testing of drivers for their ability, as well as medical and alcohol testing. Advanced driver training may be offered, including bonuses and awards. For safe driving, added incentives may also be offered to drivers by their employers. Commitment from the private sector to road safety must be visible and continuous to be effective. For example, the Shell Petroleum Company sponsored defensive driving courses for its personnel in several countries, for example, Brunei, Fiji, etc.

In Malaysia, Shell required its new drivers to undergo a three-day defensive driving course, and refresher courses were given every two to three years. Close monitoring of driving records has been maintained by classifying their accidents as preventable and non-preventable.

The automobile industry has also contributed to campaigns for improving road safety. Volvo, apart from its prestigious international safety award each year to encourage innovation in road safety, has been active in some countries in raising awareness of safety issues.
To raise road safety awareness among the public, non-government organisations (NGOs) and service groups can also play a significant role. They are in the best position to reach the local community to incorporate road safety easily into the basic lifestyle and education programmes. The private sector work can be organised and provided externally, and concurrently, the initiative for government activity should be coordinated to maximise impact.

2.6.4 Development Stages

Stage 1: It commonly happens that the government passes legislation making third-party motor insurance compulsory, but the enforcement is lacking. Sometimes, only a few motorists are insured, and most compensation claims for accident damage are settled privately.

Stage 2: Motor vehicle insurance starts receiving more priority from the government. The NRSC represents the insurance industry, and insurance companies, as well as other private sector organisations, start sponsoring events, such as conferences or safety publications.

Stage 3: Between insurance and road safety, a closer association is developed with new legislation requiring that a levy of say 10% be imposed on insurance premiums, which should be for road safety improvements. Requirements for insurance are beginning to be enforced by the police and inspection authorities. At this stage, insurance should be integrated with vehicle registration using a computerised system all over the state and the central government. Policy issues such as premium rates, policy restrictions, maximum coverage limits, hit-and-run drivers and processing times are under review. Third-party insurance premiums may often be increased slightly at this stage, and some of the additional premiums are contributed to road safety.

Stage 4: The insurance industry becomes more active in supporting road safety, rather than in its financing role. The government creates an incentive policy of encouraging private involvement in road safety and sponsorship of road safety campaigns. The accident databases developed by insurance companies are used for analysing road safety problems and are made accessible to researchers. Joint campaigns by the government and the insurance companies become more ambitious as they realise the benefits that are accruing from their initial efforts and investments. Other companies connected to the motor industry, particularly the fuel industry and motor dealers, have also become more involved in road safety.

2.6.5 Receiving Benefits and Effects

Various potential benefits result from the insurance industry contributing to road safety funding, including:

- New funding source (non-governmental).
- Marketing and business skills.
- Improved perception of the insurance industry.

The most obvious benefit resulted from the large amounts of money that became available to help a country start its road safety efforts or continue with its priority programme. Other benefits include the synergy developed when different disciplines work together in a coordinated manner. The involvement of insurance industry firms enables them to add their professional expertise to road safety in areas such as data analysis, marketing and publicity. The self-image of the insurance industry is improved as the public acknowledges the companies' social responsibility. The campaigns improve brand recognition and corporate image.

2.6.6 Scope for Action in the Asia Pacific Region

In case of failing to persuade the insurance industry to participate voluntarily, then an alternative option should be taken to impose a levy of about ten per cent on third-party motor insurance and, where possible, a smaller levy of perhaps five per cent on all other motor insurance policies. This ensures a steady stream of funds for the NRSC for promoting road safety and initiatives. Each of the countries discussed above develops a unique way of funding safety activities, and all recognise that the motor insurance industry has an important role to play in investing in safety. Each country needs to develop a funding mechanism that is appropriate for its particular social, political and economic circumstances. However, from the above case studies, it is understood that it can be done successfully.

2.7 Planning and Designing Safe Roads

2.7.1 General

Priority Actions Needed

Incorporating safety measures is essential during the planning and design stages. In most developing countries, the road networks are still being expanded and/or rehabilitated, creating opportunities to allow provisions for road safety. Some of the important components of the design process that influence the level of road safety are discussed in this section.

Developed countries adopt international design standards because these standards are generally accompanied by effective enforcement, driver training and publicity. If those provisions are not available, then by simply following those standards, any road safety improvement may not be possible in developing countries. Additionally, the traffic conditions and types of traffic using the roads will be different, and that may not allow the features of the design standard to operate as efficiently as in developing countries. It needs to be examined how to make the road network operate safely in the operating environment and traffic conditions that exist in each country.

- In schemes of rural road rehabilitation, opportunities should be taken to minimise direct access to major roads when such major roads pass through small communities, keep traffic speed relatively low, and eliminate Y-junctions.
- Inclusion of additional safety provisions, such as cycle lanes, may be considered for new roads.
- A road hierarchy may be established by the design of road networks in urban areas, by reducing traffic and speed where pedestrian and cyclist activity exists.
- Greater emphasis is to be given to the safety of the VRUs, who have a significant presence on all roads, in large proportions in developing countries.

The safety audit process is considered during the planning and DPR design stage of roads and transport networks. The World Bank, Asian Development Bank (ADB) and other development aid agencies have found that many potential safety problems need to be addressed.

Priority Actions

1 During the design stage, all proposed new and rehabilitation road schemes are to be checked from a safety perspective.
2 Existing design standards, access control, and development control are to be reviewed to ensure safety, particularly for VRU in urban and rural areas.
3 Towns and cities have localised zones for road networks, and the existing road network is classified into a hierarchy.

2.7.2 Safe Planning and Design

2.7.2.1 Safe Planning

Planning is important because on any road, conflicts can arise between different types of road users, and these may lead to accidents involving death or injury. These conflicts are typically many in numbers in town centres, but can also occur on suburban or rural roads. Conflicts between the following types of road users are common in all countries:

Figure 2.55 - Bicycle versus lorry in Hanoi, Viet Nam

Different Modes

- Motorised vehicles versus pedestrians.
- Motorised vehicles versus NMVs, particularly cycles.
- Heavy goods vehicles versus other road users.

Different Movements

- High-speed traffic versus low-speed vehicles.
- All vehicle types at road junctions; and
- Vehicles overtaking.

In developing countries, roads tend to be used more by NMVs and pedestrians compared to industrialised countries. The design standards mostly originate from industrialised developed countries. They may therefore need to be either amended or new standards developed, considering the general usage of the national road network in the developing countries. Safety features such as service roads, underpasses, railings, covered drains, pavements, pedestrian crossing facilities, motorcycle lanes, signs and road markings should be considered at the earliest stage of planning. Financing these safety features, as well as their maintenance, should also be planned at this early stage. Normally, many problems exist due to previous decisions and poor planning; it is never too late to rectify the situation or to plan. This is particularly true for those developing countries where the infrastructure and traffic volumes are growing at a rapid rate (see Reference 1).

2.7.2.2 Safe Design

Safe design is relevant to all existing roads as well as all new projects. It tends to start with a country recognising the fact that it needs to have design standards for traffic signs, road markings and specifications for the geometric design of highways and junctions to accommodate the types of vehicles that are likely to use the roads (discussed in Chapter 2). The country should then develop specific techniques for safe design and checking procedures, such as a safety audit, to ensure that standards are applied appropriately or all deviations from standards are carefully considered (discussed in Chapter 1).

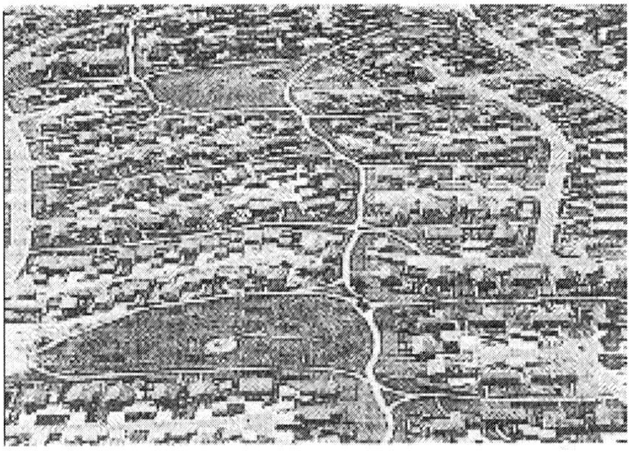

Figure 2.56 - Well-planned roads in Australia

Human behaviour is the most prevalent factor that causes or influences accidents. Engineering design is, therefore, important as it should identify and accommodate a wide range of human behaviour and encourage safer use of the roads. Sadly, this is not always considered by engineers and inappropriate engineering results in a high percentage of accidents. Engineering features are indeed easier and cheaper to change than human behaviour, and can have an immediate effect (as discussed in Section 1).

Engineering design can influence the way that a road is used at a particular location or under a particular set of circumstances. Young pedestrians often tend to behave in a relatively irresponsible and unsafe manner. In some cases, this is due to their general inability to judge speeds accurately. They can, however, be helped to survive by a set of considerations. In the urban traffic environment, the need to come into conflict with moving vehicles can be minimised by segregating pedestrian crossings, for example, underpasses or footbridges, steel railings and by controlling the speed of vehicles. NMV such as bicycles and rickshaws, together with motorcycles and three wheelers, are particularly vulnerable. These road users often constitute large proportions of the traffic on many roads in the Asia Pacific region. Due consideration should be given in the planning, design, operation and maintenance processes.

2.7.3 Key Elements

2.7.3.1 Safe Planning

The various components in the planning process, together with the key principles associated with each, thatinfluence safety are listed below.

Figure 2.57 Potentially dangerous deep V-type ditches in Papua New Guinea

Road hierarchy: The roads in a network should be clearly defined and classified into those that are primarily to be used by traffic and those that are used primarily for access to housing or other buildings and amenities, and where low speeds are required.

There should be a clear, unambiguous priority indicated at each junction so that traffic on the major (more important) road is always given priority over that from the minor road. Consideration should be given so that any road only intersects with roads of the same class or the one immediately above or below it in the hierarchy.

Catering for different transport modes: the provision of facilities for all different types of road users, including pedestrians, cyclists, other NMV, and different forms of public transport, gives the best opportunity to engineers for providing a safe traffic environment with minimum congestion and pollution.

Land use: this should be distributed to minimise conflict between traffic and pedestrians, by locating shops, workplaces and schools within walking distance of homes, including, wherever feasible, separate networks of pedestrian walkways and cycle tracks, so that travel by vehicles is minimised. All development proposals should be thoroughly examined for traffic and safety implications before approval for construction is given.

An efficient public transport system serving the various land uses can greatly reduce conflicts in an urban area by reducing the overall number of vehicles on the road and minimising the problems for pedestrians. Careful planning of the road layout around bus or rail stations and bus stops is essential, and safe routes for pedestrians should be developed near these locations.

Access control: direct frontage access onto major roads should be minimised. The construction of service roads along the major roads for shops or industrial units should be encouraged. Minor roads should be provided to access the service roads, which will then access the major roads. No access should be permitted at dangerous locations such as bends or hill crests. Access should also be minimised at or near existing junctions.

2.7.3.2 Safe Design

In general terms, good design provides conditions that are clear, simple and consistent. Many countries in the developing region have developed their standards by adapting or modifying existing standards of developed countries, such as those from the American Association of State Highway and Transportation Officials (AASHTO), the National Association of Australian State Road Authorities (NAASRA) or the UK.

Road design should present necessary information in a systematic, sequential manner. Full coverage of the technical designs is beyond the scope of this section, but more guidance is given (Reference 1). It is also important that care should be taken for the operation of roads under construction and maintenance, particularly in the developing world, where construction tends to take longer, and attention to the safety of workers, pedestrians and other road users is often inadequate.

Cross-sections: The technical design of any road will depend upon its category within the road hierarchy (see above), and will comprise lane numbers and widths, shoulders, medians and other features that are consistent with a set of national standards. It is particularly important, for example, that adequate width provision is made on hard shoulders for pedestrians and animals.

Sight distances: Adequate visibility is essentially required for a manoeuvre. A driver always needs to see ahead to stop, overtake, cross or merge with traffic safely. Thus, the criteria are needed to ensure that the road design ensures this happens and that visibility is not obscured.

At locations where visibility is inadequate, there may be a need to ban overtaking or to construct an additional overtaking lane. At road junctions, good visibility between traffic on the major road and vehicles emerging from the minor road is essential for safety. Visibility should be provided by using the appropriate speed and related stopping distance as a basis for design.

Bends and superelevation: Horizontal and vertical bends should be designed so that they can be negotiated safely at specified speeds appropriate for the category of road. This is achieved by having criteria that match design elements, including radius of curvature, surface type, and superelevation, with the speed of traffic. For example, similar radii may be provided for adjacent bends, and large variations are to be avoided. Flatter vertical curvature at the top or bottom of hills is appropriate and needs to be designed carefully to help drivers stay in control and ensure that they have adequate visibility. There should not be any sudden changes in alignment;

for example, the road should not change immediately from being straight to a series of sharp curves when moving from flat to hilly terrain.

Speed limits: Speed limits should be clearly signposted and understood, and should always be appropriate for actual traffic conditions. Slower speeds mean fewer and less severe accidents, but they can rarely be achieved using signposts. It may be necessary to provide physical features such as road narrowing, road humps, rumble strips, embedded studs, cameras and other methods to reduce speeds to the desired level, especially where such roads pass through small residential areas or villages, or where a lower speed is required for the situation, for example, on approaches to single-lane bridges. The use of road humps on principal roads needs to be carefully assessed because of noise and vibration problems and the possibility of excessive traffic delays. Warning about such devices should always be provided. Road humps have been used successfully on major roads in some countries where they pass through small communities or villages.

Road signs and markings: To inform and direct motorists, especially to warn about the hazards ahead, old integrated road network systems in all countries used to have standards for road signs and markings and in ITSs using FM radio channels to inform and guide motorists.

The signs and markings should be applied consistently within a country and should be simple to understand with minimal wording. Road signs and markings should preferably conform to international conventions to help travellers from other countries understand. Signs should be carefully sited, well-maintained and should not be obscured by trees or other features. High-quality long long-lasting materials must be used, even if the initial cost is high. Signs should be retro-reflective and road markings should be made of reflective thermoplastic where possible.

Drainage: In situations where the road is on an embankment, storm water is removed from the road surface using longitudinal gradation and cambers. In addition, drainage ditches, catch pits, channels or culverts form an essential element of any road that is not on an embankment. They are needed not only for the short-term safety of road users during rain but also for the long-term structural integrity of the road surface. Poor drainage can lead to potholes and even the complete failure of the surface. Deep drainage ditches too close to the side of the road can be a serious hazard to vehicles that run off the road while negotiating sharp curves at an interchange, or for a tired driver. They need to be designed with care, particularly on bends. L-shaped ditches are preferred if they are adequate to cope with expected levels of water runoff.

Obstacles and safety fencing: The presence of roadside obstacles, street furniture, and trees has two implications: the potential danger of collision and the obstruction of visibility. Collisions can be prevented by installing crash barriers/safety fencing or wheel guards behind high kerbs, or by removing the roadside feature. Visibility problems can be treated similarly by removing and re-siting the object, or by ensuring adequate trimming of trees or vegetation.

Figure 2.58 Movable concrete blocks and steel rail barrier in the PRC

Medians and barriers: medians or central verges with barriers can be used to segregate opposing streams of traffic, thereby helping to prevent head-on collisions while overtaking or passing parked vehicles. They can be used to restrict U-turns, segregate different road users from each other (e.g., bicycle lanes, three wheelers) and to discourage pedestrians from crossing at unsafe locations.

Proper attention is needed while designing medians and barriers to ensure that turning movements are catered for and emergency services have access. Medians must be sufficiently wide to provide safe waiting areas for vehicles while pedestrians are crossing the roads, and the ends of barriers should not be hazardous (see Figure 2.58).

Figure 2.59 Vehicle impaled on incorrectly terminated barrier

Lighting: The provision of adequate street lighting is a proven accident reduction measure in many countries. It is particularly helpful with respect to the safety of pedestrians or vulnerable vehicles operating without lights. Lighting is, however, expensive to install and maintain. Poor maintenance of lights can cause additional safety problems through uneven illumination. Lampposts should not be sited in positions where they will cause a hazard to a vehicle leaving the road. They may also be designed to collapse on impact or be protected by safety fences. In times of limited resources, lighting should first be provided where the potential conflicts or danger are greatest, for example, at junctions or at a midblock (in between two junctions) location with pedestrian crossings. Street lighting is discussed in detail in Chapter 8.

Bus stops and lay-bys: bus stopping bays and lorry lay-bys allow vehicles to stop safely and with minimal adverse effects on the ongoing traffic. They should be positioned on straight, level sections of road and should be visible from a good distance in each direction. Bus stops should be located beyond pedestrian crossings and at least 75 metres after junctions to avoid stopped vehicles blocking views of pedestrians crossing. Where goods are frequently sold at roadsides, the provision of a deep lay-bye is appropriate to accommodate the stalls and parked vehicles and thus minimising the danger.

Table 2.13 Advantages and Disadvantages of Different Junctions

Junction type	Advantages	Disadvantages
Grade separatio n - High flows	Minimal delays	Expensive
Traffic signals - Low to medium flows	Can accommodate heavy offside turns	Less space than the roundabout, many delays at off-peak times, reliable electricity supply is needed Sophisticated maintenance needed
Roundabout - Low to medium flows	Good for turning and merging with minimum delays at off-peak times	Poor safety record for cyclists. It can become congested if traffic does not give way on entry to the roundabout
Priority (give way or stop) - Low to medium flows	No delay on the major road	Possible delays to the minor road. Adequate visibility needed

Figure 2.60 – Pedestrians on the street in the PRC

Junctions: The basic principles of a good junction design are that they should allow transition from one route to another, or through movement of traffic on the main route, with minimum delay and maximum safety. The layout of a junction should be simple and obvious, catering for all types of traffic (discussed in Chapter 2).

The type of junction chosen will depend on traffic flows on the intersecting roads, considering the standards of traffic control and management within the surrounding area. The advantages and disadvantages of different types of junctions are given in Table 2.13.

Channelling and road markings should be used to position the vehicle at the correct and safest location for the driver to carry out the manoeuvre and to guide the driver through the junction.

Pedestrian facilities: Particularly in urban areas, walking is an obvious need, and arguably, at some stage of their journey, all road users are pedestrians. From metro stations, bus stops, and car parks, everyone has to walk to banks, offices, medical clinics, restaurants, shopping centres, etc. However, adequate provisions for such needs are often neglected or given insufficient attention. Pedestrians are extremely VRUs and tend to be involved in high numbers of accident deaths in many countries. Children going to school or living near the road form a large proportion of the pedestrian mass, and their road survival skills are often not as developed as those of adults. Pedestrian safety can be improved by a range of facilities, including the following:

1) Speed reduction through traffic calming measures.
2) Adequate footpath areas kept clear of obstructions.
3) Zebra or other types of crossings with or without pedestrian signals.
4) Bridges or underpasses at busy locations.
5) Barriers to prevent hazardous pedestrian movements.
6) Central refuges to assist pedestrians to cross safely in two or more stages.
7) Special crossing arrangements for school children.
8) Carefully located bus stops to minimise conflicts between moving vehicles and passengers embarking or disembarking; and
9) Special care in the design of all roads where there are large numbers of pedestrians, particularly the young or elderly.

Bicycles and other non-motorised vehicles, comprising more than 50 per cent of the traffic on many city streets, are an important part of the traffic in most developing countries in the Asia Pacific region. Separate consideration is needed due to their slow speeds and vulnerability to injury when collisions occur. Problems related to safety and congestion typically occur at major junctions, and they can be best addressed by providing grade-separated underpasses, dedicated lanes for bicycles and perhaps other SMVs.

Figure 2.61 Cyclists in the PRC

Figure 2.62 - Little provision for pedestrians in India

A less expensive solution might be provided at-grade by bringing the segregated lane crossing points ahead of the main vehicle junction. Facilities for bicycles need to be attractive and well-maintained to attract users. Facilities that are in use to a limited extent in some developing countries include:

1) Slow lanes are approximately two metres wide.
2) Separate lanes separated from the main road or service roads by a variety of barriers or fencing, ranging from a small narrow mound through various types of temporary fencing to a planted median or a concrete barrier.
3) Separate phases at traffic signals.
4) Streets for bicycles only.
5) Streets for one-way motor vehicles to accommodate two-way bicycle traffic.
6) Streets with one-way motor vehicles in one direction with one-way bicycles in the other.

7) Bridges or underpasses with steps for pedestrians and ramps for bicycles, pushchairs or wheelchairs.
8) Shared use of pavements for pedestrians, preferably with marked lanes for cycles, often associated with a ban on cyclists on the main road.
9) Pedestrianised areas for accessing bicycles, and
10) Parking bans to facilitate the movements of slow vehicles.

Providing facilities for cycles and other NMVs improves road users' safety and reduces problems related to congestion and air pollution. Many of the cities in the world that have the worst congestion and pollution have failed to make adequate provision for non-motorised traffic. More guidance on facilities for VRU is discussed in other sections of this chapter.

Traffic management

Safety and congestion within any city depend upon how well the traffic is managed. Planning can produce many benefits using some or all of the following techniques:

1) Parking restrictions and facilities.
2) Control of minor junctions through signs and road markings.
3) The use of channelling at more important junctions.
4) Streets with one-way vehicular traffic.
5) Slip roads and turning bays for traffic combined with refuges for pedestrians.
6) Central medians to restrict traffic.
7) Pedestrian crossings, bridges and subways.
8) Dedicated lanes for buses, cycles or other slow vehicles.
9) Bus stops and lorry lay-bys.
10) Traffic signals as per demand.
11) Linked traffic signals possibly using computer control.
12) Monitoring of traffic by local traffic police or by CCTV cameras.

Safety audit

It has been observed in many developed countries that even though the highway and traffic schemes were professionally designed to the latest standards, they could still lead to unexpected accidents. A common finding was that at either end of the highway, there was a transition to a road of a different standard, or pedestrians or other VRUs were not given sufficient attention.

These design deficiencies can be best reduced or eliminated by an independent road safety specialist reviewing the designs at various stages in the process. A specialist identifies any potential problems associated with the project by considering the surrounding situations and makes recommendations for improvements. Next, there should be a system that rigorously reviews such recommendations and follows up with the required design changes. Safety audits are carried out and the reports are prepared at different stages of the design process, such as:

1) Feasibility stage, conceptual design drawings on satellite images.
2) Preliminary design, with DPR drawings.
3) Final design, with detailed engineering drawings.
4) Before opening for traffic, with 'As-built drawings.
5) Inspections.

A safety audit is a quick process that improves the safety awareness of all design engineers. Safety audits are beneficial as they eliminate various potential accident situations before they are built into the network, where they would have remained unsafe for many years in the future. Opportunities also exist during road rehabilitation projects, by eliminating all Y-junctions, reducing access points and speed as the roads pass through communities. Ignoring such opportunities during the design stage will result in a less safe road than that existed before the rehabilitation.

With respect to the road standards and the road user mix, there are large differences between countries. Use of international design standards does not necessarily mean that the road will operate safely. In a developing country, there may be a completely different mix of NMV, agricultural traffic and pedestrians. Additionally, there may be many problems associated with street vendors and encroachment. These differences can be foreseen via safety audits and appropriate changes recommended to ensure that the road will operate safely in the conditions that exist in that country.

2.7.3.3 Safe Operation

Maintenance needs: As described in the preceding paragraphs, all the safety measures may fail if they are not properly maintained. This is particularly true in the case of sophisticated equipment such as traffic signals using electronic controllers or computers. It can also apply to more basic engineering techniques, such as drainage channels that become blocked and ineffective if ignored. Provision must be made for regular maintenance before any measure is adopted. Maintenance is particularly important for road markings, which can be worn out in a few months, and for signs that can be vandalised or obscured by the growth of trees and vegetation.

Enforcement needs: Traffic police are readily visible, particularly at urban junctions in most developing countries. They can help improve the effectiveness of any safety measure by ensuring that the facilities are used correctly. It is important for the police to understand any new facilities that are introduced and that their efforts to enforce regulations and penalise irresponsible road users help to improve road safety. This requires a close liaison between the engineers and the police. Traffic police need to be trained to ensure that they see their role in increasing road safety rather than just maintaining the traffic flow. It is particularly important to have trained police carrying out enforcement activity when a new scheme or junction is opened to guide and direct road users until they become familiar with the feature and how to use it.

Figure 2 .63 Cycle lane in the UK

2.7.4 Stages of Development

The key stages in developing a safer planning and design process normally include:

1) Introduce regulations covering land use, development, vehicles and road users, road signs and markings, highway design standards and safety audits.
2) Introduce a traffic act stating that the road authority must make efforts to improve safety on its road networks and to report annually on the action taken and its effectiveness.
3) Road authorities need to set up a small safety team to monitor the network and identify the problems.
4) Regulations applied and procedures developed covering access controls and uniform design standards.
5) Training to increase the skill levels of local engineers.

6) Monitoring and improving road safety on the existing road network.
7) Notes and guidance manuals prepared by safety specialists to be issued to engineers as new information becomes available.

Technical assistance to NRSCs and PRSCs in all these stages is to be provided by development banks and funding agencies.

2.7.5 Receiving Benefits and Effects

Applying the techniques listed above should provide a safer environment for all road users. A good planning and design process is expected to minimise the conflicts by:

- Better land use and public transport may reduce the number of local trips and trip lengths.
- Separate local traffic from higher-speed through-traffic.
- Providing guidance and warnings to road users about the changes in the road environment.
- Ensuring safety hazards are not included unintentionally in new schemes, for example, deep V-ditches on rural roads, keeping existing uncontrolled access, and visual traps such as troughs or unbroken sightlines across junctions with major roads.

Figure 2.64 Lane for bicycles in Hanoi, Viet Nam

Most of the techniques require funding, but can all be an investment. The techniques that lead to unacceptable conditions will have damaging financial implications for the cities and countries, and should be ignored.

Figure 2.65 Cycle lane on a minor road, PRC

2.8 Improvement of Hazardous Locations

2.8.1 General

It is essential to identify hazardous locations by a careful analysis of accident data, by studying sites, and then designing appropriate remedial measures. The safety benefits that can be derived from such procedures have proven to be particularly high. This helps implement low-cost remedial measures, and the benefits achieved can be many times higher than the cost of their implementation.

A planned programme of remedial measures, based on accident reduction targets for highway authorities, helps maximise the effectiveness of this approach. A specific annual safety budget needs to be allocated by the authorities or ensure adequate funding is set aside within the maintenance budget, for road safety.

The four main strategies are:

- Single-site or black spot programmes.
- Mass action plans.
- Route action plans, and
- Area-wide schemes.

The stages of a hazardous location improvement process are as follows:

- A good accident database.
- Agreeing on a local hazard location improvement programme.
- Analysis to identify accident black spots.
- Design of remedial measures.
- Implementing and enforcing the measures, and
- Monitoring the effectiveness of remedial measures.

Remedial measures can include better signage design and locations for installing road markings, pedestrian facilities, fencing, railings, junction modifications and improvements to visibility. Traffic calming through various speed reduction measures has proven particularly effective where VRUs are at risk.

2.8.2 Priority Actions Needed

1. To monitor the operational safety and efficiency of the road network, all road authorities must establish and train a small team.
2. According to annual targets, identify and improve the most hazardous locations on major inter-urban road networks.
3. Focus on speed reduction near schools and in residential and other areas where there are many pedestrians and cyclists.

It is the basic responsibility of all governments to ensure a safe living environment as realistically as possible and with equal concern for economic importance. Road accident costs commonly strain the resources of all countries, whether developed or developing. Experience shows that one of the most cost-effective ways of using funds in the roads or transport sector is to identify the hazardous locations and apply the funds for improvement by designing appropriate remedial measures.

This section contains a discussion on the following:

- Guidance on accident data analysis.
- Identification of hazardous locations.
- Determining priorities for action.
- Remedial measures, and
- The evaluation of results.

2.8.3 Identifying Hazardous Locations

Highway authorities should ideally be constantly working towards achieving their own annual targets for accident or casualty reduction. With budgets that are inevitably limited, they need to be sure that they are getting maximum value, in terms of accident reductions, for money spent on improving road safety.

On any road network, accidents have been shown not to be randomly distributed over the entire length but to be clustered at certain locations, i.e., hazardous locations or black spots. If accidents occurring at these locations are studied, it is often found that there are common patterns, such as driver error, to which a particular engineering feature may have contributed. Appropriate road improvement could prevent or reduce similar occurrences in the future. Black spots, by definition, are locations where many accidents occur and thus, logically, treatment of these sites should result in the best return in terms of accident prevention.

It has generally been proven that the application of low-cost engineering at hazardous sites produces high returns over many years. Simple measures can significantly reduce safety problems at such sites, and this can often be shown in a few months in terms of the monetary value. For example, the use of road signs and markings to guide and channel traffic through complex junctions, or providing safe waiting lanes for turning vehicles, can often result in a substantial reduction in accidents.

If preventive measures are based upon good accident data systems that help identify and locate hazardous sites accurately, the measures should always be most cost-effective. It is important that governments realise that the most effective use of often scarce transport resources lies in applying them to low-cost remedial work at accident black spots.

2.8.4 Key Elements

2.8.4.1 Target Setting

Experience in several developed countries has indicated that good results can be achieved towards reducing a nation's accident numbers significantly by adopting realistic accident or casualty reduction targets by all highway or road authorities. These targets are achievable and therefore to be treated seriously. The minds of engineers tend to focus on finding appropriate, cost-effective solutions regularly and on a wide scale. This assumes that a reliable road accident database for the highway authority's road network is readily available.

2.8.4.2 Identification

A good accident data system will allow easy determination of where the accidents cluster and enable the engineers to prepare priority lists of hazardous sites that need attention. This is carried out in various ways around the world, typically:

- By ranking sites using actual accident numbers.
- Account for accidents involving injuries or use a weighting system to take account of the severity, assigning using a scale with more points marked for more severe accidents.
- Account for the cost of the accidents.

There will, of course, be a need for further analysis to determine those types of accidents that would respond to remedial measures. Not all locations will have easily identifiable patterns that can be improved.

The actual definition of a black spot depends upon local conditions and should be determined after a preliminary summary of all the worst sites has been carried out. Then a cutoff can be made at the level of perhaps the 50 worst sites. Thus, in the UK, a black spot may have only five accidents in three years, while a city in Bangladesh may have defined a black spot as having more than ten accidents in a year. In the definition of a black spot it is important to use the same time period (usually accidents are expressed per year) and be consistent about the type of highway or have several different black spot groups, for example, 200 metre or one kilometre length of highway, within 50 m of a junction, such that the highway authority can really identify the worst accident clustering within a road group type.

2.8.4.3 Analysis at Hazardous Locations

In developing countries, many accident databases do not record all the information identifying the real cause of an accident, as the police collect data to apportion blame to enable prosecution and for compensation issues to be resolved. Also, the accident data systems will often need to be expanded and modified to include more and correct information necessary for the investigating engineer.

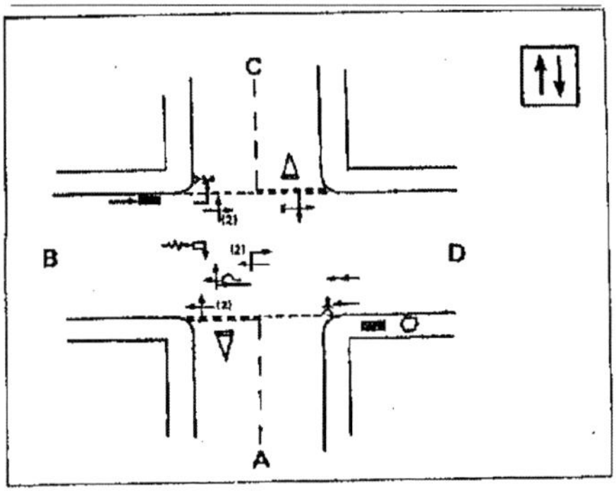

Figure 2.66 Collision Diagram

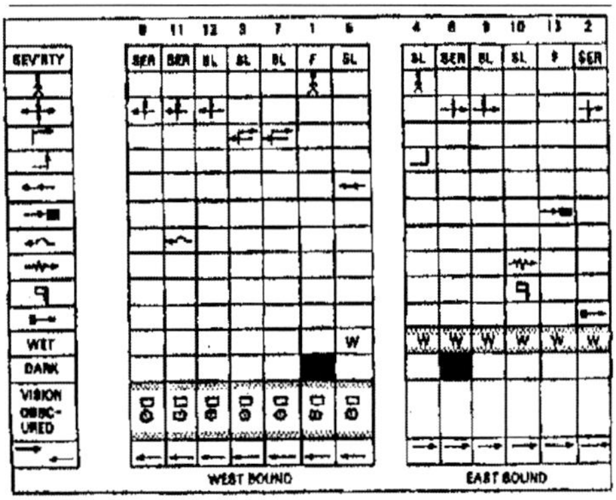

Figure 2.67 Accident Factor Grid

Most countries, at the earlier stages, commence with single-site or black spot programmes and then later move onto mass action plans, area-wide schemes and route action plans. Once the accident sites for investigation have been determined, there usually follows a three-stage investigation process, as follows:

- A preliminary analysis is carried out of the accident data to understand the accident types and conditions. Collision diagrams and accident factor grids (or stick diagrams) are useful tools here (Figures 2.66 and 2.67).
- Specialist accident investigators on site determine the exact conditions at the location. The investigation will take note of traffic flows and road user behaviour when the accidents occur most frequently, determine whether more information is required, such as speed measurements and skid resistance and start to define the problems associated with the location. Where data is limited, observation of conflicts can provide insights into the problems. This may be either simply by watching traffic informally to see typical manoeuvres and conflicts or via formal conflict studies.
- A detailed analysis of the accidents is then carried out. This will be done by simple statistical tests to determine whether the accidents have resulted from a real increase in risk or are a result of random fluctuations.

2.8.4.4 Design and Implementation of Remedial Measures

The main objective in deciding on remedial measures is to consider solutions that ideally will remove the main accident patterns identified. This can involve one or more of the following:

- removing the conflict causing the problem.
- improving the situation (e.g., give earlier warnings).
- reducing speeds, thus reducing the chance of the accident happening or its severity.

Table 2.14 lists typical accident situations along with the proven remedial measures known to have been effective in both developing and developed industrialised countries. Often, the remedial measures relate to the type and positioning of all the road furniture. This is particularly true in urban areas and at junctions. Further measures that are relevant in city centres include:

1) Turn prohibition, channelling or protected turns.
2) Traffic signals, roundabouts or revised junction designs.
3) Refuges, pedestrian crossings, bridges or underpasses.
4) Segregated cycle tracks or defined cycle lanes.

5) Parking restrictions or controls.
6) Speed limits or enforcement.
7) Traffic calming.

The most important part of this whole process is a mechanism that ensures the following:

1) Up-to-date data is available to the engineers investigating hazardous locations.
2) Their recommendations are implemented either with a specific budget or by ensuring that the remedial measures are implemented within existing maintenance or other budgets.

2.8.4.5 Traffic Calming

In recent years, speed reduction or traffic calming has developed as almost a separate science from accident reduction work, as it can produce significant environmental improvements as well as accident reduction. It has proved to be popular with engineers and residents.

Table 2.14 Accident Situations and Remedies

Situation		Remedial Measures
Skidding		Restoring surface texture Resurfacing Improve drainage
Collisions with roadside objects		Better delineation Railings or fencing Frangible posts Remove objects
Pedestrian/vehicle conflicts		Pedestrian/vehicle segregation Pedestrian crossing facilities Pedestrian fences or other protection

Situation		Remedial Measures
Loss of control		Bigger or better road signs Road markings Speed controls Safety fencing Superelevation
Nighttime accidents		Reflective signs Delineation Road markings Street lighting
Poor visibility		Trim or remove vegetation Improved sightlines Realignment
Poor driving behaviour or violating lane discipline		Road markings Enforcement Median barriers Overtaking lanes

Traffic calming can be defined as the improvement of the traffic situation by reducing traffic speeds and volume of vehicles, particularly in residential areas with emphasis on the safety of pedestrians, cyclists and VRUs, such as children or the elderly. It has been shown that when pedestrians are hit by vehicles travelling at a given speed:

- 95 per cent survive at speeds up to 32 km/h.
- 55 per cent survive at speeds up to 48 km/h.
- 15 per cent survive at speeds up to 65 km/h.

Therefore, the use of traffic calming to ensure low speeds can have a pronounced effect on reducing the severity of injury in accidents.

Traffic calming techniques commonly include:

- Speed humps.
- Road narrowing.
- Gateway features.
- Footpath widening.
- Pedestrian crossing areas.
- Landscaping and environmental features.
- Special warning signs.
- Possible reduced speed limits.
- Mini roundabouts.
- Horizontal carriageway deflection
- Priority systems.

Examples of traffic calming schemes are shown in Figure 2.68, and much more information on such measures is included in References 2 and 6. Traffic calming is particularly useful for area-wide schemes and particularly so in residential areas, as illustrated above.

It is also, however, increasingly being used to slow traffic on rural routes in many countries, particularly on the approaches to villages; see Figure 2.68. Speeding through settlements and villages often increases dramatically and results in increased deaths and injuries. Such gateways on entry to traffic-calmed sections of the route have been implemented in Fiji and are being used in Samoa. Not all the traffic calming techniques listed above will be relevant to all countries. What is important, however, is that speed reduction reduces accident occurrence, and countries need to discover what traffic calming methods work best for their road users and traffic mix.

Figure 2.68 Rural gateway, UK

2.8.4.6 Monitoring and Evaluation

While many of the countermeasures listed above are well-tried remedies that have been used for many years all around the world, it is still important that all accident sites are monitored for actual reduction, and then the true costs and benefits can be evaluated. It is important to use equivalent before and after periods (ideally two to three years) when making comparisons or evaluations, and the construction period should be treated separately.

This has at least two beneficial effects: first, it helps determine what remedial measures are most effective and appropriate for local conditions; and second, it is an excellent way of illustrating the benefits to be obtained in a manner that can readily be understood by the policymakers responsible for allocation of resources.

2.8.5 Stages of Development

The first stage is to establish an effective accident data system. As mentioned above, all these techniques rely on accurate accident data, and the development of such systems is described in the accompanying sector guidelines, *Road Accident Data Systems*. The importance of appropriate and accurate accident data cannot be overemphasised; it is the base measure of safety and is an indicator of how and where improvements can be made. The second stage is defining the staffing and financial resources needed to carry out the investigation and remedial work. It has been stressed how cost-effective these techniques are, and thus, all countries would be wise to set aside sufficient resources for the task. It does, however, assume that the costs of road accidents to the community have been established and accepted within the country.

When an accident investigation team has been established, it is often useful for it to publicise its work to the traffic police and the design and maintenance engineers. Many workers in these roles are surprisingly unaware of what can and is being achieved, and liaisons can lead to several benefits. First, accident data from the police can improve in accuracy if they understand how it is being used; and second, the police and the engineers improve their understanding of the causes of accidents and can start improving their own effectiveness in accident reduction in their own fields of enforcement, design or maintenance. In addition, following the accident analysis at a problem site, there is often a need to reinforce or publicise the recommended engineering measures. The closer the disciplines work together, the more effective the accident reduction becomes.

2.8.6 Benefits and Effects

Assuming that realistic costing for particular accident types has been established, many remedial schemes can be shown to pay for themselves in a matter of months. For any new country adopting these techniques, it is unlikely that their priority lists for action would include any schemes that would take more than one year to return the investment. Figures vary from scheme to scheme, but where there is a well-defined accident problem with many accidents of a similar type, the accident reductions that can be achieved are likely to be between 30 and 50 per cent. If, however, there is a specific problem, such as accidents on a severe bend, then in many cases low-cost signs, road markings and perhaps delineation may remove the problem altogether, giving an 80-100 per cent reduction.

2.9 Road Safety Education for Children

2.9.1 General

Twenty per cent of accidents involve people under 15 in developing countries, so education is essential.

Compared to the developed countries, the roads in developing countries are often more unsafe. Children are left with unnecessary risk due to the absence of traffic education and safety measures. Compared to the developed countries, the traffic circumstances and problems are very different, so it would be inappropriate to use teaching materials from developed countries. To improve road safety education, an incremental approach is needed. Essential components for developing and improving road safety awareness areas are as follows:

- Road safety in the school curriculum appropriate for each age group.
- Preparation of classroom materials.
- Preparation of a teacher's guide distributed to all teachers.
- Teacher training courses on road safety.
- Coordination of activity and clearly defined responsibilities.

Priority Actions Needed

- The level of existing road safety education in the school curriculum is to be reviewed. Adequacy and practicality of lessons and materials are to be assessed.

- Develop pilot projects on road safety education in high-risk areas, especially around areas where road rehabilitation is ongoing and road accidents are likely to increase.
- Strengthening road safety education, including on-road practical lessons, in the national curriculum.

This section deals with road safety education for children as formal education in schools and as non-formal education in communities. Road safety publicity and education for the public are discussed in this section.

2.9.2 Road Safety Education (RSE)

RSE is needed for the acquisition of safety knowledge and skills. These include decision-making skills and the identification and assessment of risk and strategies. RSE attempts to prepare children for different tasks at each stage of their increasingly independent use of the road network and, later, as adults.

2.9.3 Key Elements

2.9.3.1 Inclusion of Road Safety in the Curriculum

Adequate provision in the national curriculum ensures long-term prospects of RSE in schools. This is the building block on which further training is based. It is therefore preferable to teach rules and principles for crossing the road that can be applied equally to light-controlled junctions, marked pedestrian crossings, rural roads or other situations. However, classroom work alone will not improve critical skills such as road crossing. Supervised practice in an actual situation is much more effective and should be carried out for children in the community who start using the roads on their own.

Figure 2.69 - Road Safety for Children on Road

2.9.3.2 Teachers' Guide to Road Safety

To enable the teachers to teach road safety, a teacher's guide needs to be prepared and distributed to the schools.

2.9.3.3 Teacher Training on Road Safety

The teachers, through a series of local seminars, and new teachers entering the profession, through the teacher training institutes, need to be trained to teach road safety. Providing training to all teachers with basic road safety teaching throughout the country is a large commitment. It is unlikely to be achievable unless a dedicated person is appointed full-time, for at least two years, to coordinate this task. Their role should be primarily to train the teachers to run road safety seminars, initially for the head teachers, but next covering all class teachers.

2.9.3.4 Teaching Aids on Road Safety

Experience from other countries has shown that effective RSE does not rely on lectures by visiting speakers at irregular intervals but can supplement ongoing, curricular-based classroom work, and is more effective if supported by teaching resources. Some countries have found it beneficial to assist and support teachers by inducting specialist road safety officers, by providing ideas, resources and teaching materials. The following materials are useful but are not limited to classroom sessions:

1) Worksheets.
2) Posters.
3) Teaching pack.
4) Slide shows.
5) Books, games and videos.

2.9.4 Benefits and Effects

The potential benefits of RSE for children include the following:

1) Increased awareness and knowledge of the traffic environment.
2) Appropriate survival skills.
3) Understanding drivers' behaviour and attitudes that influence road safety.
4) Knowledge and understanding of how humans, vehicles and control systems interact and work.

5) Decision-making skills that will enable them to make choices and act responsibly.
6) Esteem and care for other people.
7) Knowledge and understanding of the causes and consequences of road accidents.
8) Knowledge, understanding, and skills necessary to travel safely while showing consideration for others.

Attitudes developed by people in their earliest years influence how the individual should behave on the road in later years. It is easier to teach good habits at an early age than to correct bad habits later. RSE helps to develop positive attitudes and values that have long-term benefits for the community in terms of road user behaviour. RSE must commence at an early age, in nurseries, or even earlier, through publicity initiatives aimed at parents.

2.10 Training and Testing of Drivers

2.10.1 General

Driver training is mostly used to prepare a candidate for the driving test, and therefore, the extent and quality of training will be determined by the standard of the driving test. Driving instructors must have adequate capability to provide improved training. In each country, the learner drivers should be introduced to a recommended syllabus.

For road safety improvement, it is appreciated that driver testing and training are only a small part of a whole series of actions necessary. However, due importance should be given to driver training and testing for its contribution to the reduction of road casualties and long-term reductions in accident statistics. The awareness of the inherent heavy responsibilities in possession of a driving licence is essential for all road users.

2.10.2 Priority Actions Needed

1. Ensure that the driving test examines the driver's judgement, decision-making and safe driving ability on public roads in normal traffic conditions, with stricter tests for drivers of large commercial vehicles.
2. Driving examiners should possess adequate training and be taught to drive all vehicles to the highest standards. Adequate staffing for driving tests should be ensured.

3. Driving school instructors should also be organised with their training courses. A standardised driver training curriculum needs to be developed, as well as establishing a registration system, driving instructors and driving schools.

These guidelines are prepared for the countries undergoing rapid motorisation and are restricted to standard, conventional licensing programmes. This means that the drivers passing the driving test are treated the same as all other licensed drivers. Advanced programmes on licensing include probationary or graduated licences, where certain restrictions and stricter penalties are imposed on the learner or novice driver. This requires a computerised database and an efficient monitoring system accessible from various authorised locations.

2.10.3 Key Elements

2.10.3.1 Driver Testing

In most countries, the number of vehicles and drivers is growing; a programme of structured training, followed by a valid and reliable test of competence, is therefore essential for the reduction in road accidents. In developed countries, younger drivers aged between 17 and 21 comprise 10 per cent of all licence holders, but they are involved in more than 20 per cent of all accidents. A young male driver is seven times more likely to be involved in a fatal road accident than a middle-aged man.

An effective driver testing and training system is overdue in the developing countries for the following reasons:

1) The traffic is a mixture of motorised and non-motorised modes with an inadequate road network and hierarchy, and poor traffic control methods contribute to a difficult driving environment; and
2) Unless there is a dramatic improvement in driving standards, the rapid increase in motor vehicles and novice drivers will result in a significant increase in accidents.

The driving test is the main judge of any driver training syllabus. For this reason, although driver training is experienced, the driving test is still reviewed first. Most training is intended towards passing the driving test rather than teaching the student to become a safe and competent driver. In newly motorised countries, ensuring proper training followed by a comprehensive driving test is the most practical way to improve the standards of novice drivers.

a. Legislation

The legal authorities of the state that control driver licensing and the requirements for potential licence holders must be defined by the traffic regulations. Licensing legislation is also discussed in this section, but the criteria are to be mentioned in the following:

1) Minimum driving ages by vehicle type.
2) Vehicle classifications.
3) Medical limitations.
4) Driving test procedures.
5) Provision for training, and
6) Restrictions on learners or novice drivers.

Licence categories should be the same as international guidelines to enable interchangeability and acceptance in other countries. Such classifications should be similar to those set out in European Community (EC) Directive 91/439/EEC. Licence durations, including the duration of a learner's licence, should be specified by law. In several developing countries, the learning period allowed by the validity of a learner's licence is too short and should be expanded to a 6–12-month period. Any real advantage is not known in terms of road safety by imposing such a short duration for a normal driving licence. In the UK, users of light vehicle driving licences do not require renewal until the holder is 70, after which the duration between renewals is three years. Heavy commercial vehicle licences last until the holder's age 45, after which they need to be renewed every five years until age 60 and then annually; these renewals need to be accompanied by a medical report.

Several developing countries have adopted restrictions by legislation that newly qualified motorcycle riders are not allowed to carry passengers for the first year. Such legislation should be enforced by displaying probationer plates on the motorcycle. Legislation, which can be enforced, should be enacted to promote respect for the law. Legislation should, wherever possible, make changes in the driving licensing procedures without lengthy delays. For example, any country may wish to incorporate photographs or holograms into the driving licences as security measures.

b. Medical Examination

The medical examination could, in practice, be limited to an eyesight check, by a simple test of reading a registration plate at 20.5 metres, which is still used in the UK, and clearance is issued declaring that the candidate has none of the medical conditions prohibiting driving as specified in legislation. Although the growing number of licence applications may cause backlogs and accumulation problems, in many countries, the candidate's mental or physical fitness is tested by a medical or general board.

c. Theory Examination

Besides the driving skills of candidates, their knowledge should also be tested by examination before using a vehicle. The test may be done by selecting one of several options. In countries where labour is intensive and the literacy rate is low, an oral examination is appropriate. If a formalised question paper or computer-based test is used, it should be taken before the practical driving test, and passing will be a prerequisite for qualifying for the practical test. Literacy is not necessarily a requirement for a good driver. A test covering sufficient topics will require about 50 questions for a duration of about 45 minutes.

Subjects covered by the theory test should include at least the following topics:

1) Traffic regulations.
2) Vehicle handling.
3) Vehicle manoeuvring.
4) Hazard perception.
5) Effects of weather.

Practical Driving Test

In some countries, the whole practical test is conducted on public roads, but the technical control of vehicles, i.e., emergency stops, hill starts, zigzag manoeuvres and parallel parking, is examined by using off-road facilities. Normally, the off-road test is carried out first, and this is followed by the on-road test, conducted on normal roads with light traffic. They should be as uniform as possible and include a common range of typical road and traffic conditions. The test route should be chosen to test candidates' hazard perception skills without exceeding their capabilities. Pedestrian crossings and junctions for right and left turns, including roundabouts, should be included to provide an opportunity for give way testing. The on-road test should include, but not be limited to, the following checks and exercises:

1) Proper precautions should be taken before starting the vehicle engine.
2) Make use of all controls properly.
3) Position on the road and make stops in a safe place.
4) Drive at a speed appropriate for the conditions.
5) Make effective use of rear-view mirrors.
6) Give all necessary signals using the indicators.
7) Show alertness and anticipation of the actions of other road users.
8) Overtake, meet, and cross the path of other vehicles safely.
9) Act properly at road junctions.
10) Take appropriate action at pedestrian crossings.
11) Take prompt and appropriate action on all traffic signs, road markings and traffic lights.

Tests for motorcyclists can be conducted by the examiner observing a set of exercises on a normal road system or by specially trained examiners, riding their own vehicle and following the test candidate around a set route on normal roads. Contact can be easily maintained by simple radio equipment. Several examiners at strategic locations around a route could enable many motorcycle tests to be performed simultaneously. For testing drivers of heavy goods vehicles or large buses, it should be remembered that the potential for damage by these vehicles is extremely high. It is therefore recommended that time is taken to examine a candidate for a vehicle in one of these categories should be about twice that for a driving test for a light vehicle. Large vehicles should be given an area of prescribed minimum dimensions available for manoeuvring exercises. Vehicles in these off-road areas do not interfere with one another. It is necessary to have standardised assessment forms to ensure consistency among tests, to facilitate monitoring and to provide feedback. All the errors should be pre-coded and thus standardised.

d. Driving Examiners

No matter how elaborate or strict the driving test is, it can be effective only if driving examiners are properly trained and have adequate time to conduct the test. The increase in the number of examiners has not kept up with the increase in licence applications. Lack of staffing has also had a greater impact than inadequate training. Even with forms and procedures, the effectiveness disappears if only a few minutes are available for each test due to demand.

All examiners, whether employed to test drivers of cars, motorcycles, heavy goods vehicles (HGVs), or public service vehicles (PSVs), or for the police or the military, should be subject to the same training and should be qualified to drive all categories of vehicles that they are likely to examine. Examiners should be understanding, sympathetic and make every effort to put the candidate at ease and should give clear directions and instructions. They should assess and mark driving faults as they occur. It is useful to have a standard form that examiners complete during the actual test and not retrospectively. To avoid any possible corruption, examiners should not be made aware of their test programmes before the day of the test. They should also not be permitted to give driving instruction themselves or be associated with any driving schools. Candidates with criminal records should not be permitted to become driving examiners.

e. Test Administration

Clear requirements are necessary for the administration of driver testing. These should cover all categories of driving licences and should be available in a published booklet format for sale to the public. A complete list of areas to be tested and required standards should also be published. Centres for the driving test need to be convenient for large numbers of candidates, and it is important that potential drivers do not have to travel unreasonably long distances for their test. These centres should have examiners in sufficient numbers to meet the likely demand for tests in that area. Facilities for waiting and toilets must be provided for the candidates. Parking should be conveniently located at the test centre, and provision made for all types of tests that are to be carried out at the centre. A computerised system will ease this function and the collating of test results.

Depending on the size of the region and population and the demand for tests, a senior manager has to coordinate with day-to-day responsibility for organising the driving tests. These managers should be supported by supervising examiners to monitor the testing and training systems and assess the validity of appeals. The clerical staff should also be there to support them by monitoring the test and training results for uniformity, maintaining a booking and cash handling system, preparing statistical summaries of driving test data and performing administration.

f. Cost Recovery

The provision of a driving licence should always be identified as a privilege and not a right. A considerable amount of capital cost, along with substantial running costs, is invested in testing infrastructure. To organise the funds for a testing system, as described, most countries try to recover the running costs from the driving test fees. Although administration and management staff do not usually conduct tests, the cost of their services should also be recovered from driving test fees. There is no known country that provides a driving test as a free public service. Driving instructors are charged for entry examinations to become registered instructors and will need to pay an annual fee. These charges also contribute to bearing the cost of the testing operation. Selling official guidelines to help candidates prepare for the tests can also become a worthwhile source of revenue.

2.10.3.2 Driver Training

a. Curriculum

In most developed countries, driver training is seen as a requirement for a driving licence or permit. A syllabus needs to be developed that covers sufficient elements to enable the student to pass a test for their licence. Ideally, the syllabus and the training are not just those to be followed by the examiner at the time of the test, but should aim to prepare the learner drivers for all potential hazards and situations. Driver training should not only convey this message but should reinforce it throughout the training. The learner must also possess:

1) a proper knowledge of the highway code and laws, and
2) a proper understanding of the responsibilities of a driver.

The drivers must have real concern, not only for their safety but for all road users, including pedestrians. There is no minimum number of lessons, as the requirements of each learner will depend on a variety of factors. The syllabus shall have the list of skills for which basic competence must be achieved to pass the driving test.

b. Driving Instructors

Governments are to ensure that the people who claim to be driving instructors are qualified. To receive a qualification certificate, these people must have completed a training course organised by the appropriate authorities and be qualified to drive vehicles of the type in which they wish to instruct. Once qualified, their names should appear on a website or register of driving instructors, and they are then entitled to charge professionally for driving lessons. It is understood that an official supervisory system is necessary to ensure that the standards of instruction are maintained. Qualified driving instructors may be required to pass a qualifying examination. This should cover:

1) Theory: A multiple-choice question paper on all aspects of driving and instruction.
2) Practical driving: A test of driving ability to advanced standards over one hour.
3) A practical test where the potential instructor gives a lesson to an examiner on a random driving topic.

Each of these modules must be passed before going on to the next stage. Only on satisfactory completion of all three modules shall the instructor be eligible for registration. A fee is to be paid, and continued membership is dependent on favourable assessment reports from frequent tests by government inspectors. The fees for registration should cover the costs of inclusion in the register annually. Separate schemes need to be considered to provide adequate training for drivers of HGVs and PSVs, motorcyclists and instructors.

c. Driving Schools

Although it is not essential, it has been found to be beneficial to register driving schools separately from driving instructors, to encourage driving schools to meet criteria, such as having registered instructors, adequate numbers of training vehicles, and lecture rooms for theory training. Better-quality control can be ensured because the registrations of instructors and schools can be cancelled due to corruption or other deficiencies.

There are also benefits in organising a driving schools association to aid coordination between driving schools and to establish an industry code of practice (as done in Fiji). Driving schools can also be involved in the training and rehabilitation of offenders, especially if government organisations such as the police do not have the necessary resources.

2.10.4 Development Stages

It is not necessary that all the above stages must be taken in the sequence shown. In some countries, it may be possible to miss a whole stage to improve the system. The aim must be to implement a driver testing procedure that fairly and adequately prepares new drivers for the challenges of driving.

Stage 1: tests that are conducted mainly off-road or on quiet local roads are short and simple (less than 10 minutes) and require little training for drivers. Some examiners have little or no training and have no written procedures or monitoring. The driving examiners are often required to conduct vehicle inspections. Therefore, this is a basic but ineffective stage.

Stage 2: while Stage 1 may continue to exist in provincial areas, this next level is observed in major urban areas. At this stage, the on- or off-road tests are carried out, but they are not standardised. The driver training schools begin to develop, and training may also be offered for the instructors. The examiners will focus on driving tests and routine monitoring will be conducted.

Stage 3: at this stage, assessment forms are developed and used by driving examiners. Monitoring driving instructors and driving schools has started. The focus will be from theoretical and mechanical knowledge to hazard perception and practical training. A basic theory test will be included.

Stage 4: at this stage, the candidates are given feedback, and a monitoring system is in place, publicising statistics. Quality assurance of the exam system will result in improvements in examiner consistency. An adequate number of examiners are to be available, and the test workload will be distributed to according to the number of examiners. Candidates' knowledge may be assessed with a theory test.

It is essential that improvements in the driving tests are made immediately before or simultaneously with improvements in training. Failing to do this will make it difficult to improve driver training standards.

2.10.5 Receiving Benefits and Effects

Any road safety improvements have to be seen as a long-term policy designed to accelerate the acquisition of driving experience. This, in turn, will improve the skills of new drivers and thus the safety of all road users.

Drivers can readily recognise the positive aspects of a structured training programme followed by a professional, valid test. Its effects on improving road safety and reducing accident figures are not immediate; however, the time taken will enable more highly skilled drivers to become a significant percentage of the total. Such achievements must be linked to an efficient accident data system to enable monitoring the qualifications of drivers involved in accidents. This monitoring could also extend to the available traffic citation data.

The introduction of such a modification in driver licensing will initially result in obviously lower pass rates. This is no bad sign, especially when the pass rate may have been excessively high (for example, more than 80 per cent) on the first test. One way of fine-tuning the system will be to establish the current pass rate and bring in changes gradually to the test to reduce it. As the professional driving instruction is introduced, it should then gradually counteract this reduction and tend to improve the pass rate. Monitoring driving schools should be done based on the pass rates obtained. Provisions should be there for retraining and/or retesting drivers who have committed serious motoring offences. This should have a restricting effect on any tendency towards reckless or aggressive driving and can be helpful in the reduction of accident figures.

2.11 Publicity and Campaign for Road Safety

2.11.1 General

Mass media is influential on the daily lives of people in aspects like health and well-being. Therefore, it is very possible that effective publicity can influence road user behaviour and raise awareness of road safety issues. Well-planned publicity can influence the short-term behaviour and long-term attitudes of road users. For instance, to reduce drink driving, publicity can inform people about the risks. Publicity may be highly cost-effective in that it can address the safety of large numbers of people using media and materials. However, it can also result in wastage of resources if it is not carefully planned. Based on the analysis of road accident data, publicity campaigns and selection of target groups can be planned. If the detailed data is not available, awareness-raising campaigns can be ineffective. Suitable publicity should be prepared for the new legislation, standards, enforcement initiatives, safety products and highway features.

Priority Actions Needed

1. Road accident data must be analysed to identify the nature and characteristics of the problems and the target road user group.
2. A single concise message should be the focus of publicity campaigns, and the media materials and images used must be appropriate to local conditions and the target groups.
3. Wherever possible, conducting publicity campaigns should be coordinated with engineering, legislation and enforcement and should be evaluated by surveys before and afterwards.

The potential effectiveness of publicity depends on several factors, including:

1) Credibility of the message.
2) Design of the message, and its relevance in respect of the whole planned campaign.
3) Implementation of the delivery, and
4) The extent of the country's publicity culture.

The extent to which the publicity can 'reach' the target people depends on the number of opportunities for the people to see or hear the message through printed or electronic media. Publicity can be purely subject-orientated; for example, announcing a new law on seat belt wearing, or targeting a subset of the population with information about the effects of alcohol on driving performance.

2.11.2 Need for Road Safety Publicity

In 95 per cent of all road accidents, road user error is found as the factor. Therefore, improving road user behaviour should always be a priority. Road safety publicity can educate and influence the public; it is needed to:

1) Create awareness of the vulnerability of certain road users and road accident threats.
2) Educate road users as to what constitutes safe road user behaviour.
3) Change attitudes and beliefs to a more positive approach, and
4) Inform road users of changes in traffic regulations or operating conditions.

Publicity is commonly most applicable to adults. Children in formal education may have the opportunity for detailed work on road safety, directly from a competent source in schools. Adults are difficult to reach on an individual level, and mass publicity is often the only practicable way of informing them and voluntarily modifying their behaviour. Safety must compete with other influences on the public's attention that offer more immediate and tangible benefits, but restrict the use of images that may cause mental distress.

While awareness made using posters, leaflets, radio and television has less impact than school-based education, publicity can be effective. For instance, a targeted public awareness programme on the destructive effects of alcohol or drugs can go some way to fill the gap in understanding that might otherwise be filled in school or college-based traffic education.

2.11.3 Key Elements

Road safety publicity techniques based upon posters, leaflets, billboards, newspapers, and often FM radio, television and the Internet are employed by most of the developing countries. However, to ensure success few of these campaigns appear to be planned systematically. The key components that need to be considered in publicity are discussed below.

2.11.3.1 Assessment of Road Safety Problems

Publicity campaigns need to be primarily based on the best understanding of the accident at the time. Ideally, publicity initiatives should be data-based. It is important to understand the nature of the accident or behaviour problem being targeted in some detail, if the message needs to be credible for the audience. Various types of data can be utilised, including accident data, observation, police discussions, attitude testing and knowledge testing.

2.11.3.2 Design of Road Safety Campaign

After identifying the problem, it is necessary to consider the target behaviour, target audience, what will motivate the target audience to change their behaviour, message content, the media and the type of appeal that is the most appropriate.

2.11.3.3 Campaign Message

The message must be appropriate, and the publicity should be clear, unambiguous and targeted. General exhortations such as 'drive safely' are not particularly effective, as it does not generate any alert. Campaigns should concentrate on a single concise subject, for example, 'Do not drink and drive', and should not confuse the target audience. It must be confirmed that the message is reaching the target audience. If perpetrators are targeted in the campaign, most viewers will not identify with the offender because most road users are overconfident about their safety behaviour. Positive instruction is therefore preferred to attract public attention. The timing of any single campaign or other initiative needs to be considered in terms of the length of the campaign and the most appropriate times of the day, month and year to run it. Individual campaigns should not generally be too long, as the message becomes diluted or ignored.

2.11.3.4 Selection of Target Audience

Any publicity through the mass media will result in widespread awareness; still, a publicist should select proper messages, images and media such that the target group will be influenced. It may often be appropriate to direct publicity to a limited and more manageable audience. In the case of accidents involving pedestrians, motorists are generally the target group for publicity campaigns. Likewise, in accidents involving children, the target audience is often the parents and care providers.

As road safety is a social and public issue, mass media, therefore, can be used to advance policy initiatives, and this approach is termed media advocacy. It does not attempt to change individual risk behaviour; instead, it focuses attention on changing the way the problem is understood as a public health issue. By using the media, public support for more effective policy-level approaches to public health problems may be increased.

2.11.3.5 Pilot Testing

Publicity is culturally specific and is demographically sensitive. The use of language, images and messages should be designed for precise target population groups, whether by age, sex, ethnic origin or the recently revealed importance of lifestyle. It is of prime importance in publicity to not only know your audience but also know what motivates it. However, certain general principles and techniques may be identified that will be of relevance to all regions.

2.11.3.6 Methods of Evaluation and Indicators of Success

It will not be possible to formulate effective evaluation methods unless clear objectives have been defined. According to the complexity of the subject and the period involved, the indicators vary. Effective criteria may involve one or more of the following:

1) Reduction in numbers and severity of accidents.
2) Observed changes in actual behaviour.
3) Changes in knowledge.
4) Changes in opinions.
5) Recall of the campaign.

2.11.3.7 Identifying Other 'Players'

Presenting the message from more than one source is sometimes preferable. Many other groups with the same interest include police, doctors, health promoters, interest groups, and the private sector. Different agencies have responsibilities for road safety; it is therefore important to ensure proper coordination to maximise the effectiveness and to reduce fragmentation of effort. Creating a campaign planning group to steer the initiative is preferred in some countries. Once the safety specialists have agreed upon a message, it is sometimes subjected to top-level political scrutiny. Although road safety is generally not regarded as a subject of political interest, there may still be political sensitivities in the message that may have an undesired public response. Professional agencies can be asked to submit proposals to compete for a single contract or for a longer-term association. Their proposal should include a speculative presentation of how they would tackle the brief if they won the contract. Those agencies with a proven track record in safety, health or public service publicity will obviously have the advantage.

2.11.3.8 Identifying Sources of Funding for the campaign

Most countries have direct access to radio and television time or can influence programme controllers. The campaign may be implemented through them by paid advertising, and that will form part of the publicity strategy. This may be expensive, especially if prime airtime or newspaper spots are sought. Sharing the cost of advertising with others, including commercial sponsors, may increase the buying power of the campaign planning group. Businesses, particularly in the field of insurance, will have an interest in being associated with road safety. Part of the brief (scope of work) to an advertising agency could be to locate sources of funding or value added from within the private sector and a part of their public responsibilities would be to negotiate special rates with the media to increase value.

2.11.3.9 Selecting Media for the Message

Media may be selected based on their affordability and appropriateness to the target population group. It is not to say that all campaigns must utilise paid media, though this is the obvious route to take in any country with a well-developed commercial press, and radio and television networks. A mix of suitable media for the target group and message should be created within a budget for advertising in these countries. The popular journals, radio and television programmes have a high audience of teenagers; therefore, campaigns aimed at youth might concentrate on such media.

2.11.3.10 Appeal to Public

The images chosen for the publicity campaign depend on the subject and target audience. Images or videos of shocking or horror scenes are not always appropriate or effective, though they can tackle complacency among the target population if handled effectively. The appeal of the campaign could be based on horror, fear, grief, humour, information, endorsement or responsibility. Many times, it is possible to use a celebrity with personal appeal, as road accidents have affected many people. Actors and sports celebrities in many countries are participating in campaigns for road safety after loved ones have been injured or killed in road accidents.

2.11.3.11 Implementation

The timing of the publicity should be synchronised such that it is associated with another partner measure, such as the planned introduction of a new law, an enforcement initiative or engineering initiative, or the availability of new safety equipment, for example, reflective clothing for use by pedestrians and cyclists after dark. It is particularly useful to synchronise a centrally run campaign with local initiatives on the same theme.

Formal launches designed to attract free media coverage can also be a useful opportunity to involve the appropriate minister or other figurehead, and thus further legitimise the measure. Thereafter, every effort should be made to put the message in front of as many of the target group as possible and as often as possible. There is evidence that people need to see the publicity many times before they acknowledge it, and even more times before they act upon it.

2.11.3.12 Monitoring and Documenting

It is important to collect data regarding its effectiveness once the campaign is over. The campaign should be documented, in order that it may contribute to a knowledge of what does and does not work.

The above-mentioned aspects of planning and implementing road safety campaigns represent the ideal. Developing countries should review the various campaign programmes and develop a programme that will allow campaigns to become more targeted and effective according to these principles, but in a way that is feasible for their circumstances.

2.11.4 Development Stages

For a country to make progress in publicity for road safety, there are some preconditions to satisfy, and then a logical and progressive course of action can be developed to follow. A publicity campaign can be pursued only if there are material and human resources available to make it a success, such as qualified staff who are sufficiently familiar with publicity techniques for running the campaign, or to commission and monitor a campaign run by a professional publicity and marketing consulting agency. It is important to learn from the success or failure of the campaign. The temptation can be for the campaign organisers to neglect monitoring outcomes and claim success based on output alone. Such temptation must be avoided, whereby the production of a glossy poster can be seen as a success. To justify the diversion of scarce safety funds into publicity requires more than an act of faith. It requires evidence that its objectives have been met.

2.11.5 Benefits and Effects

Publicity campaigns can create a better-informed mass of the public and contribute to safety culture, by making the demand for safety eventually generated by the population itself. Even publishing road accident data for the general public can boost the awareness among the public of the scale of the country's safety problem, provided it is permitted politically. In any nation's road safety strategy, publicity is an indispensable part, and it should not be used in isolation from other measures. Campaigns will be successful if they are implemented alongside other measures such as engineering, legislation or enforcement. The road safety publicity can be ineffective if it is mishandled, and it will be a waste of resources. To achieve success in meeting its objectives, publicity needs the same specialist attention and thorough preparation that would be given to the other technical components of the road safety plan.

Figure 2.70 Speed warning campaign using road signs

2.12 Standards for Vehicle Safety

2.12.1 General

There may be little control over the general safety of a country's vehicle fleet without vehicle manufacturing regulations that govern the safety standards, especially for systems such as braking, lighting and signalling. The specified standards of comfort, access and additional safety requirements are needed in PSVs. To ensure the safety of all road users and to minimise damage to the environment, standards of size, gross weight and maximum axle loads are essential for HGVs.

Imported used vehicles must comply with national safety standards and must always be checked on arrival in a country. To ensure that vehicles meet a minimum acceptable standard of safety, statutory testing is required at regular intervals. The following are the most important items that should be inspected:

- Braking system.
- Steering.
- Tyres.
- Lights.

There are no international fixed guidelines for initially testing vehicles. For the developing countries, it is recommended that light vehicles are to be first tested after three or four years and then annually. However, the high utility vehicles such as HGVs, PSVs and taxis should be first tested after one year, then annually, and after ten years. Inspections are to be done every six months. These are the recommendations for minimum testing frequencies. For an effective vehicle roadworthiness testing system, the components necessary in order of priority are as follows:

- A legal framework.
- A coordinating and managing authority.
- Trained and qualified staff.
- Adequate testing facilities and equipment.
- An enforcement and backup operation.

Priority Actions Needed

1. Vehicle defects are likely to contribute to accidents and should be targeted by routine and roadside vehicle inspections.
2. To encourage uniform testing standards and procedures between stations and inspectors and to minimise corruption, checklists, assessment forms, increased controls and training programmes should be used.
3. To enforce compliance with safety standards, random roadside inspection checks should be introduced by police and vehicle inspectors day and night.

The various guidelines form a set of road safety guidelines for the road safety policymakers.

2.12.2 Vehicle Safety Standards

Studies carried out in the UK indicated that between 5 and 8.5 per cent of accidents are directly caused by vehicles due to faults. Studies also indicated that 25 per cent of goods vehicles and 11 per cent of large passenger carrying vehicles involved in accidents have contributory defects. In developing countries, it is likely that vehicle defects are more often a factor in accidents as vehicle conditions are generally much worse. The vehicle fleet is usually older, with many vehicles imported second-hand from other countries, and there may be difficulty in obtaining suitable spare parts. In developing countries, lighting defects are common, and by conducting nighttime surveys of long-distance buses, it was found that two-thirds had one or no rear lights. In developing countries where road lighting, road marking and signs are inadequate and driving conditions are poor, vehicle lighting is very important.

In many developing countries, overloading is a serious problem. The strength of the chassis, suspension, braking system, tyres and engine power dictates the maximum load carrying capacity of a vehicle. A seriously compromised safety performance occurs when goods vehicles are modified unofficially to increase their load by welding reinforcements or extensions to the chassis or the addition of an extra axle. This practice is commonly found in countries that have ineffective regulation of manufacturing standards. Lorries are being modified with a combination of chassis strengthening and an extra axle to double their load-carrying capacity. Although the extra axle may only increase braking capacity by 50 per cent, unfortunately, the gross weight may have increased by two-thirds.

Very often, buses are locally manufactured on lorry chassis using materials and designs that offer little or no protection to passengers in the event of an accident. Leg space between the seats is often exceptionally insufficient to accommodate the maximum number of seats in the vehicle, and the seat frames themselves are usually formed from steel, frequently causing amputation of limbs in an accident. If provided, emergency exits are often blocked by a row of seats, making them of no use in crises. Vehicle manufacturing standards can improve road safety by reducing the extent of injuries caused by road accidents.

2.12.3 Key Elements

For maintaining the vehicles in a safe, roadworthy condition, it is necessary to implement various legal and operational systems as linked. These systems include:

1. Legislation.
2. Management and administration.
3. Equipment and facilities.
4. Training and staffing.
5. Ownership.

2.12.3.1 Related Legislation

Before any testing system can be implemented, the legal framework by which it is to be regulated should be set forth. The components or systems, standards and authority for testing are all to be included in legislation. The legislation must be a proposal that is unambiguous, clear and general. The specific requirements of a testing scheme should then be detailed in separate regulations that can be easily updated and implemented by the main legislation. The regulation on manufacturing and use may be drawn up with reference to legislation from other countries, adapted as appropriate to local needs, conditions and resources. These should define the requirements for the various types of vehicles and what initial inspection is necessary for locally built or imported used vehicles.

In many Asia Pacific countries, new and reconditioned vehicles are imported, as regional standardisation of vehicle design and manufacturing ensures safe vehicle standards. There is nothing intrinsically wrong with importing used and reconditioned vehicles if they are roadworthy.

All imported vehicles are to comply with the safety requirements of the major vehicle manufacturing standards in international use, which should be the required system of type approval.

Figure 2.71 Training of vehicle inspection staff in Fiji.

2.12.3.2 Management and Administration of Motor Vehicles

A responsible body needs to evaluate the policy implications of operating a particular vehicle testing regime or system. Continuous monitoring is required for assessing the degree to which performance targets are met. Although the staffing requirements are determined by the inspection frequency and length of inspection, the organisation of test stations is also critical. In many countries, it is the public perception about vehicle testing that illicit payments to the vehicle inspector will result in favourable treatment. Regular checking of testing premises, procedures and recently tested vehicles by selecting at random is necessary to ensure the system is open and honest. Wherever possible, during the test procedure, any direct and private contact between the tester and vehicle owner or operator should be eliminated to reduce opportunities for corrupt practices.

2.12.3.3 Equipment and Facilities for Motor Vehicles

Depending on the resources available in a particular country, the requirements will vary considerably, the stage at which the existing and proposed testing has reached, and the technical education and skills of the users. The use of low-cost, locally manufactured equipment may often be preferable to expensive high-cost equipment, which also requires skilled use and maintenance, often at a high cost. Basic testing of brakes and lights can be done even with simple equipment. The exact equipment requirements are not listed here as they are beyond the scope of this section. They would obviously be customised and relevant to the requirements and resources of each country. A manufacturer usually designs equipment for a particular market. When those are sold in a different market, the problem arises due to a lack of know-how, both in equipment operation and maintenance. Procuring equipment without maintenance contracts is, unfortunately, too common even where the inspectors have been taught how to operate the equipment.

Table 2.15 Requirements of Equipment

Level of testing	**Equipment/characteristics**
Basic No previous effective testing Below 20vehicle s/ 1,000 inhabitants GDP per capita below US$500/year. Widespread evasion of testing	Low-cost equipment Head lamp pattern marked on wall Inspecting for black smoke Brake test on road using decelerometer
Medium Testing carried out but has been restricted due to domestic situation or lack of resources Below 100 vehicles/1,000 inhabitants GDP per capita above US$500/year Evasion common	Mixture of equipment Optical head lamp meter Black smoke meter Tyre tread depth gauge [low-cost] Roller brake tester for heavy vehicles Brake test on road using decelerometer for light vehicles Steering free play meter
High Routine testing to consistent international standards Above100 vehicles/1,000 inhabitants GDP per capita above US$3,000/year	High-quality, standardised equipment Optical head lamp meter Black smoke meter CO/HC meter Tyre tread depth gauge Roller brake tester for all vehicles Steering free play and suspension checking equipment
State of the art High motorisation with need for control of vehicle numbers High-income GDP per capita above US$10,000/year	Computer controlled equipment minimises operator subjectivity and maximises throughput. Possible full control by one organisation. High security documentation and control paramount

2.12.3.4 Spot Checks and Rule Enforcement.

Random roadside spot checks, by the police and vehicle inspectors, are of crucial importance in reminding drivers and operators about the continuing need to maintain their vehicles safely. Document and weight checking can be combined with these spot checks to form a useful enforcement procedure. To control the standards of vehicle maintenance and operation effectively, regular vehicle condition checks are carried out in approved and regulated testing premises and by frequent roadside spot checks. This will make all drivers understand that they may be prosecuted if they drive a vehicle without a valid roadworthiness certificate.

2.12.4 Development Stages

Countries should be realistic about testing the number of vehicles possible by the number of trained staff available. Frequently, a country's existing inspection system may struggle due to a lack of staff, so testing cannot be done every six months. Consequently, one of the ways to reduce the number of vehicles to be tested is by increasing the time of the initial test for new vehicles. As a rule, fewer good tests of dangerous vehicles are more desirable than many tests of minimal value.

Rigorous and independent monitoring systems should be established. To identify unusual patterns in pass rates, the data should be collected frequently and analysed quickly. Fees should be set to meet the costs to be recovered, allowing investment in improved control procedures and equipment.

2.12.5 Benefits and Effects

By maintaining vehicle safety standards, the following three main benefits are achieved:

- All motorists know that they have the responsibility and obligation not to drive unsafe or un-roadworthy vehicles on the road; therefore, safety culture is promoted.
- Road accidents caused by vehicle defects are reduced significantly.
- Vehicles become safer, thus the severity of injuries to casualties is reduced.

The promotion of a safety culture, perhaps, is the most important achievement of a vehicle roadworthiness scheme. Knowing that the vehicle must meet certain minimum standards of safety will keep drivers aware of their responsibilities.

Driving safely and keeping a vehicle in good condition are closely connected. It is difficult to care about driving standards when the driver does not care about the condition of the vehicle they drive and the safety of their passengers. It is therefore advised that vehicles should be maintained to a higher standard to ensure that they do not reach a condition in which they are unsafe to use. Safer vehicles are involved in fewer accidents. Even if an accident occurs, it may result in less serious injury because of the safety features, for example, seat belts or safety glass. When all systems are working correctly, drivers may be able to reduce the speed and exercise control to take evasive action more quickly to avoid an accident or minimise the effects of an accident.

2.13 Traffic Legislation

2.13.1 General

Traffic legislation regulates the movements of people, animals and vehicles on the public highway and related activities. Where possible, the primary legislation should provide the basic features and framework, and the details should be specified within secondary legislation, through ministerial regulations, which the primary legislation should provide. Therefore, without disturbing the primary enactment, this allows flexibility for periodic revision. Legislation is to provide a framework to enforce and promote safer road user behaviour.

Legislation specific to road safety includes as follows:

- Driving licence (criteria for licence holders and the content of the driving test).
- Vehicle registration and fitness testing (roadworthiness of vehicles and registration).
- Traffic Control (speed limits, traffic signals, signs and markings, drink driving, and pedestrians).
- Road authority.

Fixed penalty points systems have been found to be effective in influencing driver behaviour in many developed countries and may be considered wherever feasible. However, these systems may not always be equally effective in the developing world. All traffic laws are to be reviewed and consolidated and the areas of legislation directly relevant to road safety should be identified. Drink and drug driving is a proven cause of accidents. Enforcement of prescribed limit legislation has resulted in reductions of associated deaths and injuries in many countries. Similar legislation prohibiting driving while under the influence of drink or drugs should be devised that specifies a quantitative upper limit.

Priority Actions Needed

1. Review of existing legislation and prosecution patterns to identify areas that need revision and identify weaknesses in the existing legal system regarding citations and prosecutions.
2. Develop and introduce legislation where it does not exist, and enforce urgently drink driving limits, use of mobile phones, introduce speed restrictions, seat belts, motorcycle helmets and compulsory third-party motor insurance.
3. Develop and introduce legislation where it does not exist, with a national road safety council (NSRC) or similar department to oversee coordination and improvement of road safety.

2.13.2 Need for Traffic Legislation

Traffic legislation regulates the use of public roads and provides a framework to promote and, where necessary, enforce safer road user behaviour. This is done by requiring compliance with specified standards for the benefit of all road users.

In developing countries where motor vehicles have increased rapidly in both numbers and capabilities, and where the existing traffic law no longer meets the needs of modern traffic conditions, updating traffic legislation is necessary. With the increase in the number of vehicles, it has also become increasingly necessary to regulate road users, to minimise conflicts and improve road safety.

The existing legislation could be reviewed, updated and strengthened first. This is best done through a systematic revision process, as mentioned below. The objective of traffic legislation should be specific and clarified to ensure:

- The legislation is written as clearly and effectively as possible.
- It is appropriate for the local environment.

Legislation is never self-enforcing; rather, it can only be effective when enforced and if the judicial system can treat offenders. All traffic legislation should be reviewed according to the following criteria:

1) Need for the regulation.
2) Regulation is acceptable in principle and applicable to the user population.

3) Unambiguous framing.
4) Consistent with other regulations.
5) Awareness of it by the user population.
6) Enforceable.

Developing countries are managing the traffic with limited policing, especially in South Asia, where many of the lower ranks of traffic police personnel are not authorised to cite traffic violations; therefore, the considerations are relevant. To increase their chances of being enforced, the initial traffic regulations should be kept to the basic minimum. Ideally, traffic legislation that is passed as a traffic act by the government should provide an umbrella framework. The details that should be specified in the regulations are under a minister's authority and, therefore, are much easier to revise. Such flexibility is beneficial for speed limits, traffic fines and penalty points, and blood alcohol limits. In these areas, the standards may change from time to time according to accident trends, public opinion or new developments in equipment, and such regulations should be revised quickly without the need to alter the main traffic act.

2.13.3 Key Elements

Several areas of legislation can have significant effects on road safety. Legislation is needed in the following key areas:

- Driver-related.
- Vehicle-related.
- Control of traffic.
- Traffic signs, road markings, surveillance cameras on a device such as FASTag.
- Fines and penalties.

2.13.3.1 Driver-related Issues

Fulfilling certain criteria relating to health, age and competence, a motorist is granted permission to drive by way of a licence issued by the state. Therefore, there is no inherent right to drive a motor vehicle on a road without a licence. Driver training and testing were covered more comprehensively in Section 2.10, and some of the key legislation concerning licensing is discussed below.

a. Driving instructors and driving schools

A registration system for individual driving instructors and driving schools exists in many countries, and no one is allowed to instruct unless they are tested to become a registered driving instructor. Thus, the competency of driving instructors is ensured not only in the knowledge of driving but also in teaching and instructing learners. The registration of schools ensures that they meet at least minimum standards and criteria on premises, vehicles, teaching aids and qualified instructors. Driving instructors are to take defensive driving courses and include some of these elements in their training of learner drivers, while driving schools are to set industry standards and controls.

b. Learner licences

In many countries, new licence holders have restrictions placed on their driving licence, such as on the types of vehicles they can drive. The engine size and power of motorcycles and cars may also be limited. Learner car drivers are usually supervised by an experienced driver until their driving test. A requirement may be considered that motorcyclists should have had at least four hours of off-road training in basic manoeuvrability skills before being issued a learner's licence. Motorcycle dealers and driving schools can often provide such training.

Table 2.16 Typical safety issues that need to be covered within traffic legislation

Driver-related	Highways-related
Driving instructors	Authority
Driving schools	Statutory responsibility
Learner licences	Access/development control
Driving tests	Signs/markings
Medical	Speed zones
Theory	Traffic calming
Practical	Safety audit
Licence holders	
Licence categories	
Professional drivers	

Vehicle-related	Fines/penalties-related	Traffic control / general safety-related
Condition requirements	Traffic accident investigation	Drink driving
Roadworthiness testing	Penalties	Seatbelts
Safety belts/equipment	Courts/hearings	Speed limits
Dangerous loads	Fines	Pedestrians
Documents/insurance		Road user education
		NRSC

c. Driving Tests

The basic purpose of a driving test is to ensure that at least a minimum standard of competence is achieved by a driver before being allowed unrestricted access to the public roads, subject to other criteria in respect of vehicles or classes of vehicle. Standards of driving tests vary throughout the world, from minimal vehicle control manoeuvring demonstrations to sophisticated multipart tests that may include a medical test to ensure physical fitness to drive and a theory test on road knowledge before the third, practical part of the test. A drive of about 30 minutes in varied road conditions accompanied by an examiner is recommended. More extensive tests produce better and safer drivers.

d. Licence Holders

In assessing the suitability for a driving licence, age is only one consideration. Setting the minimum standards of health and eyesight is appropriate on applications for a licence. A regular licence renewal system could retest health at the time of renewal.

Graduated licensing programmes have been used in advanced countries to prevent novice drivers from exceeding their limits. This is generally imposed by restricting the type of motor vehicle and driving times. In some countries, novice drivers use a ‘P’ plate to make detection of violations easier. In Japan, novice motorcyclists are forbidden from carrying passengers for the first year. For drivers convicted of certain driving offences, a retraining course is being proposed and has been introduced in several countries.

e. Licence Categories

A system already exists in most countries that categorises vehicles into classes, and the licence applies to certain types of vehicles depending on the driving tests undertaken. Commercial licences should also be incorporated within the basic driving licence system, but require additional criteria related to competence, age and driving experience.

f. Professional Drivers

A higher minimum standard for eligibility is required in most countries to become a professional driver of HGVs and PSVs. A minimum age of 25 and a minimum driving experience of at least five years are the qualifying criteria. Also, a driving test in the type of vehicle for which such a licence is also required. By passing a test for a private vehicle and then, after some time, starting to drive an HGV or PSVs should not be permissible. An additional, more stringent testing must be conducted to ensure competence to drive vehicles such as an HGV or PSV. This is not only to ensure the greater skill level required to control larger vehicles, but also the greater potential risk to public safety.

2.13.3.2 Vehicle-related Issues

To ensure the roadworthiness and safety of a vehicle at registration and throughout its working life, legislative controls are needed that are appropriate to each country. Vehicle safety standards and inspections are discussed in Section 2.12, but some of the legislation-related key issues concerning vehicles are discussed as follows:

a. Condition of the Vehicle

The traffic police, often working with the land transport department's vehicle inspectorate, hold the responsibility of enforcing laws relating to vehicle condition and operation.

Tyres: One of the most common vehicle faults that contributes to road accidents is tyre defects. It is recommended that there be a requirement of at least one millimetre (mm) of tread pattern over the entire width of the tyres and around the circumference. The limit could justifiably be raised to a higher level, say 1.5 mm, later when the general conditions of vehicles are improved. Additionally, failure of the internal ply structures and cuts that expose the ply structure weakens the tyre, eventually resulting in total failure. Radial and crossply tyres have different reactions to physical forces acting on them, and mixing them is dangerous. When cornering at speed, the crossply tyre loses grip, and the vehicle becomes uncontrollable.

These issues of vehicle defects should be incorporated into the legislation. It is recommended that, under 'unsuitable' conditions, a catchall provision be added to outlaw the use of worn tyres, and similarly, it should also be used for brakes and steering.

Lights. To see the road ahead clearly and be seen by other drivers, lights are essential. A minimum standard should be defined appropriate to each country and enforced by legislation. Subsequently, there should also be a requirement to maintain the lights in working order.

Glass. A laminated glass windscreen at the front should be maintained to allow a clear and unobstructed view, in three- or four-wheeled motor vehicles. Thus, cracked or shattered screens and toughened glass windscreens, which can cause much more serious injury in an accident, should not be lawful or permitted.

Dangerous conditions. A provision to prevent using a motor vehicle in a dangerous condition on a road should be incorporated in legislation to cover the unforeseen possibilities due to every possible safety-related defect on a motor vehicle. Such a section would cover defective shock absorbers or corroded shock absorber mountings, which are directly related to safety, and defects to the structure or chassis.

b. Vehicle Testing

In most parts of the country, to meet local standards, imported vehicles are 'type' tested. Annually testing vehicles more than three years old, supplemented by random roadside spot checks and technical examinations by government vehicle inspectors, the traffic police, or more usually both acting in partnership, general road safety can be covered. Any further use of a defective vehicle should be prohibited, either immediately or subject to conditions, until the defects have been repaired.

A vehicle defect rectification scheme is increasingly used in advanced countries, ensuring that the reconditioned vehicle is made roadworthy. In this system, by stopping a vehicle, the appropriate defect is identified. The driver is offered a chance to either repair the defect within 14 days and present the vehicle for testing or go to court. Most vehicle owners will accept the former option. They have the vehicle tested and submit the certificate of testing, or otherwise prove that the vehicle has been scrapped or broken up. Then the matter is closed.

c. Dangerous Loads

Another undesirable feature common in many developing countries is the precarious manner in which passengers and loads are carried on the roof or hanging onto the outside of vehicles. Therefore, an all-encompassing provision is essential to prevent the carriage of passengers and loads in such a way as to endanger any person. Further provisions should also be considered, referring to total gross and axle weights.

d. Safety Belts and Safety Equipment

Traffic legislation increasingly includes requirements that safety belts be fitted and sometimes that other safety equipment, for example, reflective advance break-down warning triangles, must be carried. It is important that such equipment meets appropriate local standards, and it is often best to refer to such criteria used in other, more motorised countries.

e. Documents/insurance

The rules or requirements relating to the use of vehicles are incorporated in the legislation. Besides the vehicles being roadworthy, there should also be clauses requiring that the owner hold vehicle registration documents to prove ownership. Some degree of cross-checking can be introduced if it is made compulsory that when licensing a vehicle, registration, insurance, and roadworthiness certificates are to be shown, and at the time of seeking insurance, the same vehicle registration, roadworthiness and licence documents must be shown.

2.13.3.3 Highway-related Issues

a) Authority

In the legislation, there should be a clear-cut definition of who is responsible for what in terms of traffic and use of roads. This might specify the different roles and responsibilities concerning traffic at the Ministry of Transport or Public Works, provincial governments and municipalities or the national police.

The legislation must specify, wherever possible, a statutory responsibility upon each highway or road authority (whether national, provincial or municipal) to monitor and improve road safety on their respective road networks. Sections 2.7 and 2.8 cover engineering-related matters, and Section 2.14 covers traffic law enforcement by police in more detail. Some of the important legislative issues related to roads and enforcement are discussed below.

b) Access and Development Control

The right must be given to each highway or road authority to refuse access from developments adjacent to the highways unless the access causes no undue traffic or safety problems. Anyone wishing to create access to a public road must be required to apply for permission and approval from the relevant road authority. The junction should be designed properly with acceleration and deceleration lanes, including tapers, slip roads, islands and without a median cut. A median cut may be provided at the junctions where the traffic U-turns to access the other side carriageway. In case of several closely spaced junctions, clustering may be developed by accessing a service road which merges onto the main highway.

c) Speed Zones

Each highway authority should categorise the highway or road network to specify appropriate speed zones depending upon the function that roads or sections of the network are supposed to provide. Imposition of lower speed limits, physical speed reduction devices and traffic calming in residential areas or rural highways may be considered.

d) Safety Audit

A proposed new or rehabilitation road scheme should be checked systematically by the highway authorities from a safety perspective at stages of its design and just after construction, but before opening. Road safety specialists are to be appointed for these safety checks to ensure that the needs of all road users (especially VRUs) have been taken into consideration. This formal process, termed 'safety audits', is important to ensure the development of safer road networks.

e) Traffic Calming

To provide the local highway authorities with the necessary powers to act on speed reduction and other safety measures for traffic calming, the regulations are made; otherwise, the local highway authorities are never clearly authorised. Detailed specifications for the provision may be made via ministerial regulations.

f) Traffic Signs and Road Markings

Traffic signs and road markings should be compatible with relevant international conventions and may be implemented by the relevant ministry for roads. In developing countries, many traffic signs are based on outdated standards, and the sign size is often too small to be visible to drivers while travelling at the currently permitted speeds.

2.13.3.4 Traffic Control and General Safety

Most legal systems consider the manner of driving under two categories:

- Driving recklessly. If death results from an accident, the offence may be considered more serious if the driver was reckless.
- Careless but still deserves a sanction, for the protection of other road users and the maintenance of an adequate standard of driving.

Drink driving

From a road safety perspective, any review of legislation should include drinking and driving as a priority. Road deaths can be reduced by around 15 per cent by the strict enforcement of this legislation, as observed in several countries, by prescribing the consumption of alcohol up to a limit, but exceeding which is an offence. In developed countries, blood alcohol concentration (BAC) levels between 20 milligrams (mg) and 100 mg of alcohol per 100 millilitres (ml) of blood are typically used, with most countries prescribing an acceptable alcohol level of 50 mg or lower/100 ml of blood.

It is observed that accident risk increases significantly with the BAC level. At a BAC of 50 mg/100 ml, accident risk is double that of a zero BAC level, while the accident risk at 100 mg/100 ml is almost eight times as high as at a zero BAC level.

It is strongly recommended that an alcohol limit of 30 to 50 mg or less be adopted in all countries as a matter of urgency. Random testing should also be adopted, so that any driver at any time could be required to give a specimen. Wherever possible, a roadside test using mobile devices properly disinfected before every use, and maintaining precautions against COVID-19, should be regarded as sufficient evidence for prosecution in developing countries.

Speed Limits

For urban and rural areas, maximum speed limits allowing for lower speed limits near schools, hospitals or other populated and congested areas should be set. It is important to understand that speed limits represent maximum limits and police should enforce these strictly, especially where pedestrians and other road users are at risk. Reduced speeds in urban areas and in designated area-wide zones by installing surveillance cameras has proved effective in reducing the number and severity of road accident injuries in many developed countries. Designating certain locations as reduced speed zones should be permitted under traffic legislation.

Following Too Closely

Drivers should be required to maintain a safe distance between their vehicle and the vehicle ahead, as the driver in the following vehicle will almost always be held accountable for any accidents with the vehicle in front. This distance will vary according to speed, weather and road conditions and should be clarified in driver training and the highway code. However, the law should specify that the driver behind in such an accident is automatically considered to be at fault unless extenuating circumstances can be shown. As a rule of thumb, if a series of vehicles is travelling at a speed of ‘x km per hour’, then the minimum distance should be maintained as ‘2 × metres’ between each two vehicles. It is a tentative suggestion and depends on terrain type, visibility, traffic condition, road configuration and condition, etc.

Give Way Procedures

Give way rules are required to cover junctions, turning movements, parking, approaching emergency vehicles and entering congested junctions. In legislation, the priority rules should be specified and properly signposted as a reminder to all road users.

Restriction of Mobile Phones

Use of a mobile phone while driving is common, but it is widely considered dangerous due to its potential for causing distracted driving and crashes. Due to the number of crashes that are related to conducting calls on a phone and texting while driving, some jurisdictions have made the use of calling on a phone while driving illegal. Many jurisdictions have enacted laws to ban handheld mobile phones. Nevertheless, many jurisdictions allow the use of a hands-free device. Driving while using a hands-free device is not safer than using a handheld phone to conduct calls, as concluded by case-crossover studies, epidemiology, simulation, and meta-analysis. In some cases, restrictions are directed only at minors, those who are newly qualified licence holders (of any age), or to drivers in school zones. In addition to voice calling, activities such as texting while driving, web browsing, playing video games, or phone use in general can also increase the risk of a crash.

In 2015, 660,000 drivers in the United States were estimated to use mobile phones each day while driving during daylight hours. Mobile phone use while driving has become a leading cause of vehicle crashes over the last two decades.

Falling Asleep While Driving

It is defined as criminally reckless driving in terms of the driver being without sleep for more than 24 hours. According to a National Highway Traffic Safety Administration study in the USA, drowsy driving was responsible for 72,000 crashes, 44,000 injuries and 800 deaths in 2013. Other research claims that these figures are underestimated and indicate that up to 6,000 fatal crashes caused by drowsy driving occur each year.

A lack of sleep can seriously affect your reaction time, judgement and ability to concentrate, which can be especially dangerous if you are driving. Whether you're about to embark on a long road trip or not, you need to have enough sleep beforehand to avoid drowsy driving.

The science behind drowsy driving laws is compelling, and there are some excellent reasons why Michigan and 47 other states should wake up to the very real dangers. A very critical situation is posed by excessively fatigued and dangerously sleepy drivers, causing otherwise very preventable car accidents:

- There were 795 deaths involving sleep-deprived drivers in the US in 2017.
- There were 91,000 motor vehicle crashes involving sleep-deprived drivers in the US in 2017.
- 50% of sleep-deprived drivers were 25 or younger.
- Ten to 20 per cent of large lorry or bus crashes in the US involve a tired driver.
- Sleep deprivation can impair a driver as much as drink and drug driving: A motorist who has been awake for a long time will likely experience performance deficits similar to those of someone who has been drinking. For example, a motorist who is up for a continuous 18 hours will typically exhibit performance levels similar to those of a person with a BAC level of 0.05% percent. After 21 and 24 hours without sleep, performance mimics a BAC of 0.08% and 0.10%, respectively.
- In Michigan in 2017, being "fatigued or asleep" was a factor in nine fatal crashes, 1,415 injuries and 2,164 property-damage-only crashes. In 2016, fatigue and sleep were factors in 17 fatal crashes, 1,195 injuries and 2,065 property-damage- only crashes.
- A sleep-deprived driver who has had less than five hours of sleep is nearly five times more likely to be involved in an auto accident.

Similarly, New Jersey's "Death by auto or vessel" Statute, 2C: 11-5(a), defines criminally reckless driving in terms of the driver being "asleep" or "without sleep" for more than 24 hours. Specifically, New Jersey's vehicular homicide law provides:

"Criminal homicide constitutes vehicular homicide when it is caused by driving a vehicle …recklessly", and "proof that the defendant fell asleep while driving or was driving after having been without sleep for a period in excess of 24 consecutive hours may give rise to an inference that the defendant was driving recklessly."

Source: https://www.verywellhealth.com/avoid-falling-asleep-while-driving-2224270

Pedestrian Rights and Requirements

Amongst various developing countries, mostly in the Asia Pacific region, the volume of VRUs is high, and motorisation is rapid, resulting in a large share of road accident victims being pedestrians; hence, the traffic legislation must clearly state priority. Drivers should be directed to stop for pedestrians at formal pedestrian crossings and when parking. Any overtaking near a pedestrian crossing should not be allowed. When turning, drivers must be required to give way to pedestrians already crossing. Pedestrians should be careful while crossing the road and may be legally required to use nearby pedestrian crossings and to walk on the footpath when it is provided. Provisions of railings along the footpaths restrict pedestrians from crossing the streets at any arbitrary locations.

Motorcycle Helmets

In rural areas, mandatory use of motorcycle helmets was required as far back as 1970 in Singapore, but as late as 1995 in Viet Nam. It should be introduced nationwide in all countries. It is generally evident from available accident data that motorcycle passenger head injuries are occurring and that mandatory motorcycle helmet wearing will save lives and injuries. Exemptions are usually necessary for wearers of turbans. However, the capacity for enforcement must be considered. A safe standard must be specified for helmets, and the government should make efforts to ensure such helmets are affordable and readily available from many outlets.

Seat Belts

Legislation requiring the wearing of seat belts and its effective enforcement has resulted in significant reductions in road deaths in many developed countries, for example, a 23 per cent reduction in the UK for example. In some Asia Pacific countries, seat belts have been required on a partial basis (e.g., on expressways only) for several years. The proportion of car drivers and passengers killed or injured in the developing countries is much less than in the advanced countries. Therefore, seat belts have a much less significant impact in developing countries, where pedestrians and motorcyclists cause most traffic deaths. Nevertheless, it is a measure likely to save many lives.

Seat belt legislation normally needs to be introduced in stages: the first provision of seat belts is to make them compulsory for all imported or newly manufactured vehicles, and wearing them is compulsory when seat belts are available. Then, fitting seat belts can be extended to all cars less than five years old and eventually to all cars of any age capable of carrying at least one passenger. If the full benefits of wearing seat belts are to be achieved, the police must be assigned strong powers of enforcement and exercise them.

Non-motorised Vehicles

Although traffic regulations are assumed to apply to all road users, in some countries, pedestrians, bicycles, and other NMVs, like cycle rickshaws, cycle vans, and people and animal-drawn carts, are ignored in the traffic legislation. Traffic regulations should be applicable fairly to all those using public roads and should not discriminate against SMV and NMVs. It is suggested that the legislation could be drafted to give priority to NMVs in certain situations and to provide protection for NMVs in their conflicts with motorised vehicles in the traffic stream.

Post-accident Requirements

All accidents should be immediately reported to the police, and assisting the injured persons should be a primary requirement of all those involved, and must not be declined by any medical centre. However, considering the threat in some societies to those perceived as responsible for accidents, developing countries may also want to consider making it a specific crime for bystanders to harm those involved in road accidents. The difficulties in enforcing this must be considered.

Highway Code

This must be made available. In most countries, highway codes are only advisory, whereas in the UK, the *Highway Code* refers to actual legal requirements, and the word "must" is used. In Malaysia, the highway code was published as regulatory in 1959 and regulations also support the Sri Lankan highway code. An overview of traffic rules and obligations of all road users should be in the highway code in nontechnical language.

NRSC (National Road Safety Council)

To oversee road safety improvements, a constituted NRSC should have statutory powers. Ideally, the NRSC should be established directly under the PMO because of the involvement of many ministries. Legislation should permit funding and establish a permanent secretariat to implement, follow up, and coordinate NRSC decisions. More information on NRSCs is provided in earlier sections of this chapter.

Insurance Requirements

For all vehicle owners, third-party insurance should be made mandatory. Before insurance is awarded, the vehicle owners should show registration documents and a roadworthiness certificate. Finally, for annual vehicle registration, both the certificates for roadworthiness and for valid insurance should be required.

A level of reliance on third-party insurance premiums should be contributed to fund the NRSC and its operations. The legislation should grant the authority to impose such a levy, with a specific amount. The consultation process to set the amount is determined by ministerial regulations.

2.13.3.5 Fines and Penalties

Level of Fines and Fine Collection

In developed countries, studies have shown that the risk of detection is a stronger deterrent than high penalties. If the primary objective is compliance with regulations, then it would be better to give many smaller penalties rather than fewer high-value penalties. For minor offences, on-the-spot fines are useful as quick and simple penalties, but problems with accountability can prevent them from being used in many developing countries.

To overcome such accountability problems, in several Asian countries, traffic fines are not allowed to be paid directly to the traffic police but must be paid at a separate system, such as either a national bank or a post office. The inconvenience for the time involved in the delay when being stopped and in paying a fine encourages the driver to follow the traffic regulations, and so the method is worth considering. The penalty system should consider the optimal division of responsibility between the traffic police and the courts, which will be different for each country.

Where proceedings are commenced by police but settled by payment of a fixed penalty or fine, or a combination of fine and penalty points, fixed penalty notices are used. These relieve the courts of burdensome minor matters and generate revenue. These have been used to sanction drivers for disobeying signals, excessive speed, and noncompliance with other safety-related offences. Thus, even though traffic volumes and the number of offences increase, no increases are required in judicial or prosecution resources.

Penalty Point System

Penalty point systems have existed for many years in developing countries, but have rarely been implemented effectively by the legislation. Driving convictions can be marked on the driving licence, but this practice is rarely enforced in developing countries. Penalty point systems have proven successful in developed countries where driving licence details are available on a computerised network, which is accurate and up to date. In most developing countries, accurate records on a computerised network are rarely available, but such systems should be introduced as feasible with ITS.

There is always the opportunity for corruption in charging or not charging penalties on a manual system. FASTag-based charging with an amount based on the type of traffic rule violation as detected by the ITS is more effective but involves investment.

Disqualification of the licence holder

Despite being included in traffic regulations, many countries, even at present, lack the administrative system to cancel driving licences and ensure that new driving licences are not obtained. Absence of a central driver records register available on a computerised network means the drivers simply go to another province and get a new licence, so the threat of disqualification is in vain. Efforts should be made to establish ITSs with a single central list of driving licences on a computerised network as quickly as possible, and provisions should be retained within legislation for penalty points systems and disqualification.

Fines Used to Finance the Traffic Police

A percentage of the fines collected should be used to fund traffic police enforcement equipment, police activity, or police welfare. Several countries (including Malaysia and the Philippines) have allowed this. Such funding is particularly appropriate where new and additional income is generated from the use of new equipment (e.g., speed detectors or alcohol testing devices) or from new legislation such as seat belt wearing or drink driving. It does offer a way to increase funding for traffic police enforcement without depending solely on the availability of government funds, and provides incentives to the police to carry out enforcement. In the UK, the introduction of the Sponsorship Act allows private sector funding to go directly to the traffic police rather than to general revenue.

2.13.4 Development Stages

Stage 1: Colonial / Outdated System

Traffic legislation in many developing countries is still heavily based on colonial motor vehicle codes and has been outdated and inappropriate for controlling traffic and improving road safety for many years. Such legislation was developed when there were only a few cars, traffic conflicts were not a serious issue and before the onset of modern traffic control devices such as traffic signals, radar speed detectors, FASTag, CCTV cameras and alcohol detectors. At present, NMVs are not included, insurance requirements are insufficient, speed limits are inadequate, and road signs are too small for current speeds and have unfamiliar designs. Amendments are required in all above-mentioned points; piecemeal amendments are often added to the old base as needed, and the result is often confusing and ineffective. Traffic legislation should correct these deficiencies most effectively.

Stage 2: Legislative Revision

Upon realising the inadequacy and problems of ad-hoc amendments, the traffic police and transport authorities must rewrite the draft for legislation. Sometimes, only one person, and often without relevant training, as such skills are not always readily available, is working within the Ministry of Transport. Also, the negotiations with other ministries can go on for years. Traffic legislation revision has taken more than a decade in Indonesia and Pakistan because of inter-ministerial conflicts of opinion, and revision has also been ongoing for the past few years in Bangladesh. In Nepal, traffic legislation was updated in 1993, but the traffic police were not allowed to contribute to the revision process even though they had to enforce it. Much confusion has resulted both in what is required by road users and how the legal proceedings are to be settled.

Stage 3: Working Group Committees

A working group should be formed of participants from within the government as well as outside, and should include private and public sector transport operators as well as transport specialists. This may help ensure that the traffic legislation produced will be acceptable to the public.

This is often best done under the non-partisan and neutral umbrella of the NRSC, as it minimises inter-ministerial rivalries. In revising the traffic legislation, the inclusion of traffic police on the committee is imperative to ensure practical experience is used and to obtain their support. Specialist technical assistance may be needed from abroad. Traffic legislation should specify blood alcohol limits and testing procedures, seat belt requirements, and the optimal distribution of responsibility as appropriate for that country between individual ministries and between traffic police and the courts.

Stage 4: Standardisation and Automation

If traffic regulations are streamlined with optimal distribution between legislation and regulations, then regional standardisation is achieved. Thus, the desired situation is attained. Provisions for procuring and using modern automated technologies of traffic control, such as red-light cameras and modern enforcement equipment, should be covered under the traffic regulations. A legal requirement is to be imposed on every road authority to try to improve road safety on its road network and specify traffic calming regulations.

2.13.5 Benefits and Effects

The development or revision of traffic legislation provides the proper legal framework for traffic police enforcement and specifies the requirements for improved behaviour by road users that will improve road safety. The earlier traffic legislation was old-fashioned, administrative and document-related concerns characteristic of older, outdated legislation. Modern traffic legislation should contain a safety focus only. Traffic legislation and regulations should be properly developed or revised, and should include not only information about road use but also a streamlined punishment system with adequate deterrence for offenders. Compliance with the traffic regulations results in increased predictability and uniformity of behaviour of road users, reducing the risk of road accidents.

2.14 Traffic Police and Law Enforcement

2.14.1 General

Enforcement of traffic laws is needed to encourage drivers to follow traffic rules by maintaining safe and orderly traffic flows. In the developing regions, most traffic police forces are characterised by insufficient training, minimal enforcement equipment or vehicles and many staff. Inadequate attention is given to the use of accident data, and enforcement and violations are prioritised. Despite having high staffing levels, a lack of trained and experienced officers substantially reduces their potential effectiveness.

In many areas, including traffic management, accident investigation, highway patrolling, motorcycle riding and car driving, and management skills, proper and effective training is needed. Trained police personnel are required to make the traffic disciplined on roads as well as to set an example for the public. A career structure should be available in traffic policing to allow officers to specialise and make maximum use of their additional training. Control systems should be established for empowering junior-level police officers, thus minimising potential for abuse of power by decentralising it.

Priority Actions Needed:

- Based on an analysis of accident data, traffic policing should be planned. The target should be the roads and the locations where accidents occur most frequently, and the associated unsafe driver behaviours and offences.
- Create a specialist traffic police force skilled in the use of modern enforcement equipment, tactics, and strategies, and with the ability to conduct targeted and effective enforcement campaigns. Traffic police training should be expanded and improved.
- To monitor the performance of traffic control, including the frequency of use and prosecutions resulting from modern enforcement equipment, such as alcohol testing devices and radar speed meters, efficiency and activity indicators should be adopted.

2.14.2 Need for Traffic Police and Law Enforcement

The traffic conditions become harder for the average police officer to control as the motor vehicles increase in numbers, power and capabilities. Specialised training is needed for traffic police officers to provide the best road safety and an orderly flow of traffic on the road. To provide training to all police officers is expensive and not cost-effective. The duties of a traffic officer who deals with fatal and serious road accidents and reports traffic offences do not appeal to all police officers. Also, policc driver training is a skill and is taken to high levels, and this is not necessarily within the capability of all police officers.

In developed countries, the traffic enforcement department with highly trained specialist officers has become the nucleus. They are responsible for the prevention of road accidents and the maintenance of smooth and orderly traffic flow. However, in developing countries, traffic police are mostly untrained, unequipped and poorly paid. In developing countries, there are competing demands on scarce road space, a populace largely lacking in road safety awareness, and insufficient guidance provided to road users by road signs, markings, and various other facilities, which increases the need for trained and effective traffic police.

While balancing different road use demands as diverse as child pedestrians and handcarts, overloaded lorries, and modern land cruisers, the traffic police are expected to compensate for any deficiency in traffic engineering on the roads. Specialised training becomes essential as the traffic police are held responsible for ensuring a safe and efficient road network under challenging circumstances.

2.14.3 Key Elements

The six key components are: organisation, policy, tactics, training, vehicles and equipment.

2.14.3.1 Organisation

In various developed countries, traffic police typically account for 8–10 per cent of the total police budget and in the countries where traffic policing is particularly strong (e.g., Japan), traffic police may account for 14–16 per cent of the total police budget. In most developing countries, traffic police are often underfunded and hence under-resourced. They are rarely allocated even five per cent of police budgets. Efficient organisation of the traffic police is required for such limited resources to maximise their potential effectiveness. In many developing countries, traffic police are transferred after two to three years, and there is limited opportunity to receive any advance or specialist training. In most countries, salaries are tied to the pay scales of civil servants and do not properly compensate for the hazardous and demanding nature of the work. Traffic police are required to work long shifts on point duty in tropical climates, for some time, seven days a week. Leave entitlement is often restricted, and that contributes to a high sick leave rate. Within traffic policing, too much temptation and opportunity for corruption can exist, and inadequate salaries may not prevent that.

2.14.3.2 Policy

There is a wide range of traffic regulations, and the available resources of traffic police are limited; therefore, the traffic enforcement efforts must be prioritised to suit local problems. A recommended order of offences for consideration may be as follows:

Safety: offences that could lead to an accident, i.e., speeding, signal violations, drink driving violations and pedestrian crossing violations.

Traffic management: offences that, while not inherently dangerous, do not facilitate smooth movement of traffic, i.e., illegal parking or buses loading and unloading passengers within a junction.

Equipment: offences such as poor lighting or tiredness that could contribute to an accident, but have a much lower correlation with accidents than the safety offences.

Administrative: paperwork offences such as improper vehicle registration or transfer of ownership.

Enforcement efforts should focus on safety violations that are most closely connected with accidents at high-risk sites. Traffic regulations usually specify the fine for each violation; safety violations should incur the maximum fine to highlight the seriousness of the offence.

2.14.3.3 Tactics

a) Strategic Plans and Selective Enforcement

In a policy document, the enforcement priorities need to be clearly identified, such as in a strategic plan. To help achieve the stated objectives, strategic plans, in recent years, have been used to specify the objectives of the police forces and the actions intended. To allow monitoring, deadlines should also be provided. Where safety violations are identified through accident data analysis and professional judgement, selective enforcement should provide the basis for staff deployment. Targeted enforcement campaigns should be undertaken with specified objectives and agreed methods of evaluating the effectiveness of the campaigns. Many enforcement campaigns, such as holding a Traffic Week, are too short and unfocused to have any lasting effect.

b) Verbal Warnings

In the early stages of an enforcement campaign, verbal warnings should be used when education and regulation compliance are the objectives rather than punishment. In advanced countries, most interactions between the traffic police and the public result in verbal warnings, with embarrassment and inconvenience seen as adequate deterrents. The strategy should always be to deter and not punish bad behaviour.

c) Staffing Deployment

The number of vehicles available to the traffic police force determines the actual deployment of traffic police. Point duty, mobile patrolling, accident investigation and RSE are the main functions of the traffic police.

In many developing countries, most of the traffic police are assigned to stationary point duty for traffic control and management. By posting the beat officers at high-risk sites, a lot of accidents can be prevented. Traffic police could be used more effectively at pedestrian crossings to help platoon the pedestrians, as well as requiring vehicles to stop for pedestrians. Point duties could be used more effectively through staffing in shifts. Due to the physical demands, the point duty staffing level should not be constant throughout the day and should be broken into several shifts over the day. Additional traffic police should be stationed at points of duty during the peak hours for accidents and traffic congestion, timetabled to operate during busier hours, as is now done in Nepal.

A typical level of patrol coverage in the UK is shown in Table 2.17. In developing countries, the availability of patrol vehicles results in assigning much larger patrolling routes to each vehicle, hence making police presence and creating deterrence effects on undisciplined motorists lower.

2.14.3.4 Training

The traffic police in any country reflect the living standard of that country, and in developing countries, it is no exception. Generally, traffic police are the general public's primary contact with the main police force and therefore, the image of the police force depends on how well the traffic police conduct themselves. Training programmes need to be capable of producing not only technically qualified but also professional officers with integrity who can present a good image to the public.

Training courses should be provided at junior and senior levels in traffic management and control. The senior-level courses must additionally contain traffic law, accident investigation, highway patrolling, speed control and the use of breathalysers. For all patrol officers, driver and rider training, including basic maintenance, should be provided. Traffic police training should provide courses in public relations and management skills.

2.14.3.5 Vehicles

The selection of motor vehicles for traffic patrol work should be based on their suitability for the task. A saloon motor car is thus unsuitable to undertake a general patrol on poorly maintained roads or unmade tracks. The suitability of solo motorcycles should also be considered with respect to the nature of the roads on which they are to be used. Patrol vehicles should be incident management ready and should also be equipped with accident rescue tools. To help free trapped victims, fire extinguishers, suitable articles to remove debris from the road surface, collapsible warning signs and flashing beacons should be in these vehicles. Local decisions must be taken to evaluate what is the most necessary equipment for the prevailing environment and any potential impact on the speed and handling of the vehicle.

Table 2.17 Traffic Patrol Coverage

Area	Road class	Time period	Level of patrol coverage
Urban	Trunk and A Class	08.00- 00.00	One car and two motorcycles for every 32 kilometres of carriageway
		00.00- 08.00	One car for every 30 miles of carriageway.

	Other	All	One car for every 200 miles of carriageway
Rural	Motorways	08.00-00.00 00.00-08.00	One car for every 10 miles of road (20 miles of carriageway) One car for every 20 miles of road (40 miles of carriageway)
	All-purpose trunk roads	08.00-00.00 00.00-08.00	One car and one motorcycle for every 20 miles of route One car for every 40 miles of route
	Other A and B class	08.00-00.00 00.00-08.00	One car for every 80 miles of route One car for every 160 miles of route

Figure 2.72 - Appropriate vehicles must be used for traffic enforcement.

2.14.3.6 Equipment

The safety of traffic police officers on duty is paramount. The nature of their work is such that they are always in a potentially dangerous environment in proximity to moving traffic. Proper precautions must be taken to minimise risk. A highly conspicuous uniform, or at least one with a high-visibility jacket, is essential. Other required items will vary according to local conditions, but include smog masks, illuminated or reflective batons and shading for officers engaged in lengthy periods of traffic control. Senior officers are to consider the needs and welfare of subordinates and supply the articles that are reasonably required for the conditions.

a) Speed Detection Equipment

Traffic police need to have modern technology equipment if they are to be able to do their job properly. This will include enforcement equipment, such as alcohol testers and radar speed guns. Some more sophisticated enforcement equipment is discussed next.

The enforcement of speed limits is a necessary and important road safety function, especially near locations where speed is a significant cause associated with accidents. A suitable and affordable model of speed measuring equipment is to be procured with varying degrees of sophistication. The simplest and cheapest method of measuring speed is to use a stopwatch to note the time a vehicle passes over a premeasured distance between two points. By applying the formula:

Speed=Distance Time

Either by calculator or using prepared time and distance tables, the speed can be calculated. If the vehicle passes the distance in a shorter time, it must be addressed. This method is cost-effective. It should be remembered that in this method, accuracy is increased as the measuring distance increases, and a minimum of 100 metres (m) is recommended. By using a radar gun, speed offences can also be detected; one officer can easily operate the device and obtain a direct reading of a vehicle's speed. The built-in safety features ensure accuracy and prevent the device from giving a reading when more than one vehicle is within the beam. Used in appropriate circumstances and by appropriately trained personnel, the handheld radar is an effective speed detection tool.

Laser devices produce a low-intensity laser beam that can be aimed at approaching vehicles. The pinpoint of laser light is concentrated, not spread out as in radar devices. It is also visible through the device viewfinder, removing all doubt as to the identity of the target. It is possible to use this equipment in heavy traffic flows that would render the radar system inaccurate.

Figure 2.73 Police wearing reflective jackets conduct spot checks for drunk driving.

b) Automatic Detection Cameras

These cameras use either 35-millimetre colour film format or electronic digital imaging and are presently being introduce) d in many developed countries. The use of these devices is usually a collaboration between the highway authority, which provides and maintains the installation, and the police, who operate the system.

To observe the roadway at appropriate points, the cameras are mounted in favourable locations. Offences that can be detected are the failure to comply with a red traffic signal, either at a road junction or at a pedestrian crossing, or speeding, where speed has been identified as a major cause of accidents. Such cameras can be positioned in front of or behind the site to record oncoming or receding traffic. The distance of the camera from the sensor site is determined by the camera optics used (see Figure 2.74).

Without a satisfactory method of identifying the driver, the camera system cannot function efficiently. As an alternative to prosecution, a conditional offer fixed penalty notice system may be used.

The relevance of such camera equipment requires an ability to trace vehicle owners and an efficient judicial processing system. In the ITS, the mandatory provision in the vehicle, such as FASTag, is highly effective in resolving various issues in fulfilling the above conditions of detecting speed, identifying the vehicle and its driver/owner.

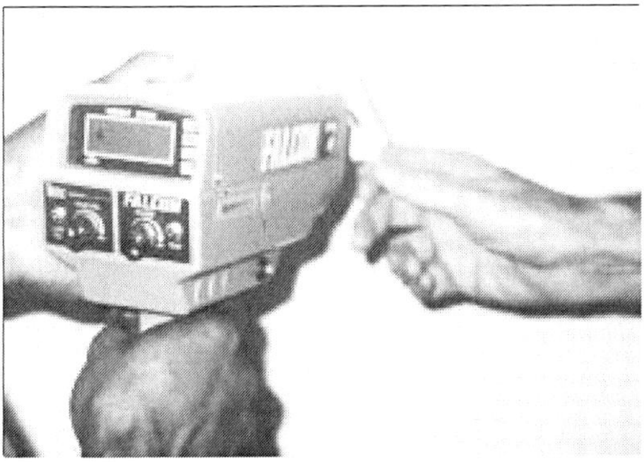

Figure 2.74 Radar being checked before use.

c) Blood Alcohol Screening Devices

It is appropriate to consider publicity as the primary means of deterrence for drinking and driving problems. However, such measures have limited effects without enforcement. In developed countries, drinking and driving has been a problem for many years. Drink and drug driving is a suitable target for traffic police enforcement, which causes significant reductions in accidents. Measurable results can be obtained, either by quantifying the number of police breath tests administered or alternatively, such as analysing alcohol-related accidents as a proportion of overall accident figures.

Considerable inroads have been made into the problem in developed countries through a combination of vivid publicity and high-profile enforcement. In the Asia Pacific region, the problem of drinking and driving varies; in some countries, it is a matter for considerable concern and in others, there is no problem. Suitable handheld breath alcohol screening devices are accurate, highly portable and easy to use at random roadside enforcement checks (see Figure 51.75). But care should be taken to ensure all possible protection against infection of COVID-19 and other diseases.

2.14.3.7 Development Stages

With motorisation and the increase in road accidents, traffic policing has evolved. Four basic stages, describing the gradual awareness of the specialised requirements of traffic policing and the need for optimal use of limited resources, are summarised below.

Stage 1: To compensate for their lack of training, equipment, and mobility, staffing strength is often used by the traffic police. Traffic control is expected from most traffic police, and they are often stationed at traffic signals to provide backup support. Finally, traffic police earn little respect from the public, whether overwhelmed or under-resourced.

Stage 2: Training begins to improve, especially in the metro cities where population pressures cause a lowering of the traffic safety concerns, and the number of mobile traffic police enforcement patrol teams is increased. A strategic plan may be developed and begin to help focus resources.

Stage 3: The mismatch between needs and capabilities is increasingly understood, and this leads to an overhaul of the national traffic police training programme. Overseas technical assistance and expert input are often required for guidance. Introducing specialised training, career tenure is upgraded, and junior police officers are empowered under trial programmes of training and auditing. The strategic plan is regularly assessed to ensure it is effectively monitored for progress. Enforcement campaigns are based on accident factors, and their effectiveness is evaluated rather than assumed.

Stage 4: Between human resources and automation, the traffic policing tasks are divided. There is minimal duplication of effort, with police rarely supplementing signals.

2.14.4 Benefits and Effects

If hazardous and high-risk locations are targeted, traffic policing should be effective to reduce the number and severity of road accidents. Also, focus and effort should be given to reduce traffic conflicts, which would not only facilitate orderly traffic flows and optimise road space use but also reduce the uncertainty and perceived risk of traffic movements at dangerous locations.

To allow analysis and engineering investigations to occur in addition to legal prosecution, the traffic police should also begin to report road accidents accurately and comprehensively.

Figure 2.75 Roadside alcohol testing site in Fiji.

2.15 Emergency Assistance for Road Accident Victims

2.15.1 General

There is general agreement on the principles of an effective emergency medical service, despite its diversity. The essential functions of such emergency medical service are as follows:

- At the roadside, the casualties are to be provided with first aid and medical care.
- Removing the casualty to a hospital.
- Provision of more definitive treatment.

In a developed country, the typical components of an ambulance service are as follows:

- Proper system for prompt notification and communication.
- Proper system for central control and coordination of operations.
- At the accident site, provide effective rescue and medical aid.
- Transport the injured persons to a hospital and the provision of definitive care in an emergency department.

In some countries, in the case of the absence of organised ambulance systems, the alternative is transporting the victims to a hospital by the first available vehicle passing the site, which is often called 'scoop and run'. In such situations, efforts should be made to educate the public on a few basic actions that can be taken to save lives, and the need to transport the victims to the nearest medical facility as quickly as possible.

It is to be ensured that accident victims get the best emergency medical treatment possible. There should be a review of the local situation to provide information on the system with available resources and current usage patterns. This is to inspect how casualties arrive at the hospital, the time taken at the site and in transit. By studying crash and injury data, and transport to hospitals, short and longer-term plans can be made for the development of a system suited to local conditions.

Priority Actions Needed

1. Provide basic first aid guidance on the treatment of accident victims, including how to stop bleeding, choking, etc., to all drivers. These may be mentioned in the back of the highway code and through targeted publicity campaigns.
2. Training police, fire, and any other emergency service personnel in basic first aid.
3. Based on the study of post-accident assistance and consequences for road traffic accident casualties, local and regional trauma plans are to be developed.

2.15.2 Need for Emergency Medical Assistance

In an average road traffic accident, about 50 per cent of deaths happen within 15 minutes of the accident because of injuries to the brain, heart and large blood vessels. A further 35 per cent die over the next 1–2 hours, having head and chest injuries, and 15 per cent over the next 30 days from sepsis and organ failure. The first 30–60 minutes between injury and initial stabilisation is the single most important factor in patient survival. The damage to vital organs, the head, spinal cord, and internal soft tissue is the most serious injury resulting from a traffic accident. Early treatment and stabilisation of these typical injuries can enhance a patient's timely and full recovery. Death or permanent disability can result from delay or well-intentioned but inappropriate first aid. Medical experience around the world has demonstrated that a patient's potential for survival and full recovery increases if they are stabilised and hospitalised within the 'golden hour'.

Ambulance services are intended to meet the following needs:

- Rapid response to life-threatening or serious incidents.
- Preservation of life at the accident site.
- Pre-hospital life support and patient stabilisation.
- Reduction in death and serious injury for accident victims.

Ambulance services may not be as appropriate or effective in developing countries, and a more flexible approach to emergency medical services needs to be considered because of:

- Lack of an effective communication network prevents ambulance services from being notified of road traffic accidents.
- Public unwillingness to give priority to ambulances, as ambulances are more usually used for nonemergency trips, such as transferring patients and hospital staff between the hospital and home.
- Lack of data prevents the need for ambulance services to be identified.

On crowded city streets, it is difficult to know how quickly an ambulance could reach a crash and transport the victim to the hospital. For instance, in New Delhi, India, the time taken for patients to reach the hospital is approximately the same as reported in cities in Europe and the US. However, the medical care for respiratory obstruction and bleeding, as provided by the first aid and transport system, is more effective.

2.15.3 Key Elements

2.15.3.1 Alternative Approaches

In emergency medical care, there are the following general philosophies. One of these formal approaches is to provide immediate first aid and emergency care at the scene of the incident, then to transport the injured person to the emergency department as fast as possible to supply skilled and definitive care. The so-called scoop and run approach is used in countries such as Australia, New Zealand, the UK and parts of the US.

Another approach is to carry the care system to the site of the incident and, there, provide appropriate treatment. This approach is taken in France, Germany, and Russia, where well-equipped vehicles carry skilled doctors to the scene. Unfortunately, there is no evidence to see which approach produces better results than the other, although both have their ardent supporters.

A third approach, an 'informal system', is found in many situations in countries in the Asia Pacific region. As there is often no organised ambulance service, the injured are picked up by bystanders or passing motorists and carried to the nearest emergency department by whatever transport is available, usually without any treatment or first aid at the scene.

This results in rapid transport to the emergency department, but without any resuscitation measures. The outcome of this system depends to some extent on the capacity of the emergency department to deal with these severe cases and to provide effective treatment on arrival.

The essential components of an organised emergency medical care system are mentioned below:

2.15.3.2 Formal Emergency Ambulance Systems

a) An Effective Notification and Communication Network

To notify the ambulance service of the location and nature of the emergency, there should be a single, convenient method. A special telephone number, such as 999, 000, or 111, which operates throughout the country and provides telephone access to all emergency services, police, fire and ambulance, mountain rescue and lifeboats, is provided to the public free of charge in many countries. Another requirement is that there must be a way of identifying the exact location of the incident. The spot can be located by GPS technology available on mobile phones.

b) Central Control and Coordination

A central control centre should be there for receiving calls from the public or from other emergency services, to coordinate the dispatching of vehicles and crew, even though the vehicles may be based at out-stations strategically placed in relation to the anticipated demands.

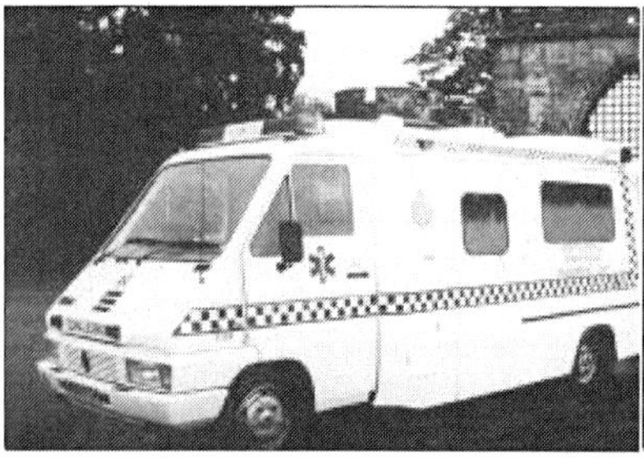

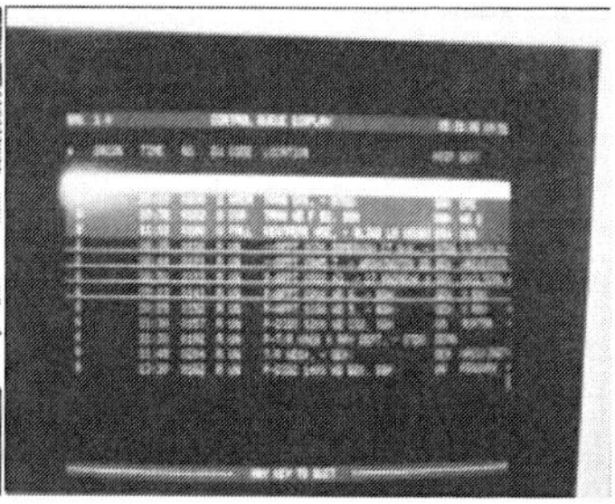

Figure 2.76 Typical ambulance used in developed countries.
Figure 2.77 Command and control Screen

Radios and telephones provide the links between the control centre, out stations and emergency departments, thus the communication by radio between the control centre and the vehicle is maintained. Details of injuries and requests for any special medical treatment at the receiving hospital can be radioed ahead. Thus, the casualties are speedily dealt with on arrival. This cooperation between the ambulance service and the hospital also helps with training staff, planning, development and an evaluation of operations.

c) Effective Rescue and Medical Aid at the Scene

The initial first aid treatment provided to the injured person within the golden hour after the injury is the most important factor relating to the successful recovery of the accident victim. Permanent injury or premature death is prevented by stabilisation of the patient and/or prevention of choking or bleeding to death.

d) Effective and Appropriately Equipped Transport Vehicles

Ambulance vehicles should be clean, comfortable, and have enough room for treatment to be provided to the patient. The vehicles used for patient transport should be capable of travelling over the local terrain. A range of basic first aid equipment, as appropriate, should be available in the vehicles depending upon the training of the crew.

e) Training and Evaluation of Staff Performance

For emergency medical services, training for personnel is commonly provided at basic and advanced levels. Basic level training includes control of bleeding, preservation of a clear airway, cardio-respiratory resuscitation and the stabilisation of fractures. Intravenous therapy and cardiac defibrillation are included in advanced training.

f) Documentation

For medico-legal reasons, and to provide a basis for the evaluation of the efficiency and effectiveness of the service, records must be kept of all operations and treatment given. To determine the overall efficiency as well as the time associated with each facet to identify potential time savings, records should include the time of initial response, time spent at the scene and the time of transport to the hospital.

It is also important to use the E codes as per the ICD-9 coding system, to record that the victim was injured or killed in a road accident and to provide statistics for planning purposes and to permit comparisons with police data to assess underreporting.

2.15.3.3 Informal Systems

There are three basic requirements for such systems to operate effectively. These are:

- Drivers of private vehicles must be willing to transport road accident casualties.
- Road users need knowledge of basic first aid.
- To help achieve the first two requirements, publicity campaigns and programmes need to be undertaken.

In many developing countries, a regulation is commonly found in the traffic legislation that passersby must assist road accident victims, but it is rarely enforced. People fear being blamed for the accident by the police and crowds who may become hostile. Someone who is bleeding badly may cause damage to car upholstery and clothes, hence the reluctance of passersby to carry the victim.

Timely transport of victims to hospitals is important, but according to recent research conducted in India, in many cases, only basic first aid is required to sustain life. Therefore, basic first aid teaching should be incorporated into as much driver training as possible, including highway codes.

To remind the driving public of their legal responsibilities as well as the importance of helping, as it could mean the difference between life and death for the victim, publicity measures should be used. The public will also need to be convinced that, apart from the inconvenience, there are no risks or disadvantages to assisting the injured.

Sometimes, well-meaning but misguided local perceptions can aggravate a patient's condition, for example, making an injured person drink water in the belief that this will help. Publicity campaigns should consider current local practice, but if local practices need to be discouraged or modified, recommended practices should be considered. Hospital locations also need to be publicised, and countries should consider requiring private hospitals and clinics to treat road accident victims.

In many developing countries, even when other hospitals are closer, road accident victims will often be taken to the main public hospital, where treatment is free. The delay thus occurs in treatment may result in a more serious situation.

2.15.4 Development Stages

2.15.4.1 Committee for Emergency Medical Services

To review the provision of emergency medical services and the fate of victims of road accidents, an inter-sectoral committee should be set up under the control of the NSRC. Representatives from the departments of health, transport, public works and police departments, hospitals with emergency departments, existing ambulance services and possibly the insurance industry should be included. The committee would review the present situation and the resources available.

2.15.4.2 Short-term Action Plan

In countries where there is no emergency ambulance, the short-term action plans, for one or two years, may be more effective. Given limited resources, as the first step, to encourage informal scoops and runs by:

- Provide basic first aid training to police, fire service, and other rescue personnel.
- Provide information using highway code, publicity, etc., to all private or professional drivers and passengers on the four or five basic steps to assist a road accident victim from bleeding or choking to death.
- Provide information and guidance on how to transport injured persons to the nearest hospital.
- Provide a mechanism through insurance companies to cover any minor expenses incurred in bringing the injured victim to the hospital, for example, the cost of taxi fares (here, insurance companies could save money if the victim reaches a better hospital earlier).

With maintained medical equipment, trained nursing or paramedical staff, hospital emergency departments can be kept upgraded.

2.15.4.3 Medium-term Action Plan

During this stage, the focus should be on the following:

- Improving the communication system.
- Training personnel.
- Procuring and upgrading vehicles and equipment.

- Developing a notification, dispatch, and control system.
- Developing a record system to monitor and evaluate overall performance.

The action plan, for three to five years (or in the short term, where some components of an emergency medical system already exist), starts with the appointment of a person by the emergency services committee. The person will have the responsibility to coordinate the activities of the agencies to provide the necessary resources to maintain an improved system. For example, it may be felt that the ambulance system should be improved; this is best done by building up a new ambulance service by appointing a professional manager, whose duties could include:

- Defining the need for ambulances.
- Responsibility for obtaining finances.
- Maintaining staff resources that are adequate in numbers.
- Coordinating with all interested parties.
- Financing the manager's post and the ambulance service will depend upon government policy, but possible sources include:
- Legislation to obtain a levy on vehicle insurance, fuel or highway tolls.
- Improved central funding to existing agencies.
- Development banks and aid agencies for certain aspects, which include studies and pilot projects.

2.15.4.4 Long-term Action Plan

The manager should prepare plans for the next one or two years, together with a longer-term strategy over perhaps five years.

The following areas could be improved within a one or two-year period in the whole country (if small) or in a trial area (if a large country), including:

- Funding mechanisms.
- A suitably equipped fleet of vehicles.
- Trained crews.
- Training for all levels of staff.
- Equipment for first aid.
- Coordination between agencies.
- Communication systems.
- Response times.

The development of plans to upgrade communication and notification systems, to train personnel, and to obtain effective and appropriate vehicles can proceed in the longer-term strategy. In urban areas and where distances to medical centres are short, sufficient skills may be developed to apply effective respiration management, control of bleeding and immobilisation of the spine using a basic level of training.

2.15.4.5 Benefits and Effects

The availability of levels of care and their effectiveness, the methods of evaluation employed and the level of services available in the region under study help estimate the potential number of lives that an improved emergency service system could save.

Coordinating with the actions of hospital emergency departments, the first aid and emergency transport services, are the developments which may take place regionally. It has been found that improvements in pre-hospital care, a systematic approach, categorisation of hospitals into levels of trauma care capability and regionalisation of emergency care decrease the number of deaths.

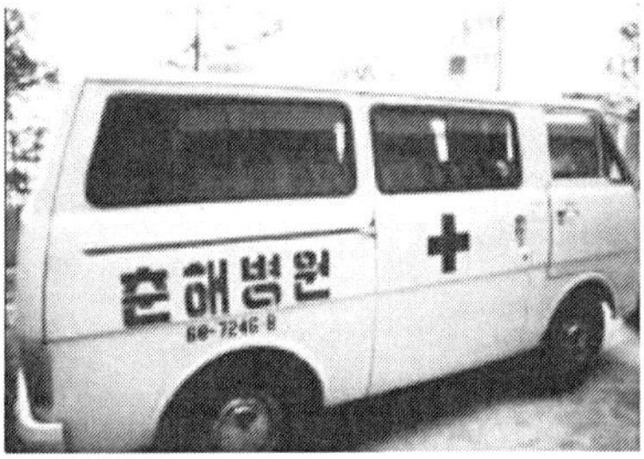

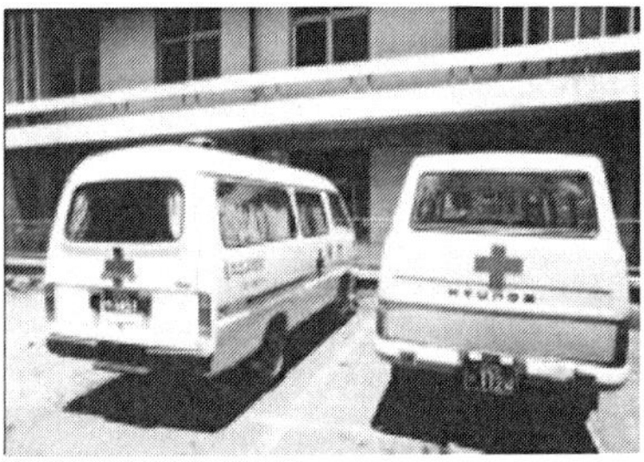

Figure 2.78 **Figure 2.79**
Typical ambulances used in Asian countries

2.16 Road Safety Research

2.16.1 General

Road safety research is essential to understand problem areas and priorities for improvement. Research provides the framework of knowledge against which policy decisions can be taken and countermeasures devised.

To identify problems, evaluate any changes and assess the effectiveness of any countermeasures adopted, accurate and comprehensive accident data is required. Consequently, when seeking to establish a road safety research programme, improving the accident database is frequently one of the priorities. Road accidents are very complex, and different sectors are involved in the operation of road safety; therefore, to reduce the suffering and losses caused by road accidents, local research is required to provide a scientific and objective approach. Research on road accidents is usually best carried out by specialist researchers in universities or road research institutes, but can also be done by others who have an interest in road safety. Lots of research has been undertaken internationally in road safety, and many of the findings of such research can be of value and assist researchers and practitioners in all countries.

Studies should be undertaken to identify the factors involved in various road accidents and to monitor the effectiveness of any countermeasures implemented. To provide immediate improvements at known hazardous locations, the development and monitoring of low-cost engineering countermeasures are of particular importance.

Priority Actions Needed

1. Identify and prepare a consolidated list of all road safety research undertaken at various institutes in the country.
2. All future research should be from accepted priority areas and in improving the accident data system, where the accident data is inadequate to provide an accurate baseline to assess the road safety situation.
3. To ensure road safety research is relevant and findings can be applied, researchers should coordinate with the NRSC and the lead road safety agency in the country.

Safety research is a scientific and objective study with the aim of reducing the suffering and losses due to road accidents. It has the following main objectives:

- Better understanding of the situation and the identification of problem areas.
- Development of countermeasures for problem areas.
- The evaluation of the effectiveness of any remedial action undertaken.

The guidelines define the need for road safety research, a possible framework for research, the stages of development, and the benefits and examples of good practice.

2.16.2 Need for Road Safety Research

Prevention of road accidents is justifiable not only on social and humanitarian grounds, as road accidents are a major cause of death and injury, but also on an economic basis, and therefore, the study of the causes is essential. Road accidents may result in between one per cent and three per cent of a country's GDP per annum.

It is to be noted that, even when the international agencies were providing substantially increased support for funding specific safety projects or ensuring that major highway or urban transport projects contained a safety component, relatively little funding was provided for pure research. The UK's ODA has been one notable exception, which for more than 20 years has funded much of the TRL road safety work in developing countries.

As motorisation is increasing dramatically in many developing countries, road accident numbers are bound to increase, and the need for road safety research becomes stronger. Country-specific research is to be carried out to identify measures that may be appropriate in the developing world.

2.16.3 Key Elements

2.16.3.1 Programme Content

Road safety research in developed countries is usually carried out as typical sectors of activities, and their aims are usually as follows:

1) **Accident data:** To develop the database and analyse it to improve the background knowledge relating to road safety and traffic engineering.
2) **Road users:** To develop and apply human performance characteristics and behaviour patterns in different traffic situations. This helps improve the background knowledge that shapes decisions concerning vehicles and the traffic environment, and measures relating to road users, particularly drivers, pedestrians and cyclists.
3) **Roads:** To facilitate the achievement of the desired standard in the planning, design, construction and operation of safe roads, methods and features are developed and applied.

4) **Vehicles:** To improve the background knowledge that shapes codes concerning vehicles, vehicle components, and the traffic environment, methods are to be developed and applied for studying vehicles and vehicle components in different traffic situations.

These components are in every transport problem and so offer a convenient framework against which a comprehensive research and development programme may be planned. To suggest specific areas for comprehensive research work in this field, the problems found in each sector can be used to consider the general area of road safety, as shown in Figure 2.98.

2.16.3.2 Staffing and Funding

There should be several members in a road safety research unit, working together to maximise the research's potential impact on road safety policy. In India, the Indian Institute of Technology (IIT) and the Central Road Research Institute in New Delhi (India) have several road safety researchers.

Training should be covered by university courses, short in-house courses and by overseas experts. A career path is also needed to ensure that road safety researchers are motivated to remain in the field by creating positions for road safety engineers in every road department of the government.

At TRL, the fund allocated to road safety research is one-third of the total turnover of approximately £32 million (US$51 million) per annum. However, even in developed countries, road safety research is not always adequately funded. In the US, transport advisory boards raise the demand so that the road safety research budget should be increased by 50 per cent because, in terms of years productive life lost, road safety research receives only one-eighth of what is allocated for heart disease research and only one-seventeenth of what is allocated for cancer research.

2.16.3.3 Dissemination and Application

Road safety research findings need to be shared, discussed, and applied for the full benefits to be achieved. Here, failure provides as many lessons as success. Despite the inevitable disappointment resulting from such failures occurring, research institutes should publicise all results. Most research institutes publish the work being carried out and completed. The Central Road Research Institute in India publishes annual reports that summarise the research and development work for the previous year. IIT also produced a review of its injury control research efforts for 1991–1996. Research findings can also be disseminated through seminars, training courses and international conferences.

2.16.4 Development Stages

A country needs to proceed through several stages to develop an effective road safety research capability. In the development of road safety research, the major milestones and the activities involved are typically as follows:

1. Identify previous research conducted. Early road safety research works conducted by individuals from the academic or medical sector need to be centrally collected and organised. Individuals, organisations, engineering universities and teaching hospitals involved in road safety research work should be listed on a central database.
2. Target priority areas identified and ensure a reliable accident database. An accurate and comprehensive accident database is required to adopt an objective and scientific approach to road safety. Road safety problem assessment requires scrutiny of accident data.
3. NRSC to (i) guide research, (ii) promote dissemination, and (iii) apply findings. Road safety research should focus on reflecting the lead agency's (NRSC) priorities and needs to be incorporated into any action plan or strategy developed.
4. Effectiveness of road safety remedial measures evaluated by road safety researchers. By undertaking evaluation work of road safety remedial measures, both foreign-funded and locally financed, the road safety research could be promoted and expanded.
5. Establish links with other road safety research institutes in other countries. While research findings are not always sharable by countries for varying conditions, the lessons learned and approaches used should be shared to help minimise the repetitions, which no country would like to finance. Annual research publications lists should be circulated between research institutes in various countries, and twinning arrangements, along with exchange programmes developed.
6. Solicit private sector-commissioned road safety research. Central government funds were not available for road safety research in more than half the countries surveyed. Transport-related industries, i.e., insurance companies and automobile manufacturers, need to be approached for research funding. Road safety research will need to be directly relevant.
7. Establish a road safety research centre within the existing institutes, with full-time researchers. Road safety research will eventually need

to be undertaken by specially trained professionals to make the output relevant to the needs.

2.16.5 Benefits and Effects

Road safety research produces many benefits, which can be categorised as below, by providing specific examples of the benefits.

2.16.5.1 Problem Assessment

An accurate assessment of the road safety situation is required before developing any action plan. Road safety research objectively evaluates the data available, identifies any data deficiencies and appraises the relative situation with respect to accident trends, high-risk road user groups. Research should cross-check accident data available from the police against the data collected from hospital and insurance records. The research should query any discrepancies and the police will be responsible for tallying reported road accident numbers and casualties.

So far, research on road safety undertaken by TRL's Overseas Centre has been instrumental in documenting the growing problem of road accidents in developing countries and even the accident rates were found to be many more times than those in motorised countries. Despite low levels of motorisation, the road accident casualties tended to affect the younger population and pedestrians in a much larger proportion of total road accident casualties than in countries with a much higher rate of motorised vehicle travel.

TRL's Overseas Safety Research Team developed a MAAP to address the lack of accurate and comprehensive accident data.

2.16.5.2 Development of Countermeasures

To enable safety authorities to plan effective countermeasures and campaigns, a good accident data system is essential. Many measures have been developed in the area of road environment, related to road traffic management and increased understanding of road user behaviour. These include the following:

- Improved junction design techniques.
- Urban safety management studies.
- Better road surfaces.

The most significant results of the research, however, are perhaps linked to issues such as not wearing seat belts, drinking and driving and the use of mobile phones while driving. The financial benefits of these research studies are listed below in Section 2.16.6.4.

2.16.5.3 Evaluation

A report by TRL estimated the monetary benefits obtained from examples of its research work. These were selected as places where it was considered possible to improve on most of the benefits. These estimates were understandably approximate and were expressed as the monetary benefits to the community in the year of the report. Below are the benefits shown from UK road safety research-based projects, and illustrate the value of carrying out effective research programmes:

Cost of research annual benefit (As at 1993)

	£million	**£million**
Seat belt wearing	8.5	650
Junction accident studies	1.1	5
Urban safety management schemes	6.3	17
Drink and drive studies	3.0	60
Road texture and accident rate	16.0	15

The production cost of the highly acclaimed safety-conscious road design guide for developing countries, *Towards Safer Roads*, has been estimated at £0.025 million (US$0.034 million) while the annual benefits from the safe design could reach £10 million (US$14 million) per annum. This has been developed into a textbook and is available from CRC Press of Taylor and Francis Publishers, USA, *Computer-Aided Highway Engineering*.

2.16.5.4 Scientific Approach

Policies and countermeasures are visible but superficial and with little ultimate effect on the expectation of road safety. Because of the complexity of accident-causing factors and the emotional nature of road accidents, the organisations involved are tempted to embark on policies and countermeasures. Research should help road safety policy to be based on an objective and informed basis. The framework, against which informed decisions can be made, is provided by effective road safety research, and every developing country should have some local research activity on road safety issues to aid decision-making.

2.16.6 Examples of Implementation

National transport research centres exist in several countries within the Asia Pacific region. This region is under the coverage of the ADB and the United Nations Economic and Social Commission for Asia and the Pacific (UN/ESCAP). They invest a reasonable amount of their time and budget in safety-related issues. Various locations are identified where the institutes are gaining wider recognition for their work in the safety field:

- Central Road Research Institute, New Delhi, India.
- Institute for Road Engineering, Bandung, Indonesia.
- Korean Transport Research Institute, Seoul, Republic of Korea.
- Institute Kerja Raya Malaysia, Kuala Lumpur, Malaysia.
- National Transport Research Centre, Islamabad, Pakistan.

Although details are not available for all these organisations, the following few examples illustrate that some excellent research work and potential exist in the region.

1) In India, the Central Road Research Institute's traffic and safety environment unit was established in 1986, as a technical wing of the traffic and transport department to give road safety greater emphasis. The institute's main objective of the road safety research was to create road safety awareness among the decision makers in India, with much of the early work dedicated to analysing accident data. At present, the institute's road safety research covers a wide range of areas, including non-motorised vehicle safety, motorcycle helmet usage, design standards and construction practices, provisions with the medians, RSE in schools and driver evaluation methods.

2) In Indonesia, the Institute of Road Engineering, founded in 1984, is the primary research organisation for the highway sector. Six years later, it began collaborating with SweRoad on road safety research. It began to work on the identification of hazardous locations and came to realise that an improvement of the accident database would be its priority.

In addition, several universities in the Asian region carry out valuable road safety research. These include:

- Bangladesh University of Engineering and Technology, Dhaka, Bangladesh.
- University of Bangalore, Bangalore, India.
- IIT-Delhi, New Delhi, India.
- Accident Research Unit, University Pertanian, Serdang, Malaysia.
- Chulalongkorn University, Bangkok, Thailand.

Over the past two decades, TRL has collaborated on research with several countries, including Egypt, Indonesia, Malaysia, Pakistan and Papua New Guinea, while the Republic of Korea has sent several road safety researchers to work at TRL. The Institute of Transport Economics in Norway, as part of its overseas research programme, has also produced several standardised publications for regional use in Southern Africa. Although these were produced for application in Africa, they are also useful for other developing countries. Consolidating and disseminating such information and research results were earlier made within ADB and UN/ESCAP regional studies to improve road safety in the best possible way.

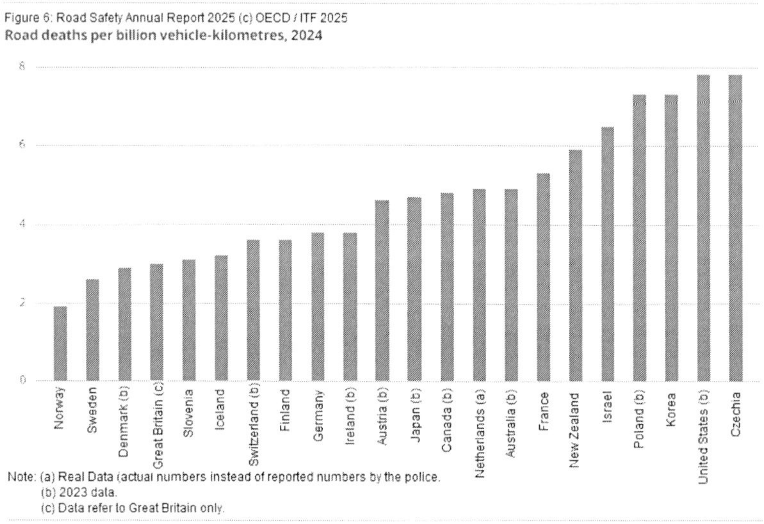

Figure 2.80 Road deaths per billion vehicle-kiloetres, 2024, Source: ITF (2025), Road Safety Annual Report 2025, OECD Publishing, Paris,

2.17 Costing Road Accident

2.17.1 General

With the high growth of vehicles in terms of volume and technology, there has been a dramatic increase in road accidents throughout the developing world. It is essential that adequate sums of money are spent to improve the situation. In the absence of an estimate of accident-related developments, it is difficult to identify the sums of money that should be invested each year on road safety countermeasures. The basic measures include construction of pedestrian and vehicular underpasses, widening roads to provide driving comfort, segregation of fast and SMV along with motorised and non-motorised vehicles, provision of cycle tracks, steel railings, covered drains, pavements, crash barriers, elevated corridors, FOBs, road signs, road markings and ITS devices like FASTag, etc. To ensure that road safety is given adequate priority in terms of investment in its improvement at the level of national resource planning, the accident cost valuations are a primary requirement.

A second need for road accident cost figures is to ensure that the best use of funding is made, and that the best and most appropriate safety measures are provided in terms of safety improvements by using the latest technology. The gross output or human capital approach is the method currently recommended for use in the developing world, despite having various methods available for estimating the cost of road accidents. The loss of current resources, such as vehicle damage, medical treatment, police and administration costs and damage to street furniture, is considered when a person is killed or injured. There is a loss to society of the person's output, so the cost of any loss in the future is also considered in the estimation.

An amount that reflects pain, grief, and suffering of the accident victim and those who care for the victim is usually included, using the gross output method. The degree of severity is considered in estimating the accident costs, so that separate values are determined for fatal, serious, slight and damage-only accidents. The cost of road accidents on a national basis is then obtained by multiplying the costs by accident severity by the number of those accidents that occurred in the year.

Once the total extent of the human casualty toll, and hence the economic costs of road accidents, is known, the road safety situation is more likely to be accepted by decision makers in the Ministry of Road Transport and the Ministry of Social Welfare to invest an increased amount in road safety. Road accident costs can then begin to be used to justify safety measures and be considered along with construction and maintenance costs in the cost-benefit analysis of road safety improvement projects.

2.17.2 Priority Actions Needed

- Until local estimates are available, assume one to two per cent of national GDP is lost annually through road accidents.
- Prepare interim local estimates of the costs of road accidents ranked by severity using the approach recommended by the TRL in the UK.
- Set a research project in motion at a university or economic research institute to prepare accurate valuations of road accident costs ranked by severity using the gross output method.

Road traffic accidents (RTAs) are commonly viewed in terms of individual personal losses or as general statistics. Unlike aviation or rail accidents, where frequently, many persons die in a single accident, the large-scale economic and social impact of road accidents is rarely observed, because the deaths and casualties in a road accident normally happen in smaller numbers.

Various methods exist for costing road accidents, including court awards and life insurance contracts, but the two most common methods are the gross output method and the willingness to pay method. The gross output method is also referred to as the human capital approach and is based on assessing the economic consequences of road accidents, usually supplemented by a notional sum to reflect pain, grief and suffering for those involved and for family and friends of those killed and injured, as a proxy for accident costs.

The willingness to pay method estimates the amount of money for a particular number of people affected who would pay to avoid an accident and produces a much higher valuation of accident costs.

2.17.3 Need for Road Accident Costing

The socioeconomic burden of road accidents is highlighted by accident costing. Developing countries face many challenges, and they need many resources. Not all road accidents and casualties are reported to the police, and there is usually no other system for estimating road accidents and the corresponding casualties nationwide. Therefore, road safety doesn't receive due consideration.

The perception that road accidents are random, unintentional, and predestined, i.e., unavoidable, is also a problem that needs to be resolved by making people understand the presence of a reason, direct or indirect, behind every accident. The reason, therefore, needs to be resolved by investing either money or planning.

Knowledge of accident costs allows safety impacts to be economically justified. In cost-benefit analyses, road safety measures have been frequently ignored or downplayed on the grounds that the associated costs and benefits are too intangible. When road safety is included in cost-benefit analyses, it is often only factored in on a subjective basis and not applied consistently for comparison purposes.

Average wage rates are frequently used with assumptions that are required to reflect the amount of agricultural labour and employment rates. Surveys have been conducted on accident victims in some countries, but are generally on a small scale and often produce conflicting results. It is often assumed that urban pedestrian deaths are of persons of low income. However, an accident costing exercise in Kerala, India, estimated the victims' income to be three times that of the local average per capita income.

The 'lost output' resulting from RTA deaths is calculated as the average earnings multiplied by the number of working years lost, which is the average retirement age minus the average RTA fatality age, and then weighted at the accepted government rate to a present-day value. No cost is normally assigned to the amount of increased travel time caused by road accidents. Unlike other major causes of death in developing countries, road accidents strike down those in the prime of life; the average RTA fatality age is usually between 28 and 31 years, the most productive age, and thus of value to society.

2.17.4 Key Elements

When examining this sector, the key components that need to be considered relate largely to the various cost components. The components can be classified as follows:

- Casualty-related costs that include lost output, medical costs, and the pain, grief, and suffering value.
- Accident-related costs that include property damage and administration.
- Accident data.

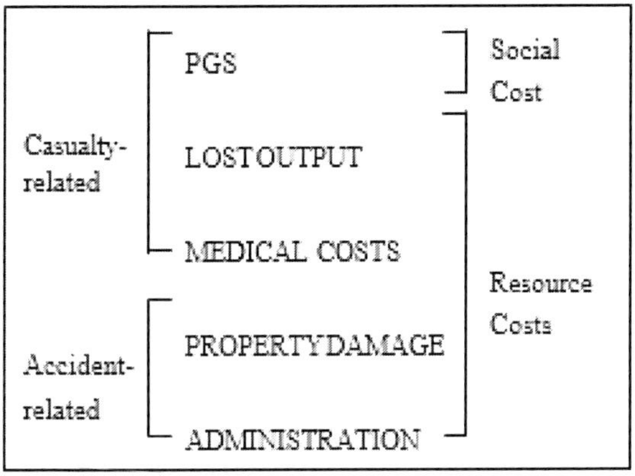

Figure 2.81 shows all are resource costs, except for the pain, grief and suffering component.

Figure 2.81 Road accident cost components

2.17.4.1 Accident Data

The accident data is often taken as authentic in developed countries, but may be incomplete and inconsistent in dcveloping countries, which have little road safety awareness. To make cost estimation possible, casualty figures need to be provided separately for each accident severity type by the traffic police, as shown in Table 2.18.

Table 2.18 National Accident Data Requirements

Type	(No.)	Casualties		
		Fatality	Serious Injury	Slight Injury
Fatal	✓	✓	✓	✓
Serious	✓	0	✓	✓
Slight	✓	0	0	✓
Damage-only	✓	0	0	0

As per the internationally accepted definition of a road accident death includes all related deaths within 30 days of the accident while many countries report only deaths that occurred at the spot or within a few days. Adjustment factors should, therefore, be applied while making international comparisons.

To ensure that an accurate assessment of total road accidents and casualties is achieved, the published Ministry of Health figures, when available, should be compared with police reported figures.

Serious injuries are identified as those that require hospitalisation for at least one night, while slight injuries require medical treatment but no overnight stay in a hospital. In many countries, the number of police-reported road accident injuries is often unrealistically low, and, in such cases, hospital records should be collected to obtain the actual number of serious and minor injuries or use police or insurance company data.

Damage-only accidents are even less well recorded than injury accidents. Only a few developing countries require damage-only accidents to be reported. For accident costing in the UK, a damage-only accidents are six times more likely than accidents with injuries. Subsequently, insurance companies calculated that this ratio was more than 17 times that of accidents with injuries in urban areas of the UK. A nationwide ratio of 15 damage-only to accidents with injuries is now used in the UK for cost estimation. It is important to estimate the number of damage-only accidents as in the US, the cost of damage-only accidents is estimated to be higher than the cost of all fatal accidents.

2.17.4.2 Vehicle Damage Costs

In developing countries, vehicle damage costs are the only property item valued, while all property-related damage costs should be valued, e.g., street furniture, guard-rails and walls. Vehicle damage costs include the repair or replacement cost, minus salvage value, insurance claims, surveyor fees and any business lost due to the vehicle being out of commission (although this is rarely counted in developing countries). Vehicle damage costs should be calculated net of import duties and sales tax. Vehicle damage costs are often the largest cost component in analyses, in the gross output method. Insurance claim surveyor reports, although still limited to insured vehicles, have the advantage of estimating vehicle damage costs regardless of coverage limitations, and often they will report the present value of the car, any remaining salvage value and will itemise the cost of labour and replacement parts separately. Where they are available, they provide a good source for estimating damage costs.

Average vehicle damage costs should not be assumed to apply equally to all vehicles involved in accidents. While accidents involving pedestrians are a serious problem in many developing countries and contribute significantly to casualty figures and costs, accidents involving pedestrians often involve little or no vehicle damage.

Accident reporting systems, such as TRL's software, MAAP, can be configured to include data on vehicle damage and can provide the number of reported accidents incurring no vehicle damage.

2.17.4.3 Medical Costs

Inpatient and outpatient medical care, prescription costs, service fees (X-rays and operations) and rehabilitation costs (including artificial limbs) are medical costs, including emergency medical services. Medical costs are a difficult area in which to collect data. As they rarely account for more than five per cent of accident costs, data collection can be kept to a minimum, as even gross errors have relatively little effect on the overall accident costs. The cost of an inpatient stay (per night) or an outpatient visit is never estimated either by the health ministry or by the individual hospitals. As the bed charges refer only to the patients' charge (and for partial service) and do not reflect the amount of subsidy received or the true cost incurred, those should not be used in costing.

2.17.4.4 Administration Costs

The police and the insurance companies normally incur administration costs for road accidents. The number of accidents not reported to the police and those accidents not involving insured vehicles makes the situation complicated. The responsibility for accident reporting should not be divided between several police forces or divisions. In Bangladesh, the traffic police report damage-only accidents and the general duty police report injury accidents.

For convenience, many developing countries adopt the cost ratios determined by research in the UK and assume administration costs to represent 0.2 per cent of the total resource costs in a fatal accident, four per cent of serious accidents, 14 per cent of slight accidents and ten per cent of damage-only accidents.

Table 2.19 Road Traffic Accident Cost Ratios

Road Traffic Accident Cost	Per cent
Administration cost to represent	0.2
Total resource costs in a fatal accident	4
Serious accidents	14
Slight accidents	10

In an alternative method, a token amount is assumed for those accidents concerning police and/or insurance involvement.

2.17.4.5 Pain, Grief and Suffering Component

Early estimates of road accident costs did not attempt to consider pain, grief and suffering but were focused exclusively on the direct economic costs. With the onset of social cost-benefit analyses in the 1970s, a notional value for pain, grief and suffering was included in developed countries. They reflected society's and the individual's aversion to death. The original value for pain, grief and suffering of an RTA death was determined by the amount required to ensure all lives received a positive value under the net output method, no matter what age and remaining productivity. Pain, grief and suffering values were increased in the UK, at various times during the 1970s and 1980s and ended at 38 per cent of the resource costs of an RTA death, 100 per cent of a serious injury, and 10 per cent of a slight injury. Many developing countries have used these values obtained from the UK, and the accident costing in India and Nepal has used 20 per cent of lost output costs.

The human costs are used at almost twice the resource costs in the UK, with the increasing introduction of the willingness to pay method. This method is inappropriate for developing countries, and indicates that the calculation and use of a value based on the gross output method, perhaps still underestimates the real cost of road accidents in a country.

2.17.5 Development Stages

To produce accurate estimates of road accident costs and to incorporate these estimates into cost-benefit analyses, a developing country needs to proceed through several stages, as mentioned below, along with the activity involved:

1. Accept one per cent of annual national GDP as a minimum estimate of annual road accident costs. To highlight the costs of road accidents incurred annually and to emphasise the overall national cost of road accidents that are usually considered as individual cases with separate financial and personal losses, this working estimate should be used.
2. Conduct an accident costing exercise based on the gross output method. With economic indicators available for lost output calculations and vehicle damage costs derived from insurance companies and surveyor data, the accident cost components can be estimated. As a small percentage of total costs, the medical and administrative costs can be assumed. A conservative estimate of ten to 20 per cent of lost output is initially used for estimating pain, grief and suffering.

Pain, grief and suffering estimates should be agreed upon by the NSRC or those involved in road accident consequences, i.e., insurance company representatives, police, doctors and legal experts. While some technical assistance may be required for initial accident costing calculations, a local research institute could be trained to conduct accident costing updates and revisions. TRL Road Note 105 provides guidance on how to carry out costing in developing countries.

Table 2.20 Hospital RTA Casualty Survey Format

	Inpatient	**Emergency Room Only**	**Brought in Dead**
January			
February			
March			

1) Assess the extent of underreporting of injury and damage in accidents. Hospital surveys should be conducted to produce a more accurate estimate of accident injuries, in case the RTA injury is felt to be underreported by the police. Key hospitals along major highways should be surveyed using their admission registers, with the understanding that the RTA is listed as the cause of admission. Table 2.25 shows the minimal data required by such hospital surveys. Surveys of insurance and fleet companies should also be undertaken to determine the frequency and average cost of vehicle damage in damage-only accidents.
2) Estimate the total annual road accident costs and average accident costs. Reflecting both reported and unreported accidents, these estimates can be determined locally. An average reported accident cost would also include the costs of unreported injury and damage-only accidents as well. These costs would replace the GDP percentage estimate of the annual national cost of road accidents and should be used in road safety programmes and campaigns.
3) Conduct accident victim surveys. To check the accuracy of cost estimates, accident victim surveys should be conducted. By following up on a random sample, say 100 or so, further research on accident costs can be conducted. The researchers obtain from police records the names of those involved in the selected accidents. These individuals or their families can be interviewed to find out how long the victim was in the hospital, how many days' work were missed, the average salary and the cost of vehicle repairs.
4) Introduce accident costs in project cost-benefit analyses for the effects of road widening on accidents; the safety impacts should be evaluated based on accident costs and engineering judgement. According to before and after- accident findings, the cost savings should be refined.
5) Target road safety funding to a percentage of national road accident costs based on the national road accident cost estimate, anywhere

between 10 and 50 per cent. Japan allocated 0.6 per cent of its annual GNP to road safety after calculating that accidents cost 4.3 per cent of annual GNP, i.e., safety funding allocation was about 50 per cent of annual losses.

6) Regularly revise estimates and consider the willingness to pay method. Every three to four years, the road accident costs should be recalculated and updated annually during the interim period by the per capita GDP growth. If the countries are considering moving towards the willingness-to-pay method, they should first carry out a pilot project, with the complex surveys that are involved.

2.17.6 Benefits and Effects

2.17.6.1 Road Safety Awareness

All accidents and related casualties, i.e., reported and unreported injuries, and damage-only accidents, need to be considered to estimate the total cost of road accidents nationally. Reported RTAs are expected to be only an underestimate of the actual total accidents that occur each year. Considering the underreporting that exists in all countries, the severity indices found in developing countries indicate that a much more serious problem exists in the developing world. Road accident costing is an opportunity to overcome underreporting, as hospital surveys often identify casualty totals much higher than police RTA figures. In the PRC, Ministry of Health statistics showed 111,000 road accident deaths, whereas official police statistics for the same year of 1994 showed 66,362 road accident deaths. Similarly, with so few vehicles insured in developing countries, insurance claims for accidents may represent only a fraction of the total vehicle damage costs incurred in the country, in that year. Road accident costing reveals the true extent of the problem in both the human casualty toll and in economic terms.

2.17.6.2 Road Safety Funding

After estimating the total cost of road accident losses, an optimal safety budget can be determined, and adequate resources can start to be allocated to road safety. A road accident costing procedure should also combine public and political support and generate private sector support. Road accidents will be shown to have many more victims as well as serious economic consequences than is commonly thought. Compared to the thousands of persons killed or injured each year, the economic losses to the country are often viewed more seriously by the governments. Being able to estimate the economic losses provides a means of attaining political interest in and support for road safety initiatives. It also tends to galvanise press and media support for safety initiatives to reduce such losses.

2.17.6.3 Promotion of Cost-effective Road Safety

Once the average accident cost is estimated, road safety can now begin to be factored in a consistent manner on the same basis as construction and maintenance in a cost-benefit analysis of road projects and vehicle safety products, e.g., seat belts, motorcycle helmets and reflective stickers. Safety engineering measures in developed countries have proven to have high rates of return, and road safety expenditure can be promoted as an investment rather than purely as an expenditure. Safety measures can be shown to be cost-effective, a basic requirement in this era of scarce resources.

2.17.6.4 Controlling Measures

The rapid and high growth of motorisation is a concern for the governments of various countries, as this is resulting in more road accidents. The loss of lives and assets is significant in road accidents and is a very serious socioeconomic issue. Lives are not only important for families but also for the countries. The governments or road authorities must exercise some measures to control the growth of motorisation, including the growth of motorcycles. The use of private vehicles, taxis and motorcycles should be discouraged, resulting in the use of public transport. To discourage road users using private vehicles, taxis and motorcycles, some measures are proposed as follows:

- During the peak hours, restrict the entry of private cars, taxis and motorcycles on main streets in the cities to central business districts, allowing public transport vehicles only. Private vehicles and motorcycles are to be redirected to some other routes, which are much longer and outside the city.
- Provide efficient public transport 24 hours a day. Destinations such as hospitals, airports and railway stations must be covered adequately.
- Impose a limit on the sale of fuel to private vehicles and motorcycles.
- Impose a high levy on the sale of fuel to private vehicles and motorcycles; the amount may be allocated for road safety measures.
- On non-urban/rural roads, impose random checking for speeding, alcohol consumption, mobile phone use, vehicle surveillance, with reports being sent by patrolling police, etc. Impose strict penalties on

offenders by detaining them in custody for one, two or three days, supported by legislation and charging fines.

- To control corruption by police, all drivers should be prosecuted at some strategic location like a toll plaza equipped with CCTV and audio recording.
- Overloaded vehicles are sent back by the weighbridge.
- Stopping on the carriageway causes many accidents during the day, night, fog and rain and must not be allowed. Broken down vehicles should attract a large fine if they remain on the carriageway and should be removed as soon as possible.

Imposing such measures will be of great help to communities across the world. The following are web links to important sites on this crucial social issue:

1. United Nations on Traffic Accidents Takes Toll on Social, Economic Progress, General Assembly Agrees, Adopting Resolution Improving Road Safety https://www.un.org/press/en/2014/ga11498.doc.htm
2. World Bank: "Road crashes are the biggest safety challenge for youths". https://blogs.worldbank.org/endpovertyinsouthasia/road-crashes-are-biggest-safety-challenge- youths
3. ADB on Technical Assistance for Socioeconomic Impact of Road Crashes (Financed by the Poverty Reduction Cooperation Fund) https://www.adb.org/sites/default/files/project-document/69005/tar-stu-38081.pdf
4. ResearchGate on Social Justice Approach to Road Safety in Kenya: Addressing the Uneven Distribution of Road Traffic Injuries and Deaths across Population Groups https://www.researchgate.net/publication/45389690_Social_Justice_Approach_to_Road_Safety_in_Kenya_Addressing_the_Uneven_Distribution_of_Road_Traffic_Injuries_and_Deaths_across_Population_Groups
5. Clinical Epidemiology and Global Health: despite flows in transport policy, it is noticed that many more proven wider social factors are distressing the number of RTAs, including injuries and deaths. A wider social determinant among low socioeconomic groups, including identity crises, largely contributes towards an increased rate of road accidents in India. https://cegh.net/article/S2213-3984(20)30159-7/fulltext

References

Some figures/text from "Road Safety Guidelines" by the Asian Development Bank.

"Road Safety Guidelines" was largely written and developed by Ross Silcock Ltd. in association with TRL Overseas Centre, the UK. It was produced under the ADB regional technical assistance (RETA 5620: Regional Initiatives in Road Safety), which has also resulted in the production of a comprehensive set of guidelines for decision makers on road safety policy.

CHAPTER 3

TRAFFIC SAFETY STUDIES AND MEASURES

Abstract: This chapter covers general discussions on traffic safety: studies and measures, traffic safety as a public health priority, conceptual framework, mediating factors, traffic volume, traffic speed, traffic conflicts, considerations for traffic safety in roadway design, traffic calming measures, access management, pedestrian countermeasures, traffic characteristics, human factors, driver characteristics, designing roads for young drivers, designing roads for the elderly drivers, pedestrians, pedestrian crossings, signals and signs, school zones, measures for bicycles and motorcycles and road safety planning and audit.

3.1 Traffic Safety Studies and Measures

Road safety is a major economic, social and health issue globally, and has been a critical concern of developing as well as developed countries across the world. Forecasted road accidents are going to be the fifth leading cause of death in 2030, but currently stand in eighth position. Referring to the WHO (World Health Organization) report in 2018, an estimated 1.35 million people are killed in road crashes and up to 50 million are injured worldwide every year. That means an average of nearly 3,700 people die on the world's roads every day.

Before we declare a victory in the war against highway deaths and injuries, we should take a closer look at the factors to overcome such a grave issue. The relationship between the built environment and traffic safety needs proper attention as to how planning can either help or hinder. We need to examine broad impacts on traffic safety at the macro levels of the region and community, and then examine impacts at the micro levels of the street and site.

3.2 Traffic Safety as a Public Health Priority

Public health and safety concerns have always been central to planning. Recent research has focused on active living, specifically how the built environment affects physical activity and hence overall health (Transportation Research Board (TRB), 2005). Significantly less attention has been paid to the important health outcome of the built environment, traffic-related deaths and injuries.

Worldwide, roughly 50 million traffic-related injuries and 1.35 million transportation-related fatalities occur each year. It has been observed that more than half of these injuries and fatalities involve pedestrians, and the majority occur in developing countries. Because of the rapid increase in the use of personal automobiles in developing countries, these numbers are projected to increase by 65 per cent by 2020, thus raising traffic fatalities to the sixth leading cause of death worldwide (WHO 2004), which is preventable.

Traffic fatalities in the United States are already the sixth leading preventable cause of death (Mokdad et al. 2004). This roughly causes suffering from 40,000 fatal crashes and an additional 800,000 injuries each year, according to the Fatality Analysis Reporting System & the General Estimate System.

The United States used to have the second-safest transportation system in the developed world; its traffic safety record has since fallen behind every other developed country, including the UK, Australia, and the entirety of continental Europe. As per the WHO (2004), whether the safety is measured in terms of traffic fatalities per capita or traffic fatalities per vehicle mile travelled, this still stands as a fact. Pedestrians are the victims of 11 per cent of traffic fatalities in the United States. Referring to the Surface Transportation Policy Project 2004, far out of proportion to walking, pedestrian injuries outnumber pedestrian fatalities by a ratio of almost fifteen to one.

3.2.1 Conceptual Framework

A conceptual framework is presented in Figure 3.1. The built environment affects crash frequency and severity through the mediators of traffic volume and traffic speed.

Safety is primarily affected by the roadway design, the traffic speeds it allows, and secondarily by the traffic volumes it generates. In turn, the increased traffic volumes are the primary determinants of crash frequency, while traffic speeds are the primary determinants of crash severity.

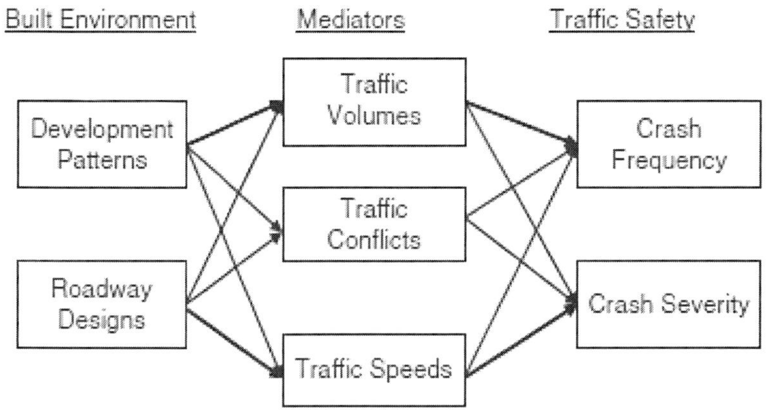

Figure 3.1 Conceptual framework linking the built environment to traffic safety

3.2.2 Mediating Factors

Fatalities along individual roads vary seasonally, with crashes increasing during periods that experience seasonal increases in vehicle miles travelled (VMT). Conversely, reductions in annual mileage during economic recessions often reduce per capita crash rates (Balkin and Ord 2001).

A consistently significant factor influencing the risk of a motor vehicle crash was the quantity of kilometres driven, as found by a study on young drivers (Bath 1993, 5). A lower crash rate, similarly observed for female drivers, is approximately equal to their lower average driving mileage (Butler 1996).

3.2.3 Traffic Volume

Human error is a key factor in traffic safety. Traffic crashes are caused by failure to notice a potential hazard, delayed response to a perceived hazard, or unexpected behaviours by other road users. Every trip, whether as a motorist, pedestrian or bicyclist, involves an element of risk. With other conditions remaining the same, the more vehicular travel, the greater the risk of crashes. The relationship between per capita traffic fatalities and VMT for urban and rural areas in the United States was examined by Litman and Fitzroy (2005). The relationship is roughly linear, which means that as

VMT increases, so do traffic fatalities. In urban areas, each 1 per cent increase in travel is associated with a 1 per cent increase in traffic fatalities and in rural areas, each 1 per cent increase in VMT is associated with a 1.5 per cent increase in traffic fatalities (Litman and Fitzroy 2005).

3.2.4 Traffic Speed

Traffic speed is the other main mediating factor. We all know that higher speeds give drivers less time to react to unforeseen hazards, resulting in increased force of impact when crashes occur. Driving at a speed of 65 km per hour typically needs more than 86 metres to stop on a wet surface; at 48 km per hour, the emergency stopping distance drops to just over 40 metres; and at 32 km per hour, it is about 19 metres (see Figure 3.2).

Lower vehicle speeds are associated with other general safety benefits: lower speeds have a profound effect on pedestrian safety. A pedestrian has an 85 per cent chance of being killed if struck by a vehicle travelling at forty miles per hour. As per the UK Department for Transport 1997, the fatality rate drops to 45 per cent at thirty miles per hour and to 5 per cent at twenty miles per hour or less (Zegeer et al. 2002a). The crash severity increases exponentially with vehicle speed (see Figure 3.3), and the relationship is nonlinear.

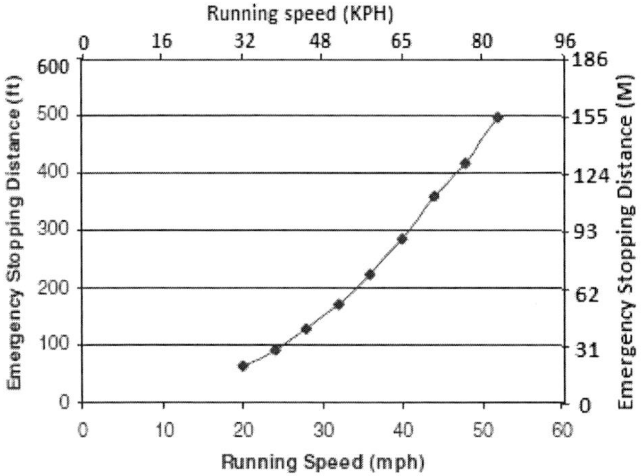

Figure 3.2 Typical emergency stopping distance on a wet road for various running speeds (Transportation Research Institute 1997)

Thus, it is very likely that a car/pedestrian collision appears to increase with vehicle speed. Generally, main street type designs, with low speed, experience the lowest rates of vehicle–pedestrian crashes, and city areas with wide travel lanes and higher operating speeds experience the highest rate of vehicle–pedestrian crashes.

This is the reason that European engineers design roadways for lower vehicle operating speeds, at least in developed areas (Federal Highway Administration 2001; Lamm, Psarianos and Mailender 1999; Organisation for Economic Co-Operation and Development [OECD] 1998; UK Department for Transport 2007).

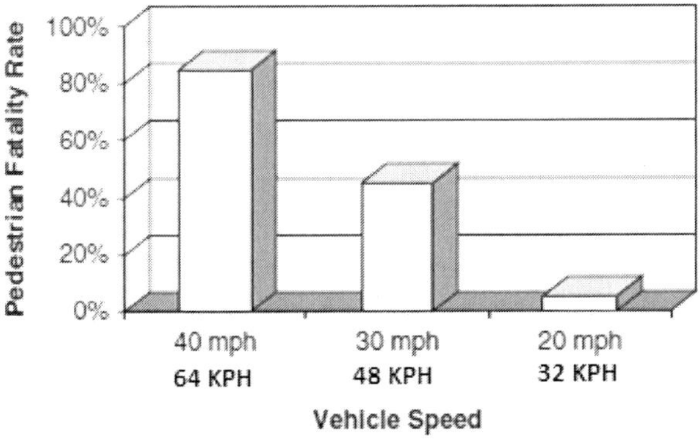

Figure 3.3 Pedestrian fatality rates for collisions at different speeds (Zegeer et al. 2002a, 13)

3.2.5 Traffic Conflicts

Traffic speed alone is not the cause for crashes; rather, it is speed differentials among vehicles in the traffic stream. Similarly, traffic volume alone is not the cause for crashes, but rather conflicting movements when the traffic volumes are high. The conflicts also arise due to on-street parking, access management, traffic calming, junction control, and pedestrian countermeasures. An extra box, representing the mediating effect of traffic conflicts, has been added to Figure 3.1 to make this point explicit.

3.2.6 Traffic Safety and Roadway Design

From the viewpoint of traffic safety, the conventional theory for roadway design is that wider, straighter, flatter and more open is better. It is presumed that these designs are more forgiving of driver error, and thus result in reduced incidence of crashes and injuries. Referring to the American Association of Highway and Transportation Officials (AASHTO) *Green Book* (2004a, 67), "every effort should be made to use as high a design speed as practical to attain a desired degree of safety." Two facts and two concurrent trends support this view.

One of the two facts is that in rural areas, highway safety is improved by high-speed design features such as wide shoulders and gentle bends, particularly on two-lane rural roads (Zegeer and Council 1995). The other fact is that the interstate highway system, designed for high speeds, has experienced lower crash rates than other roadway classes.

The two concurrent trends are (1) the sharp fall in crash rates over the past 40 years, at the same time (2) lanes and shoulders have been widened, bends straightened, and the design for speed generally raised. Concurrent timing has led to the assumption of causality, specifically, that the use of design for higher speeds enhances roadway safety (Dumbaugh 2005a).

To account for an array of factors that influence the safety of highways fails to account for the conventional engineering wisdom. In an urban context, the land-use and vehicle-operating conditions are entirely different from rural areas. The lower crash rates on main roads are at least in part attributable to controlled access, and that eliminates the turning manoeuvres and speed differentials that cause most urban crashes (Dumbaugh 2005a; 2006b). In addition, pedestrians and cyclists and vulnerable road users are banned from main roads.

Accounting for the changes in the demographic mix of the driver population and increased seat belt use and improvements in emergency services, we see the main concurrent trends. Some national studies of crash performance mentioned that changes in highway infrastructure that have occurred between 1984 and 1997 have not reduced traffic fatalities and injuries, and have even had the effect of increasing them. Other factors, primarily the changes in the demographic age mix of the population, increased seat belt usage and improvements in medical technology, are responsible for the downward trend in fatality from accidents. (Noland 2001). This study found that "changes in infrastructure have actually led to increased accidents and fatalities" (Noland and Oh 2004, 532).

3.2.7 Roadway Width

To relieve congestion, there is constant pressure to add lanes and widen roads. Research studies have shown that road widening may have various operational benefits, but those occur at the cost of safety reduction, even after controlling traffic volumes (Dumbaugh 2005b; Harwood 1986; Milton and Mannering 1998; Noland and Oh 2004; Sawalha and Sayed 2001; Vitaliano and Held 1991; Hummer and Lewis 2000. Also see Table 3.1). Conversely, eliminating lanes or reduction in width of the carriageway appears to reduce accidents. Studies on projects that converted four-lane undivided roadways into two-through lane roadways with a centre turn lane find that traffic crashes decrease as lanes are eliminated (Huang, Stewart, and Zegeer 2002; Knaap and Giese 2001).

Table 3.1 Collision rates by cross section, development type and development density

	Collisions per One Hundred Million Vehicle Miles (161 million Vehicle kilometres)			
Development Density:	Medium		High	
Development Type:	Residential	Commercial	Residential	Commercial
Cross Section				
Two-lane	110	270	230	a
Three-lane	180	210	190	230 b
Undivided four-lane	230	260	370	1500 b

a) No data.
b) Very small sample sizes.
Source: (Hummer and Lewis 2000).

In urban areas, wide lanes can adversely affect traffic safety. It is found that wider lanes were associated with statistically significant increases in fatal crashes (Noland and Oh 2004). It was also ascertained that while wide lanes reduced the probability of run-off-roadway crashes in rural settings, they were associated with increases in the same crash types in urban areas (Lee and Mannering 1999). Re-examination of the findings on lane widths and traffic safety has consistently shown an increase in crashes as lanes exceed eleven feet in width (Hauer 1999).

The root cause may be speed, which declines somewhat as individual lanes and street sections are narrowed (Harwood 1990; Farouki and Nixon, 1976; Heimbach, Cribbins, and Chang 1983; Clark 1985; Gattis and Watts 1999; Gattis 2000; Fitzpatrick et al. 2001). On narrow streets, the drivers seem to behave less aggressively and go through fewer red lights. Drivers may feel less safe and drive more cautiously on narrow streets (Mahalel and Szternfeld 1986). Prudent drivers set the pace on two-lane roads, and others must follow. High-speed drivers set the prevailing speed on multilane roads, where passing is possible (Burden and Lagerwey 1999). However, one should be careful not to give too much credit to narrow junctions alone. Dumbaugh (2005b) concluded that it is not narrow lanes by themselves that reduce speeds, but narrow lanes combined with other design elements, such as roadside streetscape elements, that reinforce the message to slow down.

3.2.8 On-Street Parking

On-street parking is on shopping streets and most residential streets. Between the traffic and pedestrians, the parked cars act as a buffer and are convenient to shoppers and residents. However, these benefits are at the expense of safety. It has been suggested that on-street parking accounts for a significant proportion of urban crashes (Seburn 1967; Humphreys 1978; Texas Transportation Institute 1982; McCoy et al. 1990; McCoy et al. 1991; Box 2000, 2002, 2004; ITE 2001). For children, this is especially true, as many child injuries and fatalities from motor vehicle crashes occur when children dart out from between parked cars. Conflicts with parked cars in permitted parking zones produce about 40 per cent of total crashes on two-way major streets, 70 per cent on local streets, and a higher percentage on one-way streets. With the increase in parking turnover rate, the number of crashes increases, meaning that land uses that generate high turnover will also generate more traffic crashes (Humphreys 1978). Angled parking increases the danger compared to cars parked parallel to the kerb (McCoy et al. 2001; ITE 2001; Box 2002).

Interestingly, no study of crash rates was found on comparable roadway sections with or without kerbside parallel parking, which is the ultimate test of on-street parking's safety impact. One study measured the residential street typology and the rate of crashes with pedestrians and found that the existence of parking did not affect crash rates (Swift et al. 2008). It is also quite possible that where parking is provided, parked cars account for a large proportion of crashes, and where parking is not provided, the overall crash rates are about the same as on sections without parking.

The effect on bicycle safety is another consideration with on-street parking. One of the main causes of vehicle–bicycle incidents is 'dooring'. This occurs when a vehicle occupant suddenly opens a door into the path of a cyclist. Designers considered creating facilities that place cyclists out of the door zone. Researchers in Norway suggested that prohibiting on-street parking resulted in a 20 to 25 per cent reduction in vehicle–bicycle collisions. So, while parking acts as a buffer for pedestrians and provides 'friction' to slow down the vehicles, it causes challenges for cyclists and can hide children from drivers.

3.2.9 Traffic Calming Measures

Providing speed humps, roundabouts, etc., used for traffic calming, are perceived by some traffic engineers, residents and members of the media as obstacles in the roadway. Had such measures been truly obstacles, they would have increased crash rates. However, the fact is, they do just the opposite by slowing down the traffic.

Several international traffic calming case studies were summarised by The Insurance Corporation of British Columbia (Geddes 1996). Crash frequencies declined from 8 to 100 per cent with traffic calming. Collision frequencies were compared before and after traffic calming measures were installed (Ewing 2001). From the sample, collisions declined significantly after traffic calming, the difference being statistically significant at the .001 probability level. Adjusting for changes in traffic volumes and removing cases for which volume data was not available, collisions still declined significantly at the conventional 0.05 probability level.

For individual traffic calming measures, all reduced the average number of collisions, and hatched junctions and roundabouts produced differences that were statistically significant.

In these statistics, the mitigating role of traffic conflicts is implicit. Compared to speed humps, hatched junctions are believed to have a better safety record because their design enforces deceleration on entry and less acceleration on the exit (Ewing 1999). The likelihood of rear-end collisions is thus reduced. By reducing the number of conflicting movements at uncontrolled four-way junctions, safety was improved in Seattle (US) on roundabouts, which overcame the disadvantages of a traditional urban grid (see Figure 3.4).

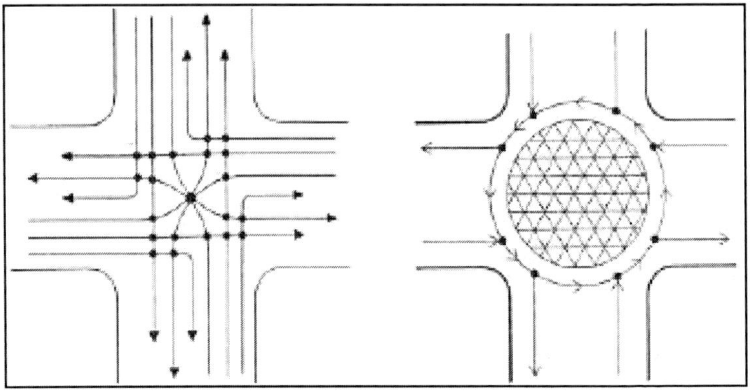

Figure 3.4 (a) Conflict points for an uncontrolled junction with and without a roundabout for the right-hand drive system (USA, Europe, Australia, Japan, etc.)

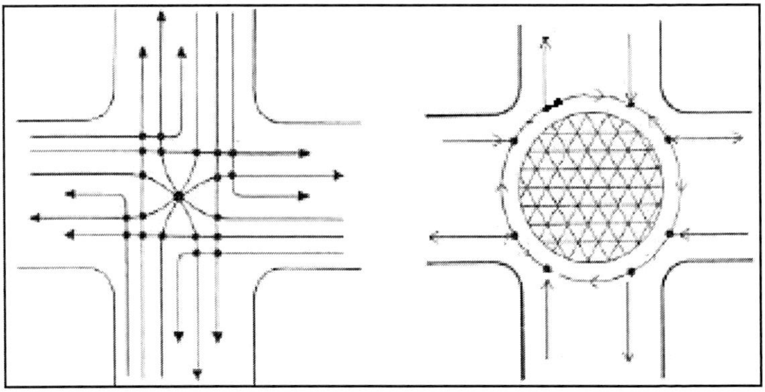

Figure 3.4 (b) Conflict points for an uncontrolled junction with and without a roundabout for the left-hand drive system (the UK, India, Malaysia, Singapore, Nepal, etc.)

European and British traffic calming treatments are more intensive and more integrated with their surroundings than US measures. The average speed drop is almost eleven miles per hour or 30 per cent in Britain (County Surveyors Society 1994) compared to under seven miles per hour or 20 per cent in the US (Ewing 2001). As shown in some studies, the presence of speed humps on a street was associated with lower odds of children being injured (Tester et al. 2004).

3.2.10 Access Management

In urban traffic crashes, speed is not the only culprit; conflicts between through-moving vehicles are created by the presence of driveways and side streets along arterials and those attempting to turn in and out of them. Occurrences of rear-end crashes take place when drivers decelerate to negotiate turns or enter the traffic stream from driveways or side streets at lower-than-prevailing speeds. Occurrences of angle crashes are commonplace as drivers attempt to turn into driveways or side streets, but have insufficient time to clear opposing traffic lanes.

There are two strategies commonly followed for moderating access-related crashes. The first is to reduce the speeds of through-moving vehicles, thereby minimising speed differentials with turning vehicles (Dumbaugh 2006a). The second is to control turning movements while maintaining higher speeds for through-moving vehicles through access management. Access management is exercised for controlling the location, spacing and operation of driveways, median openings and street connections to a main roadway.

The planning engineers can summarise the traffic safety benefits associated with access management techniques. They range from a 20 to 67 per cent reduction in crashes associated with the addition of right/left-turn bays, and with the addition of left/right-turn dividers, respectively.

In up to about 40 access points per mile, the crash rates appear to vary with the square root of access density (Committee on Access Management 2003). On roads with unlimited left/right turns, the crash rates are higher (Gluck, Levinson, and Stover 1999). The two variables—access point density and non-traversable medians—have dual effects and are described in Table 3.2.

Table 3.2 Crash rates on urban and suburban roads with different levels of access control (per million vehicle miles)

	Median Type		
Access Points per Mile (1.61 Km)	Undivided	Two-Way Left-Turn Lane	Nontraversable Median
<20	3.8	3.4	2.9
20-40	7.3	5.9	5.1
40-60	9.4	7.9	6.8
>60	10.6	9.2	8.2

For operational reasons, raised medians are favoured by highway agencies and pedestrian advocates. For pedestrians, who can cross the road in stages, the raised medians provide refuge areas. A study found that the pedestrian–vehicle crash rates were about half as high on arterials with raised medians compared to undivided roadways or roadways with central two-way left/right-turn lanes (see Figure 3.5).

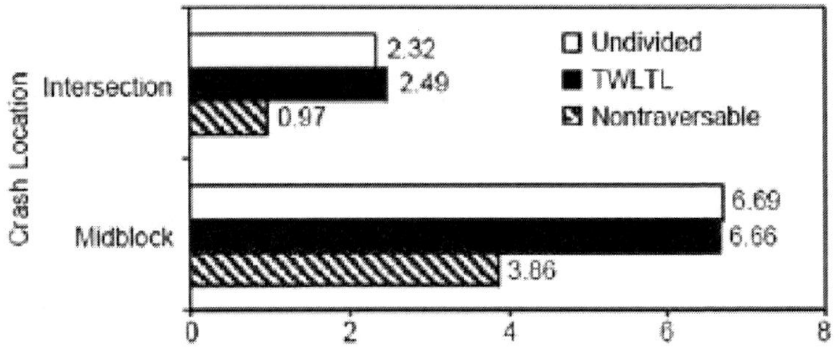

Figure 3.5 Pedestrian crash rates for suburban arterials with different access control (Bowman and Vecellio 1994)

Safety benefits appear to vary by type and width of medians. In a study, it was observed that pedestrian collisions fell by 23 per cent when a six-foot painted median was replaced with a wide raised median (Claessen and Jones 1994). In another study, the narrowest medians (four feet wide) had four times the pedestrian crash rate of the widest medians (ten feet wide) (Scriven 1986). Very narrow medians do not affect pedestrian crashes, although they may reduce vehicle-to-vehicle crashes (Johnston 1962; Leong 1970). Painted medians and two-way left/right-turn lanes do not reduce vehicle–pedestrian crashes on multilane roads, while raised medians and raised crossing islands may do.

Cycling could be safer and more comfortable without the distraction of constant driveways and traffic, and this can be instigated by suitable access management.

Two warnings should be noted before declaring that access management is a win-win for motorists and pedestrians. First, it is not clear that access management strategies, if considered, may enhance pedestrian safety, but the medians do. The concept of access management considers the wide spacing of signalised junctions, preferably with distances of one-quarter mile or greater (Florida Department of Transportation 2006; Minnesota Department of Transportation 2002; Nevada Department of Transportation

1999). Such spacing makes the number of opportunities limited for pedestrians to cross with signal facilities, and encourages hazardous crossings. Second, access management may involve the provision of service roads for local traffic adjacent to the main line or parallel reliever roads. This may partly reduce the safety benefits currently attributed to access management when access-related crashes are transferred from a main arterial to parallel roads.

3.2.11 Junction Control

Vehicle–vehicle and vehicle–pedestrian conflicts are concentrated at junctions. Junction control in some form is more effective for reducing conflicts than others. Although all-way stops have never been a favourite with traffic engineers, they produce lower vehicle speeds near junctions than those produced by traffic signals or two-way stops. From a safety standpoint, with moderate traffic volumes (say, up to 10,000 vehicles per day on major streets), all-way stops appear to outperform signals (Bissell and Neudorff 1980; Ebbecke and Schuster 1977; and Syrek 1955). One study found that when traffic signals at low-volume urban junctions were converted to all-way stops, collisions declined by 25 per cent (Persaud et al. 1997).

Traffic engineers in the US have never been in favour of roundabouts, as they considered them to have many defects. With modern roundabouts, a reduction in travel speeds is achieved by giving way to circulating vehicles, deflection at entry and the curvature of the travel path through the junction. The number of conflict points is reduced by circulation around the centre island, largely eliminating certain types of collisions such as right-angle and left/right-turn head-on crashes.

Various studies have found that roundabouts are superior to other junction control devices with respect to safety (Persaud et al. 2002; Jacquemart 1998; Maycock and Hall 1984; Robinson 2000; Schoon and Minnen 1993, 1994). Crash severity is lessened at roundabouts, even if crash frequencies are comparable to other junctions (Brown 1995). In the United States, the change in crash rates was evaluated following the conversion of 24 junctions to modern roundabouts and in crash rates significant overall reduction of 39 per cent. For crashes involving injuries, the reduction was 76 per cent and crashes involving deaths or incapacitating injuries fell by about 90 per cent (Persaud et al. 2002).

Compared to large or multilane roundabouts, small and medium capacity roundabouts are safer (Maycock and Hall 1984; Alphand, Noelle, and Guichet 1991). Substantially lower pedestrian collisions have been reported at single-lane roundabouts than at comparable junctions with traffic signals (Brude and Larsson 2000). Depending on the study and design, crash reductions are most pronounced for motor vehicles, less pronounced for pedestrians, and uncertain for cyclists (Robinson 2000; Schoon and Minnen 1993, 1994; Brown 1995). Comparative crash statistics are presented in Table 3.3.

In the United States, there is a preference for traffic signals at locations with heavy pedestrian and bicycle traffic, while the European experience envisages roundabouts as relatively safe for pedestrians and cyclists. For crossing by pedestrians, signals provide a periodic gap in traffic, while roundabouts do not do so because of the continuous flow of vehicles. When crossing a junction, signals require no deflection of motor vehicles, while roundabouts may cause motorists to cross paths with cyclists. For pedestrians with disabilities, there are particularly serious issues of access. Low levels of cycling and walking are the cause of some of these, creating a more hostile relationship between drivers and other roadway users.

Table 3.3 Crash rates for pedestrians at roundabouts and signalised junctions in the UK

Intersection Type	Pedestrian Crashes per Million Trips
Mini-roundabout	0.31
Conventional roundabout	0.45
Flared roundabout	0.33
Signals	0.67

3.2.12 Roadside Design

Most of the pedestrian amenities, including pavements, trees and lighting, are located at the roadside. In case the motorists lose control and leave the roadway, a wide 'clear zone' is safer; therefore, conventional engineering design practice encourages placement of such features as far away from the roadway as possible (Transportation Research Board 2003, V-43).

It is observed that most utility pole crashes occur along curves and within 3.5 metres of the road, (Hall et al. 1976), that 85 per cent of tree-related crashes occurred within thirty feet of the road, (Zeigler 1986) 60 per cent of trees involved in crashes were located along horizontal curves, and that 80 per cent were within twenty feet of the road (Turner and Mansfield 1990).

The above statistics only tell us where roadside crashes occur, but not whether roadside crashes are more likely to occur or are more severe when fixed objects are near the roadway. Also, any conclusions will not necessarily apply to low-speed urban streets, but are only related to clear zones on high-speed rural roads. The statistics found that urban roadways with trees located in the nominal 'clear zone' have fewer roadside crashes than locations where trees were not present (Lee and Mannering 1999). The safety effects of urban streetscape improvements along arterial roadways, with the addition of roadside features such as trees and concrete planters, reduced crashes by five to 20 per cent (Naderi 2003). The frequency of injurious roadside crashes against the actual percentage of road segments that had clear zones indicates that the probability of a roadside-object-related crash was largely independent of the roadway's fixed-object offset (Dumbaugh 2005b) (see Figure 3.6).

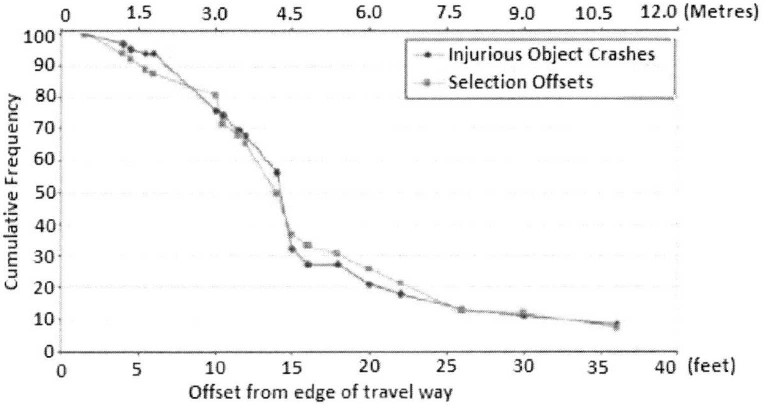

Figure 3.6 Injurious roadside crashes and offset (Dumbaugh 2005b)

3.2.13 Pedestrian Countermeasures

To improve the safety of roadways for pedestrians, engineering can provide solutions: (i) separating pedestrians from vehicles by time and space; (ii) increasing the visibility of pedestrians and (iii) reducing vehicle speed. The last of these measures is already discussed in the section on traffic calming (Retting, Ferguson, and McCartt 2003). A list of 47 such measures is described in *The Pedestrian Facilities Users Guide* (Zegeer et al. 2002a).

The proxies for traffic safety to document impact are used in most of the studies of pedestrian countermeasures. In some cases, the travel speeds have been measured, and in others, the conflict counts and giving way behaviour are considered. In such studies, the actual crash rates are seldom measured. As would at first appear, this may not constitute as big a shortcoming; however, to provide an accurate estimate of multiyear crash rates, the conflict counts have been shown (Hauer and Garder 1986).

In urban areas, the pavements are an absolute necessity along all streets serving developed areas. Where there are no pavements, vehicle–pedestrian collisions are more likely than those with pavements; the collisions are one-and-a-half to two times more likely, according to a study (Knoblauch et al. 1988). Provisions of pavement clearances, vertical kerbs, trees between the street, pavement and parked cars can increase safety.

Traffic is forced to stop for pedestrians at signalised and stop-controlled junctions, with or without marked crossings. Nevertheless, whether to mark crossings at uncontrolled junctions and midblock locations is the issue.

It was observed in a study that at uncontrolled locations, drivers were found to approach pedestrians somewhat slower at a crossing and after markings were installed, resulting in an increase in crossing usage (Knoblauch, Nitzburg, and Seifert 2001). However, no changes in driver behaviour or pedestrian assertiveness were found in this study. Still, it would be good practice to mark pedestrian crossings at relatively low speed, low volume, unsignalised junctions.

Driver speeds before and after installation of crossing markings at uncontrolled junctions were evaluated in another study (Knoblauch and Raymond 2000) by collecting the vehicles' speed data under three conditions: (i) no pedestrians present, (ii) pedestrians watching, and (iii) pedestrians not watching. A considerable reduction in speed was observed under both the no-pedestrian and the pedestrian-not-looking conditions. Drivers on relatively low-speed arterials became more cautious and more aware of pedestrians where the crossing markings are made.

A study was done for comparison over five years of crashes involving pedestrians by selecting 1000 marked crossings and 1000 matched unmarked crossings at uncontrolled locations. In this study, all sites lacked traffic signals or stop signs on the approaches (Zegeer et al. 2002b). It was revealed that on two-lane roads, at an uncontrolled location, the presence of a marked crossing had no difference in the collision rate compared to an unmarked crossing. Furthermore, on multilane roads with traffic volumes above 12,000 vehicles per day, having a marked crossing without other substantial improvements was associated with higher pedestrian collision rates compared to the same site after providing controlling measures for other site factors. Finally, the crossing issues were mitigated by providing raised medians.

Comparing different engineering schemes, it was found that a particular crossing mechanism has a dramatic effect on motorists' propensity to give way to crossing pedestrians (see Figure 3.7). A statistically significant advantage is received by equipment that shows a red signal to motorists as opposed to devices that do not show a red indication. On busy, high-speed arterial streets, midblock signals, half signals, and high-intensity activated crossing (HAWK) signals have compliance rates greater than 95 per cent. In prompting motorists to give way, achieving 65 and 87 per cent compliance, pedestrian crossing flags and in-street crossing signs were also effective, respectively. It is to be noted that most of these crossings were installed on lower-speed and lower-volume, two-lane roadways. High-visibility signs and markings, and overhead flashing beacons, had much lower compliance rates. The *Manual on Uniform Traffic Control Devices* (MUTCD) recommended such changes.

In many countries, the safety benefits of pedestrian-activated signals at uncontrolled crossing points are applied. By installing the Pelican signals in Australia, the quarterly crash rate fell by 90 per cent, and it was found to be highly effective in reducing crashes (Geoplan 1994). Similar to a standard mid-block pedestrian signal, the Pelican signal acts, but during the pedestrian clearance phase, the display facing motorists changes to a flashing yellow, indicating that vehicles may proceed cautiously through the crossing but are required to give way to pedestrians, intending to cause less delay for motorists than standard pedestrian-activated signals. Installing standard pedestrian-activated signals at midblock locations also showed significant reductions in crashes, and the observed reduction was 49 per cent.

Site average and range for motorist yielding by crossing treatment. [Msig = midblock signal; Half = half signal; Hawk = HAWK signal beacon; InSt □in-street crossing signs; Flag = pedestrian crossing flags; OfPb = overhead flashing beacons (push button activation); Refu = median refuge island; HiVi = high-visibility signs and markings; OfPa = overhead flashing beacons (passive activation)]

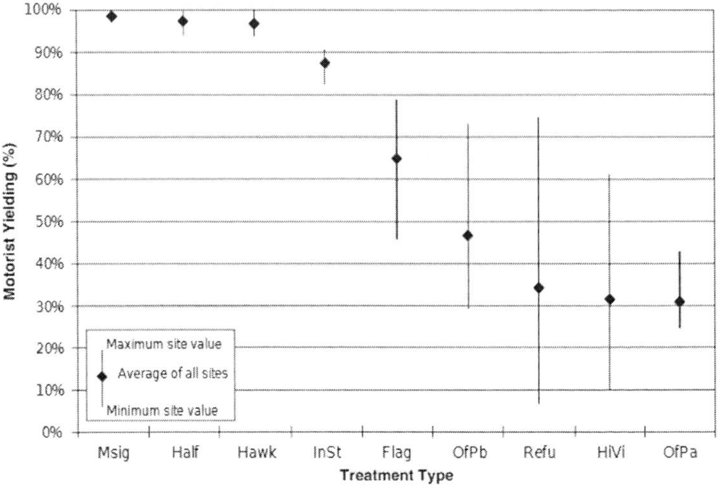

Figure 3.7 Average percentage and range of motorists giving way to pedestrians with different crossing types (Fitzpatrick et al. 2006, 49)

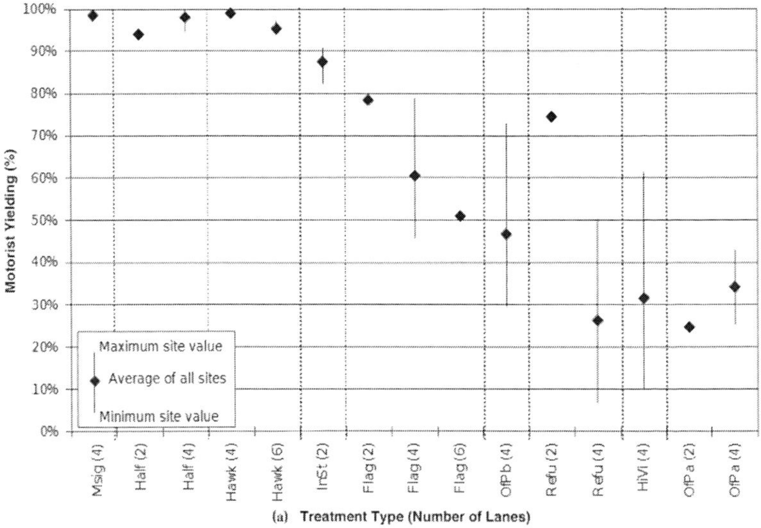

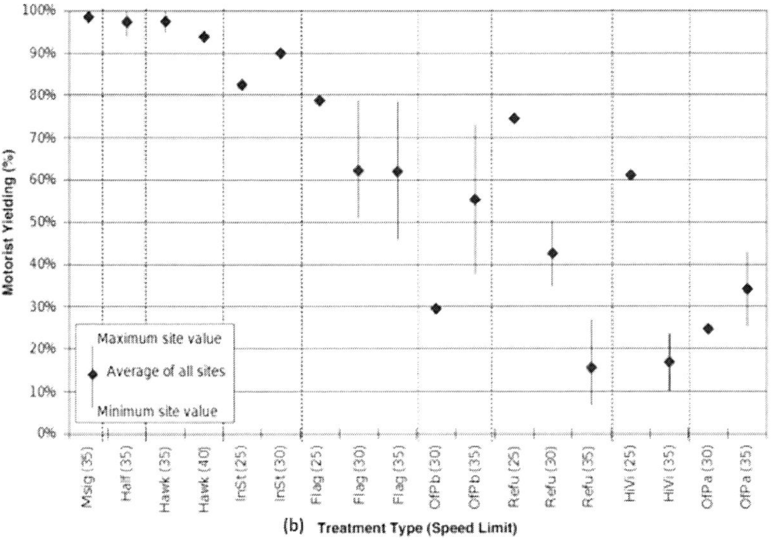

Figures 3.8 (a) and (b) Percentage of vehicles stopping more than ten feet, twenty feet, thirty feet, forty feet and fifty feet from the crossing for each placement of the stop line (Van Houten and Malenfant 1992)

In the area of pedestrian safety, some Canadian research has focused on six countermeasures:

- Interventions to prompt pedestrians to look for turning vehicles when crossing at signalised crossings, including modification of the pedestrian signal head.
- Modification of pedestrian signals to increase the clarity of the indication for the clearance interval. A countdown may be provided to inform the pedestrians about the time left for crossing.
- The use of pedestrian-activated flashing beacons at mid-block crossings and at crossings on major junctions not controlled by traffic signals.
- The use of advance stop lines to get motorists to stop upstream of crossings.
- Interventions to increase the conspicuousness of crossings.
- The use of multifaceted programmes that focus on engineering, enforcement and education (the three Es) to increase the giving way to pedestrians in crossings.

These countermeasures have shown changes in the behaviour of motorists and/or pedestrians (Van Houten and Malenfant 1999). For example, placing advance stop lines fifty feet upstream of a crossing rather than the standard four feet causes most of the drivers to stop well in advance of the crossing rather than encroaching on it (Figure 3.8). At signalised junctions, exclusive pedestrian intervals stop all vehicle traffic for all or part of the pedestrian crossing have shown significant reduction in conflicts between pedestrians and motor vehicles (Van Houten et al. 2000). Flashing warning lights that are automatically activated by the presence of pedestrians have shown reductions in both vehicle speeds and conflicts at uncontrolled crossings (Hakkert et al. 2002; Prevedourous 2001).

Finally, the best suggested countermeasure for pedestrian and bicyclist safety is simply that more people should adopt the habit of walking and bicycling, and this can be viewed as another positive effect of compact development patterns (Jacobsen 2003). There is a direct relationship between the number of cyclists and pedestrians and their safety. For a 100 per cent increase in walking, the attendant increase in injuries is only 32 per cent. So while there might be more injuries, there are fewer per capita.

3.3 Traffic Characteristics and Human Factors

The basic parameters describing a traffic stream are:

- Volume (or flow rate), q, is the number of vehicles passing a point per unit of time. Typical units for volume are veh/day, veh/hour, and veh/sec.
- Speed (or velocity), V or v, is the distance travelled by a vehicle per unit of time. Typical units for speed are km/h (symbol V) or m/s (symbol v). A comprehensive discussion of speed parameters is provided in Chapter 6.
- Density (or concentration), k, is the number of vehicles per unit length of lane or roadway at a given instant of time. The typical unit for density is veh/km.

These are related to each other by the equation:

$$q = kv_S$$

(v_S is the space mean speed). This equation applies only to uninterrupted flow conditions.

Space mean speed, Vs, for a length of lane, L, can be defined as either:

- the average of the simultaneous spot speeds, Vi, of each vehicle in L. If speeds are constant, this is the same as the average speed of the vehicles passing some point in L (refer to Austroads 1998a); or
- an estimate made from measurements of the time the vehicles take to traverse L (refer to Lay, 1998).

Other traffic flow parameters frequently used to describe conditions are (Austroads, 1998a):

- Headway, th is the time gap between successive vehicles in the traffic stream
- Spacing, s, is the distance between the same physical point on two successive vehicles in the traffic stream; and
- Occupancy, Ω, is the proportion of time that vehicles cover a designated point in the traffic lane. Occupancy is directly related to density.

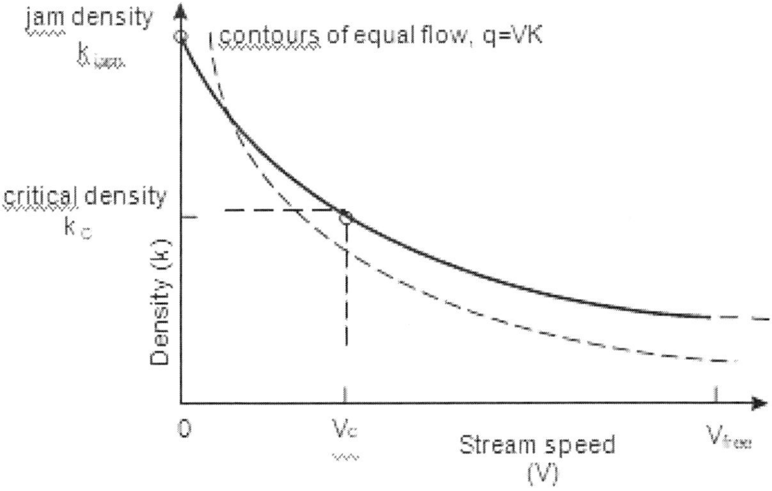

Figure 3.9 (a) Speed-density relationship (Lay, 1998)

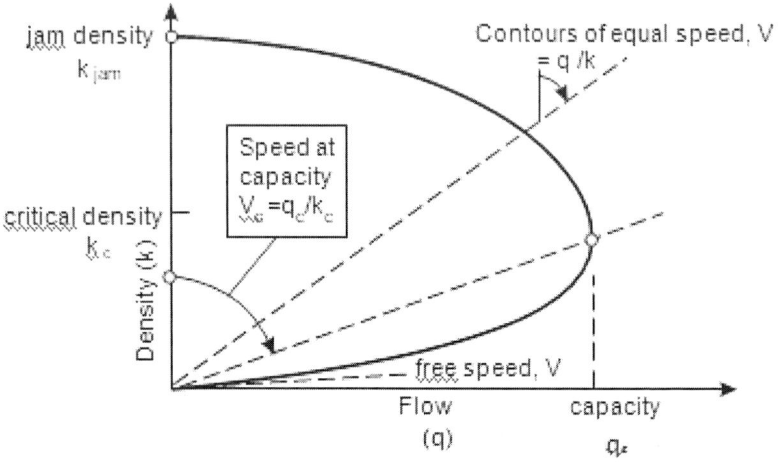

Figure 3.9 (b) Density–flow relationship (Lay, 1998)

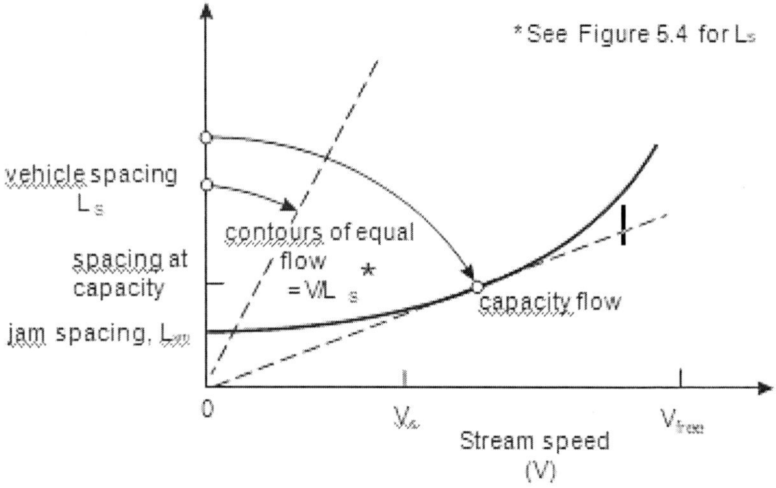

Figure 3.9 (c) Speed–spacing relationship (Lay, 1998)

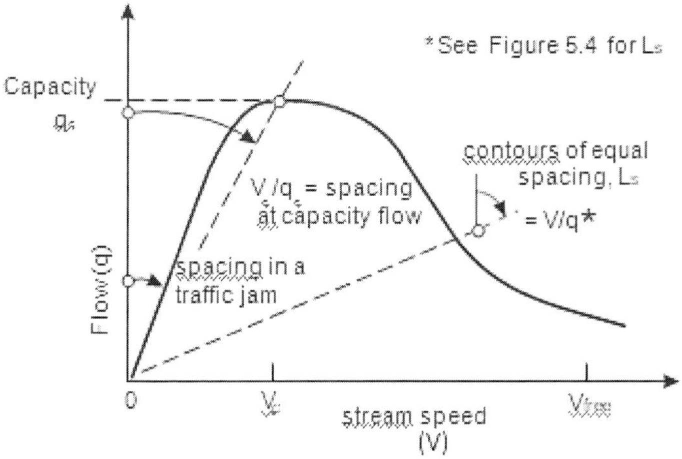

Figure 3.9 (d) Speed–low relationship (Lay, 1998)

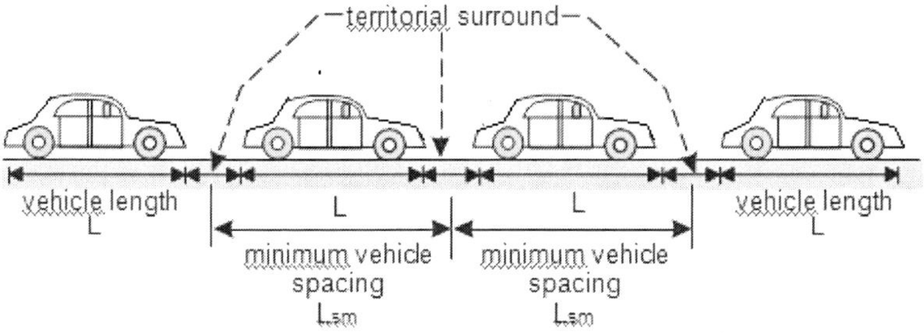

Note that the minimum vehicle spacing, Lsm, which exceeds L, defines the maximum lane occupancy

Figure 3.10 Stalled traffic (Lay, 1998)

Density increases as vehicles travel closer together. This occurs as the volume increases and is accompanied by a reduction in speed (which is required for drivers to feel comfortable with the distance from the vehicle in front). Figure 3.9 (a) shows the relationship between speed and density, and Figure 3.9 (b) shows the relationship between density and flow. Figure 3.9 (c) shows the relationship between speed and flow.

Traffic volumes vary with density from points of zero flow to a maximum at a certain density, as illustrated in Figure 3.9 (b). The zero points represent no vehicles at all, or so many vehicles that the vehicles have come to a stop.

Figure 3.9 (d) illustrates the relationship between speed and spacing. Figure 3.10 represents stalling. It shows the maximum density possible (jam density) and illustrates the maximum lane occupancy.

The speed of the traffic stream is affected by:

- Weather conditions.
- Cross traffic.
- Weaving traffic.
- Disability vehicles.
- Accidents.
- Other marginal conditions.

As speed decreases, vehicles travel closer together (i.e., density increases). As interference increases, the volume can be maintained within certain limits, but with reduced speed and increased density. When the interference becomes so great that the average speed drops below the value at which stable flow can be maintained, there is a rapid decrease in speed and traffic flow, and severe congestion sets in.

Bottlenecks occur when the roadway capacity is reduced in a single area. Provided the approach flow does not exceed the outflow capacity of the section, the flow will be stable. Once the inflow exceeds the outflow capacity, queuing begins, and a breakdown of flow occurs. Speeds are reduced, and the queuing continues until the inflow to the section is equal to the outflow capacity. Bottlenecks can be avoided by designing a road link with consistent roadway capacity (i.e., consistent level of service).

For safety improvements, information on road user characteristics is essential. We can broadly consider three categories of road users: (i) drivers' characteristics, (ii) pedestrians' characteristics, (iii) cyclists' characteristics and (iv) motorcyclists' characteristics.

3.3.1 Driver Characteristics

Information specific to drivers of four-wheeled vehicles is discussed in this section. Almost all the decisions required in the design task are based on the driver's behaviour. The interaction between the driver, the vehicle, other drivers and vehicles and the road itself is the governing criterion for the efficient operation of the road system. Failure of some part of the system causes crashes. Improper use of the road system has devastating consequences as it is a complex, dynamic and fundamentally unstable system (Fuller et al, 2002). The road designers are to be responsive to the needs of road users and understand the psychological factors that make them vulnerable, which helps create a safer road environment.

The performance of drivers is influenced by many human factors that are beyond the ability of road designers to overcome. However, road designers are assisted by some principles for responding to human factor problems:

- Physiological and psychological state:
- Make the roads physically easy to use.
- The needs of fatigued drivers are to be considered.
- Stimulus-driven high physiological states are to be avoided.
- Information processing:
- When critical information is being processed, avoid distractions.
- Informational overload is to be avoided.
- Memory-related errors are to be avoided by placing the necessary information close to the vulnerable phases of the task. Do not rely on the information being stored in the driver's mind.
- Design road features to accord with the driver's expectations.
- Use speed guidance at critical road segments.
- When coming off the highway, a driver may feel that the change in speed makes them think that the car is going much slower than it is.
- Error management:
- Adopt practices for error management.
- Increase feedback to drivers regarding the quality of their performance.

To determine a range of design parameters, response or reaction time is a major component. This reaction time represents the period taken by the driver to convert perceived information into action. The components of response time for the average (50th percentile) driver are:

- Visual detection (refer above).
- Identification.
- Decision making.
- Driver response (or volition).
- Vehicle response.

3.3.2 Designing roads for Younger Drivers

Many younger drivers (18 to 24 years) are significantly overrepresented in the crash statistics. The following factors influence the characteristics of young drivers:

- Risk Lifestyle: 15–20% of young male drivers deliberately take risks; this is a lifestyle factor they have been conditioned to or have adopted. They are more likely to be involved in traffic violations and to crash, and may be considered 'thrill seekers'.
- Risk Exposure: an environment in which the younger drivers operate that creates greater exposure to risk, such as older cars providing less protection, more passengers causing more casualties in a crash, and night driving.
- Risk Seeking: attitudes play an important part in peer pressure. A demonstration of superior skills is in high-risk driving, for example.
- Overestimating competence: ability is often overestimated by younger male drivers, and thus their ability to correct a situation:

  ✓ Identifying distant hazards may be poor.
  ✓ They see less risk in various driving scenarios.
  ✓ They put themselves in driving situations (for example, leaving shorter gaps) where they come into conflict with other drivers.
  ✓ They find it difficult to manage and control their speed.
  ✓ They are prone to driving too fast for the conditions.

3.3.3 Design Recommendations

Younger drivers have less skill and do not find it easy to select the appropriate action when encountering various situations. A young driver often uses inappropriate speeds. Younger drivers can suffer from information overload, leading to some information being ignored. Designs must consider the attitude of younger drivers and their performance can be improved by:

- Providing adequate and well-located signage to ensure that the driver's attention is directed to the hazards ahead.
- High contrast, clear and dynamic warning signs.
- Allowing longer reaction times for complex situations.
- Where drivers are likely to respond too late, use high skid resistance surfaces, such as when approaching traffic signals.

- Control driver behaviour in urban areas using traffic calming devices.
- Give drivers heightened awareness and induce greater vigilance by modifying perceptions of the roadway. Apply devices to give an illusion of higher speed, such as rumble strips with decreasing longitudinal spacing.

3.3.4 Designing roads for older people

Ageing can present as the slowing down, alongside the following characteristics: (Fuller et al, 2002):

- Both cognitive and motor behaviour result in the loss of the ability to perform complex tasks. Coordination is influenced by the changes in the neuromuscular system.
- Loss of muscle strength, endurance and tone and a decrease in joint movements.

For older drivers, their needs and corresponding areas of difficulty are described below (Fuller et al, 2002):

- Vehicle control:
- Road-related tasks, such as difficulty in various manoeuvres.
- Traffic-related tasks, such as car following, overtaking, etc., and
- Roadside services, for example, difficulty in contacting the breakdown service, changing a wheel, etc.
- Journey information:
- Map reading, identifying visual displays, understanding announcements, and
- Information about diversions due to road blockages caused by road works, traffic jams, etc.
- Environmental conditions:
- Weather that causes difficulties in seeing, reading and understanding signs and audible information, and
- Night driving, which causes difficulties resulting from visual deficits and sensitivity to glare.
- Identifying parking areas and manoeuvering into spaces.

3.3.5 Improving the Road Environment for Older Drivers

Enhancing the road can be achieved as follows (Fuller et al, 2002; Jennie Oxley et al, 2000):

- Motorways:
- Ensure enough information with adequate size, lighting and glare protection; and
- Where lane changes are required, provide increased distances to enter the required traffic lane.
- Junctions:
- Signal-controlled junctions are better than those without signals.
- Adequate roadside information.
- Clearly defined permissible vehicle paths.
- Right/left-turn movements are separate.
- Traffic signals have greater prominence, and
- Sight distances to junctions controlled by "Stop" and "Give Way" signs should be improved (longer perception/reaction times—three seconds is desirable).
- Road works: advance and clear information is required. Fluorescent signs have been claimed to improve the legibility and recognition distance for older drivers by a significant margin—3.2s earlier at 100km/h (Dunne, 2001).
- Railway crossings: clear and early warnings are required before railway crossings.
- Traffic manoeuvres: car following, overtaking, entering and leaving a traffic stream and lane changing require:
- Good visibility.
- On two-lane roads, extended use of overtaking lanes; and
- Clear information.
- Traffic-related tasks: information in time, correctly located, with adequate dimensions and contrast and designed to accommodate environmental conditions are required by older drivers.

3.3.6 Design Recommendations

To improve the performance of older drivers and fulfil their needs, the roadway can be designed as follows, which has the potential to reduce crash risk, as indicated by studies (Oxley et al, 2000):

- At junctions controlled by 'Stop' and 'Give Way' signs, increase sight distances.
- Right/left-turn movements should be separated from opposing through movements at junctions.

- At signalised junctions, enhance the prominence of traffic signal displays.
- At junctions, define clearly permissible vehicle paths to prevent wrong choices or the use of traffic lanes.
- For older drivers involved in crashes, the gap selection is a significant problem. Appropriate decisions in the design process can reduce the risk for older drivers. For older drivers to address the problem, some countermeasures are as follows:
- Wherever appropriate, replace 'Stop' and 'Give Way' signs with fully controlled traffic signals. This lessens the decision-making task for older drivers.
- A roundabout means that drivers only need to select a gap in one direction of traffic at a time.
- Gap selection and dealing with high traffic volumes and high speeds can be made easier by providing fully controlled right/left-turn phases.

The environment is improved for all drivers by most of the measures designed specifically for the older drivers' environment. An enhanced driving environment at minimal cost with potentially large benefits can be provided by some additional thought at the design stage.

3.3.7 Pedestrians

Pedestrians are the most vulnerable road users, and they form a large proportion of road fatalities and injuries. For pedestrian safety, simple safety measures can often provide significant improvements.

- At potential points of conflict, warn road users and pedestrians of each other's presence.
- At areas where pedestrian activity is likely to be heavy, inform the drivers.
- To minimise points of conflict, guide drivers and pedestrians.
- By use of special facilities (such as ITS), driver and pedestrian movements can be controlled.
- Avoid isolated and unexpected pedestrian facilities to ensure no surprises for drivers.
- For all users, clear paths are to be defined.

Pedestrian facilities should be designed considering their needs of disabled people. The Commonwealth Disability Discrimination Act 1992 (DDA) has the following objectives:

- Discrimination against persons on the grounds of disability is to be removed, as far as possible, in the areas of:
- Access to premises, accommodation, education, work, clubs and sports facilities.
- The provision of goods, services, land, and facilities.
- Adherence to the existing laws of the country, and
- The administration of Commonwealth laws and programmes.
- As far as practicable, to ensure that persons with disabilities have the same rights to equality before the law as the rest of the community, and
- Within the community to promote recognition and acceptance of the principle that people with disabilities have the same fundamental rights as the rest of the community.

Disabilities may be physical, intellectual, mental, emotional or sensory and are listed as follows:

- Total or partial loss of sight.
- Total or partial loss of hearing.
- Total or partial loss of speech.
- Disfigurements.
- Difficulties in walking, including total or partial loss of legs.
- Difficulties in using arms fully, including gripping.
- Difficulties learning and orientation.
- Malfunctions caused by chemical leaks into the environment, such as reduced vision or breathing difficulties.
- Poor mental health.
- Neurodiversity.

3.4 Design Implications of Roadway Elements on Traffic Safety

3.4.1 Footpaths

A general minimum width of 1.2 m is required for a basic walkway. For a wheelchair and a pram to pass a clear width of 1.5 m and for two wheelchairs to pass comfortably 1.8 m may be provided. To meet the specific needs of those with a disability, including those in wheelchairs, footways should possess all of the following:

- A width between 1.5 m and 1.8 m (minimum).
- A level longitudinal grade as much as practicable—appropriate longitudinal grades are summarised in Table 3.4.
- Slopes or cambers not steeper than 1 in 50, with an absolute maximum of 1 in 40.
- Over the full width of the footway, a vertical clearance not less than 2.4 m, with 2.7 m preferred.
- No manhole covers with grates, to prevent wheelchair wheels and walking sticks from being trapped in the grating. If this is not practical, they should be of a nonslip surface and flush with the footpath. The openings of drainage grates should not be more than 13 mm wide and not more than 150 mm long, and arranged perpendicular to the direction of pedestrian movement.
- An unobstructed passage free from all obstacles, including trees, signs, posts and bicycles.
- A smooth, obstacle-free surface is to be maintained in this condition.
- Guidance devices and tactile surfacing as integral parts of the design.

3.4.2 Surface Treatments Including Tactile Indicators

The surface of footpath pavements should be smooth and not slippery due to the decreased mobility of some pedestrians, and the sensitivity to jarring and vibration of some people in wheelchairs. Some people may stumble and fall over even small ridges and protrusions as low as 6 mm. Surfaces should not deviate more than 5 mm from a 500 mm long straight edge laid anywhere on the surface. For visually impaired persons, the location of tactile indicators must be chosen with care. The merits of commonly used surface treatments in relation to those with a disability are considered below (Austroads, 1995).

Table 3.4 Longitudinal grade requirements for paths

Aspect	Steep Ramp	Ramp	Walkway	Footpath
Grade	1 on 8	1 on 14 to 1 on 19	1 on 20 to 1 on 33 or flatter	Grade of road[3]
Maximum elevation	190mm	<1.0m preferred	Not relevant	Not relevant
Spacing of Landing's	Not applicable	Grade 1:14 -6m Grade 1:19 - 14m[1]	Grade 1:20 — 14m Grade 1:33 — 25m[2]	Not usually required[3]
Kerb	Not applicable	Required	Not required	Not required
Hand rail	Not applicable	Required	Not required	Not required
Fence	Not applicable	As required	As required	As required
Usage	Kerb crossings	Pedestrian only path	Pedestrians only and shared facilities	Pedestrians (bicycles in some circumstances)

Notes:
1. For grades in between 1 on 14 and 1 on 19, at intervals obtained by linear interpolation.
2. For grades flatter than 1 on 33, no landings are required. For grades between 1 on 20 and 1 on 33, landings should be placed at intervals obtained by linear interpolation
3. Where the grade of the road exceeds 3%, and access by wheelchair bound people is required, consideration should be given to applying the standard for walkways. This is particularly applicable in the vicinity of facilities attracting these users.

The minimum widths are for locations where pedestrians gather, such as at the entrance to schools and crossings, recreation facilities and bus stops; widths up to 5m may be appropriate.

Concrete and Asphalt

Hard smooth surfaces are generally quite appropriate, such as concrete and asphalt. Expansion/contraction joints in concrete should not be wider than 13 mm, and to provide an increased coefficient of friction, concrete surfaces should be brushed perpendicular to the direction of movement.

Paving Blocks and Bricks

Paving slabs and bricks used on footways in external areas should not be glazed because these surfaces can become very slippery when wet. They may be laid on a firm well well-compacted base, preferably concrete. As people with slight visual impairment frequently use differences in pavement colour as a means of guidance, the paving blocks should be uniform in colour.

Loose Surface Material

It is very difficult for some physically disabled people to walk on loose surface materials such as exposed aggregate, gravel, sand, soil, grass and mulch. They can also cause severe difficulties for people in wheelchairs. Such surfaces are only appropriate on some recreational/jogging tracks.

Tactile Indicators

For assisting vision-impaired pedestrians, tactile ground surface indicators have been designed to provide directional guidance and warnings. They are detected through contact with a foot or a walking stick. They are made of synthetic rubber, ceramic and clay tiles and stamped concrete. Obviously, rather than outdoor footpaths, these are more suited to indoor and/or lightly trafficked areas. On footpaths and in shopping complexes, commercially produced guidance strips and tactile tiles are commonly used.

3.4.3 Pedestrian Crossings

The approach sight distance (ASD) and the crossing sight distance (CSD) are to be considered for pedestrian crossings:

- To see the crossing and alert them to a possible pedestrian hazard, next allowing time to react and brake, stopping the car before the crossing, if necessary, a driver requires a minimum distance, the ASD. It is the distance from a driver's eye of 1.15 m to the road surface at the crossing.
- To see approaching vehicles and decide if they have time to cross the road safely is CSD. This distance may be considerably longer than the ASD for wide crossings.

At horizontal curves or immediately past the crest of a vertical curve, pedestrian crossings should not be provided because of the restricted CSD. Pedestrians may have walking disabilities or be visually impaired. These factors are to be considered in designing pedestrian crossings and signals.

Crossings Provisions:

- Should be clearly marked as per the applicable design standard.
- Should have a clear delineation between road and footpath with at least colour contrasts between the two (e.g., asphalt and concrete) and perhaps, tactile guidance (see pavement design below).
- Should be well-lit.
- Allow a smooth transition from a raised footpath to road level with dropped kerbs. This should be on the safe side of stop signs and should not be located on a kerb radius unless there is no left/right-hand turn traffic.
- Adequate ASD at the pedestrian crossings and the crossings must be clear of parked vehicles. This may require banning parking for some

distance on each side of the crossing, the distance being determined for each case to ensure that parked vehicles will not obscure the required sight lines.
- Shorten the travel path across the road where practicable, and include a kerb extension.
- On wide carriageways, there should be a pedestrian refuge island.

3.4.4 Signals and Signs

Special considerations are needed for physically challenged persons. Visual impairment could include colour blindness, partial blindness or zero vision. The inability to distinguish red or green is the most common form of colour blindness, although there are those who have less than average perception of one or more colours. To overcome a deficiency in red/green differentiation, the standardisation of traffic signals with the red signal always on top is very effective. To facilitate sign recognition, international symbols and different shape signs can be installed. This emphasises the need to use the signs as defined by the standard of the country. Some helpful, general principles for signals and signs include:

- Using audible signals: these are of help to the visually impaired, but also benefit the general public.
- Signs placed lower down: these are more easily read by people in wheelchairs.
- Push buttons for pedestrian lights: these are to facilitate use by manually or visually limited people.
- Locating traffic signal posts and buttons: installed at standard positions for the same reason.
- Coloured signs: contrast with the background colours, such as light-coloured letters on a dark background.
- Large lettering on signs: compatible with the viewing distance using the recommended typeface.
- National and international symbols: useful for those who cannot read, and
- Signs in braille: for public facilities such as toilets, telephones and service information at major public transport stops.

To cross a road safely, pedestrian signals assist pedestrians. The common minimum walking speed of 1.2 m/s covers most cases, but at some locations, the needs of older or disabled pedestrians need to be considered.

A crossing can be scanned using a device such as an electronic pedestrian detector, which allows a longer time for a disabled pedestrian to cross in safety. Conversely, the green signal would stay on for a shorter period than average at a crossing that is used only by faster walkers. To detect wheelchairs and steel-framed prams at pedestrian crossings, loop detectors at the rear of kerb ramps, similar to those on roadways, may be used. If used, detecting a wheelchair or pram would override the normal signal phase. Such loops may also detect bicycles.

3.4.5 Physical Pedestrian Aids

By providing pedestrian refuges, medians and safety zones, a safe physical separation of pedestrians from vehicles within the roadway may be achieved. In addition to these aids, the provision of street lighting should also be considered. Physical pedestrian aids may be used in conjunction with traffic signals to reduce the exposure of pedestrians to vehicular traffic. On wide, heavily trafficked roads, these provisions may be used if the number of pedestrians crossing the road is insufficient to warrant traffic signals.

Narrowing the road is sometimes caused by pedestrian refuges, medians and kerb extensions, reducing the crossing time so pedestrians are less exposed to traffic when crossing the road. Therefore, provisions of these devices can also have the added benefit of helping to reduce the speed of passing motorists. It is important to ensure that the device will not create a hazard for other road users when designing physical pedestrian aids.

In the absence of a dedicated track for motorcycles and other two-wheelers, it is seen that to avoid traffic congestion, motorcyclists are travelling on the footpaths. To restrict such habits, sometimes posts are installed on the footpaths, but such posts also restrict the movements of wheelchair users. This essentially needs some effective measures for serving both purposes.

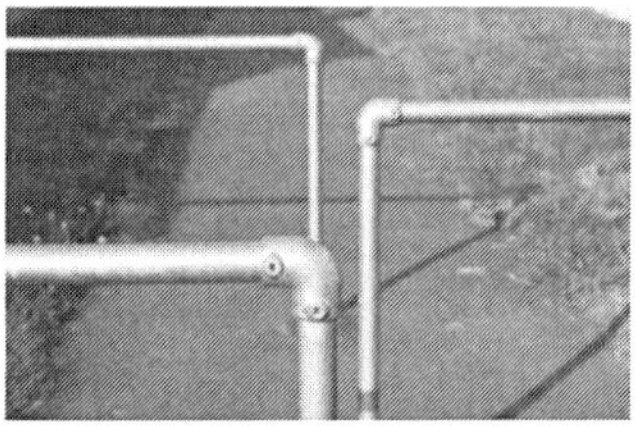

Figure 3.11 Barriers to restrict motorcycles and allow wheelchair users

3.4.6 School Zones

A school zone can have a time-based speed zone; therefore, a VMS (Variable Message Sign) may be installed to regulate vehicle speeds in the vicinity of schools. Additional engineering, such as the reduction of road width, etc., can improve driver compliance with the reduced speed limit.

A typical school zone may have the following speed limits:

- 40 km/h for zones normally with a speed of 70 to 80 km/h.
- 30 km/h for zones normally with a speed of 50 to 60 km/h.

On a road where the speed limit is greater than 80 km/h, if a school zone is required, then a permanent 80km/h speed zone must be installed to provide buffer zones for the 60km/h school zones. Times of operation for school zones are to be considered as mentioned in the relevant guidelines.

The length of the school zone should extend along the front of the school and should be no less than 200 m at either end. The school zone should not extend more than 100 m beyond the limits of the school frontage, preferably centred. The signs and markings are to be in accordance with the relevant standard. If the length of the school zone is longer than one minute's drive or where there are major junctions within the school zone, at appropriate locations, repeater signs should be installed to remind drivers of the speed limit in the school zone. On a road where motorists can readily perceive the presence of the school, the school zones are provided on roads adjacent to schools, with the following exceptions:

- On multi-lane roads.
- On roads where no kerb-side parking is permitted in the proposed school zone and pedestrian-actuated signals, or grade-separated facilities have been installed, and
- At pre-schools, nurseries or day-care centres, with the exception of a pre-school next to a school, in which case the school zone may be extended to include the pre-school.

For ensuring the safety of school children, at the excluded locations of the school zones, other traffic engineering schemes are to be considered.

3.4.7 Bicycles

Cyclists comprise 2 to 10 per cent of the traffic in developed countries and more in developing countries. As the use of bicycles resolves several issues of traffic, pollution, parking, and energy consumption etc., a range of measures should be considered:

- Improve the cycle network.
- Improve safety and security.
- Integrate cycling with the public transport system.

The reasons for convenient and safe bicycle facilities on roads need to be understood. Main roads policies are to be adopted to provide bicycle facilities on state-controlled roads. Bicycles have a right to use the road system, as bicycles are defined as vehicles in road regulations, unless specifically excluded in situations such as on some motorways and controlled access highways. Bicycles are also allowed to travel on footpaths in many places unless specifically prohibited by a local law.

Characteristics

The following people may be defined as cyclists: school children, commuters, athletes and recreational users. Primary school children do not have developed road skills and awareness of dangerous situations. They should preferably be provided with off-road facilities. Secondary school children are more adventurous and may prefer public roads to off-road paths, particularly if the latter requires a longer journey. Recreational cyclists are usually not in any hurry to reach their destination and prefer to do most of their travelling on the quieter paths and streets. However, for longer journeys, they will use the road system. Commuter cyclists have varying needs. Some of them want to reach their destination in the shortest time, irrespective of traffic conditions, and some are prepared to take longer on less stressful routes. At the end of the journey, secure facilities are required, especially where stops occur for long periods. Athletes travel long distances for training and are found on arterial urban roads and highways. Many of them also commute to work.

Facilities Required

To consider special or separate facilities for cyclists, the following definitions apply:

- Full integration—in this, the motor vehicles and cyclists share the same lane.
- Partial integration—a side lane shared by motor vehicles and cyclists is to be widened to allow motor vehicles to overtake cyclists without changing lanes.
- Partial separation—a separate lane or sealed shoulder is provided for cyclists and parked vehicles.
- Full separation—a separate path away from the carriageway is provided for cyclists. This path may be exclusively for cyclists or shared with pedestrians.

The degree of integration or separation to be adopted depends on vehicle volumes, bicycle volumes (12-hour two-way), the presence of parking, and the design speed of the road.

Road Design Criteria for Cyclists

At speeds up to 30 km/h, a desirable lateral clearance between bicycle operating spaces of 1.0 m is required on cycle ways. The required clearance to rigid obstacles beside the cycle way is 1.0 m

Side wind force on cyclists is exerted by passing heavy vehicles, and it is desirable to provide adequate clearance between the bicycle envelope and these vehicles. At speeds of 60, 80 and 100 km/h of motor vehicles, the clearance between the cyclist envelope and a lorry of 1.0, 1.5 and 2.0 m, respectively, is desirable for cyclist safety. These clearances are not always achievable for the site and other situations, and the designer must decide on the maximum clearance.

Cyclists typically travel at 20 to 30 km/h, and downhill, the speed may be up to 50 km/h. To allow travel at these speeds, the roads and paths should be ideally designed; Provisions for compulsory stop or give way controls should be minimised except at locations where it is safe, possible and practical to provide them.

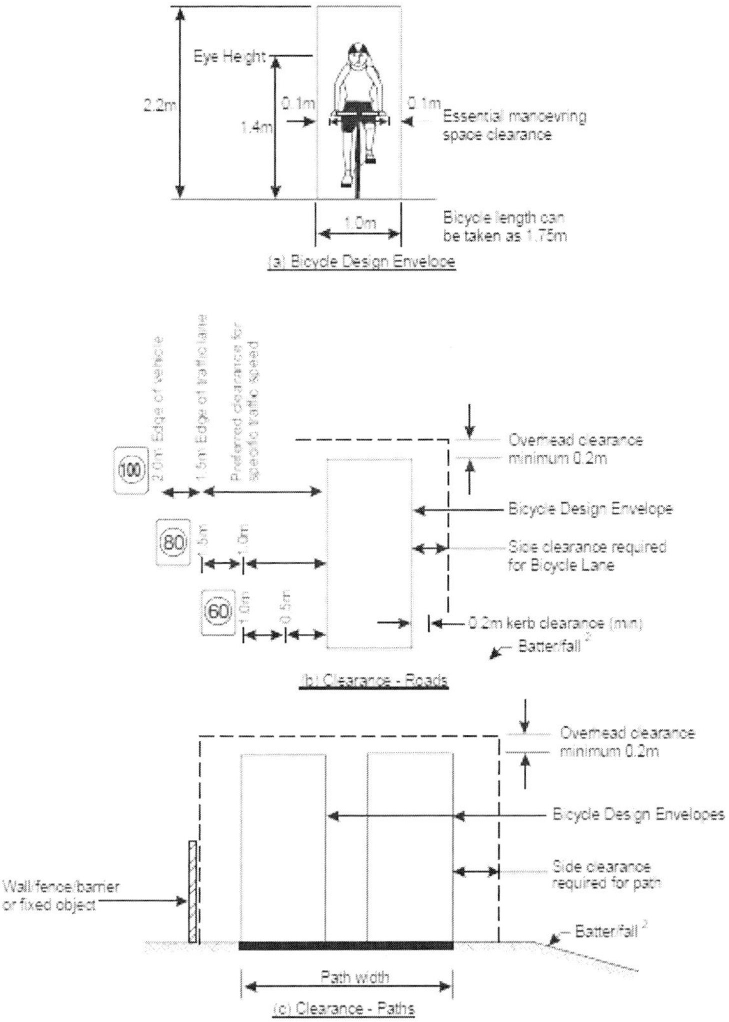

Figure 3.12 Bicycle operating space (Austroads, 1999a)

Gradients

To avoid the hazard of downhill riding and to accommodate ease of riding uphill, the grades should be as flat as possible. The applied desirable grade should be 3% with a maximum of 5%. The length must be limited if steeper grades are adopted where unavoidable, and by using flatter sections at regular intervals. At sharp horizontal curvature (i.e., curves < 200m), steep grades must not be applied.

Table 3.5 Minimum Horizontal Curve Radii (Austroads, 1999a)

Speed (km/h)	Superelevation (%)					
	0	2	3	4	5	6
20	10	10	9	9	9	9
30	25	24	23	22	21	21
40	50	47	45	43	41	41
50	94	86	82	76	73	73

Sight Distance

Cyclists require adequate sightlines to negotiate horizontal and vertical curves safely in the same way as car drivers. Parameters used to calculate stopping distance are:

- Perception/reaction time of 2.5 s.
- Eye height of 1.4 m.
- Object height of 0 m.
- Coefficient of longitudinal deceleration of 0.25.

Stopping distance is needed:

- In junction design.
- In setting out the alignment of paths.
- In positioning terminals and handrails.
- At entries to underpasses.
- For landscaping in the field.
- Ensure the safety of cyclists.

The chart in Figure 3.12 provides information on minimum stopping distances, the chart in Figure 3.13 provides the minimum length of crest curvature to meet the design requirements and the chart in Figure 3.14 sets out the lateral clearances on horizontal curves.

Partial Separation

- Partial separation may be achieved by:
  1. Sealed shoulders.
  2. Exclusive bicycle lanes, or
  3. Shared bicycle/car parking lanes.

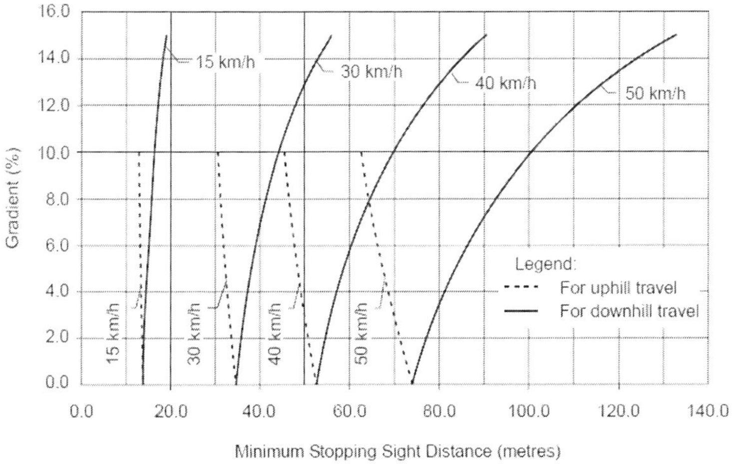

Figure 3.13 Minimum stopping sight distance (metres) (Austroads, 1999a)

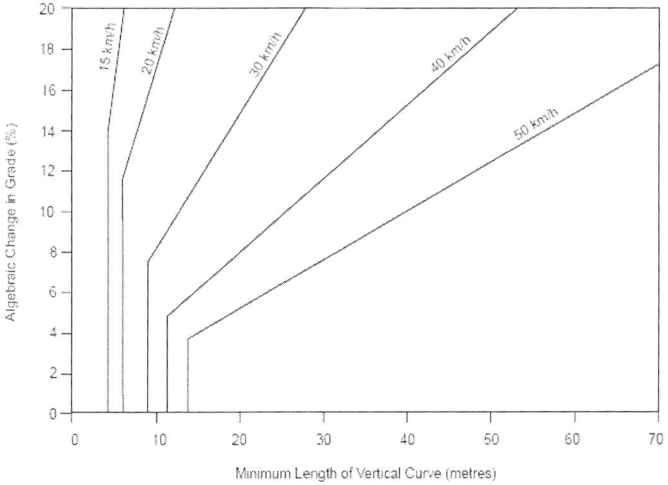

Figure 3.14 Minimum length of crest curvature (Austroads, 1999a)

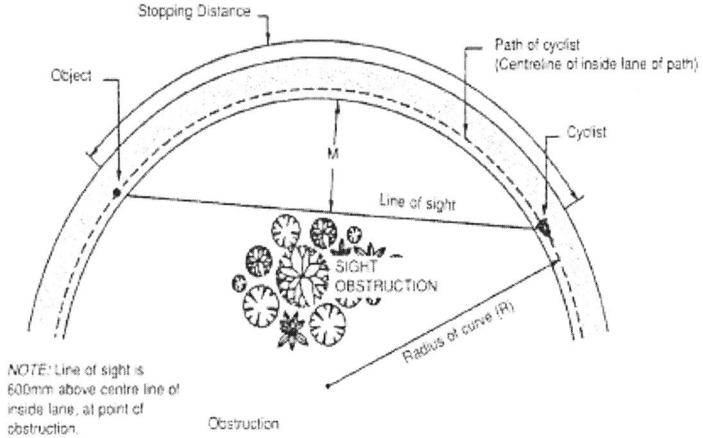

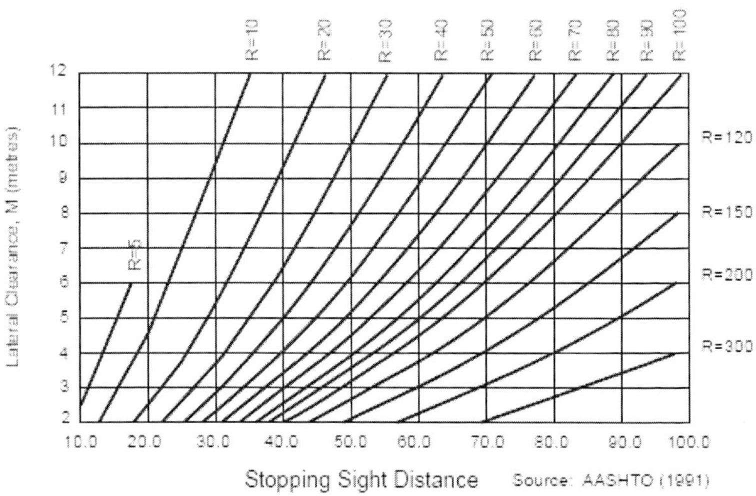

Figure 3.15 Lateral clearances required on horizontal curves (Austroads, 1999a)

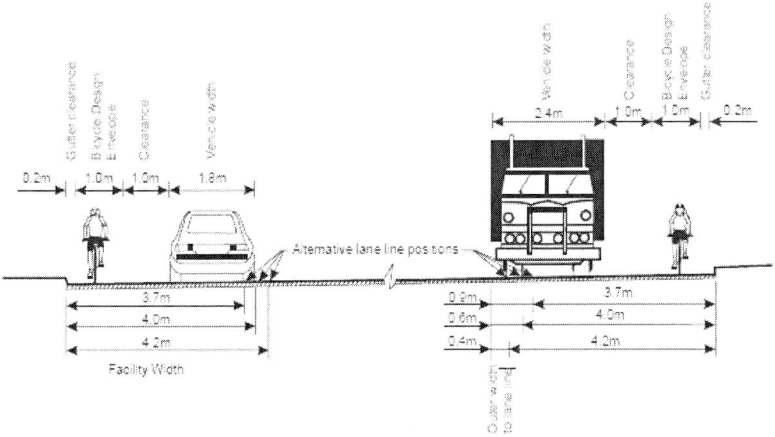

Figure 3.16 Vehicle positions on carriageways associated with wide kerb side lanes (Speed ≤ 60 km/h) (Austroads, 1999a)

Exclusive Bicycle Lanes

Where motor vehicle speeds exceed 80 km/h and/or bicycle traffic is concentrated, such as near schools or along major routes near cities, an exclusive bicycle lane is always preferred, and the following should be considered:

- To make the flow of bicycles in the same direction as the traffic, a bicycle lane should be provided on either side of the road.
- As there is no escape for cyclists, should a car door be opened suddenly, the bicycle lane should not be between the kerb and parked cars.
- Where there is little demand for parking throughout the day or where parking can be prohibited during certain designated hours, bicycle lanes may be provided to suit the peak travel demand of cyclists and motor vehicles, such as at rush hour and school drop-off/collection times.
- Raised markers or barriers are hazardous to cyclists; therefore, bicycle lanes are not to be delineated by them.

It is important that dust and debris are regularly swept as part of routine road maintenance, as these come from the adjacent lanes and tend to accumulate in bicycle lanes, as they are not swept away by motor traffic. The widths of bicycle lanes and sealed shoulders may be provided as per the design standard. The absolute maximum width may be 3 m, to avoid the lane being attractive to vehicular traffic. When adjacent motor traffic is moving at high speed (e.g., 100 km/h) and large vehicles form a significant proportion of the traffic stream, this is the desirable width to provide an adequate level of service to the cyclists, where demand for cycling is high. The width of 3.0m allows cyclists to overtake each other without encroaching into the adjacent traffic lane.

Shared Bicycle/Parking Lanes

Cycle and car parking lanes may be shared, and are appropriate in speed zones that are 60 km/h or less. They are generally required where there is a significant parking demand. These lanes are given legal status with pavement marking and signs. Unless it is impracticable, cyclists are required to use these lanes. Cyclists may use the general traffic lanes to make a right/left turn or to avoid hazards in the bike lane. Vehicles may only use these lanes for parking, property access or turning.

The desirable minimum width of a combined and shared bicycle/parking lane is 4.0 m and the desirable maximum width is 4.5 m. A greater width may tempt car drivers to use it as a vehicle lane. If the lane is a combined bus/bicycle lane, the minimum width should be 4.1 m. If the kerbside lane is a combined car/bicycle lane, the minimum width should be 3.7 m.

Other Design Issues

- Shared bus/bicycle lanes.
- Provisions for bicycles at junctions.
- Provisions for bicycles on roundabouts.
- Regular maintenance of the roads for bicycles, and
- Schemes for local area traffic management.

3.4.8 Motorcycles

Motorcycles, although a small proportion (1%), are still an important part of the traffic stream and require specific attention because of the differences in operating characteristics with other vehicles. The over-representation of motorcycle casualties in road crashes in terms of deaths and hospitalisations makes special considerations even more important for motorcycles.

The design process is complex, in which the demands of various road users are to be fulfilled and compromises made to strike a balance to obtain the best result for all the road users. Designers need to make judgements on the proper mix of treatments to achieve the best result. In this context, provisions for motorcycles must be undertaken, and this section is intended to guide good industry practice.

Characteristics

Motorcycles are not to be considered as smaller cars or fast cycles:

- Motorcycle handling characteristics are grossly different from cars.
- Having only two tyres, motorcycles are far more dependent on good, consistent traction on the road surface.
- Motorcycles are more manoeuvrable than cars and use different parts of the carriageway (for example, bends).
- Protection for riders is not provided by the motorcycles in the event of an accident.
- Motorcycles are driven by their own power source, resulting not only in greater speed but also have greater opportunities to negotiate varying traffic conditions.

The Road Surface

The road surface should have a uniform and predictable surface friction, and that is critically important to the motorcyclist for staying upright and in control of the vehicle.

In wet weather, the skid resistance of various materials, including markings on the surface, is changed. A slippery surface material can lead to inconsistent surface friction and unexpected loss of control. Motorcycles also require more distance to stop in wet weather. The existence of gravel or debris on the road surface is another major hazard. It results in loss of traction and leaves no time to act. Despite being a maintenance issue, design should also be carried out to ensure, as far as practicable, that debris will not accumulate on the road surface by treatments, such as avoiding overland low paths spreading onto/across the road. Grooving on the pavement surface, along the direction of travel, can cause difficulties for motorcyclists as some tyres can be trapped in it. When going around a bend, a motorcyclist may not track parallel to the lane lines, causing the tyres to become trapped in the grooves, resulting in loss of control. This practice of grooving should be avoided.

Unresolved issues are:

- Inadequate skid resistance of most surface marking materials.
- Slipperiness is caused without a friction additive in rubberised crack sealant.
- Inability to warn motorcyclists about adverse superelevation.

Road Surface Furniture

Consistent surface friction is interrupted by road surface furniture when struck by a motorcycle and includes:

- Temporary steel plates over trenches are a major hazard to motorcyclists.
- Curvature on approaches and evenness of surface at rail level crossings.
- Tactile or profiled edge lines when used with sealed shoulders.
- Tram tracks in unexpected locations—warnings are required.
- Services access covers in carriageways.
- Temporary kerbs in high-speed areas must be secured.
- Low kerbing to separate traffic streams; a motorcycle's foot peg may hit the kerb while cornering.
- Road humps are a hazard, particularly at night.
- Rumble strips on the approach to junctions cause uneven braking ability. They should be located away from the braking area.
- Lateral rumble strips on sealed shoulders can look like paint, give a considerable jolt, causing loss of control, and
- Iron bars used in the construction of roads may become dislodged and cause accidents.

Road Layout and Alignment

It must be recognised that motorcyclists may not be as conspicuous as other vehicles; therefore, visibility is a critical issue for all road users.

Adverse superelevation is particularly hazardous. The friction between the tyre and surface controls the stability of a motorcycle, and when the motorcycle is leaning over in the curve, this stability is minimised. A localised uneven road surface or low-friction area can be catastrophic.

On any alignment where curves of varying radius are used, the curvature changes in parts way around the curve, causing similar problems for a motorcyclist as described above. Compound curves, therefore, should be avoided. The alignment should not surprise the motorcyclists.

Figure 3.17 Introduction of a new traffic signal in India, giving a head start for two-wheelers

In India, the city of Chennai has introduced a new traffic signal giving motorcyclists a 30 to 40-second head start to improve the flow of traffic. This has also reduced the conflict between the motorcycles and the larger vehicles.

Junction layouts should be kept obvious and simple, with particular attention given to the following:

- Visibility between vehicles—motorcycles may be difficult to detect and may take a longer time to stop.
- It is best to keep the layout simple, as nighttime visibility is critical for motorcyclists, and the junction layout should be as per expectations.
- Right/left turn visibility—right/left-turning vehicles may find it difficult to see oncoming motorcycles.
- Parked vehicles may obstruct a motorcycle.
- All surface treatments must ensure adequate skid resistance.
- Islands need to be equipped with lights, because motorcycle headlights have limited effectiveness.
- To detect motorcycles, signal detector loops should be tuned, and
- Zebra crossing markings must have adequate skid resistance at slip lanes when motorcycles lean into a corner.

Additional Issues to Be Considered for Roundabouts Include:

- Visibility of the central island, particularly at night.
- Visibility of splitter islands.
- The camber must be consistent on the approach and around the roundabout. Adverse cambers are not ideal.
- The roundabout design should ensure early recognition.
- Lane marking materials should be skid-resistant, similar to that of the road surface.
- Sight lines should not be blocked by landscaping.

Delineation, Signing and Lighting

Although good delineation, signing and lighting are required for all road users, delineation may be more critical for motorcyclists for the following reasons:

- They cannot consult a map or street directory while riding.
- The headlights of the motorcycle have limited effectiveness.
- Visors of the helmets may be obscured as no wipers or cleaning are available while riding.

The following issues must be addressed:

- Clear delineation of vehicle paths.
- Road bends must have consistent and credible advisory speed signs.
- Adequate direction information.
- Effective street lighting aligned with the road layout provides lighting in areas of complex decision-making.
- Glare from street lighting—riders wearing closed-face helmets may have a problem.

Roadside Layout

Issues for motorcyclists include:

- An area clear of obstructions in the event of a rider falling from the vehicle.
- In the event of veering or being forced from the traffic lane, sealed shoulders are essential.
- Barrier kerbing, which is particularly hazardous to motorcyclists.

- Kerbs—these can hit/snag the foot pegs of a motorcycle, creating instability and may cause an accident.
- Kerb colouring—this should blend in with the road surface—a contrasting colour is beneficial.
- Landscaping located in areas where the risk of running off the road is high should use frangible planting.

Roadside Furniture

Safety devices are designed to minimise the consequences of crashes. However, sometimes these devices are major obstacles for motorcyclists. When a motorcyclist falls from the vehicle, the safety barriers pose a significant hazard to that rider. Any solid object in the path of the rider, if struck, will cause considerable damage to the motorcyclist. If possible, it is therefore better to keep the roadside clear of all such objects. Some related issues to address are:

- Poles and posts on the roadside—these are a hazard to all users, particularly motorcyclists. These must be located on the inside of bends sufficiently away from the edge of the lane to ensure that the rider will not impact them when leaning into the bend.
- Watering systems—these should not be allowed to spray water onto the road surface, as it alters the friction characteristics and introduces sudden variability in the friction value.
- Visibility should not be obstructed by signs at the riders' eye height, which can vary between 1.3 and 1.6 m.
- Guideposts at the road edge should be flexible—brittle posts can spear a motorcyclist, and rigid, sharp posts can cause loss of limbs if struck.
- The use of safety barriers should only be there when the likely damage or injury from hitting the barrier is less than that of hitting the hazard it shields.
- Funding is required to provide solutions that are expensive and therefore divert resources to more urgent situations.
- Reflectors on a barrier should be soft, not sharp and should not be located inside the W-shape of steel beams.

For motorcyclists, there are many unresolved safety issues. Adopting good practice in minimising the incidence of crashes and their consequences will go a long way. Careful design can remove unnecessary hazards and make the road environment safer for motorcyclists and other road users.

3.5 Road Safety Planning and Auditing

A road safety audit (RSA) can evaluate the accident potentiality and the performance of the road in terms of safety. A safety audit is a preventive measure. It is derived from the principle that 'prevention is better than cure'. A carefully planned RSA can provide invaluable benefits by identifying possible hazards and, in turn, assist in evolving appropriate corrective and improvement measures to minimise the hazards. In many developed and developing countries, RSAs are mandatory for all highway and road projects funded by international funding agencies. Urban roads are typically of lower priority for safety studies. In fact, at the metropolitan level, the authorities are not very worried about evaluating the functioning of urban roads in terms of safety.

In terms of the heterogeneous characteristics of traffic, urban roads are more prone to accidents, whereas on non-urban/rural highways, homogeneous traffic is observed. Moreover, on urban roads, mixed traffic including pedestrians, cyclists, carts, animal-driven vehicles, etc., with motorised traffic is observed. The road users' behaviour also differs on urban roads from that on non-urban highways.

Road development has two typical applications: one is 'green field alignment' for the construction of a new road or bypass, and the other is 'brown field alignment', which is the widening of existing roads involving pavement rehabilitation and upgrading from two-lane to four-lane to six-lane, etc. Road safety planning is done for green field roads, and an RSA is done for the existing roads. In road safety planning, various safety requirements are to be envisaged and incorporated. In RSAs, the shortcomings of existing roads are identified alongside the additional requirements with respect to road safety. In this section, the RSA is discussed in relation to the rehabilitation and upgrading of existing roads.

An RSA includes appraisal of the audit brief by the design team, appointing an audit team, qualification of the personnel, drafting an audit report, the final report, responses and meetings, responses by the design team, etc. The team should be independent. For ready reference by the team, checklists should also be prepared specifically targeting possible issues observed on existing roads. In many industrialised countries, the central and local road authorities have made the improvement of road safety a statutory duty. Under such legislation, each level of local authority that acts as a road authority is required to undertake road safety activities on its road network. This often includes the collection of accident data, accident analysis, identification of black spots and the design of remedial measures. In recent years, RSAs have been made mandatory in the list of responsibilities for road developments.

The RSA is a formal procedure. The audit proposes the safety requirements for the improvement and rehabilitation of existing roads and their inclusion in the maintenance of existing roads. An objective approach to the assessment of accident risk is required in the process of RSA. The safety audit is to be undertaken by a team that is independent of the design team. The prime objectives of this audit are broadly accident investigation and prevention.

3.5.1 Objectives

An RSA's objectives are as follows:

- To minimise the number and severity of accidents on the existing or upgraded road.
- To avoid the possibility of the scheme giving rise to accidents elsewhere in the road network.
- To identify and meet the needs and perceptions of all types of road users and to achieve a balance between safety and mobility of traffic.
- To reduce the future cost, considering that unsafe designs may be expensive or even impossible to correct in the future.
- To create awareness of safe design practices by all involved in the planning, design, construction and maintenance of roads.

3.5.2 Evaluation of the Effective Functioning of Various Road Elements

The RSA must visualise the effectiveness of various road elements and their evaluation. Any safety audit has to start the study with collection of traffic data, accident data and assessing of the road elements such as, horizontal alignments, vertical alignments, service roads, junctions lanes, geometric layout, pedestrian facilities, traffic signs, pavement markings, signals, unsignalised junctions, high embankments, bridge railing system, marker posts and delineators, median cuts, arrangement of street lighting, ends of guardrail sections, visibility obstruction by trees and other vegetation, hazardous positions, degree of safety of school children, bus stop locations, FOBs, pavements, covered drains, vehicular and pedestrian underpasses, railway crossings, side drains, median drains, utility corridors, planting in medians and avenues.

Table 3.6 Tasks at various stages of the RSA

Preliminary /Feasibility Stage	**Draft design stage**	**Detail design stage**	**Construction Stage**	**Pre-opening Stage**	**Operation and maintenance stage**
Upgrading scheme Impact Alignment Profile Service roads Underpass Median cuts Junction Layouts Land acquisitions	Alignment Profile Service roads Underpass Bridges Culverts Flyovers Median cuts Junction Layouts Land acquisitions Bus bays Truck lay byes	Geometrics Signals Signs Markings Lighting Junction Service roads Underpass Bridges Culverts Flyovers Median cuts Junction Layouts Bus bays Truck lay byes	Signs Markings Crash Barriers Pavement Lighting	Driving study in the day and night	Management Maintenance Infrastructure Overloaded and Unfit vehicles Undisciplined driving Surveillance Visibility and Weather

3.5.3 Procedures and Practices

Schemes that should be audited cover a wide range of types and sizes, on different classes of roads, in urban and rural areas. These include:

- Major new road up-grade schemes.
- Major new road rehabilitation schemes.

- Minor improvements.
- Traffic management schemes.
- Development schemes.
- Maintenance works.
- Evaluation of pedestrians, parking, access, tolls and facilities.

Although addressing all the schemes may be desirable in an RSA, considering limitations in financial and personnel resources, it may not be possible. Therefore, it is important to prioritise schemes and have well-defined tasks, which will result in the appropriate level of audits required and their acceptance by the authorities.

3.5.4 Benefits of a Road Safety Audit

The benefits of RSAs extend from reduced accidents to improved economics if properly incorporated in policy making and design. Some of these benefits include:

- Safer functioning of the roadway system.
- Reduced future cost in the improvement and maintenance of road schemes.
- Reduced need to modify existing schemes.
- Safety improvements, standards and specifications.
- Effective consideration of the safety needs of vulnerable road users.
- Insurance cost savings, lower health care and assets lost costs due to reduced collisions.
- Design improvements.

Acknowledgements

Road Safety Guidelines was largely written and developed by Ross Silcock Ltd. in association with the Transport Research Laboratory (TRL) Overseas Centre, the United Kingdom (UK). It was produced under the Asian Development Bank (ADB) regional technical assistance (RETA 5620: Regional Initiatives in Road Safety), which has also resulted in the production of a comprehensive set of guidelines for decision makers on road safety policy.

CHAPTER 4

HIGHWAY SAFETY AT JUNCTIONS AND ROUNDABOUTS

Abstract: This chapter covers various topics including a general overview, junction safety on high speed rural main roads, crash data methodology and findings, conventional and innovative safety improvements, highway safety at junctions on superelevated bends, junction safety analysis, results of past research, process of research studies, econometric modelling, crash frequency analysis, severity analysis, selecting junctions for studies, crash frequency model, crash severity model, vehicle paths, speed-bend relationships, test vehicles, geometric elements, highway safety at roundabouts and conclusions and recommendations.

4.1 General

The safety of traffic and pedestrians at a junction depends on various factors, such as:

- Junction type—a two-lane undivided major road with a two-lane undivided minor road, a four-lane divided major road with a two-lane undivided minor road and a four-lane divided major road with a four-lane divided minor road.
- Whether the junction has signals or not.
- Whether the major road is on a straight tangent or a superelevated bend.
- There are visible speed limits.
- Visibility available on major and minor roads at night.
- Effect of weather.
- The surface condition.

Sometimes, at a junction, the major road is located on a superelevated bend; in such cases, the designers have to treat such road crossings with extra care. The AASHTO Policy on Geometric Design of Highways and Streets

(AASHTO, 2001) advised that at junctions, the alignment should be as straight as possible, and the gradient should be as flat as practical. Such a statement allows the design of junctions on bends, but with maximum possible measures, if other solutions are economically prohibitive. The policy, however, warns that for the turning vehicles, such bends may introduce the disadvantage of adverse superelevation. This needs proper data collection and very effective studies where the bends on major roads have high superelevation rates, and the minor road approaching the junction has adverse grades with a sight distance restriction due to the grade line. This is to ensure that the combination of vertical and horizontal curvature shall allow adequate sight lines at a junction. National policy does not forbid locating junctions on bends on the major road if alternatives are expensive, but suggests avoiding them wherever practical.

So far, several such junctions with bends on a major road have been built. Indications about the safety concerns have been raised for many of them. Following a series of recurring fatal events, the DOTs of the US and the road authorities in other countries have decided to remove exits from some junctions.

Due to bends with a steep superelevation, designing junctions on such segments is currently avoided. Expensive alternatives are called for because of design restrictions, such as realigning the roads or adding grade separations (bridges). For some locations, bridges are currently deemed appropriate for construction to allow minor roads to cross over main ones. These are the design issues associated with locating a junction on bends. Specifically, the main points of concern are:

- To determine whether superelevated junctions are truly more dangerous compared to similar junctions located on tangents or straight roads.
- If the junctions on bends on major roads are more dangerous, then assess what geometric characteristic or combination of characteristics makes them more dangerous.
- Identify cost-effective safety improvements for existing junctions on superelevated bends.
- Propose effective design recommendations for cases where a proposed junction is being considered for design on a superelevated bend.

Typical four-lane, median-divided rural highways at junctions have 42 possible points of conflict between vehicles, leading to more crashes than alternative median designs like J-turns, as shown in Figure 4.1.

Median-separated highways provide distinct advantages over undivided roadways by separating traffic, providing a recovery or stopping area for vehicles, and providing space for vehicles turning left, see Fig. 42 (a) and vehicles turning right, see Fig. 42 (b). In many cases, they also provide the same safety and travel time benefits as rural roads but at a lower cost. However, these safety benefits can be diminished by an increase in the frequency and severity of junction crashes, especially right-angle crashes that occur while a vehicle from a minor road is making a turn through the median and onto the highway.

As interchanges and traffic lights are not always the most effective or cost-efficient solutions to these problems, transportation agencies need innovative, low-cost designs that can be used to improve their safety. To help establish design guidance and safety data for these treatments, the National Cooperative Highway Research Program (NCHRP) undertook a project.

Many of the junction designs examined can significantly improve safety at a lower cost than constructing an interchange. In doing so, an expanded set of options is available for dealing with problem junctions. In the past, options for improving safety at high-speed rural junctions were limited. The options include a new interchange, which involves major new spending, or installing a traffic signal, but the improve safety is not guaranteed and may even worsen it.

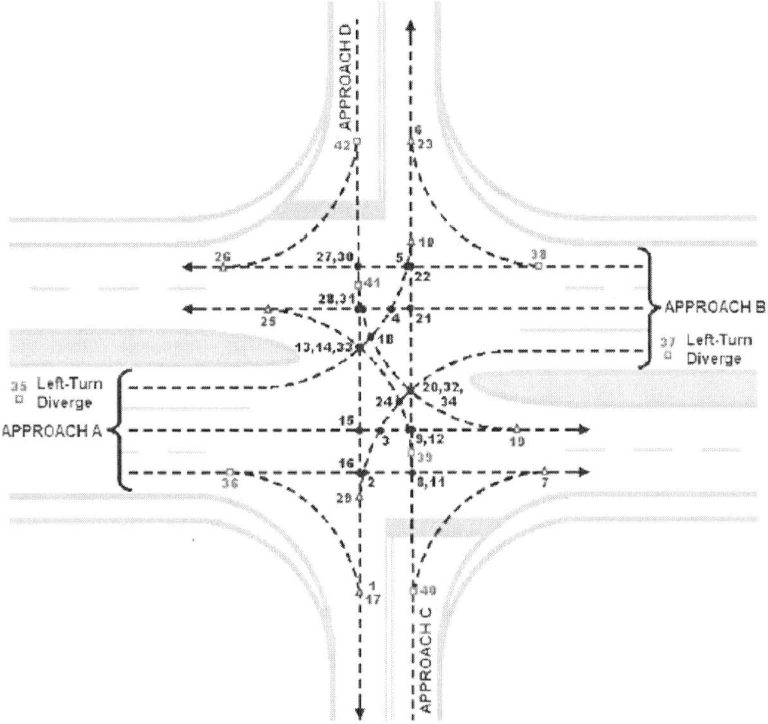

(a) 42 Conflicting movements at a right-hand drive junction (USA, Europe, etc.)

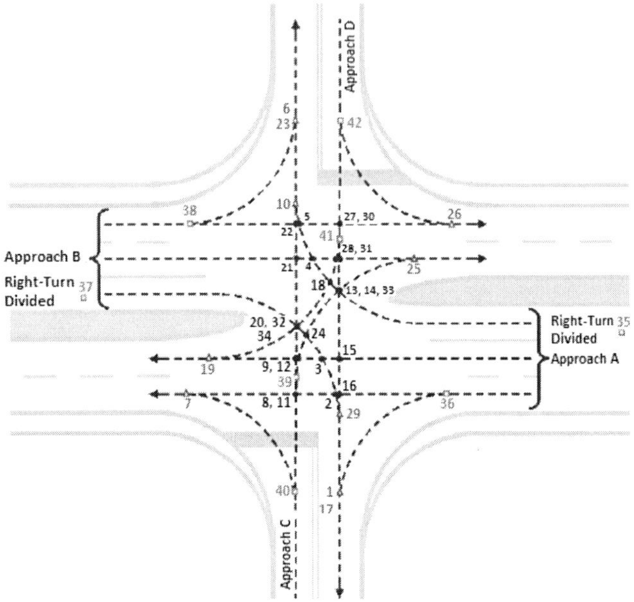

(b) 42 Conflicting movements at a left-hand drive junction (UK, India, Malaysia, etc.)
Figures 4.1 (a) and (b) 42 Conflicting movements on a four-lane median divided ruralhighway junction

The need is to make an informed decision about how to handle a problematic junction and how to design it to achieve a 100 per cent reduction in more severe right/left-angle crashes.

4.2 Junction safety on high-speed rural roads

In this section, we shall be considering main roads that are multi-lane and median-divided, where most junctions are at grade. The typical junction on a rural main road does not have signals but is a two-way stop-controlled (TWSC) junction, where a two-lane roadway (the minor roadway) meets a four-lane main road (the major roadway).

The severity of crashes at most TWSC rural junctions is more concerned with traffic volume on the minor road than with traffic volume on the main road. As traffic volume on the minor road increases, that results in an increase in both crash rates and crash severity.

TWSC junctions on rural commuter routes are those where the main road has moderate volumes of traffic of 10,000 to 12,000 vehicles per day and where the minor roadway has high volumes of traffic of 2000+ vehicles per day. The volumes are highly peaked in situations where a rural commuter route leads to a central business district, and the main road is at or near vertical or horizontal bends. For such problematic at-grade highway junctions, in addition to conventional safety improvements, several innovative safety improvements are also available.

Converting two-lane roads to four-lane or six-lane main roads has become a popular improvement in many states and countries. In a country's highway system, main roads are now the fastest-growing component. Main roads are less expensive to construct, with fewer interchanges than main roads, yet they support similar traffic speeds and capacities and, where there are fewer than five access points per mile, have similar safety records. Still, the specific safety performance of main roads is relatively little known. Therefore, it is essential to provide an analysis of crash characteristics for rural main roads, TWSC junctions, with a synthesis of safety strategies at these junctions. Researchers have focused on determining the relative impact on crash rates of traffic volumes on both the major and minor roadways, and junction geometry with respect to median width, presence of turning lanes, etc. Various researchers surveyed strategies currently being applied by state transportation agencies to improve safety at TWSC main road junctions.

4.2.1 Crash Data Methodology and Findings

The project team for planning and design of junctions is to examine five years of crash data for selected TWSC main road junctions. Many of the junctions may have very low volume, and at some of them, the minor roads may be unpaved with an aggregate surface.

First, crash rates, crash fatality rates, and crash severity rates relative to traffic volumes on minor roadways are to be examined at each junction.

- The crash rate is the total number of crashes divided by the millions of entering vehicles (MEV) over five years.
- The fatal crash rate is the total number of fatal crashes divided by a hundred million entering vehicles (HMEV) over five years.
- The crash severity index assigns a weight of five to fatal crashes, four to major injury crashes, threeto minor injury crashes, two to

possible injury/unknown crashes, and one to property damage only (PDO).

Having the rates for crash and crash fatality plus the crash severity index, the rates are averaged over five years, stratified by increasing traffic volumes on the minor roadways, increasing all three rates. This indicates that as traffic increases on the minor roads, the crashes are occurring more frequently and becoming more severe. Next, relative to both major and minor roadway traffic volumes, the crash types are examined, whether head-on, right/left angle or rear-end, etc. The right-angle crashes are the most severe, and these happen when drivers on the minor roadway fail to leave an adequate gap between vehicles when crossing the main road.

Next, it is to be presented that the percentage of right/left-angle crashes increases as minor roadway traffic volume increases and that traffic volume on the main road itself does not seem to have any impact on crash type. This trend should be examined another way, by selecting a number, say, 20 of the TWSC main road junctions, and observing that ten are with the best safety performance and ten with the poorest safety performance. Then, by comparing each junction's expected crash-severity rate, based on a statistical model, with the actual, observed crash-severity rate, the relative safety performance may be determined. The junctions whose actual crash-severity rates were lower than expected are considered the 'best' junctions, which means they performed better than predicted, and the junctions whose actual performance was worse than predicted are considered the 'poorest' junctions in terms of safety.

Table 4.1 shows a sample of average traffic volumes for minor and major roadways at the ten highest-severity junctions and ten lowest-severity junctions.

Table 4.1 Sample of average daily traffic (ADT) on approaches of high- and low-severity TWSC main road junctions

	Average Traffic Volume (ADT) on the main road	**Average Traffic Volume(ADT) on minor roadways**
Low-crash-severity junctions	20,350	428
High-crash-severity junctions	11,480	2,400
Average for all junctions that experienced at least one crash	10,860	2,400

As expected, the average traffic volumes on the minor roadways were 'low' and 'well below' the overall average, at junctions with low crash-severity ratings, and the average traffic volumes on the minor roadways were well above the overall average, at junctions with high crash-severity ratings. Sometimes it may be surprising to observe that the low crash-severity junctions are on rural main roads with the highest volume of traffic. The difference in crash type distribution may be presented between high crash-severity and low crash-severity junctions. The high crash-severity junctions are expected to have very high occurrences of right/left-turn crashes, while the low crash-severity junctions have low occurrences of right/left-turn crashes.

Intersections having other potential problems include the following:

- Where minor-roadway drivers' ability to judge gaps in traffic is obstructed and hindered by horizontal or vertical bends on the main road.
- Where commercial developments, for example, petrol stations, shops, fast food, etc., where additional turning movements and higher volumes create more opportunities for crashes.

4.2.2 Survey Methodology and Findings

Email queries may be sent to the traffic departments of various states in the country, focused on the following:

- States' experiences with main roads (lane miles, criteria for upgrading a junction to an interchange, etc.).
- Safety performance of the main roads in various states.
- Innovative geometry layout of at-grade junctions on main roads, and
- General respondent comments.

Responses to such surveys normally indicate that safety at rural TWSC main road junctions is a growing concern for state transportation departments. Remedial actions to improve safety at these junctions have already been applied by various states. The effectiveness of these countermeasures is supported by subjective evidence in several states,but very few states had objective evidence only. Several conventional and innovative improvements for main road junctions are being implemented in various countries. A few of these improvements are summarised and illustrated below.

4.2.2.1 Conventional safety improvements

Improving signage on minor roads and in the median, adding a double yellow centre line in the median and stop/give-way bars, adding advance rumble strips for minor roadway traffic in advance of the stop, and then, depending on the crash type experience, adding left and right turn lanes on the main road are the various conventional strategies for safety improvements at these junctions. Installing traffic signals is considered if a junction continues to have crash problems. Signalisation, whether it improves the safety of high-speed main road junctions, may be verified by statistical analysis. However, it can be understood that signalisation can convert a more severe right/left-angle crash problem into a less severe rear-end and red-light-running crash problem. If a right-of-way is available, the ultimate conventional solution is, of course, converting an at-grade junction to an interchange, but that could be prohibitively expensive.

4.2.2.2 Innovative safety improvements

Two-movement cues

It is possible to misjudge a gap in the far-lane traffic while crossing a median-divided main road junction. Various low-cost strategies focus on providing cues and guidance to encourage drivers on the minor roads to move through the junction in two stages. At the first stage, the lanes on the near carriageway are crossed, stopped or slowed down in the median crossover. By the second stage, turning left/right or crossing the set of lanes on the other side carriageway is achieved.

Wide medians

Several studies have shown that, with the increase in the median width, crash frequency reduces. The drivers are encouraged by wider medians to treat the lanes on two carriageways of the divided main road as two separate junctions; this results in a two-movement crossing or left/right turn. For long vehicles, such as lorries and school buses, an ample refuge is also provided by a wide median. However, it is to be remembered that, in the median crossover lanes, very wide medians can result in more crashes.

Left/right-turn acceleration lanes

Unfortunately, for large vehicles, not all medians on main roads are wide enough. In situations where widening of the median is not feasible, a left/right-turn acceleration lane in the median may be created. This is an effective and innovative alternative and is shown in Figure 4.2.

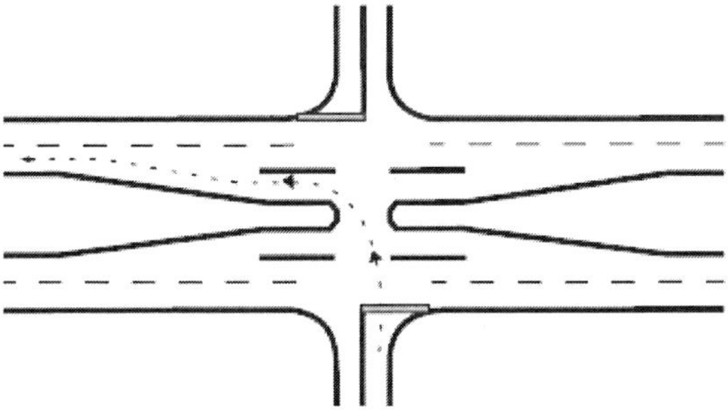

Figure 4.2 Median left-turn acceleration lane (Graphic from Harwood, D.W., "Innovative Junction Improvements," Midwest Research Institute, Kansas City, MO)

Wherever this strategy has been applied, it has been very popular with the drivers and has reduced delays and conflicts considerably. Compared to a direct left/right turn into a high-speed lane, a left/right turn using an acceleration/merge lane is much safer. The acceleration lane provides a refuge along the median for large, left/right-turning vehicles. The vehicles in the acceleration lane attain the required speed before they merge onto the main road traffic and which reduces the likelihood and severity of rear-end crashes.

The conflict between a left/right-turning vehicle from the minor road that is stopped in the median and another vehicle turning left/right from the main road is also reduced by the acceleration lane created in the median. When both vehicles are large lorries, this conflict can be particularly problematic. Ideally, to provide heavy lorries with enough distance to accelerate and merge with main road traffic, an acceleration lane should be 450 m long. This requires some driver education because most drivers are not familiar with left/right turn acceleration lanes. To encourage proper use of the acceleration lane, enhanced markings are also needed.

Indirect left/right turns

When driving on the right, left turns are a factor in a large percentage of crashes. The authorities in some states in the US are implementing strategies to replace left turns, and sometimes cross-traffic movements, with indirect movements. To eliminate left turns from the main road onto a minor road, the jug-handle style of movement (as shown in Figure 4.3), and loops are sometimes used.

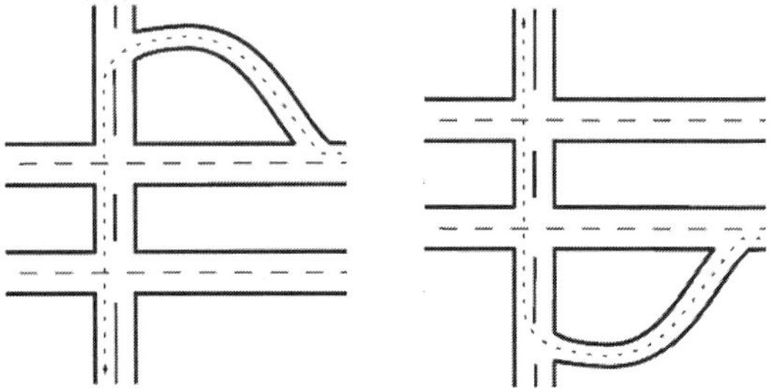

(a) (b)

Figure 4.3 Jug-handle layout of traffic movements (a) for right-hand drive junctions (USA, Europe, etc.) and (b) for left-hand drive junctions (UK, India, etc.)
(Graphic from Harwood, D.W., "Innovative Junction Improvements," Midwest Research Institute, Kansas City, MO)

The left turns from the minor road may be eliminated by closing the median crossover. For the USsystem of traffic, left turns are allowed from the main road, but left turns from the minorroad are blocked by a raised median, as shown in Figure 4.4. The drivers from the minor road who want to turn left are forced to turn right and make a U-turn through the median, moving for some length after passing the minor road joining the main road.

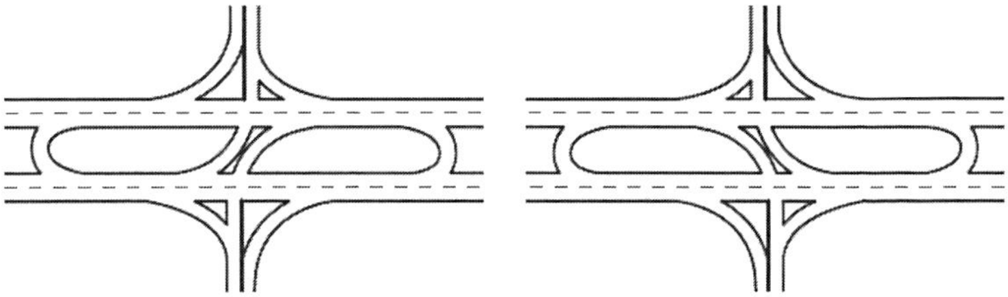

Figure 4.4 Directional median opening (a) for right-hand drive junctions (USA, Europe, etc.) and (b) for left-hand drive junctions (UK, India, etc.)

4.2.2.3 Implementation Benefits

Currently, innovative improvements of at-grade, TWSC main road junctions, like those discussed in this section, are not widely implemented. It is therefore difficult to predict the potential safety benefits of implementing such treatments more widely. It is already observed that actual crash-severity rates at grade-separated interchanges are about one-third compared to crash-severity rates at at-grade TWSC junctions. It is, therefore, not unreasonable to estimate that implementing innovative treatments at the ten worst-performing at-grade TWSC junctions could reduce the crash-severity rate by half.

Using data from the Iowa DOT, for an average of five years of crash experience, estimates for fatal, severe-injury, minor injury, and crashes resulting in property damage, the economic loss from ten junctions is currently about $1,633,500 per year. If the distribution of crash severity stays the same, reducing the crash severity rate by half would save a little over $800,000 per year in crash losses, which is $80,000 per junction per year.

To equip a junction with a directional median and a median U-turn cross-over (similar to Figure 4.4), a conservative (high) cost estimate is about $250,000; the estimated life for such an improvement would be 15 years.

In general, adequate information is required to identify locations where innovative main road junction improvements, like those described herein, may be beneficial. The details on implementing such improvements, as well as information about states that have used these strategies, are essential.

4.2.3 Highway Safety at Junctions on Superelevated Bends

In this section, methodologies and results of a few studies related to junction safety are reviewed. This is to determine whether certain junction geometric characteristics, particularly superelevation, adversely affect safety in terms of both accident frequency and severity.

Highway safety is to be examined for TWSC where the main road is located on a superelevated bend. The focus is on high-speed divided highways; however, two-lane roads were also examined to assess the potential safety factors. The studies were to determine and explain the underlying causes of the crashes and provide general countermeasures. The desired output should be a set of design rules that address safety at new and existing superelevated junctions.

Reviews are made of the methodologies and results from research on highway safety and the methodology followed in those studies. The next section provides an analysis of the bend effect using a sample of junctions where both routes are two-lane roads. Next, a comprehensive analysis is discussed for junctions along four-lane divided highways. Finally, the details are given on the most frequent crash type, occurring between vehicles on the outside of the major road and vehicles attempting to cross through the median, along with a summary of the research findings and providing design recommendations for cases where a junction is being considered for design or improvement on a superelevated bend.

4.2.4 Methods of Junction Safety Analysis

Several alternative methods of safety analysis have been pointed out in recent research studies. The various methods considered for this research include safety audits, collision diagrams, direct traffic observation, and statistical analysis.

For investigating hazard problems with possible application to existing roads, road safety audits, and safety reviews, have evolved as the emerging methods (Pietrucha et. al., 2000). Safety audits are applied to existing roads, and a team of safety experts carries out a comprehensive field review of each location on the highway. However, in the present case, the safety audit method is not useful, because the research studies explore unknown problems, while safety audits refer to expert knowledge about known issues and identify the missing measures.

Evaluating the compliance of roadway design to the design standards has been the focus of the traditional safety analysis based on collision diagrams. Human factors, such as visibility issues, are typically not considered in these safety reviews. Hence, the traditional approach is not an appropriate method for the research.

Direct observations of traffic operations may provide clues about the potential causes of crashes. Insight can be gained about human behaviour to complement the geometric design characteristics by observing the driver's behaviour. Detecting the link between crash occurrences cannot be ascertained by this approach, which is resource-demanding.

Based upon the complexity, diversity, and stochastic nature of transportation problems, the analysis of statistical modelling is a promising choice and an appropriate method of analysis.

4.2.4.1 Statistical Modelling

To identify the geometric characteristics that make a junction safe in terms of both accident frequency and severity, statistical modelling techniques have been used. For isolating such characteristics, several forms of statistical models can be used.

Frequency, or count data models, are discussed first. For determining safety factors that affect the frequency of accidents at a given location, the count data models are appropriate.

Discrete outcome models are the second type used to determine safety factors that increase the probability of an accident being of a particular severity, given the fact that the accident has occurred.

4.2.4.2 Count Data Models

Over the span of decades, many types of accident frequency models have been developed. Conventional linear regression was used in earlier models. Later, for modelling accident frequencies, these types of models were found to be inappropriate, and because of the random nature of crashes, it was concluded that Poisson and negative binomial regression models provided a more reasonable approximation of crash counts. Many researchers have developed models of these types in recent years.

The Poisson or negative binomial may be selected based on the presence of overdispersion in the data. When the variance of the predicted variable is greater than the mean, this results in 'overdispersion'. In transportation safety analysis, such a case occurs quite often. The negative binomial distribution is appropriate in case overdispersion exists. The negative binomial distribution is changed to a Poisson in the case of no overdispersion.

The negative binomial is preferred to the Poisson for modelling purposes, because incorrectly specified parameters in the model may be caused by the exclusion of the overdispersion parameter. Otherwise, the variability explained by overdispersion may be incorrectly incorporated into other variables. The number of zeroes in the sample is another factor, in addition to overdispersion, that affects the selection of an appropriate model. A procedure for determining the appropriate model specification for crash data was explained by Shankar et al. (1997). It was arguedthat the traditional Poisson and negative binomial models do not address the possibility of a zero-inflated counting process. It was shown that in many cases, a zero-inflated version of the modelis more appropriate by distinguishing the truly safe road section from a zero-accident state. The probability processes that are zero-inflated allow one to isolate independent variables that determine the relative accident likelihoods of safe versus unsafe roads.

A test was developed to determine whether the zero-inflated state was justified (Miaou, 1989). If the mean and variance of the accident frequencies are approximately the same, the Poisson model is recommended to be used. The negative binomial and zero-inflated Poisson (ZIP) models were found to be more appropriate if overdispersion is present, where the variance is significantly greater than the mean. Due to under-reporting accidents, there may be other reasons for excess zeroes. For rural junctions, underreporting may be a particular problem. This reason provides clear justification for selecting a zero-inflated model over the traditional negative binomial model.

4.2.4.3 Discrete Outcome Models

From the various categories, such as PDO, injury, or fatality, the severity of an accident can be classified as a discrete outcome.

The multinomial logit (MNL) formulation is an appropriate method of modelling such data (Washington et al., 2003). Nested logit models are used in the evaluations of recent applications. The nested logit model accounts for shared characteristics of severity levels; otherwise, it would result in a model that is incorrect.

4.2.5 Results of Research

This section looks at the research on highway safety, showing relationships between traffic volume, along with crashes, geometric characteristics and weather.

4.2.5.1 Traffic Volumes

Traffic is obviously the primary contributing factor relating highway variables to crashes. Undisciplined driving, whether intentional or unintentional, is the root cause of many crashes. Unintentional undisciplined driving is caused by poor road conditions, absence of road signs, non-functioning traffic signals and insufficient visibility due to adverse weather conditions. Over the decades, various predictive models have been developed relating crashes to traffic volume.

Studies have been done for crashes at three-point junctions on two-lane roads (Pickering, Hall, and Grimmer, 1986). The mean number of crashes per unit time was predicted by the Poisson model in the form below:

$$N=K*(ADT_1*ADT_2)^{0.5},$$

where:
N = number of crashes per unit time
K = constant

ADT_1 = Average Daily Traffic (ADT) on a major road
ADT_2 = Average Daily Traffic (ADT) on a minor road.

The product of the traffic volumes on each road in the preceding model was found to be most appropriate for modelling purposes. Studies were also carried out on four-point junctions on non-urban freeways (Bonneson and McCoy, 1993). This also found that the basic cause of accident frequency is the ADT values, which are the only significant variables. Separate variables were created in this case, for the ADT on each road and are shown below:

$$N = K*(ADT_1)^{0.256}(ADT_2)^{0.831},$$

where:
N = mean number of crashes per unit time
K = constant

In many other studies, traffic was found to be the only significant factor in the analyses. Traffic is the major variable present in most of the models under research studies, and it is also the only factor included in the models presented here. However, this factor is outside the direct control of transportation authorities and agencies. One must have something which can be the basis of their decisions related to safety. A variety of geometric design elements have a wide range of effects on the number of crashes at an junction, and this was found in various studies. Transportation engineers are to apply the design elements correctly and avoid potential safety problems.

4.2.5.2 Geometric Characteristics

Horizontal alignment and superelevation are the design elements of primary concern in this research. Accident black spots, which are mostly seen on horizontal bends, showed an increase in the crash rate by 1.5 to 4 times that of a similar tangent section (Zeeger et al., 1992).

The relationship between curvature and safety was further explained by McGee et al. (1995) and Vogt and Bared (1998). In their studies on accidents on rural roads, it was noted that increasing curvature has a negative impact on safety (Shankar et al., 1995). High superelevation rates, as are common with sharp horizontal bends, also lead to increases in accidents (Zegeer et al., 1992), and it was concluded that a reduction in accidents of up to 11% would be achieved by improving the superelevation of bends by following the AASHTO guidelines. For any given deflection angle, the design with the larger bend radius is always safer than a similar junction with a smaller radius bend (Hauer 1997). The change in accident rate was also observed to be proportional to the change in value of the radius of the bends.

The next important aspect is the presence of vertical bends (i.e. those that are present on hill or mountain roads) that also appear to increase crash frequency (Shankar et al., 1995; Vogt and Bared, 1998). The studies on a model showed that the presence of auxiliary turning lanes is likely to decrease the number of accidents (David and Norman, 1975). Various other studies have also shown similar results, particularly for the presence of left/right-turn lanes (Bauer and Harwood, 1996). It was also found that limited sight distance at both signalised and unsignalised junctions results in an increase in crashes (Hanna et al., 1976), and shorter sight distances result in higher crash rates (Bared and Lum, 1992).

It was further determined that crashes increase at a junction with an angle of 90 degrees (McCoy, Tripi, and Bonneson, 1994). Right-angle junctions happened to be more dangerous than those that are only slightly skewed (Bared and Lum, 1992). This was also verified for rural junctions. Further, it was observed that wider lanes and shoulders result in fewer multiple-vehicle crashes, and wider medians also result in fewer crashes for rural unsignalised junctions. (Bauer and Harwood, 1996 / 1998).

Signals typically benefit junctions with higher volumes, but for a low-volume junction, signals may increase the number of accidents. It was also observed that signals do not reduce the overall number of crashes, but instead cause more rear-end and fewer right-angle crashes (King and Goldblatt, 1975).

Studies revealed that drivers had greater difficulty detecting bend characteristics at night, particularly in sections with no lighting (Hagiwara et. al. 1999). The absence of lighting increased crash frequency in a study of rural junctions (Bauer and Harwood, 1996). The roadside hazard rating was developed by Zeeger et al. (1987). It was found that lorry crashes are more prevalent in rural environments and during the night. This may be due to factors such as tiredness, design, and improper lighting (Blower, Campbell, and Green, 1993).

4.2.5.3 Weather

Studies have found that regional weather is insignificant in crash prediction (Vogt and Bared, 1998). However, extreme weather conditions combined with extreme horizontal or vertical alignment are a factor (Shankar, Mannering, and Barfield, 1995).

It was found that crash rates were the highest due to fog during the winter months of December to March, although the crashes were found to be the least severe. This is possibly due to greater care being taken by drivers due to the inclement weather conditions (Brown and Baass, 1997).

4.2.6 Research Methodology

Based on an extensive study of research, a research methodology has been developed. Focusing on the topics such as selecting a junction, data collection and statistical analysis, the methodology is described in this section.

4.2.6.1 Selecting a Junction

The first step is to develop a method for selecting junctions for analysis. Junctions with two lanes on all approaches and another with four lanes on major approaches are the two sets of junctions that were required. The junctions of state routes were selected for the two-lane analysis. This can greatly reduce the data collection, as a large amount of information should be readily available directly from the authorities. The junctions of local and county roads involve more rigorous data collection and are to be examined for the four-lane analysis. As the sample size of state-state junctions that fits this criterion was very small, it may be necessary to include these roads.

4.2.6.2 Data Collection

Data may be obtained for the project in either of the two ways. (i) From a road authority or agency, all accident data and traffic volumes may be obtained. (ii) By visiting the site, the remaining data may be directly collected. Measuring the geometric characteristics of the roadway and counting the traffic for local and county roads may be done. The amount of required data varies in each of the two ways mentioned above.

4.2.6.3 Statistical Analysis

To determine those characteristics which have an impact on junction safety, the above-mentioned data collection is followed by a statistical analysis. To determine the effects of variables for which complete information is available, such as geometry, etc., a statistical analysis is to be carried out, involving the development of econometric models. Various effects were quantified by comparing the proportion of crashes between two samples: one sample of junctions on bends and another with junctions on tangents. Complete information may not be available for the variables, such as weather.

4.2.7 Econometric Modelling

The two models and the focus points to determine what geometric characteristics tend to make junctions more or less safe are: (i) a frequency model to predict the number of accidents at a given junction per unit time, and (ii) a severity model to predict the damage caused in a particular accident. For modelling, estimating and analysing econometric models, the modelling program LIMDEP 7.0 is commonly used along with STS, SAS, and LIMDEP. LIMDEP 7.0 is well-suited for this type of project because it is primarily orientated towards cross-section and panel data.

LIMDEP 7.0 provides maximum likelihood estimates and standard error values for each coefficient. The p-values are provided, which test the null hypothesis that the true value of a regression coefficient is zero. The estimated coefficient value divided by the estimated standard error is the z-score of an estimated coefficient. A normal random variable has an absolute value larger than the z-score obtained, and the p-value is the probability of that.

There is good evidence to suggest that the corresponding variable is statistically significant, in case the p-value is small. To determine statistical significance, a threshold P-value of 0.10 was used for the research. In the final models, all parameters with p-values below 0.10 are included. However, the parameter estimate may have some practical significance, even if the p-value is above the threshold. For instance, the estimated coefficient may indicate a significant impact and could become more significant if the sample size were increased, or if a variable has a P-value of 0.25. As such, in the final models, the parameters demonstrating a high level of practical significance were included where that is appropriate.

By using LIMDEP and a stepwise modelling approach, models were developed for both the two-lane and four-lane cases, with each case for the junctions on the bend and at a tangent. Including all variables, an initial 'full' model is developed for each case. The initial problems, such as multicollinearity, are addressed, and the affected variables are removed, as appropriate. The effects of the variables on junction safety are explained most completely in the resulting 'full' model. If the sample sizes are increased, although not all variables are statistically significant in the initial model, many still displayed practical significance and would likely become statistically significant.

In the next step, based on the p-values, the variables with the highest p-value were removed from the initial model. The coefficients and the remaining variables are examined for p-values for changes due to multicollinearity. Until all variables had p-values of 0.10 or less, the number of vehicles in the models was reduced to obtain the final 'reduced' model. Some coefficients of other variables changed as variables were removed. The closer the coefficients are, the more accurate the estimate for a variable between the full and reduced models.

4.2.8 Crash Frequency Analysis

In accident analysis, Poisson and negative binomial regression count models are the most appropriate methodological techniques for frequency modelling, which is the consensus of all contemporary empirical work. ZIP and zero-inflated negative binomial (ZINB) regression models are extensions of standard Poisson and negative binomial regression and have gained considerable recognition in accident frequency analysis. These models considered that the traditional application of Poisson and negative binomial models does not address the possibility of zero-inflated counting processes. Some junctions have accident probabilities that are very low over some time, and they can be virtually safe, which is identified as zero-inflation, which is said to be in a zero-accident state. For accident frequency, other junctions may follow a normal count process, in which non-negative integers, including zero, are possible outcomes over the same period. Each of these modelling forms is discussed in this section, starting with the Poisson model.

The following equation for the Poisson model gives the probability of a junction, i, having yi accidents over a period of time:

$$P(Y_i = y_i) = P(y_i) = \frac{EXP(-\lambda_i)\lambda_i^{y_i}}{y_i!},$$

Where,

$P(y_i)$ = the probability of junction 'I' having 'y' accidents over the time period

λ_i = the expected value (Poisson parameter) of 'y_i'.

The Poisson parameter is specified as a function of explanatory variables and is equal to theexpected number of accidents at the junction over period i, it is denoted by E(yi). The variables for traffic volumes, roadway geometry, weather conditions, and other related factors are to be included for crash analyses. The log-linear model is the most common relationship between the Poisson parameter and the explanatory variables, and is given below:

$$E(y_i) = \lambda_i = EXP(\beta X_i)$$

Where:
X_i = vector of explanatory variables
b = vector of estimable parameters.

In the Poisson distribution, the variance and mean of the predicted variable must be equal, which is a severe limitation. A biased, inefficient coefficient estimate may result from the possibility of over-dispersion, or having a variance greater than the mean. A negative binomial distribution is commonly used instead of a Poisson distribution to settle this over-dispersion constraint.

By adding a gamma-distributed error term, EXP(ei), with mean 1 and variance α2, to the Poisson model, the negative binomial model is obtained as given below:

$$\lambda_i = EXP(\beta X_i + \varepsilon_i)$$

This error term allows the variance to differ from the mean as given below:

$$Var[y_i] = E[y_i][1 + \alpha E[y_i]] = E[y_i] + \alpha E[y_i]^2$$

As 'α' approaches zero, the Poisson model is regarded as a limiting case of the negative binomial model. Consequently, depending upon the value of 'α', the selection between thetwo models is made. The negative binomial distribution has the form given below:

$$P(y_i) = \frac{\Gamma((1/\alpha) + y_i)}{\Gamma(1/\alpha) y_i!} \left(\frac{1/\alpha}{(1/\alpha) + \lambda_i} \right)^{\frac{1}{\alpha}} \left(\frac{\lambda_i}{(1/\alpha) + \lambda_i} \right)^{y_i}$$

By the standard maximum likelihood methods, λi can be estimated.

In the Poisson model, $(y - E[y])^2 - E[y]$ has mean zero, where E[y] is the predicted count λ_i. Null and alternative hypotheses are generated (Cameron and Trivedi 1990). On this assumption, a test for over-dispersion is based, such that

$$H_0 : VAR[y_i] = E[y_i]$$

$$H_A : VAR[y_i] = E[y_i] + \alpha g(E[y_i]),$$

where g(E[yi]) is a function of the predicted counts that is generally given the values of

$$g(E[y_i]) = E[y_i] \text{ or } g(E[y_i]) = E[y_i]^2$$

To conduct this test, a simple linear regression is estimated where Z_i is regressed on W_i, where

$$Z_i = \frac{(y_i - E(y_i))^2 - y_i}{E(y_i)\sqrt{2}}, \text{ and}$$

$$W_i = \frac{g(E(y_i))}{\sqrt{2}}.$$

After running the regression, Zi = bWi, H0 is rejected for the associated function 'g', if 'b' is statistically significant in either case, then. In the present instance, it may be concluded that a Poisson model is rejected in favour of a negative binomial model as random sampling does not satisfactorily explain the magnitude of the over-dispersion parameter (Washington et al., 2003).

The ZIP model can address the limitations imposed on the traditional Poisson model because of the possibility of zero-inflated count processes. Two different processes are at work for some zero accident count data, which is assumed in the ZIP. The events assumed in the ZIP model, Y = (Y1, Y2,..., Yn)', are independent and create the model:

$$Y_i = 0 \text{ with probability } p_i + (1 - p_i)e^{-\lambda_i}$$

$$Y_i = y \text{ with probability } \frac{(1 - p_i)e^{-\lambda i}\lambda_i^{y}}{y!}, \ y = 1, 2, \ldots$$

where y is the number of crashes, and the mean and variance of Y_i can be shown to be:

$$E(Y_i) = (1 - p_i)\lambda_i^{y_i},$$

$$Var(Y_i) = E[Y_i] + \frac{p_i}{1 - p_i}E[Y_i]^2$$

A similar formulation to the ZIP is followed in the ZINB regression model. It is again assumed that the events, Y=(Y1, Y2,..., Yn), are independent and the model is as given below:

$$Y_i = 0 \text{ with probability } p_i + (1 - p_i)\left[\frac{1/\alpha}{(1/\alpha) + \lambda_i}\right]^{1/\alpha}$$

In cases where the zero-accident state may exist, the choice of an appropriate form of the model is critical. A direct test cannot be done to determine if the zero-accident state and non-zero accident state are totally different; therefore, selecting an appropriate model is problematic. This is because the traditional and zero-inflated models are not nested. Vuong

(1989) developed a test statistic for non-nested models that is well-suited

where $\mu_i = \dfrac{\left(\dfrac{1}{\alpha}\right)}{\left(\dfrac{1}{\alpha}+\lambda_i\right)}$.

for this setting when the distribution can be specified.

$$m_i = \log\left(\frac{f_1(y_i \mid x_i)}{f_2(y_i \mid x_i)}\right)$$

Y_i=y with probability $(1-p_i)\left[\dfrac{\Gamma((1/\alpha)+y)\mu_i^{1/\alpha}(1-\mu_i)^y}{\Gamma(1/\alpha)y_i!}\right]$,

For Vuong's statistic, let fj(yi|xi) be the predicted probability that the random variable Y equals yi under the assumption that the distribution is fj(yi|xi), for j=1,2, and let

where:

f1(yi|xi) is the probability density function of the zero-inflated model, and f2(yi|xi) is the probability density function of the Poisson or negative binomial distribution.

Vuong's statistic for testing the non-nested hypothesis of the zero-inflated model versus the traditional model is:

$$v = \frac{\sqrt{n}\left[(1/n)\sum_{i=1}^{n} m_i\right]}{\sqrt{(1/n)\sum_{i=1}^{n}(m_i-\overline{m})^2}} = \frac{\sqrt{n}(\overline{m})}{S_m}$$

where:

$\overline{m}$ is the mean,

Sm is the standard deviation, and
n is the sample size.

Vuong's value is asymptotically standard normally distributed, and the test does not indicate any other model if |v| is less than 1.96 (the 95 per cent confidence level for the t-test). However, if the 'v' value is greater than 1.96, the zero-inflated regression model is favoured, while a 'v' value of less than -1.96, the Poisson or negative binomial regression model is favoured.

Negative binomial models were developed in all cases when developing the models for accident or crash frequency. Due to the inherent danger of incorrectly specified coefficients due to the lack of an over-dispersion term, the Poisson models were not used. The parameter still has practical value, even if the 'a' term is relatively small.

By using Vuong's test statistic, the presence of a zero-count state was tested. A zero-altered form of the model is favoured when Vuong's statistic value is greater than 1.96, whereas the traditional model form is favoured when the value is less than -1.96.

4.2.9 Severity Analysis

The level of injury sustained by the most severely injured passengers inside the vehicle is the measure of the severity of an accident (Chang and Mannering, 1999). In some countries and states, the accident severity is classified as PDO, injury (I), or fatal (F). For such considerations, the severity level is a discrete outcome. The multinomial logit model (MNL) is an appropriate method of modelling such data.

To estimate the probability that vehicular accident n has severity i, multinomial logit models are used by determining the likelihood of discrete outcomes given that an accident has occurred. Mathematically, the expression is as given below:

$$P_n(i) = P(S_{in} \geq S_{In}) \quad \forall I \neq i \,,$$

The probability is Pn (i) that a discrete outcome 'i' (accident severity category i) occurs in an accident 'n', where 'P' denotes probability and 'Sin' is a function that determines the severity of accident 'n'. The linear form of the severity function is given below:

$$S_{in} = \beta_i X_n + \varepsilon_{in} \,,$$

Where,

βi is a vector of statistically estimable coefficients,
Xn is a vector of measurable characteristics that determine severity, and
εin is a disturbance term influencing accident severity independent of each severity category.

A multinomial logit (MNL) model can be derived to estimate the probability of accident severity by assuming that the disturbances have a generalised extreme value (GEV) distribution (McFadden, 1981).

$$P_n(i) = \left[\sum_I \exp[\beta_I X_n]\right]^{-1} \exp[\beta_i X_n],$$

Where all variables are as previously defined, and the coefficient vector βi is estimable by standard maximum likelihood techniques.

4.2.10 Binomial Test for Comparing Proportions (by case study)

Referring to the research study "Safety of Junctions on High-Speed Road Segments with Superelevation" by Savolainen and Andrzej P. Tarko, for the Joint Transportation Research Program, Civil Engineering, Purdue University, we get the following: because of a lack of complete information, several variables may not be included in the accident or crash frequency models. The lighting, weather, and road conditions are such variables.

There is no way to accurately determine the same parameters during periods when there are no crashes, because these parameters are known only at the time of each crash. However, using the crash-specific information available in the crash database is an alternative way, and the effect of such elements in relation to curvature can be obtained.

A comparison can be made by selecting two sample junctions, between the proportions of crashes in each sample that had occurred under certain conditions, and these proportions can then be compared. If proportions of crashes are significantly different, it can be claimed that the sample with the higher proportion is over-represented. The crash database is to be used to select the two separate samples, one for junctions on bends and another for junctions on tangents.

At a junction on a bend, the first sample consists of 246 crashes that occurred along four-lane highways. This same sample may be used for the safety analysis on the four-lane major road, later in this section. The complete four-lane sample consisted of 260 junctions. In the second sample, the remaining 14 crashes from this sample were used. This second sample was for all crashes that occurred at tangent junctions along the same divided four-lane highways. By searching the crash database for the major road and county, these junctions were selected and included in the sample. To determine whether the selection criteria were met (four-lane divided highway, rural, unsignalised), aerial photos of the junction were examined.

To extract lighting, weather, and surface conditions during all crashes, in the two samples, the database was then used. It can be determined if a variable is over-represented or under-represented for the junctions on bends by comparing the proportion of crashes related to each variable between the two samples. For example, the proportion of crashes on the bends is significantly greater than the proportion on the tangent sample during dark conditions. The bend case is overrepresented. It can then be concluded that the combination of curvature and darkness makes the junctions from the first sample more hazardous. For these types of variables, the appropriate comparison is made by testing whether the proportion of crashes between the two distinct groups is equal.

Using the binomial distribution, the appropriate statistical test was performed. A good estimate of the true proportion of crashes occurring at junctions on bends, or the likelihood of success in the binomial sense, is given by,

$$S = C / (C + T)$$

Where

C = the total number of crashes at junctions located on bends
T = the total number of crashes at junctions located on tangents

4.2.10.1 Safety Analysis of Junctions along Two-Lane Roads

To determine whether junctions on bends experience more crashes than similar junctions on tangents for two-lane roads, a safety analysis is needed. This analysis should focus on selecting a junction on high-speed rural two-lane highways. Junctions should be chosen along the state and freeway routes only, ensuring more complete data is available from road authorities and agencies. This approach avoids the need to collect traffic and crash data from the site.

Selecting Junctions for Studies

By using a state atlas, the selection of an initial group of junctions can be done. An attempt should be made to pair groups of junctions on tangents and bends with similar traffic and geometric characteristics located along the same major roads wherever possible. This would produce two sets of samples of an equal number of junctions and with similar characteristics in most respects, the only exception being the radius of curvature.

At the site, it is difficult to find such suitable pairs along the same major road because in rural areas, such junctions are typically far away from each other. For this reason, two separate groups of junctions should be selected instead, one on bend sections and the other on tangent (straight) sections.

This will enable direct comparisons to determine if there was a significant differencebetween those on bends and tangents in general. If the initial search produces, for example, 27 possible junctions of state-state roads on bends, then, after initial selection, the junctions are to be verified to ensure that they meet the study criteria using images obtained from satellite imagery. From the selected junctions located on bends, some may be removed for any one of the two reasons:

- The junctions would have occurred on a major road that was a four-lane highway. Such junctions were excluded from this sample because four-lane divided highways are inherently different from two-lane highways. The four-lane case is examined in a separate analysis in the next part of this chapter.
- The junctions that have been realigned during the study period. Since crash data is only available for this period, such junctions should be removed from the sample if it is not possible to identify which crashes occurred with which alignment.
- Crash and traffic data for many junctions may be obtained from police, hospitals, insurance agencies or from a previous research project. The final sample for the two-lane study may consist of 50–100 junctions.

Volumes

After selecting the junctions, traffic volumes for each junction on major and minor roadscan be obtained. Traffic counts are conducted by DOTs in the US every three to five yearsalong all state routes. At least two years of traffic volume data should be made available for each junction in the sample. Annual average daily traffic (AADT) values for each leg of the junction are also required for each corresponding year. The volumes for each junction were derived by summing the volumes of each approach and dividing by two. The resulting volume represents the total volume of vehicles entering the junction per day. This assumes that traffic was evenly distributed in both directions on each approach. This same procedure should be used for each year for which traffic data is available at each junction. Linear interpolation may then be used to bring all traffic volumes to the same year.

The traffic for a particular year is to be used in the analysis, deriving it from the middle of the period covered by the data. An example calculation is illustrated in Table 2-2 for the junction of State Road 37 and State Road 54/58 in Lawrence County, the US. An aerial photo of the junction is shown in Figure 2-5. The table shows AADT values for each of the three legs for the years 1997 and 2001 available from the INDOT flow maps. By interpolating between the two values, estimates are obtained for the target year, 1999.

Table 4.2 AADT Values for SR-37 & SR-54/58

	AADT BY YEAR		
APPROACH	**1997**	**2001**	**1999**
SR-37 NB	18750	19880	19315
SR-37 SB	24000	24840	24420
SR-54 WB	5930	6320	6125

The AADT value for SR-37 in this case would be the average of the northbound and southbound volumes for the forecast year. Through interpolation, the forecast year volumes for SR-37 are 19,315 vehicles per day for the northbound approach and 24,420 for the southbound approach. Averaging the volumesof each approach gives a final value of 21,868 vehicles per day. For the minor road, SR-54, theAADT would be 6,125. For four-point junctions, the values for each approach of the minor road would be averaged to obtain the appropriate minor road AADT, as done for the major road.

Figure 4.5 Junction of SR-37 & SR-54/58 (Source: http://www.mapquest.com)

Crashes

Next to traffic volume estimation, crashes are extracted from the crash database for the period under consideration. At each junction, the number and severity of all crashes were obtained up to a threshold distance of thirty metres from the junction. Similar data for the period should have been obtained already for several junctions in the study, leaving only one additional year of data to be collected in these cases.

Degree of Curvature Approximation

To obtain geometric design data for each junction, ideally, the construction plans of the junctions located on bends should be used. However, for most junctions in the sample, such plans may not be available. Satellite imagery may be used as an alternative.

Next, for each of these junctions, the degree of curvature is to be approximated using geometric design templates by laying over the imagery. The approximation may also be done by taking aerial photographs of each site and scaling them up to a 1-inch equals 100-feet scale, and the degree of curvature for each was then measured using the design templates to the nearest degree per 100-feet cord length. The list of junctions on bends and the corresponding degree of curvature values are described in a tabular form.

Safety Evaluation

After obtaining traffic volume data, crash data and curvature, to determine the effect of curvature on crash frequency and severity, the statistical models are developed.

Crash Frequency Model

After obtaining traffic volumes, crash data and curvature, a negative binomial model can be developed to determine whether curvature had a significant impact on crash frequency. The form of the negative binomial model is as below:

$$C = K \cdot AADT_1^{\alpha_1} \cdot AADT_2^{\alpha_2} \cdot \exp(\beta_1 X_1 + \beta_2 X_2 + \ldots + \beta_N X_N), \quad (4.1)$$

where:

C = expected number of crashes
AADT1 = average annual daily traffic on a major road
AADT 2 = average annual daily traffic on a minor road
K, α_1, α_2, β_1, β_2, b_N = constants
X_1, X_2, X_N = vectors of explanatory variables.

To test for the presence of a zero-count state, Vuong's test statistic is to be used, and the statistic may be, for example, 1.851. This indicates that the unaltered negative binomialregression is more appropriate than a zero-inflated negative binomial model. The final model for the bend effect is shown in the equation below.

$$C = 0.0004 * AADT_1^{0.66} * AADT_2^{0.54} * \exp(0.37FLASH - 0.03D) \quad (4.2)$$

In Table 2-3, the sample modelling results, including variable explanations from LIMDEP, are shown.

Table 4.3 Sample Results of Negative Binomial Model for Bend Effect

Negative Binomial Regression
Maximum Likelihood Estimates
Negative Binomial Regression
Maximum Likelihood Estimates

	Dependent variable	CRASH			
	Weighting variable	ONE			
	Number of observations	104			
	Log likelihood function	-267.6686			
	Restricted log likelihood	-314.6434			
Variable	Explanation	Coeff.	Std. Err.	t-ratio	P-value
ONE	Constant	-7.7388	1.1846	-6.5329	0.0000
ADT1	AADT on Major Road	0.6639	0.1668	3.9801	0.0001
ADT2	AADT on Minor Road	0.5365	0.1262	4.2512	0.0000
FLASH	Flasher Indicator Variable	0.3744	0.1637	2.2874	0.0222
D	Degree of Curvature	-0.0353	0.0185	-1.9068	0.0566
Alpha	Over-dispersion Parameter	0.3477	0.0819	4.2435	0.0002

The sample results of the model show that an increase in crash frequency is associated with increasing ADT (Average Daily Traffic) and the presence of a flasher. Clearly, as traffic increases, due to the increase in potential conflicts, the crash frequency will increase as well. For the FLASH variable, the positive coefficient may seem counterintuitive. However, the results do not mean that at a location, flashers increase the number of accidents. As the flashers are typically installed at high-crash locations to warn drivers of a potential hazard, the coefficient is positive. Also, counterintuitive results are provided by the degree of curvature variable. The junctions on bends had fewer crashes on average than those on tangents, as indicated by the negative coefficient on this variable.

This sample result says that, where both routes are two-lane roads, there does not appear to be a safety problem associated with curvature for junctions. Because of the limited number of junctions on bends, this finding has to be taken with caution, if a large number of junctions are not studied.

Crash Severity Model

Besides examining the crash frequency to determine whether curvature played a role in increasing the severity of an accident, the crashes in the sample were also examin n injury or fatal accident, by developing a multino , is given by the equation below:

$$P(I/F) = \frac{e^{S_{I/F}}}{1+e^{S_{I/F}}} \quad (4.3)$$

Where, SI/F is the severity function.

$$S_{in} = \beta_i X_n + \varepsilon_{in} \,. \tag{4.4}$$

The parameters βi and εin are constant terms, and Xn is a vector of explanatory variables. The likelihood of a severe crash increases as the severity function increases. For the 815 crashes in the sample, the modelling process resulted in the equation below:

$$S_{I/F} = -0.02D + 0.54FOURLEG + 0.02FLASH - 0.85 \,, \tag{4.5}$$

Where,

D = degree of curvature (degrees per 100-ft cord length),
FOURLEG = four-point junction indicator variable
FLASH = flasher presence indicator variable.

Positive coefficients indicate a variable, and as it is increased tends to make crashes more severe. Conversely, a negative sign indicates that as the variable is increased, crashes tend to become less severe. The crashes are to be less severe at three-point junctions, as shown by the model. This may be because there are fewer possible conflict points there are fewer right/left-angle collisions compared to four-point junctions. The installation of a flasher has a slight tendency to cause more severe crashes. Crashes tend to be less frequent and less severe at the junctions in the sample, as indicated by the degree of curvature, which again has a negative coefficient.

The results of the study indicate that it is not possible to confirm a negative impact of curvature on junction safety for the case where both intersecting routes are two-lane roads. However, additional research may further explore this impact, with a larger sample size, as the sample size with junctions located on bends for this study was relativelysmall. Due to the small sample size, the results may be an indicator of randomness within the data rather than an actual trend.

4.2.10.2 Safety Analysis of Junctions on Four-Lane Roads

A safety analysis may be conducted to determine whether, under similar situations, the number of crashes is higher at junctions located on bends than located on tangents for the case where the main road is a four-lane divided highway. The secondary roads are two-lane with stop-control on each leg. County and local roads may be included in this sample of junctions, as there are not enough state-to-state junctions to fit the study criteria.

For example, if at a junction of a major road with a four-lane divided configuration that is on a bend and the minor road with a two-lane undivided configuration has 103 severe accidents, out of which 87 are right-angle crashes over seven years, this should obviously raise concerns for the road users and the authority.

Table 4.4 Sample Crash Data for Junctions

Year	**Total Crashes**	**Right-Angle Crashes**
1986	15	14
1987	21	17
1988	21	14
1989	21	19
1990	15	13
1991	6	6
1992	4	4
Total	103	87

Due to the recurring accidents, the following typical measures are implemented at the junction:

- Flexible delineators were added to the islands on the right/left turn lanes off the junction.
- Flashing beacons are installed.
- Rumble strips are added on the approach to the junctions.
- STOP is painted on the roads.
- To reduce the required crossing time, the stop bars on the minor road wererelocated closer to the main road.

These changes are expected to result in a considerable reduction in crashes. However, safety concerns remain. A comprehensive engineering investigation should be conducted at the junction to determine whether signalisation is warranted. If the minimum volume of traffic (portion of MUTCD Warrant 7 - USDOT, 2001) is not satisfied, signalisation is denied. To restrict left/right turns on the southbound approach, the junction could be channelled. Crossing movements between the minor roads are also restricted. Only the northbound vehicles are now allowed to enter the median at the junction.

A significant reduction in the number of crashes, particularly right-angle collisions, is expected to result from the channelling and median treatment. However, this median treatment requires traffic on the southbound, eastbound, and westbound approaches. To fulfil the requirements to accommodate these movements, a grade separation may be programmed for construction soon. However, such a solution is expensive either to both authorities and road users. The purpose of the study for such junctions on the four-lane major roads was to determine whether more dangerous on bends than on tangents.

Selecting Junctions for Studies

Junctions are to be selected by the study team in coordination with the road authority. The analysis of junctions on two-lane major roads focused exclusively on state-to-state junctions. A preliminary list of junctions is to be prepared from a state atlas and the county flow maps, as was done in the analysis of two-lane roads. To verify whether each junction listed meets the criteria for the study, the list is then sent to each of the districts. For a junction to be selected, a few criteria must be met, such as the major road has to be a rural, divided highway with limited priority located on a bend and the minor road is required to have two-way stop-control.

The district offices may make some additions and deletions to the initial list, and a final list may be compiled and used to plan for field data collection. This list should include various junctions located on bends and on straight sections of the road.

The authority should have a database that contains crash, roadway inventory and traffic volume data for a select group of states. Due to underreporting and classification problems, some of the crash data may be unreliable. For instance, a highway may have many crashes at the junctions, but they are reported without an officer on the scene. Additionally, some cases exist where crashes are attributed to the wrong junction. Sometimes, the same information is not available in all states and other sections of highways, and some necessary geometric details could not be obtained from the database. Due to these potential complications, the database from these authorities is not always usable for research.

Volumes

For the primary roads, volumes are collected from traffic flow maps as described before. For the secondary roads, volumes may fall into either of two categories: these are state roads and local/county roads. Data for state roads was obtained in the same manner as for the primary roads.

For non-state roads, similar count data is generally not available. For junctions where the minor road was a local or county road, two-hour traffic counts on the non-state roads should be done for three months. The counts are to be conducted during peak traffic periods wherever possible. The number of vehicles is recorded at each junction for each of the ten traffic movements entering and exiting the minor road, as illustrated in Figure 4.6. For each minor approach, through, left-turning and right-turning traffic counts are to be done. From the major road, left-turns and right-turns onto the minor road are also to be recorded. The number of heavy vehicles may be very little, as observed at the sample junctions, particularly for the local roads, and their number is not recorded.

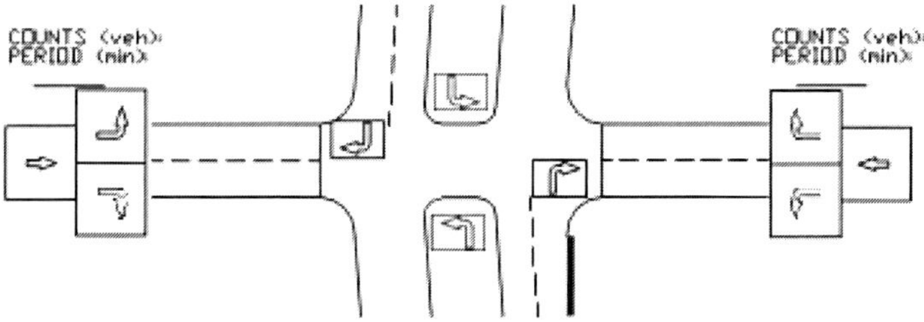

Figure 4.6 Traffic movement counted

In a three-step process, the final counts are converted to AADT values. By using the hourly factors, the two-hour counts were first converted to 24-hour volumes. By selecting a sample of 60 minor roads from the road authority, the hourly adjustment factors were determined. The research team should record 24-hour volumes at each minor road within the system at least once every five years. Junctions are selected with similar hourly volumes of data for this study. To convert one-hour counts to AADT counts, the hourly factors are used. The averages of consecutive hourly factors are the bi-hourly factors. An approximate AADT may be obtained by multiplying the counted number of vehicles by the bi-hourly factor.

Next, for the day of the week on which the counts were taken, the daily counts were adjusted. The department uses the data provided by the study team to adjust its count data based on the day the count is taken. The value obtained from the previous step is multiplied by the appropriate weekly factor.

The final step is to adjust for the month in which the count was taken in converting the traffic volumes. The adjusted AADT from the previous step is multiplied by a monthly factor to arrive at the final estimated AADT value. The equation for converting the two-hour counts to AADT counts is then:

AADT = (two-hour count data) x (bi-hourly factor) x (weekly factor) x (monthly factor)

Geometry

To obtain information on junction geometric features, field data is collected for a specified period at each of the groups of selected junctions under analysis. Information for each junction was entered into a field data collection sheet similar to the one shown in Figure 4.7. For completeness, all data collection sheets for all junctions are to be included in the appendix of the report.

By using a measuring wheel, the carriageway, lanes, shoulders, and median widths are to be measured and taken to the nearest half foot for each approach. Lane widths are generally fairly consistent for the major road, majority of junctions having lane widths of 12 feet.

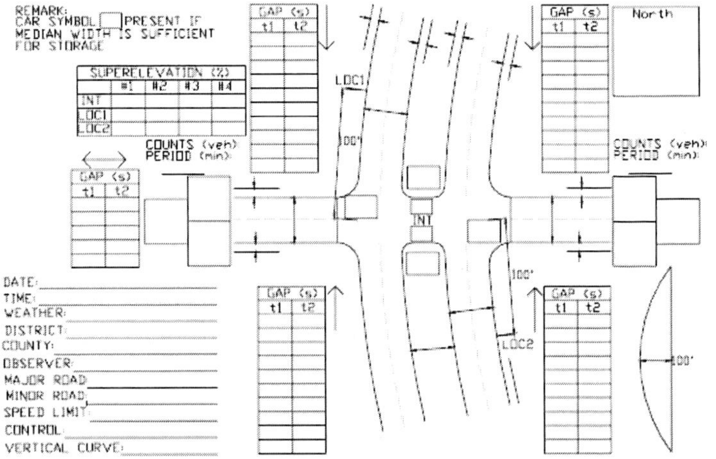

Figure 4.7 Data collection sheet

Lane widths vary from 8 to 13 feet on the minor roads, with the wider lanes typically being found on the roads with higher volumes. On each approach, the shoulder widths are measured from the edge of the outside lanes. On the major road, shoulder widths were between 2 and 10 feet. In cases where there are no auxiliary lanes, larger shoulder widths are typically found. Median widths range from 17 to 250 feet, with most of them falling between 36 and 48 feet.

As the TAC (TAC) was measured, therefore, the sight distance was not measured directly. The actual crossing time required is compared with the TAC. The time required by a vehicle to clear the major road is the crossing time measurement, and this is similar to the time gap (tg) in AASHTO (2001).

All junctions should have a usable median in this study. For that reason, the time required for a car to safely pass from the stop bar on the minor road to the median of the major road is definedas the crossing time. By using a stopwatch, the crossing times are measured. The times were recorded to the nearest hundredth of a second.

From each approach to the median, five measurements are to be taken. One crossing time value is there at three-point junctions to cross half of the width of the major road to merge with the traffic of the major road, and two crossing time values at four-point junctions to cross two halves of the width of the major road to go to the minor road on the other side. In general, both crossing times at four-point junctions were fairly close. Figure 4.8 shows the distances over which the crossing times were measured.

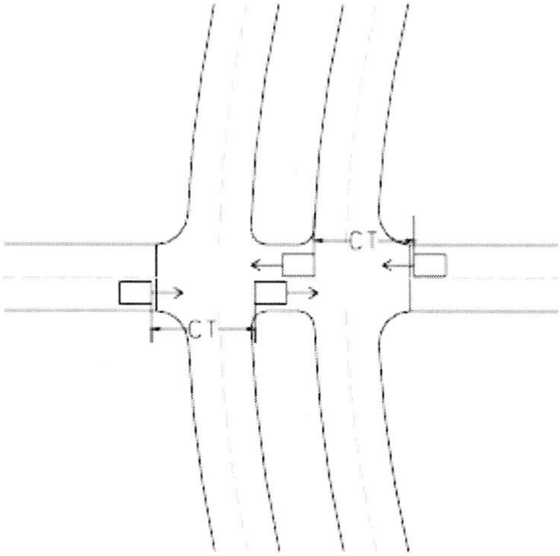

Figure 4.8 Crossing times

The time a driver has to cross from the stop bar to the median safely is the TAC and is based on sight distance at each of the four possible stopping points. By using a stopwatch, the times are recorded. The moment when a vehicle first comes into the stopped driver's field of view is the beginning of the TAC, and the moment when the oncoming vehicle crosses the path between the stopped vehicle and the median is the end of the TAC. Figure 4.9 and Figure 4.10 detail the measured times. At each junction, the TAC is to be measured ten times. At each stop bar, two measurements are to be taken, one for traffic approaching from the left/right side and the other for traffic approaching from the right/left side, which is on the other side of the median. Four-point junctions have 60 total measurements, and three-point junctions have 40.

For highway design, sight distance is a major concern for engineers. Due to the presence of curvature and superelevation, sight distance may become restricted. Safe design requires a minimum length for the leg of a clear sight triangle along the major road. As per AASHTO (2001), "The sight distance should be equal to or greater than the minimum value for specific junction conditions."

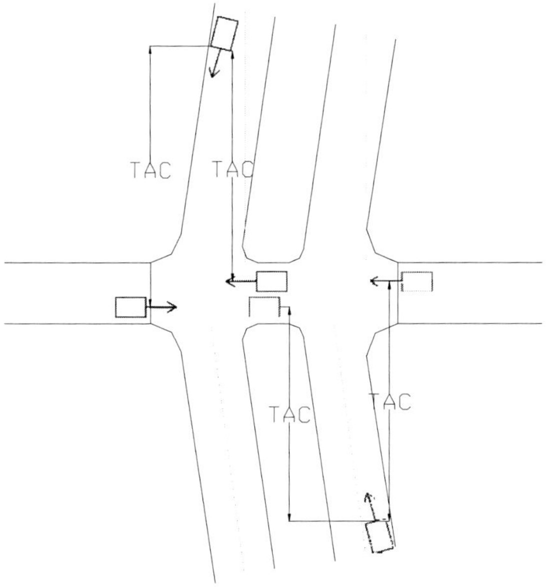

Figure 4.9 Time Available for Crossing to the median

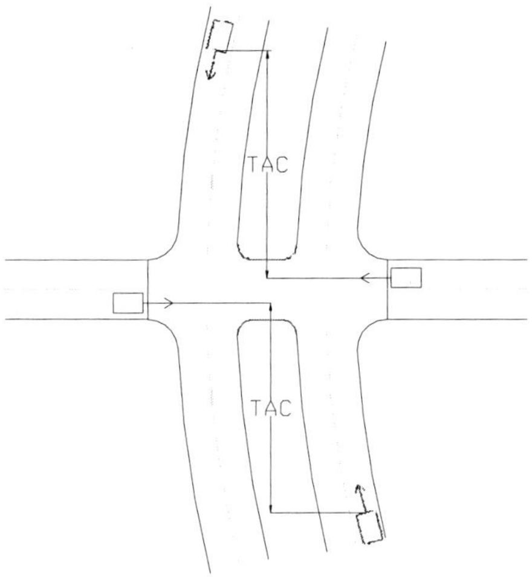

Figure 4.10 Time Available for Crossing entire roadway

To determine if the sight distance requirements as per the design standard for specific speed are sufficiently met at each junction, they are compared with the measured times available for crossing at the junctions. The difference between the measured crossing time and the time allowable for crossing is labelled as the marginal TAC (MTAC). All junctions in the sample should meet this minimum sight distance requirement for traffic crossing to and from the median. However, there may be some cases where the sight distance requirement is not met for vehicles attempting to cross the entire junction. These junctions result in negative MTAC values, indicating the available sight distance is less than the required specified sight distance. For modelling purposes, the MTAC values for each case were transformed to develop the MTACINV variables shown below:

$$\text{MTACINV} = \frac{1}{MTAC} \quad \text{for crossing to and from the median,}$$

$$MTACINV\ 2 = \frac{1}{MTAC+5.5} \quad \text{for crossing to and from the median,}$$

For the second case, an additional 5.5 seconds are added to the MTAC to make the resulting value greater than zero (positive). It is assumed that, rather than a direct linear relationship, this relationship is better explained using the inverse function.

Studies were also carried out on the available gap versus the degree of curvature for each case where a vehicle attempts a crossing manoeuvre. The available gaps, as expected, are shortest on the approach inside the bend and longest on the approach outside the bend, with a few exceptions. However, no clear relationship is available between the degree of curvature and the available gap shown for any of the cases. Based upon these findings, curvature of large radius, say more than 400 metres, does not appear to be a cause of restricted sight distance along four-lane divided highways.

For a major road, the radius of curvature may be obtained by staking out a one-hundred-foot cord along the horizontal bend using wire and chaining pins. As shown in Figure 4.11, the distance from the middle ordinate of this cord to the edge of the road surface was then measured.

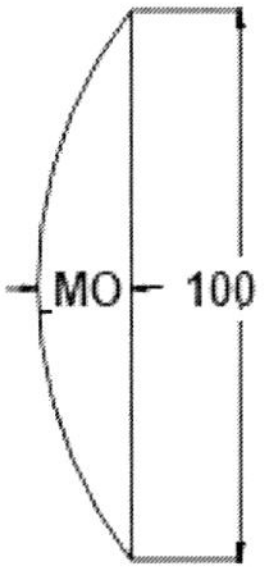

Figure 4.11 Middle ordinate approximation of a bend

To compute the radius, the length measured from the middle ordinate to the cord was then used in the equation as below.

$$R = \frac{10000}{8 * MO} + \frac{MO}{2},$$

Where,

R = radius of curvature (feet),
MO = length of middle ordinate (feet).

It is necessary to convert the radius values to the degree of bend, for modelling purposes, because, as the sharpness of the bend increases, the bend radius decreases, which differs on a tangent section. The radius is infinite on a tangent section, and such values cannot be used in the modelling process. Thisproblem is corrected by the degree of bend, because it has a finite value without exception. The radius values were converted to the degree of curvature using the equation:

$$D = \frac{5729.6}{R},$$

Where,

D = degree of curvature (degrees per 100-ft cord length),R = radius of curvature (feet).

Superelevation is also to be measured at some locations at each junction. At the same three locations for each of the four lanes, the measurements are to be taken using an electronic level. Along a major road, one measurement is to be taken directly in the middle of each junction, and another measurement is to be taken one hundred feet in each direction. These measurements are to be obtained to the nearest tenth of a per cent. Curvature and superelevation information at the studied junctions is to be shown in a tabular form. The variables, such as the speed limit on the major road, the presence or absence of flashers, and whether the junction was located on a vertical bend or not, are to be collected at each junction. The speed limits at the junctions along the major road are to be noted. The junctions with speed flashers and the junctions that are located on crest bends are also to be noted. Junction angles are to be measured using a drone or aerial photographs. A protractor may be used to determine the skew (difference from 90°) for each leg of the junction.

Crashes

The crash database available from the road authority may be used to extract the crash records for each junction. To obtain copies of each crash report, photos of these crash records may be used. Referring to the crash reports, the issues that arise when assembling data from the database are to be cleared up, and mistakes are to be corrected; otherwise, they would have gone unnoticed. Crash reports containing ambiguous location information are to be removed from the sample. This usually happens where two routes overlap, forming two different junctions. Various other problems are also to be identified, such as cases where the wrong coding was entered into the database. It is recommended that the original crash reports be obtained wherever possible to remove such errors in future studies.

Reviewing the full set of crash reports, the final sample may consist of all crashes over the period of four to five years. For each crash, for use in the development of crash frequency and severity models and in the binomial comparison test, the following information is to be extracted:

- Severity (fatal, personal injury, PDO)
- Light condition (daylight, dawn/dusk, dark/streetlights on, dark/streetlightsoff, dark/no streetlights)
- Weather (clear, cloudy, rain, snow, sleet/hail/freezing rain, fog/smoke/smog)
- Road surface condition (wet, muddy, slush, snow/ice)
- Primary contributing circumstance.

There may be different contributing factors listed on the crash reports for the sample junctions. Failure to give way and driver inattention are the causes of over 75% of these crashes. Unfortunately, these characteristics are dependent on individual drivers and are beyond the direct control of the transportation authority. Statistical models are required to explain why these mistakes are made and what measures can be taken to correct them.

4.3 Highway Safety at Roundabouts

The safety of traffic and pedestrians at a roundabout is inherent in its planning and design. Once the construction of the roundabout is over, the safety measures are only left with pavement marking, signage, lighting, etc. The planning and design are therefore critical for a safe roundabout. In this section, the discussion is mainly on the various considerations for the design of a roundabout or rotary.

Choosing between safety and capacity is involved in designing the geometry of a roundabout. When the geometry forces traffic to enter and circulate at slow speeds, roundabouts are safer. To produce this reduced-speed environment, the horizontal curvature and a narrow lane are used, and hence, the capacity of roundabouts is negatively affected by these low-speed design elements. Furthermore, the manoeuvring requirements of the largest vehicles expected to travel through the roundabout govern the geometric parameters. Thus, an optimal balance between safety provisions, operational performance, and large vehicle accommodation isrequired to determine in the process of designing a roundabout.

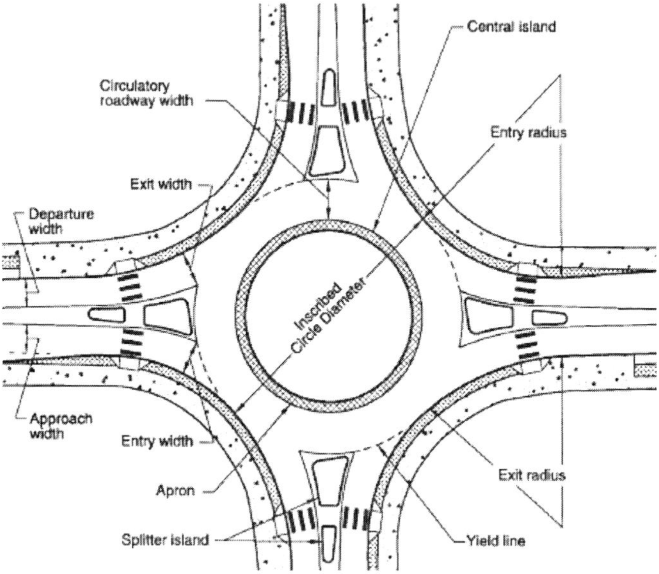

Figure 4.12 Basic geometric elements of a roundabout (a) for right lane junctions (USA, Europe, etc.) and (b) for left lane junctions (UK, India, etc.)

The process of designing roundabouts requires a considerable amount of iteration among geometric layouts, operational analysis and safety evaluation, and so the process is more comprehensive than for other forms of junction. The designer often needs to revise and refine the initial layout and attempt to enhance its capacity and safety. It is rare to produce an optimal geometric design on the first attempt. The process of designing and evaluating a roundabout is described as a graphical flowchart in Figure 4.13.

The operating speeds of typical vehicles approaching and negotiating a roundabout are 40, 55, and 70 km/h at 100 m from the centre of the roundabout. Drivers can choose their desired paths safely and efficiently due to the relatively uniform speed of all drivers on the roundabout.

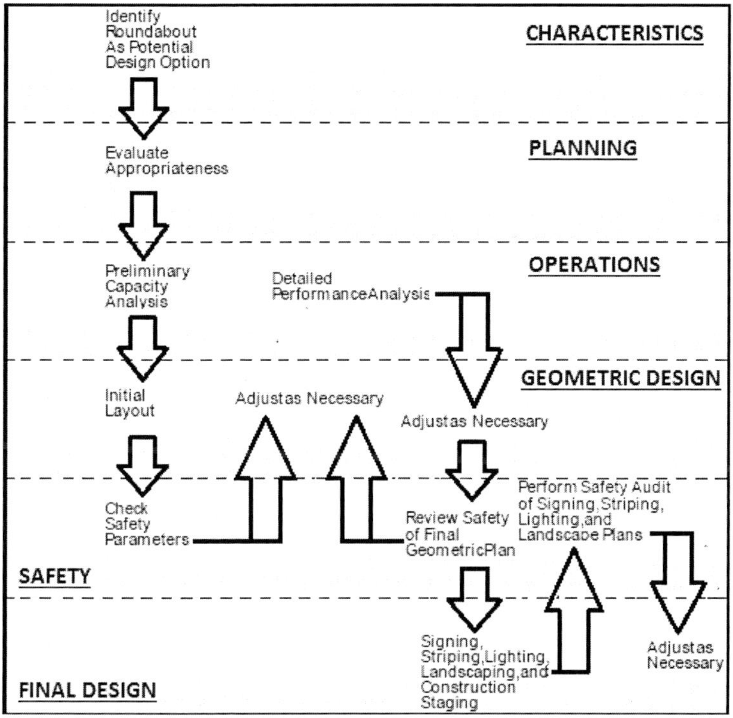

Figure 4.13 Flowchart of a roundabout design process

Table 4.5 Maximum entry speed with respect to site categories

Site Category	Maximum Entry Speed
Mini-Roundabout	25 km/h
Urban Compact	25 km/h
Urban Single Lane	35 km/h
Urban Double Lane	40 km/h
Rural Single Lane	40 km/h
Rural Double Lane	50 km/h

4.3.1 Vehicle paths

The fastest path allowed by the geometry is drawn to determine the speed of a roundabout. This is the smoothest, flattest path possible for a single vehicle, in the absence of other traffic and ignoring all lane markings, traversing through the entry, around the central island, and out the roundabout exit. Usually, the fastest possible path is this 'through movement', but in some cases it may be a right/left turn. A vehicle is assumed to be 2 m wide and maintains a minimum clearance of 0.5 m from the kerb. Thus, the centreline of the vehicle path is drawn with the following distances to the particular geometric features:

- 1.5 m from the kerb.
- 1.5 m from a roadway centreline.
- 1.0 m from a painted edge line.

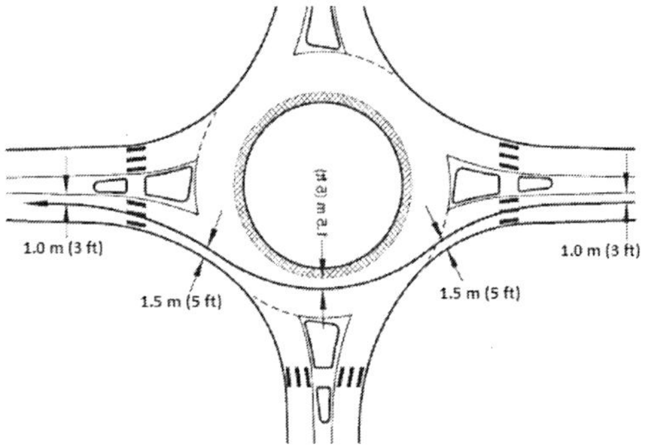

(a)

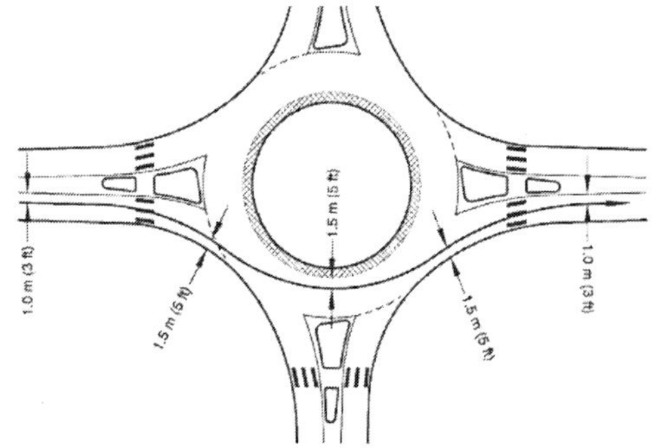

(b)

Figure 4.14 Fastest vehicle path through single-lane roundabout (a) for right-hand drive junctions (USA, Europe, etc.) and (b) for left-hand drive junctions (the UK, India, etc.)

The fastest path for the 'through movement' is a series of reverse bends (i.e., a bend to the right/left, followed by a bend to the left/right, and next followed by a bend to the right/left) as shown in Figures 4.16 and 4.17. To account for the time it takes for a driver to turn the steering wheel, a short length of tangent should be drawn between consecutive bends. It is suggested to draw the path freehand, rather than using drafting templates or a computer-aided design (CAD) program. The way a driver negotiates the roundabout, with smooth transitions connecting bends and tangents, is more naturally represented by the freehand technique. After the fastest path is sketched, the minimum radii using suitable bend templates or by replicating the path in CAD and using it to determine the radii may be measured by the designer.

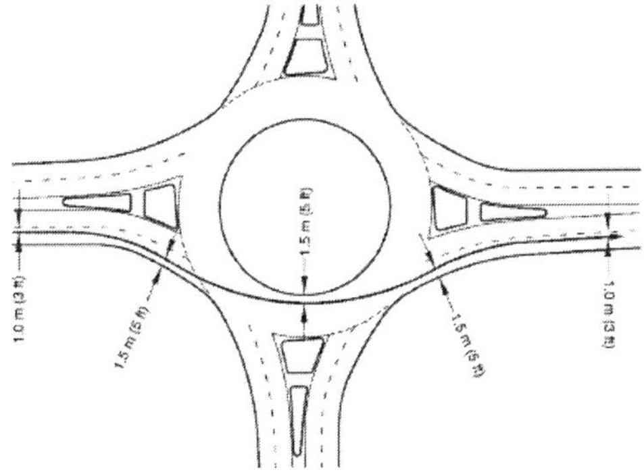

(a)

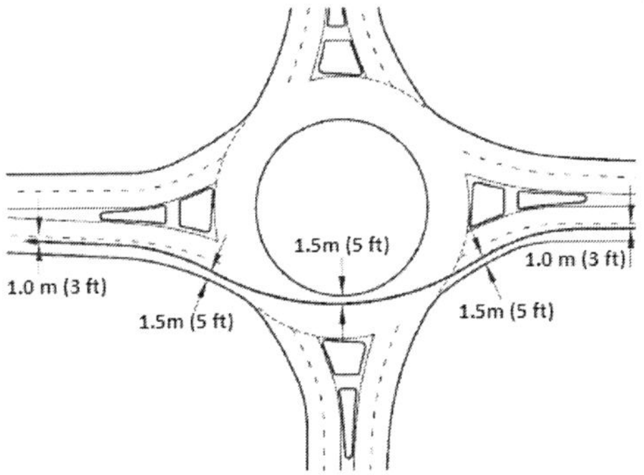

(b)

Figure 4.15 - Fastest vehicle path through double-lane roundabout (a) for right-hand drive junctions (USA, Europe, etc.) and (b) for left-hand drive junctions (the UK, India, etc.)

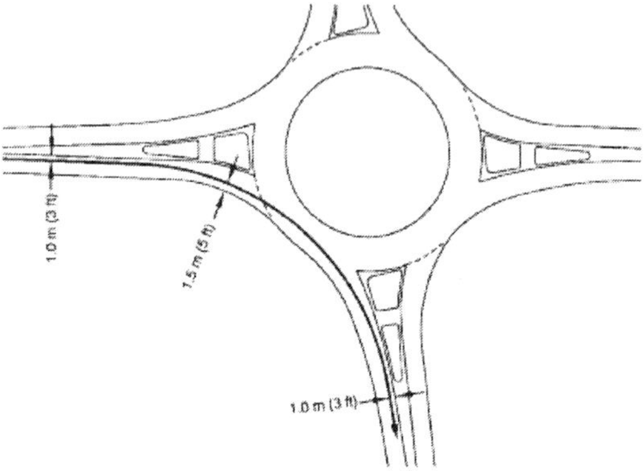

(a)

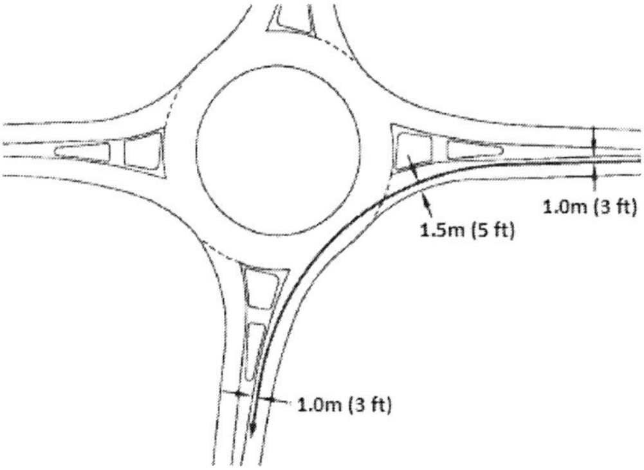

(b)

Figure 4.16 Example of critical right/left-turn movement (a) for right-hand drive junctions (USA, Europe, etc.) and (b) for left-hand drive junctions (the UK, India, etc.)

From the smallest radius along the fastest allowable path, the design speed of the roundabout is determined. As the vehicle bends to the left/right around the central island, the smallest radius usually occurs on the circulatory roadway. The radius of the entry path (i.e., as the vehicle bends to the right/left) should not be significantly larger than the circulatory path radius, which is important when designing the roundabout geometry.

For all approaches to the roundabout, the fastest path should be drawn. Determining the fastest path is a subjective process and requires a certain amount of personal judgement; it is advisable to obtain a second opinion.

4.3.2 Speed-bend relationship

A Policy on Geometric Design of Highways and Streets, AASHTO (4), documents the relationship between travel speed and horizontal curvature. Equation 4.20 can be used to calculate the design speed for a given travel path radius.

$$V = 127R(e + f) \quad \text{(4.20a, metric)}$$

where:

V = Design speed, km/h, R = Radius, m, e = superelevation, m/m, f = side friction factor

$$V = 15R(e + f) \quad \text{(4.20b, U.S. Imperial Units)}$$

where:

V = Design speed, mph, R = Radius, ft, e = superelevation, ft/ft, f = side friction factor

Superelevation values are usually adopted to be +0.02 for entry and exit bends and -0.02 for bends around the central island.

4.3.3 Speed consistency

Although achieving the fastest movements, an appropriate design speed is essential; however, another important objective is to achieve consistent speeds for all movements. Therefore, with reductions in overall speed, speed consistency is necessary to minimise the crash rate and severity between conflicting streams of vehicles. The task of merging into the conflicting traffic stream, minimising critical gaps, is also simplified, thus the entry capacity is optimised. Two implications of this principle are noted below:

1. Between successive geometric elements, the relative speeds should be minimised; and

2. Between conflicting traffic streams, the relative speeds should be minimised.

As shown in Figure 4.17, for each approach, the radii for the five critical paths must be checked.

- The minimum radius is R1 at the entry path, on the fastest ‘through path’ prior to thegive-way line.
- The minimum radius is R2 of the circulating path, on the fastest ‘through path’ aroundthe central island.
- The minimum radius is R3 at the exit path, on the fastest ‘through path’ into the exit.R4.
- The minimum radius is R4 at the left-turn path, on the path of the conflicting left-turnmovement.
- The minimum radius is R5 at the right-turn path, on the fastest path of a right-turningvehicle.

It is to be noted that these vehicular path radii are not the same as the kerb radii. The basic kerb geometry is to be laid out first, and then the vehicle paths are to be drawn by following the procedures as described before.

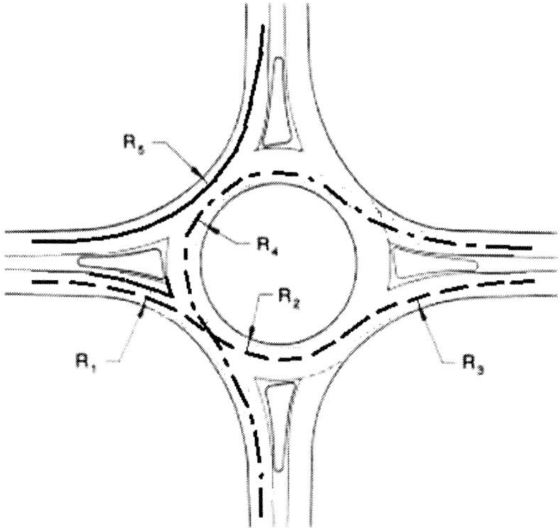

(a)

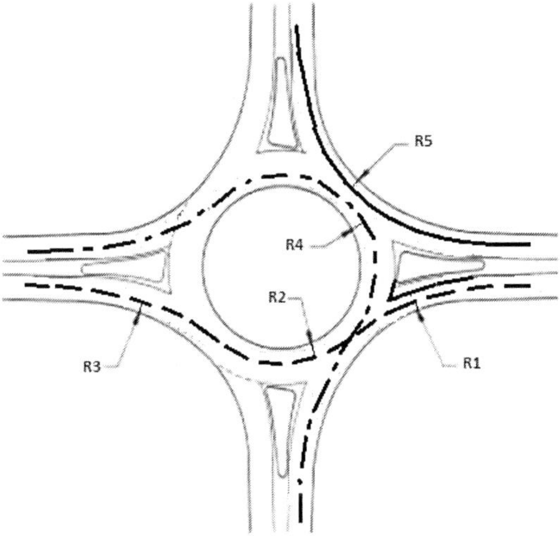

(b)

Figure 4.17 Vehicle path radii (a) for right-hand drive junctions (USA, Europe, etc.) and (b) for left-hand drive junctions (the UK, India, etc.)

It is desirable that on the fastest path, radius R1 should be smaller than R2, which in turn should be smaller than R3. This ensures that at the roundabout entry, the speed will be reduced to its lowest level and the likelihood of crashes will be reduced. However, within given right-of-way or topographic constraints, in some cases, it may not be possible to achieve an R1 value less than R2. In such cases, the radius value R1 is greater than R2, provided the relative difference in speeds is preferably less than 10 km/h or 20 km/h.

It is relatively simple to reduce the value of R1 on single-lane roundabouts. To achieve a slower entry speed, the kerb radius at the entry can be reduced or the alignment of the approach can be shifted further to the left. However, this has potential for higher exit speeds that may put pedestrians at risk. It is generally more difficult on double-lane roundabouts, as the natural path of adjacent traffic streams may overlap, caused by overly small entry bends. The geometry leads a vehicle in the left/right approach lane to naturally sweep across the right/left approach lane just before the approach line to avoid the central island, and this occurs when the paths overlap. The same is also observed within the circulatory roadway when a vehicle entering from the right-hand/left-hand lane naturally cuts across the left/right side of the circulatory roadway close to the central island. At double-lane roundabouts, when the paths overlap, it may reduce capacity and increase crash risk. Therefore, care must be taken when designing double-lane roundabouts to achieve ideal values for R1, R2 and R3.

To minimise crashes, the exit radius, R3, should not be less than R1 or R2. With pedestrian activity at single-lane roundabouts, exit radii may be the same or slightly larger than R2 to minimise exit speeds. Additional care must be taken on double-lane roundabouts to minimise the likelihood of exiting paths overlapping. When a vehicle on the left/right side of the circulatory roadway, which is next to the central island, exits into the right-hand/left-hand lane, this may cause paths to overlap. In situations where no pedestrians are expected, the exit radii may be just large enough to minimise the likelihood of exiting paths overlapping. Nevertheless, lower radius exit curvature may be necessary to ensure sufficiently low speeds at a pedestrian crossing.

To ensure that the maximum speed differential between entering and circulating traffic is not more than 20 km/h, the radius R4 of the conflicting left-turn/right-turn movement must be properly determined. Because of the lowest circulating speed, the left-turn/right-turn movement is critical for the traffic stream. Between the entry and circulating speeds, a large differential may result in an increase in single-vehicle crashes due to loss of control. By adding 1.5 m to the central island radius, the radius R4 can be determined. Based on this assumption, the approximate values of radius R4 and corresponding maximum values of radius R1 for the roundabout diameter are given in metric and US imperial units in Tables 4.6 and 4.7.

4.3.4 Design of Vehicles

The need to accommodate the largest motorised vehicle likely to use the junction is an important factor determining a roundabout's layout. Many of the roundabout's dimensions will be dictated by the turning path requirements of this vehicle. During the design process, the designer must be aware of the vehicle and possess the appropriate vehicle turning templates or a CAD-based vehicle turning path program to determine the vehicle's swept path.

Table 4.6 Approximated R4 values and corresponding R1 values (metric units)

Inscribed roundabout diameter(m)	Approximate R4 values		Maximum R1 values	
	Radius (m)	Speed (Km/h)	Radius (m)	Speed (Km/h)
Single Lane Roundabout				
30	11	21	54	41
35	13	23	61	43
40	16	25	69	45
45	19	26	73	46
Double Lane Roundabout				
45	15	24	65	44
50	17	25	69	45
55	20	27	78	47
60	23	28	83	48
65	25	29	88	49
70	28	30	93	50

Table 4.7 Approximated R4 values and corresponding R1 values (imperial units)

Inscribed roundabout diameter (ft)	Approximate R4 values		Maximum R1 values	
	Radius (ft)	Speed (mph)	Radius (ft)	Speed (mph)
Single Lane Roundabout				
100	35	13	165	25
115	45	14	185	26
130	55	15	205	27
150	65	15	225	28
Double Lane Roundabout				
150	50	15	205	27
165	60	16	225	28
180	65	16	225	28
200	75	17	250	29
215	85	18	275	30
230	90	18	275	30

The approaching roadway types and the surrounding land use characteristics govern the choice of vehicle. The road agency should be consulted to identify the vehicle at each site. The AASHTO guide on Geometric Design of Highways provides the dimensions and turning path requirements for a variety of common highway vehicles (4). Commonly, WB-15 (WB-50) vehicles are the largest vehicles along collectors and arterials. Larger lorries, such as WB-20 (WB-67) vehicles, may need to be addressed at junctions on main roads. For local street junctions, smaller vehicles are generally chosen.

To accommodate large vehicles, larger roundabouts need to be used while maintaining low speeds for passenger vehicles. However, land constraints may sometimes limit the ability to accommodate large combinations, but achieve adequate deflection for small vehicles.

A lorry apron may be used at such times to provide an additional traversable area around the central island for large vehicles. The apron provides a lower level of operation than standard non-mountable islands and should be used only when there is no other means of providing adequate deflection while accommodating the test vehicle. Determining the vehicle's path through the critical turning movements by using a CAD-based computer program is described in Figures 4.18 and 4.19.

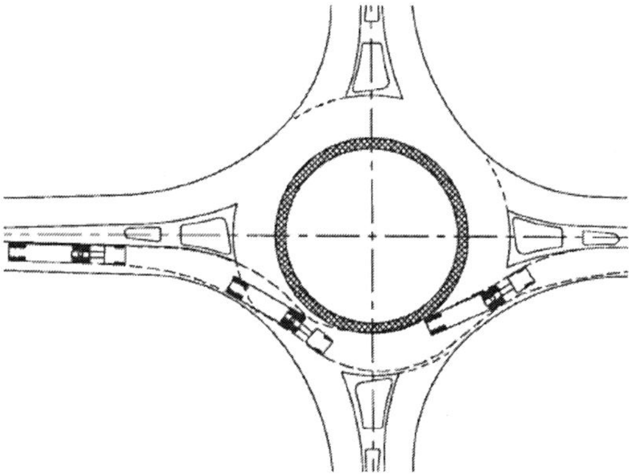

(a)

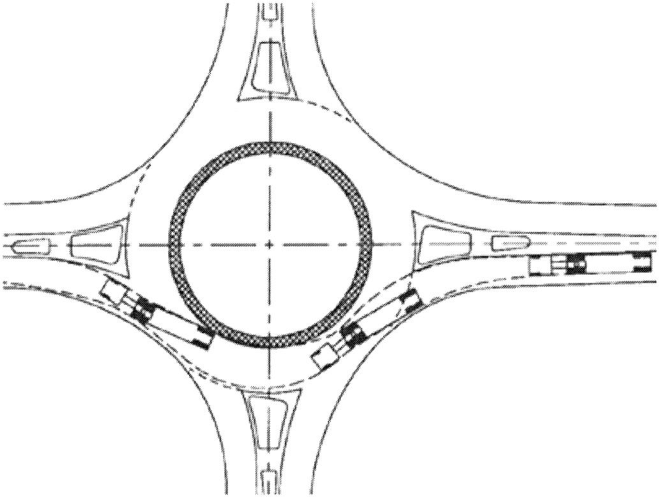

(b)

Figure 4.18 Through movement of WB-15 (WB-50) vehicle (a) for right-hand drive junctions (USA, Europe, etc.) and (b) for left-hand drive junctions (the UK, India, etc.)

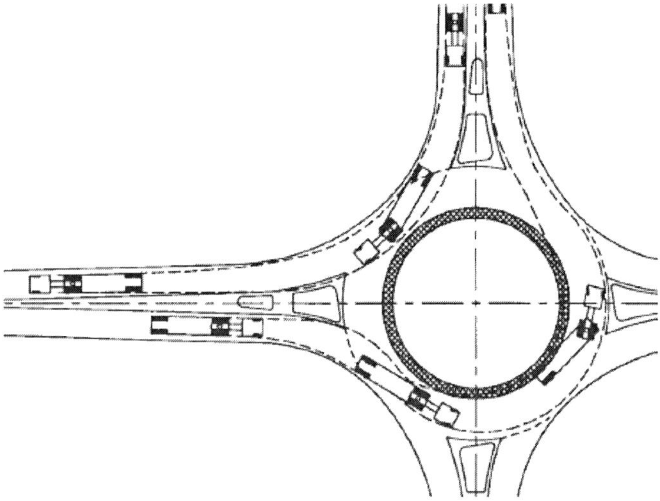

(a)

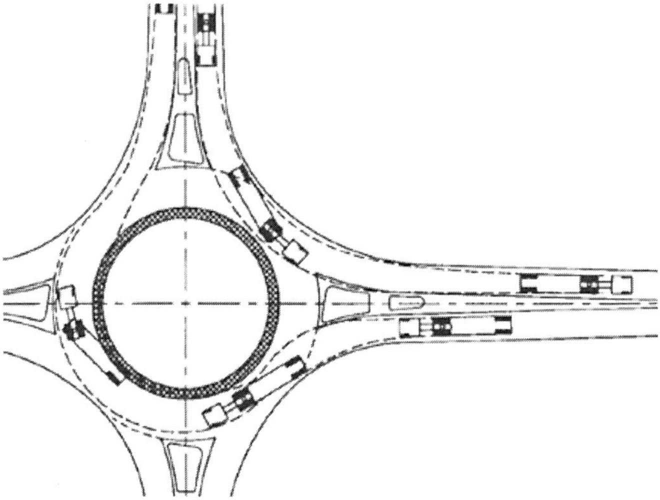

(b)

Figure 4.19 Left-turn and right-turn paths of WB-15 (WB-50) vehicle (a) for right-hand drive junctions (USA, Europe, etc.) and (b) for left-hand drive junctions (the UK, India, etc.)

A roundabout is optimally located means the centrelines of all approaches pass through the centre of the inscribed circle. The location should support an adequately designed geometry so that vehicles will maintain slow speeds through both the entries and the exits. To the approaching drivers, the radial alignment also makes the central island more conspicuous.

Alignment at Approaches and Entries

If it is difficult to align the legs through the centre point, a slight shift to the left/right (i.e., the centreline passes to the left/right of the roundabout's centre point) is acceptable. At the entry, this alignment will still allow sufficient curvature, which is highly important. When the roundabout is relatively small, it may be beneficial to introduce a slight offset of the approaches to the left/right in order to enhance the entry curvature. Care must be taken to ensure that an excessively tangential exit is not produced with such an approach offset. In urban situations, to keep vehicle speeds low and reduce the risk for pedestrians, the exit geometry must produce a sufficiently curved exit path. For an approach alignment, it is almost never acceptable to be offset to the right/left of the roundabout's centre point. This alignment brings the approach in at a more tangential angle and reduces the opportunity to provide sufficient entry curvature, resulting in vehicles being able to enter the roundabout too fast, causing more crashes and higher crash rates between entering and circulating vehicles. It is also desirable to equally space the angles between entries, providing optimal separation between successive entries and exits. This results in optimal angles of 360/4=90 degrees for four-point roundabouts, 360/5=72 degrees for five-point roundabouts, and so on. This is consistent with the findings of the British accident prediction models. Figure 4.20 illustrates the preferred radial alignment of entries.

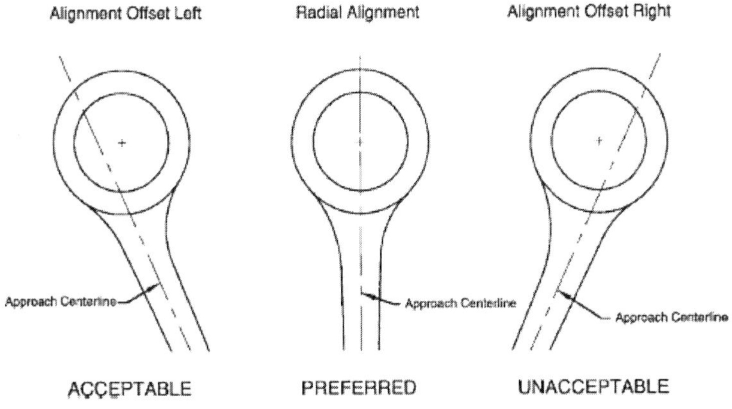

Figure 4.20 Radial alignment of entries

4.3.5 Geometric Elements

4.3.5.1 Roundabout diameter

The size of the roundabout is largely dependent upon the turning requirements of a test vehicle on single-lane roundabouts. While maintaining adequate deflection curvature to ensure safe travel speeds for smaller vehicles, the diameter must be large enough to accommodate the test vehicle. In accommodating the test vehicle and providing deflection, the circulatory roadway width, entry and exit widths, entry and exit radii, and entry and exit angles also play a significant role. In constrained situations, a judicious selection of these geometric elements may allow a smaller roundabout to be used. Broadly, to accommodate a test vehicle WB-15 (WB-50), the roundabout should be a minimum of 30 m (100 ft). For some local street or collector street junctions, where the test vehicle may be a bus or single unit lorry, smaller roundabouts can be used.

Double-lane roundabouts are not usually a constraint for accommodating the test vehicle. To determine the size of the roundabout, either the need to achieve deflection or the need to fit the entries and exits around the circumference with reasonable entry and exit radii between them is chosen. Generally, for a double-lane roundabout, the roundabout should be a minimum of 45 m.

4.3.5.2 Entry width

The entry width determines a roundabout's capacity, not the number of entering lanes. That means, with incremental increases to the entry width, the entry capacity increases steadily. Entries are striped to designate separate lanes which have sufficient width to accommodate multiple traffic streams, which is at least 6.0 m. However, even when more than one lane of traffic is expected to circulate, the circulatory roadway is usually not striped. The entry widths should be kept to a minimum to maximise the roundabout's safety. Each entry should have a certain width, with the capacity requirements and performance objectives dictating a number of entry lanes. The test vehicle may still require a wider entry to meet the turning requirements. However, crash frequency is increased with larger entry and circulatory widths. Therefore, a trade-off between capacity and safety is involved for determining the entry width and circulatory roadway width. The minimum width necessary for capacity and accommodation of the test vehicle to maintain the highest level of safety should be provided by the design. Entry widths for single-lane entrances typically range from 4.3 to 4.9 m; sometimes, for site-specific test vehicles and speed requirements for critical vehicle paths, values higher or lower than this range may be applied.

The capacity requirements can be met by increasing the entry width in either of the two ways:

1. By adding a full lane upstream of the roundabout and maintaining parallel lanes through the entry geometry, or
2. By widening the approach gradually (flaring) through the entry geometry.

4.3.5.3 Circulatory Roadway Width

From the width of the entries and the turning requirements of the test vehicle, the required width of the circulatory roadway is determined. In general, circulatory roadway should always havea minimum width equal to the maximum entry width, i.e., up to 120 per cent of the maximum entry width and should remain constant throughout the roundabout (3).

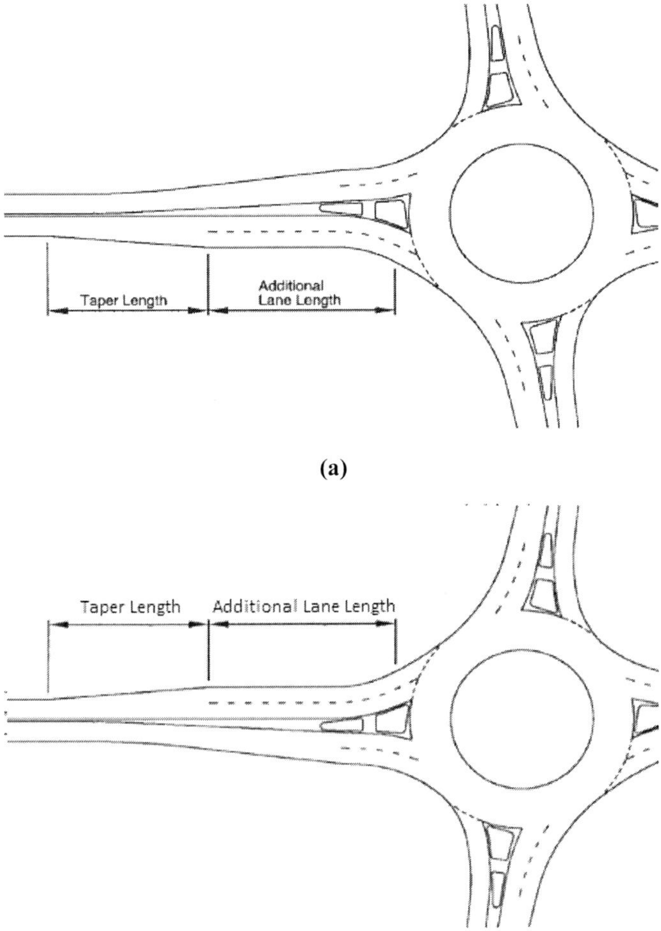

(a)

(b)

Figure 4.21 Approach widening by adding a full lane (a) for right-hand drive junctions (USA, Europe, etc.) and (b) for left-hand drive junctions (the UK, India, etc.)

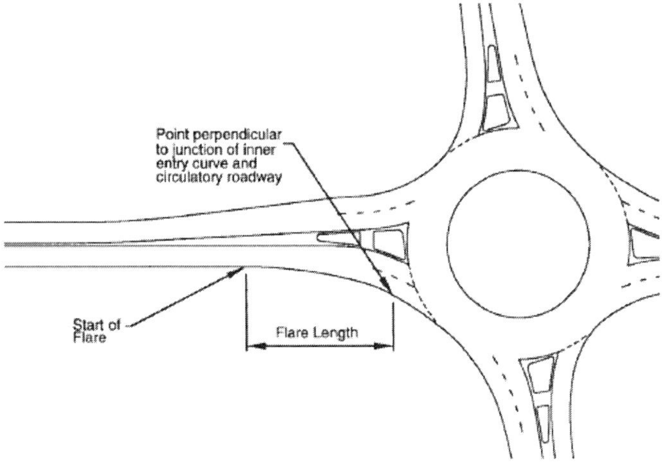

(a)

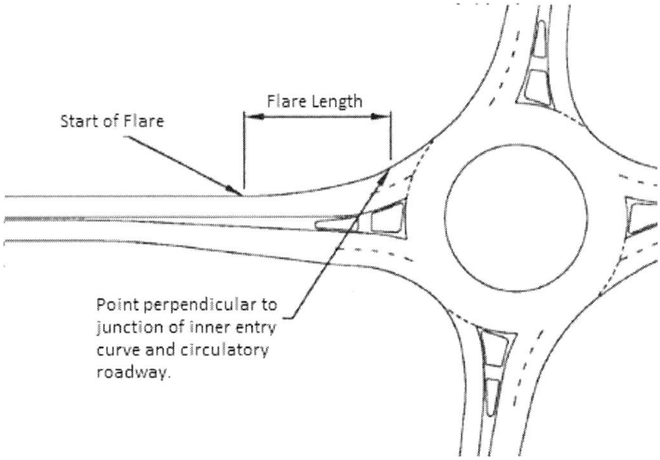

(b)

Figure 4.22 Approach widening by entry flaring (a) for right-hand drive junctions (USA, Europe, etc.) and (b) for left-hand drive junctions (the UK, India, etc.)

4.3.5.3.1 Single-lane roundabouts

At single-lane roundabouts, the test vehicle should be accommodated by the circulatory roadway. The path of the test vehicle through each of the turning movements should be determined either by using appropriate vehicle-turning templates or by a CAD-based computer program. Usually, for determining circulatory roadway width, the left-turn/right-turn movement is the critical path. As per the AASHTO policy, a minimum clearance of 0.6 m should be provided between the outside edge of the vehicle's tyre track and the kerb line. Derived widths required for various radii for each standard test vehicle are provided in AASHTO Table III-19 (1994 edition).

Where the inscribed diameter is small or the test vehicle is large, the turning requirements ofthe test vehicle may dictate that the circulatory roadway be so wide that the amount of deflection necessary to slow passenger vehicles is compromised. To accommodate larger vehiclesin such situations, the circulatory roadway width can be reduced, and a mountable kerb on the central island, placed behind a lorry apron, can be used. Lorry aprons generally provide a lower level of operation than standard non-mountable islands. Lorry aprons are sometimes driven over by four-wheel drive automobiles, may surprise inattentive motorcyclists, and can cause load shifting on lorries. Therefore, while accommodating the test vehicle, the use of lorry aprons should be restricted when there is no other means of providing adequate deflection.

4.3.5.3.2 Double-lane roundabouts

At double-lane roundabouts, the test vehicle does not usually govern the circulatory roadway width. The width required for one, two or three vehicles, depending on the number of lanes at the widest entry, should be used to establish the circulatory roadway width to travel simultaneously around the roundabout. Accommodating the combination of vehicle types side-by-side depends on the specific traffic conditions at each site. It may be appropriate to design the width for two passenger vehicles or a passenger car and a single-unit lorry side-by-side, if the entering traffic is predominantly passenger cars and single-unit lorries (P and SU vehicles, AASHTO), where articulated traffic is infrequent. It may be necessary to provide sufficient width for the simultaneous passage of an articulated lorry in combination with a P or SU vehicle, if semitrailer traffic is relatively frequent having presence of greater than 10 per cent.

4.3.5.4 Central Island

The central islands are typically circular in shape, although exceptions exist. A constant speed around the central island is promoted by its shape with a constant-radius circulatory roadway. On the other hand, oval or irregular shapes are more difficult to drive on, as these cause higher speeds on the straight sections and reduced speeds on the arcs of the oval. For the vehicles entering the roundabout, judging the speed and acceptability of gaps in the circulatory traffic stream becomes harder due to this speed differential. It can also be deceptive and lead to more crashes for the circulating drivers. Noncircular central islands, as they get larger, have a rapidly increasing degree of the above disadvantages because circulating speeds are higher. Oval-shaped islands generally do not face such a problem. Raindrop-shaped islands may be used at locations where certain turning movements cannot be safely accommodated, such as roundabouts with one approach on a relatively steep grade or at junctions.

In case aprons are used, they should be designed so that they are traversable by lorries, but discourage passenger vehicles from using them. Their width should generally be 1 to 4 mwide with a cross slope of 3 to 4 per cent away from the central island. The outer edge of the apron should be raised a minimum of 30 mm above the circulatory roadway surface (6) to discourage passenger vehicles from using the apron. To differentiate the apron from the circulatory roadway, the apron should be constructed of coloured and/or textured paving materials. Care must be taken to ensure that delivery lorries, as their rear trailer wheels track across the apron, do not experience load shifting.

4.3.5.5 Entry Bends

For single-lane roundabouts, the entry speed objectives are relatively simple to achieve. There is noconflict between traffic in adjacent lanes with a single traffic stream entering and circulating. Thus, to produce the desired entry path radius, it can be reduced or increased as necessary.The approaching vehicles will adjust their path accordingly and negotiate through the entry geometry into the circulatory roadway, provided sufficient clearance is given by the test vehicle.

On urban single-lane roundabouts, the entry radii typically range from 10 to 30 m. Use of larger radii should not result in excessive entry speeds. For roundabouts on local streets, the entry radii may be below 10 m if the test vehicle is small.

Consideration should be given for rural and suburban locations; the speed differential between the approaches and entries should not be greater than 20 km/h. It is desirable to reduce the speed of approaching traffic prior to the entry curvature by incorporating approach bends or some other speed reduction measures. At double-lane roundabouts, the entry curvature is more complicated to design. Conflicts between adjacent traffic streams may be caused by overly small entry radii, ultimately resulting in poor lane utilisation of one or more lanes and significantly reducing the capacity of the approach. Sideswipe crashes may increase as the safety performance is degraded. For avoiding conflicts between adjacent entry lanes at double-lane roundabouts, the techniques and guidelines are to be investigated.

4.3.5.6 Exit Bends

Exit bends are to be provided with bends of larger radii than entry bends to provide the desired speed to minimise the likelihood of congestion at the exits. A balance is needed to maintain low speeds at any pedestrian crossing on exit. The exit bend should have a radius (R3 in Figure 4.19) not smaller than the circulating path radius (R2); otherwise, vehicles will be travelling too fast to negotiate the exit geometry and may crash into the splitter island or into oncoming traffic in the adjacent approach lane. Likewise, to ensure low speeds at the downstream pedestrian crossing, the exit path radius should not be significantly greater than the circulating path radius.

The design of the exit bend is to be curvilinear, tangential to the outside edge of the circulatory roadway. Similarly, the projection of the inside edge of the exit roadway is to be curvilinear, tangential to the central island.

At single-lane urban roundabouts, to maximise safety for pedestrians crossing the exiting traffic stream, exits should be designed to enforce a bend exit path with a design speed below 40 km/h. Usually, exit radii should not be less than 15 m. However, the exit radii may be as low as 10 to 12 m at locations with pedestrian activities and no large articulated lorry traffic. A very slow design speed achieves maximum safety and comfort for pedestrians.On urban compact roundabouts smaller than 35 m, such low exit radii may be used in conjunction with similar or smaller entry radii. At double-lane roundabouts, the design of the exit curvature, like the entries, is more complicated than at single-lane roundabouts.

4.3.5.7 Splitter Islands

Splitter islands are essential on all roundabouts, except those with very small diameters, when the splitter island would obstruct the visibility of the central island. Splitter islands are also called separator islands or median islands, and their purpose is to provide shelter for pedestrians, wheelchair users, cyclists, pushchairs, etc. On the roundabout, they assist in speed control, guiding traffic, separating traffic streams, deterring wrong-way movements and provide a place for mounting signs.

As shown previously in Figures 4.23 and 4.24, the splitter island envelope is formed by the entry andexit bends on a leg. In general, the total length of the island should be at least 15 m to provide sufficient protection for pedestrians and to alert approaching drivers to the roundabout geometry. To prevent exiting traffic from accidentally crossing into the path of approaching traffic, the splitter island should extend beyond the end of the exit bend. For the design of the splitter island, standard AASHTO guidelines for island design should be followed. Using a larger nose radius at approach corners maximises island visibility and offsets kerb lines at the approach ends to create a funnel effect. The funnel helps to reduce speed as vehicles approach the roundabout significantly.

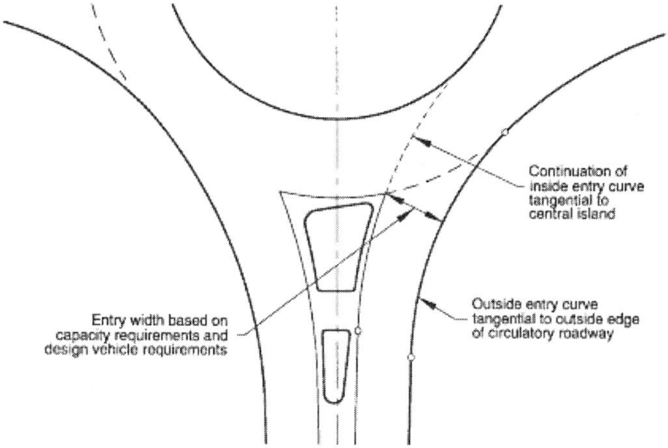

(a)

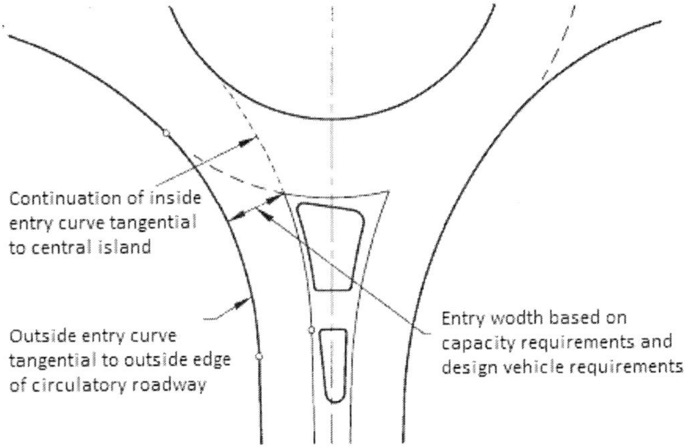

(b)

Figure 4.23 Single-lane roundabout entry design (a) for right-hand drive junctions (USA, Europe, etc.) and (b) for left-hand drive junctions (the UK, India, etc.)

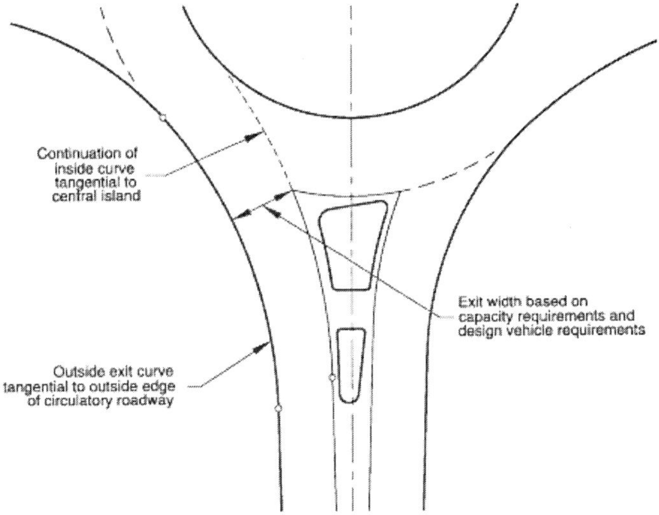

(a)

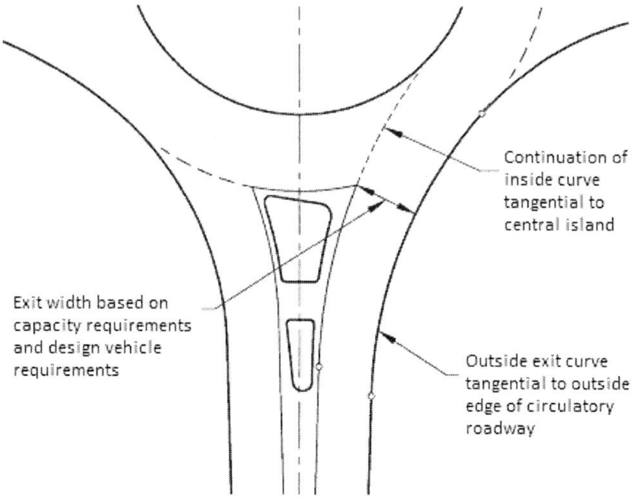

(b)

Figure 4.24 Single-lane roundabout exit design (a) for right-hand drive junctions (USA, Europe, etc.) and (b) for left-hand drive junctions (the UK, India, etc.)

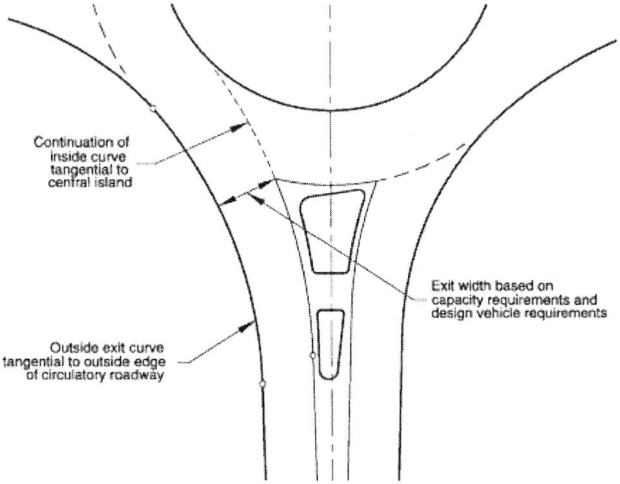

(a)

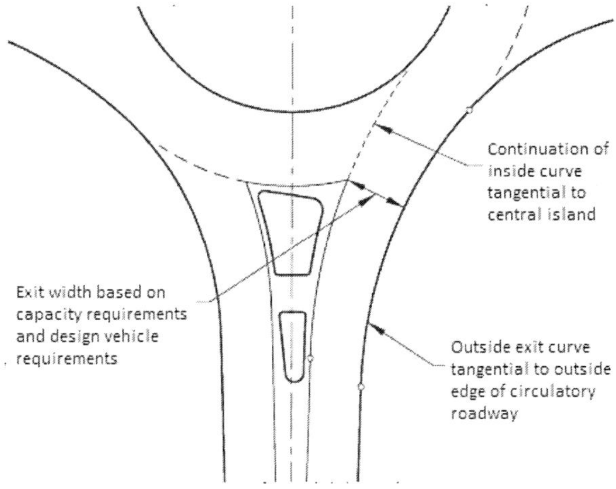

(b)

Figure 4.25 Minimum splitter island nose radii and offsets (a) for right-hand drive junctions (USA, Europe, etc.) and (b) for left-hand drive junctions (the UK, India, etc.)

4.3.5.8 Pedestrian Crossings

Pedestrian crossings at roundabouts are designed with a balance between pedestrian convenience, pedestrian safety and roundabout operations:

- Pedestrian convenience: pedestrians need the crossing locations as close as possible to the roundabout to minimise the deviation of travel. A distant crossing from the roundabout causes pedestrians to choose a shorter route other than the zebra crossing, and that may put them in greater danger.
- Pedestrian safety: the crossing location and distance are important and are to be selected judiciously. A minimum crossing distance reduces exposure of pedestrians to vehicles and hence reduces the possibility of conflict. As the driver's attention is directed to the left/right to look for gaps in the circulating traffic stream, pedestrian safety may also be critical at the give-way sign. At crossings, the advantage of the splitter island should be availed; a longer splitter island is required when crossings are located too far from the give-way line. To reduce the chance that vehicles will be queuing across the crossing, they should

also be located at distances away from the give-way line measured in increments of approximately one vehicle length.

4.3.6 Stopping Sight Distance

Stopping sight distance is the distance along a roadway required for a driver to perceive and react to an object in the roadway and to brake to a complete stop before reaching that object. Stopping sight distance should be provided at every point within a roundabout and on each entering and exiting approach.

A NCHRP Report 400, "Determination of Stopping Sight Distances (8)", recommends the formula given in Equation 4.2a or 4.2b for determining stopping sight distance (presented in metric units, followed by a conversion of the equation to imperial units).

$$d = (0.278)(t)(V) + 0.039\frac{V^2}{a} \qquad \text{(4.20a, metric)}$$

Where: d = stopping sight distance, m;
t = perception-brake reaction time, assumed to be 2.5 s;
V = initial speed, km/h; and
a = driver deceleration, assumed to be 3.4 m/s^2

$$d = (1.468)(t)(V) + 1.087\frac{V^2}{a} \qquad \text{(4.20b, imperial units)}$$

Where: d = stopping sight distance, ft;
t = perception-brake reaction time, assumed to be 2.5 s;
V = initial speed, mph; and
a = driver deceleration, assumed to be 11.2 ft/s^2.

Table 4.8 gives recommended stopping sight distances as computed from theabove equations.

4.4 Conclusions and Recommendations

This research study is formulated on junctions where the major road is located on a superelevated bend and discusses an analysis of accident frequency and severity. The study aimed to determine whether junctions on bends are prone to a greater number of accidents than similar junctions located along tangent (straight) sections and, if so, to determine the measures to improve the relative safety of such junctions.

Table 4.8 Design values for stopping sight distances

SSD (ft)	Speed (mph)	SSD (ft)	Speed (mph)
10 (3 metres)	8.1 (10 kph)	10 (3 metres)	46.4 (75 kph)
20 (6 metres)	18.5 (30 kph)	15 (4.5 metres)	77.0 (125 kph)
30 (9 metres)	31.2 (50 kph)	20 (6 metres)	112.4 (180 kph)
40 (12 metres)	46.2 (75 kph)	25 (7.5 metres)	152.7 (245 kph)
50 (15 metres)	63.4 (100 kph)	30 (9 metres)	197.8 (1765 kph)
60 (18 metres)	83.0 (130 kph)	35 (10.5 metres)	247.8 (398 kph)
70 (22 metres)	104.9 (168 kph)	40 (13 metres)	302.7 (487 kph)
80 (25 metres)	129.0 (207 kph)	45 (15 metres)	362.0 (582 kph)
90 (28 metres)	155.5 (250 kph)	50 (16 metres)	427.2 (687 kph)
100 (30 metres)	184.2* (270 kph)	55 (17 metres)	496.7 (800 kph)

*Assumes 2.5 s perception-braking time, 3.4 m/s^2 (11.2 ft/s^2) driver deceleration.

In the study, two separate junction samples are examined. The major route and minor route are both two-lane highways, which were represented and examined in the first sample of junctions. The junctions between state routes only were taken in this sample to reduce the amount of field data collection required to conduct the statistical analysis. The road curvature at the studied junctions did not appear to have a significant impact on either crash frequency or severity, as shown by the results of the study. Due to the limited number of junctions studied, these results have to be viewed with caution.

The second sample is of junctions where the major route was a four-lane divided highway, and the minor route was a two-lane road. In selecting such junctions, the state-state restriction was removed because very few such junctions could be found within the state. In addition to those junctions on bends, some additional junctions on tangent (straight) sections were included to obtain a meaningful sample size. The junctions on the tangent (straight) section were selected for being in proximity to the superelevated junctions and had similar geometric characteristics.

For the study where the major road is a four-lane divided highway, curvature appears to have restricted visibility and has a clear impact on junction safety. Crashes are found to increase in both frequency and severity at junctions where the four-lane road was on a superelevated bend with high-speed traffic. Additionally, several other findings provide additional insight into safety at these junctions, as found in the results.

Allowing adequate sight distance for vehicles attempting to enter or cross a major road is a primary concern for the geometric design of junctions. In the past, the reductions in sight distance were caused by extreme curvature and superelevation rates.

As a possible cause for an increase in crashes at junctions located on bends, the sight distances were examined. At each of the junctions in the sample, the sight time, or the time from when an approaching vehicle is visible to a vehicle at the stop bar to when the approaching vehicle reaches the collision zone, was determined.

Furthermore, the time required was determined experimentally to perform the crossing manoeuvre at each junction. To determine the effects of sight distance on junction safety, these results were then used in the modelling process. It has been found from the results of these field measurements and statistical analysis that the reduced sight distances do not appear to be directly related to curvature, and there was no clear pattern observed between sight distance and crash occurrence within the sample. This is because the sight distance provided at each junction in the sample is sufficient and therefore was not the reason for the crashes. However, this does not mean that sight distance is not an important safety factor.

The crash patterns were also examined to gain insight into what makes junctions on bends more susceptible to crashes, in addition to sight distance considerations. To identify which types of crashes were overrepresented for the sample on bends, the number of crashes by collision zone and type was used. Compared to junctions on tangent segments, the junctions on bends experience a significantly higher percentage of right/left-angle and single-vehicle crashes and a significantly lower percentage of rear-end and sideswipe collisions.

To analyse the effects of nighttime driving and different weather conditions, these crash patterns were broken down further. The junctions on bends experience a higher proportion of crashes during night conditions, but a lower proportion of crashes during adverse weather conditions, as compared to tangent junctions. Studying the nighttime crashes, only 20 per cent of crashes in the tangent sample occurred during darkness, and for the junctions on bends, 32 per cent of the crashes occurred during darkness. The statistical difference between the two proportions is highly significant, indicating thisresult did not occur by chance; rather, the combination of curvature and darkness appears to make junctions particularly susceptible to crashes. Insufficient lighting is a likely cause of this problem. It is recommended that lighting be installed where a junction is located on a superelevated bend.

The weather effect contrasts with the previous result; the number of crashes that occur during inclement weather conditions was greatly underrepresented for the case of junctions on bends. The junction sample on the tangent experienced 11.2 per cent of its crashes during rain events and 4.6 per cent during snow events, and conversely, the junction samples on bends experienced only 3.3 per cent and 1.2 per cent, respectively. Intuitively, one would expect the opposite to be true. Changes in driver behaviour under adverse weather conditions may be the reason, as the weather conditions worsen, drivers may begin to drive more cautiously.

Although the effects of weather and night conditions can be reasonably explained, some results of the study may require further research. Theoretically, drivers on a minor road may overestimate the amount of time available to cross the major road. Some investigation was done to examine the difference between the distance a vehicle travels during the sight time and the straight-line distance between a stopped vehicle and one entering the sight zone.

In some cases, junctions which are skewed to the left (from the inside of the bend) tended to experience a significantly larger number of crashes. While statistically significant results were not obtained from modelling, it appears that there may be a hazard associated with a combination of curvature and skew. Further research on this issue may prove helpful in explaining why this type of crash is more prevalent. The study may be able to determine the relationship between crashes and curvature. As the degree of curvature is increased, crash frequency increases significantly within the sample. These results are not random and are found to be highly statistically significant. There appears to be a relationship between the two variables. Based upon the results of the statistical analysis, critical design values for curvature may be determined.

In summary, the important findings and recommendations for designers and safety managers from this study are:

1. Crashes were found to increase in both frequency and severity at junctions where a four-lane major road was on a bend.
2. Sight distance in the studied sample does not appear to be affected by the horizontal bends. Furthermore, no significant relationship between sight distance and crash frequency was found.
3. Junctions on bends experience a significantly higher percentage of right-angle and single-vehicle crashes and a significantly lower percentage of rear-end and sideswipe collisions than junctions on tangent segments.
4. In comparison to tangent junctions, the junctions on bends experienced a higherproportion of crashes during night conditions. It is recommended that lighting installation be considered in cases where a junction is located on a bend, particularly where severe superelevation is present.
5. The study indicated the relationship between junction safety and the horizontal curvature of the main road and indicated a considerable increase in crash frequency when the horizontal bend radius is shorter than 3,000 ft. These design values would apply only to four-lane high-speed rural junctions with STOP signs on crossing roads.

References

Credit for using some figures/text: Highway Safety at [Intersections and Roundabouts]

IRC:65-2017 on "Guidelines for Planning and Design of Roundabouts"

Visvesswaraya National Institute of Technology, Nagpur, For circulation to participants of NRIDA Training Programs, Rural Road Safety Course, Safety in Junction Design, https://pmgsy.nic.in/sites/default/files/RSA%20%28Session%207%20_%208%29.pdf

AASHTO_2001_A_Policy_On_Geometric_Design

CHAPTER 5

SPEED LIMITS

Abstract: This chapter covers a general overview, speed limits, design speed, operating speed, posted speeds, legal speeds, traffic characteristics overview, turning roadways and junction corner radii, road safety improvement programme (HSIP), context-sensitive design, design speed selection process, comprehensive, complete, and logical procedure to select design speed, the anticipated operating speed (85th per centile speed), directional distribution (d), and standards based on design speed.

5.1 General

The primary objective of road design and operations is to make roads safe; it is necessary to build the various features according to the operating speed. Therefore, vehicle speed is a crucial control element in the road design process.

The greatest dangers are listed below, while the road users are teenage drivers:

- Distracted driving (texting while driving).
- Impaired driving (alcohol and drugs).
- Inconsistent or no seat belt use.
- Number of passengers.
- Speeding.

Figure 5.1 Real-time vehicle speed message (VSM)

The speed at which road alignment should be designed is called design speed. Road plans and construction that are based on the design speed concept are supposed to result in consistent alignments. However, over time, there has been growing concern about the design speed selection procedure. The main concern is that the design speed is being exceeded by many drivers on roads that have lower design speeds (below 90 km/h).

The assumption that consistency can be achieved by building individual alignment features upon the same design speed is incorrect. Furthermore, the design speed concept fails to provide alignment consistency, as is seen in the research.

AASHTO policy also assumes that drivers operate uniformly on a roadway that has all its alignment features based on the same design speed. This assumption is highly unrealistic. Australia's McLean states that "design speed is no longer the speed adopted by the faster driving group but has become a design procedural value" [15], or "a speed which is 'safe' from the designer's point of view" [9]. As a result, it seems that a more flexible approach should be used to select the design speed for a given roadway.

In 1991, a study of 28 horizontal bends in Maryland, Virginia, and West Virginia in the US revealed a disparity between the design speed and operating speeds [9]. One bend had a design speed of 100 km/h, and the other 27 bends had design speeds ranging from 30 to 80 km/h. All bends with design speeds less than or equal to 80 km/h had 85th percentile speeds that exceeded the design speed. The bend of 100 km/h (design speed) was the only one that had an 85th percentile speed lower than the design speed. This implies that desired speeds are being under-estimated even though AASHTO guidelines advise that "the speed selected for design should fit the travel desires and habits of nearly all drivers" [1].

Per centile = (Number of Values Below "x" / Total Number of Values) × 100
Per centile Formula, $P = (n/N) \times 100$

Where,
n = ordinal rank of the given value or value below the number
N = number of values in the data set
P = per centile

One of the reasons for this is that AASHTO guidelines advise that "above minimum design values should be used where feasible", while design speed controls just the minimum values. This approach results in a situation where drivers would speed up on a few successive, relatively flat bends, not knowing that the next bend is a sharp one. On the other hand, the presence of a bend with a smaller radius than the minimum radius may not be a cause of inconsistency if that bend follows and precedes bends with slightly larger radius.

An 85th percentile design speed is normally adopted, which is defined as the speed that is greater than the speed of 85 per cent of drivers. In some countries, this is as high as the 95th to the 98th percentile.

Traffic engineers rely on the 85th percentile rule to help establish speed limits on national roads. Typically, the speed limit is set to the speed that separates the bottom 85 per cent of vehicle speeds from the top 15 per cent. For example, if the speeds of 100 vehicles are measured and 85 vehicles are travelling at 60 kph (37 mph) or less, the speed limit for the subject street could be set at 56 kph (35 mph). Statistically, the 85th percentile speed is slightly greater than the speed that is one standard deviation above the mean of a normal distribution. Here, the 85th percentile speed as per the survey is 60 kph (37 mph), and the speed limit posted at the site is 56 kph (35 mph).

5.2 Speed Limit

Speed is one of the most important factors considered by travellers in selecting routes or transportation modes. The speed of vehicles on a road depends, in addition to the capabilities of the drivers and their vehicles, upon five general conditions: the physical characteristics of the roadway, the amount of roadside interference, the weather, the presence of other vehicles and speed limitations (established either by law or by traffic control devices). Although any one of these factors may govern travel speed, the actual travel speed on a facility usually reflects a combination of these factors.

The objective for the design of any engineered facility used by the public is to satisfy the public's demand for service in an economic manner with efficient traffic operations and with low crash frequency and severity. The facility should, therefore, accommodate nearly all demands with reasonable adequacy and should only fail under severe or extreme traffic demands. Because only a small percentage of drivers travel at extremely high speeds, it is not economically practical to design for them. They can use the roadway, of course, but will be constrained to travel at speeds less than they consider desirable. On the other hand, the speed chosen for design should not be that used by drivers under unfavourable conditions, such as inclement weather, because the roadway would then be inefficient, might result in additional crashes under favourable conditions, and would not satisfy reasonable public expectations for the facility.

There are important differences between design criteria applicable to low- and high-speed designs. For design purposes, the following definitions apply:

- Low speed is 70 km/h and below.
- High speed is 50 km/h and above.

Several tables and figures for high-speed conditions will show values for 70 km/h to provide information for transitional roadway sections.

Different types of speed terms have been defined and are used in the design process. They are listed below:

5.2.1 Design Speed

The current AASHTO definition is:

"The maximum safe speed that can be maintained over a specified section of road when conditions are so favorable that the design features of the road govern".

Design speed is applied to individual geometric elements rather than to sections of a road. The main factors by which design speed can be determined are:

- Sight distance.
- Radius.
- Superelevation (e = V2/(225xR).
- Friction.
- Topography: the physical landscape of the area.
- Anticipated operating speeds: what speeds drivers are likely to use on the road.
- Adjacent land use: the surroundings of the roadway.
- Functional classification: the road's intended purpose (e.g., a major road vs. a local street).

To provide a logical procedure for selecting a design speed, it is necessary to define additional speed terms. Another definition of design speed was adjusted by AASHTO while discussing the policy on road classification: "the maximum approximately uniform speed which probably would be adopted by the faster group of drivers but not, necessarily, by the small percentage of reckless ones." In AASHTO's "Policy on Geometric Design of Rural Roads", the design speed concept was changed from the behavioural measure to the minimum standard for all design elements related to the chosen design speed: "A speed used for the design and correlation of the physical features of a road that influence vehicle operation" [1].

5.2.2 Operating Speed

Operating speed is the speed at which drivers operate their vehicles. The 85th percentile speed of a sample of observed speeds is usually regarded as the operating speed on a particular section [12].

5.2.3 Posted Speed

Posted speed refers to the maximum speed limit posted on a section of road using the regulatory sign [12]. Speed limits are typically set based on a roadway's design. For example, whether it is a narrow two-lane road or a modern controlled access motorway and whether it is in an urban, suburban or rural setting. The posted speed limits, if perceived as reasonable by most drivers, should result in uniform traffic flow. Advisory speed plates are often used in cases where design and operating speeds are different [9].

Speed limits should also give the motorist an idea of what is a reasonable speed to drive in an unfamiliar location. Advised speeds generally seem low to most drivers because AASHTO's maximum side-friction factors are based on 1940s measurements. These values may not reflect the performance and tolerance of today's vehicles and drivers.

It has been pointed out that measuring the speed distribution of a roadway and setting the speed limit to the 85th percentile may reduce the need for enforcement and reduce crash risk by narrowing vehicle speed variation. However, numerous studies of travel speeds on rural roads have shown that 85th percentile speeds increased when speed limits were raised to 65 mph. Therefore, the 85th percentile speed is not stationary; rather, it is a moving target that increases when speed limits increase.

"In states where speed limits were raised to 65 mph in 1987, there was a 15-20 per cent increase of fatal accidents on rural interstates each year" [10]. Two assumptions that follow from basing a design speed on the 85th percentile speed are: motorists can decide on an acceptable travel speed, and the speed limit is set at the 85th percentile speed. TxDOT suggests using the design speed as the initial speed limit when a new facility is designed [22]. After a reasonable operation period, an 85th percentile speed can be used to set the new speed limit.

5.2.4 Legal Speed

The legal speed is the minimum or maximum speed limit that can be posted on a particular road class. Under normal roadway conditions, it is illegal to go below the minimum speed limit or to exceed the maximum speed limit.

5.2.5 Traffic Characteristics Overview

Information on traffic characteristics is vital in selecting the appropriate geometric features of a roadway. Necessary traffic data includes traffic volume, traffic speed, and the percentage of lorries or other large vehicles.

5.2.6 Traffic Volume

Traffic volume is an important basis for determining what improvements, if any, are required on a road or street. Traffic volumes may be expressed in terms of average daily traffic or hourly volumes. These volumes may be used to calculate the service flow rate, which is typically used for evaluations of geometric design alternatives.

5.2.7 Average Daily Traffic

ADT represents the total traffic for a year divided by 365, or the average traffic volume per day. Due to seasonal, weekly, daily, or hourly variations, ADT is generally undesirable as a basis for design, particularly for high-volume areas. ADT should only be used as a design basis for low and moderate-volume areas, where more than two lanes are not justified.

5.2.8 Design Hourly Volume

The design hourly volume (DHV) is usually the 30th highest hourly volume for the design year, commonly 20 years from the time of construction completion. For situations involving high seasonal fluctuations in ADT, some adjustment of DHV may be appropriate.

For two-lane rural roads, the DHV is the total traffic in both directions of travel. On roads with more than two lanes (or on two-lane roads where important junctions are encountered or where additional lanes are to be provided later), knowledge of the directional distribution of traffic during the design hour (DDHV) is essential for design. DHV and DDHV may be determined by the application of conversion factors to ADT.

5.2.9 Computation of DHV and DDHV

The percentage of ADT occurring in the design hour (K) may be used to convert ADT to DHV as follows:

$$DHV = (ADT)(K)$$

The percentage of the design hourly volume that is in the predominant direction of travel (D) and K are both considered in converting ADT to DDHV, as shown in the following equation:

$$DDHV = (ADT)(K)(D)$$

5.2.10 Directional Distribution (D)

Traffic tends to be more equally divided by direction near the centre of an urban area or on loop facilities. For other facilities, D factors of 60 to 70 per cent frequently occur.

5.2.11 'K' Factors

'K' is the percentage of ADT representing the 30th highest hourly volume in the design year. For typical main rural roads, 'K' factors generally range from 12 to 18 per cent. For urban facilities, 'K' factors are typically somewhat lower, ranging from 8 to 12 per cent.

5.2.12 Projected Traffic Volumes

Projected traffic volumes are provided by the Transportation Planning and Programming (TPP) Division upon request and serve as a basis for the design of proposed improvements. For high-volume facilities, tabulation showing traffic converted to DHV or DDHV will be provided by TPP if specifically requested. Generally, however, projected traffic volume is expressed as ADT with 'K' and 'D' factors provided.

NOTE: If the directional ADT is known for only one direction, the total ADT may be computed by multiplying the directional ADT by two for most cases.

5.2.13 Service Flow Rate

A facility should be designed to accommodate the design traffic volumes (ADT, DHV, DDHV). The necessary capacity of a roadway is initially based on a set of 'ideal conditions'. These conditions are then adjusted for the 'actual conditions' that are predicted to exist on the roadway section. This adjusted capacity is termed service flow rate (SF) and is defined as a measure of the maximum flow rate under prevailing conditions. Adjusting for prevailing conditions involves adjusting for variations in the following factors:

- Lane width.
- Lateral clearances.
- Free-flow speed.
- Terrain.
- Distribution of vehicle type.

The SF rate is the traffic parameter most used in capacity and level-of-service (LOS) evaluations. Knowledge of road capacity and LOS is essential to properly fit a planned road or street to the requirements of traffic demand. Both capacity and LOS should be evaluated in the following analyses:

- Selection of geometric design for a junction.
- Determining the appropriate type of facility and number of lanes warranted.
- Performing ramp merge/diverge analysis.
- Performing weaving analysis and subsequent determination of weaving section lengths.

All roadway design should reflect proper consideration of capacity and level of service procedures as detailed in the TRB's *Road Capacity Manual*.

5.2.14 Terrain

Level or rolling are the two types of terrain often presented when choosing appropriate design criteria since these are the predominant terrains in Texas, US. Some areas of the El Paso District and some areas of other western districts may be considered mountainous. Whenever mountainous conditions are encountered, refer to AASHTO's *A Policy on Geometric Design for Roads and Streets* for appropriate design criteria and design considerations.

5.2.15 Turning and Junction Corner Radii

Traffic volume and vehicle type influence the width and curvature of turns and junction corner radii. Minimum designs and turning templates for various vehicles are to be achieved as per applicable design standards.

5.2.16 Safety

TxDOT is developing additional strategies to incorporate safety into its system, contributing to the goal of eliminating traffic deaths statewide by 2050 (Vision Zero). Towards this end, the department uses several initiatives related to developing and operating a safer road system.

5.2.17 Highway Safety Improvement Programme (HSIP)

HSIP is a federally funded programme administered by TxDOT's Traffic Safety Division that allows road safety improvements through strategic safety planning and performance measures. HSIP requires states to develop and implement a strategic road safety plan (SRSP). The purpose of the SHSP is to identify and analyse road safety problems and correction opportunities, including projects or strategies to address them, evaluate the accuracy of data, and prioritise the proposed improvements. The SHSP establishes five target areas, including the number and rate of fatalities, serious injuries, and non-motorised fatalities and serious injuries.

5.2.18 Safety Analysis

Safety analyses use crash data, traffic volume and roadway geometrics. There are various analytical tools and methods available for analysing the safety of potential improvements, including historical crash data analysis, the highway safety manual (HSM) predictive method, and a crash modification factor (CMF) evaluation.

A historical crash data analysis involves a review of three to five full calendar years of crash data with respect to characteristics such as severity, crash types, frequency, rates, patterns, clusters and contributing factors. Crash diagrams, such as heat maps, bar charts and other maps graphically showing the crash locations, are used to help interpret the data. A crash rate is the number of crashes that occur at a given location during a specified time, divided by a measure of exposure. Crash rate is calculated per 100 million VMT using the following formula: crash rate = 100,000,000*A/(365T*V*L),

Where:
A = Number of reported crashes (in section or at location)
T= Time frame of the analysis, years
V = AADT, vehicles/day
L = Length of section, miles

The HSM predictive method provides procedures to analyse safety performance in terms of crash severity levels and collision types. There are various spreadsheets and software developed to automate predictive analyses.

CMFs are used to estimate the anticipated impact of a countermeasure or mitigation on safety performance. A CMF is an index of the expected change in safety performance following a modification in the traffic control strategy or a design element. It can be used to estimate the safety effectiveness of a given strategy, compare the relative safety effectiveness of multiple strategies. The CMF Clearinghouse (ref. www.cmfclearinghouse.org) offers a repository of CMFs.

Design departments could create a new sub-department for traffic simulation and safety analysis. The purpose of the new department could be to provide guidance and support for safety analysis. The design department could also develop a safety tool which can be used to estimate the safety score of a particular roadway segment by selecting various design elements. Rural roads tools (two-lane and multi-lane rural) are available for use, with development currently underway for future junctions and an urban roads tool. Use of the applicable tools should begin during project scoping to evaluate the safety impacts of design decisions.

5.2.19 Safety Analysis Data

The DOT's State-Wide Traffic Analysis and Reporting System is a good resource for traffic data, while the Crash Records Information System is utilised for crash data. Historical crash data is analysed to identify the potential safety problems that might be corrected. CRIS generates detailed crash data used to determine unsafe locations, crash types, and contributing factors.

Statewide average crash rates are used in the crash rate analysis method and are useful to compare against the crash rates of a particular road segment/junction. TxDOT maintains ten years of crash data. Summary reports of crash data are available at the following link: https://www.txdot.gov/government/enforcement/annual-summary.html
TxDOT uses crashes per year, the level of crash severity and the crash type. Severity is defined on the KABCO scale as follows:

K. Fatal injury.
A. Incapacitating injury.
B. Non-incapacitating injury.
C. Possible injury.
O. No injury.

Where accident frequencies include a wildlife-vehicle collision as a contributing factor in the CRIS records, consult the District Environmental Coordinator or the Environmental Affairs Division to determine if a wildlife crossing structure could improve safety at these hot spot areas. The ENV Natural Resource Management Section and District Environmental Coordinators can provide information to conduct a hot spot analysis and details on types of crossings, including schematics used within TxDOT and other states.

5.2.20 Context-Sensitive Design

The Intermodal Surface Transportation Efficiency Act (ISTEA) and the National Road System legislation of 1995 are the first steps toward context-sensitive design. During the design process, engineers must cope with different aspects of design (environmental, etc.) that will sometimes conflict with the engineering design concept. Context-sensitive design could be described as design in harmony with the community and the environment. Traditional training and education have been challenged by the need to approach this problem from various perspectives.

5.2.20.1 Land Use

Because of a lack of coordination between land use and transportation plans, roadway facilities are not designed to be of optimum service to the community. To result in an appropriate road design speed selection and be of maximum benefit to the community, closer coordination is essential between land use and transportation planners. The selection of design speed should consider the expected change in a facility and the existing scenario over time, for instance, from a rural environment to a suburban one [22]. The road design process must recognise and incorporate the phenomena of growing pollution and other environmental issues in the region.

5.2.20.2 Environmental Issues

Environmentally sensitive areas such as wetlands, animal habitats, nature conservation areas, etc., impose certain limitations when choosing locations for transportation facilities. Aesthetic issues must be considered. When designing a road through a scenic area, the general view of the area should not be obstructed. Preserving the natural environment and ensuring liveability are essential considerations when designing/building a new roadway.

5.2.20.3 Community Liveability

Roads are no longer being designed only for use by cars. There are a variety of other users that include pedestrians, cyclists, buses, farm equipment and heavy lorries, etc. Therefore, approaches along with solutions to road design must be creative and flexible, and in turn, to the design speed selection process. In designing a transportation facility, an equally important goal is to preserve the attributes and characteristics of the community because it is a major 'client'.

5.2.20.4 Public Involvement

Citizen participation should be a significant part of the road design process. In the design process, the needs and desires of the local and public community should be incorporated. Organising public workshops and other events should be encouraged to bring the public together with engineers to discuss the road design process.

5.2.20.5 Summary of the Context-Sensitive Design

A compromise between engineering judgement (safety, geometric design) and environmental aspects is entailed while designing roads in a sensitive context. Therefore, context-sensitive design aspects are successful for roadways that provide the best mobility while ensuring the highest safety standards and address the natural and human environmental aspects.

5.2.21 AASHTO and Federal Guidelines

In the *Policy on Geometric Design of Roads and Streets* (AASHTO 1994), the federal guidelines and codes are found, which are the most important and most used resources for designing roads and urban streets.

- When the conditions are favourable, the design features of the road govern the design speed, which is the maximum safe speed that can be maintained. The topography, adjacent land use and type of road are directly related to the assumed design speed.
- The selected design speed should be the highest that is allowed for maximum safety, mobility and efficiency under constraints of environmental quality, economics, aesthetics, and social and political impacts.
- The selected design speed and the operating speed should be similar. Before enabling driving at relatively higher speeds, the topography conditions must be considered. As per the physical road limitations and traffic, the drivers adjust their speeds. The influence of road hierarchy is almost negligible.
- Based on the travel desires and habits of most drivers, a suitable speed is selected for design. A cumulative distribution of vehicle speeds when plotted as a percentage of vehicles versus observed

speeds has the typical ‘S’ pattern. In this speed distribution curve, the design speed should be a high-percentage value, i.e. almost all typically desired speeds should be considered.

If there is a straight section, wide bends or rolling terrain, drivers will naturally use higher speeds, and such road sections should be designed for them. The average trip length is a pertinent consideration in selecting design speed; it follows that the longer the trip, the greater the desire for expeditious movement.

Changes in terrain and other physical features may dictate a change in design speed in certain sections. In such conditions, to guarantee safety, a low design speed should be implemented gradually. Generally, with respect to elevation, the design speed does not change significantly. Elevated structures are subject to rapid freezing in northern climates.

Thus, rates of superelevation may have a greater impact on design speed. To ensure a balanced design, all new road features should be based on the selected design speed. A more consistent design speed selection can result if the above-stated guidelines are followed. However, as no quantitative guidance for these guidelines is provided by AASHTO, this will eventually result in inconsistent design.

5.2.22 Design Speed Selection Process

Applications of standards in various states in the US are based on AASHTO design criteria: “most of the AASHTO design criteria are expressed as minimum values or as ranges of values for particular conditions. There is some degree of flexibility in the specific practices that might be adopted by an individual agency”.

It is normally required that the minimum design standards be applied, but occasionally there may be conditions that warrant consideration of a lower standard. In such cases, departure from design standards must be thoroughly documented for review.

5.2.23 Standards Based on Design Speed

The following standards are based on design speed:

- Curvature and super elevation.
- Required sight distances.

Both these road design parameters are directly related to the design speed. Since they are also closely related to traffic safety, minimum values for these standards must be achieved. The design speed selection procedure is performed by extracting design speed values from the following applications:

- Two-lane rural roads.
- Rural multi-lane roads.
- Urban multi-lane roads.
- Two-lane urban arterial and collector roads.

5.2.23.1 Comprehensive, Complete and Logical Procedures to Select Design Speed

Design Speed is defined as: "The maximum safe speed that can be maintained over specified section of road when conditions are so favorable that the design features of the road govern" [1]. Also, it has been defined as "The theoretical value of speed governing the design of a road elements in conditions of safe and comfortable travel in a free traffic flow" [18].

"Although clearly identified as design criteria policies or guidelines, the AASHTO publications entitled 'A Policy on Geometric Design of Rural Roads' (Blue Book) and 'A Policy on Design of Urban Roads and Arterial Streets' (Red Book) have been consistently accepted by the courts as the nationally organized standards for road and Street Design [2]". The general opinion of design engineers of the Department of Transportation, Delaware (DelDOT), is that the design speed selection procedure, based on AASHTO guidelines, does not provide enough flexibility.

Several sources show that design speed does not provide consistency in road design. Many drivers often exceed it, thus increasing the likelihood of accidents. "A tendency exists for some drivers to travel faster than the design speed on which the original design of the road section was based by substantial amounts, especially at the lower design speed levels" [20]. Basic design goals include safety, efficiency, economy, comfort, ease of maintenance, environmental quality, social issues, usage of the facility, etc. Commonly, most of the goals mentioned above should be attained with limited resources. The design engineers need information of a certain type to start the process of selecting the design speed.

5.2.24 Necessary Information for Selecting Design Speed

Selecting a suitable road and its appropriate design speed requires the following details:

- Location, to classify the road as arterial or collector or local, etc.
- Area characteristics.
- Topography.

Based on such details, an initial road alignment can be prepared and should be accompanied by an estimate of the project cost to obtain its approval by the initiator (government, state agency or private). Further details will be needed next, toward the beginning of the design speed selection process.

In the case of reconstruction or upgrading of an existing facility, the current problems and desired objectives should specify the necessary resources. If the information on design speed is not available, it should be estimated from geometric features like sight distance and horizontal curvature along road sections.

Once funding for the project is available, a detailed design of the road can be taken up. The designer must be familiar with the project to maintain an accurate cost-benefit analysis and selected design speed. During that process, the designer needs to decide which part of the project is going to be reconstructed (or rehabilitated) and which part requires a new design and rebuilding, to establish a reasonable level of safety and service.

The following sections present essential information that an engineer must understand. These include the whole methodology for selecting an appropriate design speed, and, at the same time, place the design speed within the design process.

5.2.24.1 Terrain Characteristics

- Level or plain terrain.
- Rolling terrain.
- Mountainous terrain.

Level or plain terrain is considered any combination of vertical and horizontal alignment that allows heavy vehicles to maintain approximately the same speed as passenger cars [16]. It generally includes short grades of no more than two per cent. An alternative second definition for level terrain is an area on which the grade ranges between 0 to 8 per cent [9].

Rolling terrain is considered any combination of vertical and horizontal alignment that causes heavy vehicles to reduce their speeds substantially below those of passenger cars, but does not cause heavy vehicles to operate at crawl speeds for any significant length of time [16]. An alternative second definition for rolling terrain is an area on which the grade ranges between 2.1 to 15 per cent [19].

Mountainous terrain is considered any combination of vertical and horizontal alignment that causes heavy vehicles to operate at crawl speeds for significant distances or at frequent intervals [16]. An alternative second definition for mountainous terrain is an area where the grade is above 15 per cent. Once terrain characteristics have been specified, the designer should select the type of road (urban/rural).

5.2.24.2 Type of Road for the Projected Year

1. urban
2. rural

Urban areas, within boundaries that are set by the state or local officials, are defined as having populations of 5000 or more people. Urban areas are further classified as urbanised if the population is larger than 50,000. They are classified as small urban areas if the population is between 5000 and 50,000. [1].

Rural roads do not fall into the classification criteria for urban roads. Urban areas should be further classified by several attributes: residential, commercial, school zones, resort areas, high-density, cyclists, etc. Similar classification might be important for rural areas: wetlands, industrial zones, zones of economic interest, etc.

The next step is determining the function of the road.

5.2.24.3 Functional Classification with Predominant Services and Access

Urban roads

- Interstate—restricted and fully controlled access.
- Freeway or expressway—controlled access.

- Other principal arterial—partial or no control of access.
- Minor arterial—has more access than major roads, and has local bus routes.
- Collector—has land access and traffic circulation within residential, commercial and industrial areas, serves local bus routes.
- Local—permits direct access, and usually does not have bus routes.

Rural roads

- Major or minor arterial—this is a freeway with full access, or other major arterial.
- Major or minor collector—provides approximately equal levels of access.
- Local—has access to land adjacent to the facility.

5.2.24.4 Projected Traffic Volumes for the Target Year

Forecasting the number of vehicles is supposed to be used for road design. It needs the base year's actual traffic count, annual growth rate, seasonal variations and projected estimated value at the target year.

5.2.24.5 Acceptable Traffic Flow Quality

This is a level of service which reflects social costs such as travel time delay and energy consumption. The lower limit of design speed (Vdl) should be determined after considering:

- Traffic flow quality (level of service).
- Terrain characteristics.
- Functional classification of the road.

To determine the number of lanes required for the desired level of service, anticipated traffic volumes should also be used. The policy mentions that planning a new road is to provide as many lanes as necessary for the anticipated volume. The minimum acceptable level of service is C or D. However, in the design speed selection process, the traffic volume and level of service do not play a role.

5.2.24.6 Traffic Composition

- Passenger car (percentage)
- Single unit truck (percentage)

- Single unit bus (percentage)
- Semi-trailers (percentage)

For safety, speed differentials must not exceed 15 km/h. The presence of a significant number of lorries or vehicles that have six or more wheels can result in non-uniformity. A potential solution is adding a lane for slow-moving vehicles. If this is not practical, lower design speeds should be chosen.

Speed Variance

It has been shown that speed variance on a road segment is minimal if the difference between the design speed and the posted speed limit is between 8 and 16 km/h. However, technology is much more advanced than in the past. Therefore, speed differences are much smaller. Nevertheless, DOT designers do not see traffic composition as a vital factor during the design speed selection process.

5.2.24.7 Constraints That Prevent Achieving Higher Design Speeds

Several difficulties might limit the magnitude of the design speed:

- High construction costs.
- Difficulties in acquiring right-of-way (physical objects: such as buildings, mountains, houses, etc.)
- Social (public concern, residential areas, etc.)
- Environmental (wetlands, etc.)
- Energy-related (higher speeds, higher fuel consumption).
- Presence of pedestrians, cyclists and school zones.

These constraints can result in posted speed limits lower than the desired design speed. To avoid over-design, sections that encompass some of those constraints (or part of them) should be considered primarily. Hence, a more adequate value of design speed might be maintained.

In most design guides, the selection of design speed for a specific road section of a particular classification is primarily influenced by the nature of the terrain (level, rolling, mountainous). In addition, the selection of design speed is influenced by motorists’ expectations in relation to the free speed at which it is safe to drive (in rural areas), or legal to drive (in urban areas). In urban areas, many disruptions to flow (e.g., from junctions) result in free speeds that are less than those on rural roads. Drivers are also more likely to adjust their desired speeds to the imposed speed limits. Consequently, the design speed for the new road or a major road improvement in an urban area is usually selected after considering the likely speed limit regime for the road. [3].

Speed limits are often based on the limitations of drivers' sight distances or road conditions like the surrounding environment (trees, objects), poor pavement, school zones, etc.

5.2.24.8 the Anticipated Operating Speed (85th Per Centile Speed)

The 85th percentile speed is generally considered the most appropriate choice for design speed [3]. Anticipated operating speed for a new road can be determined from operating speeds on similar roads. Several analyses have shown that the angle of a bend is the best predictor for determining operating speeds. "Relative to other alignment properties, road curvature had the greatest influence on driver speed behaviour". Lamm and Choueiri [16] developed a model that relates the 85th per centile speed to the degree of horizontal bend: V85=52.656-1.135D (D-degree of bend). "For new roads, operating speed V85 can be selected using the general range of bend radii likely to be adopted. For reconstruction of the existing facility and realignment the existing V85 can either be estimated from Table 2 or be measured directly from speed studies on existing straights" [21]. "In practice, designers obviously cannot take direct account of actual speed expectations on proposed alignment sections for new roads or major improvements and they rely instead on typical speed distribution data previously obtained on similar existing roads" [3].

5.2.24.9 the Anticipated Posted Speed Limit for the Road

For each road section, the anticipated posted speed should be considered in terms of the constraints discussed in this chapter. The initial design speed value (approximately Vdl+20km/h) should be chosen in a way that reflects the anticipated operating speed or the anticipated posted speed limit that is generally larger. The upper limit of the design speed range (Vdu) should be used as the speed limit for the appropriate road class, plus 20 km/h for additional safety.

5.2.25 Process of Selecting Design Speed

The expected 85th percentile speed should be obtained before starting the design speed selection process.

5.2.25.1 Expected 85th Percentile Speed for Each Road Section

The computation of the radius or the degree of bend is based on the initial design speed value. Each bend and each tangent (DC=0) has its expected 85th per centile speed: V85=52.656-1.135D (D-degree of bend). Once the initial design speed has been selected, which is the 85th percentile speed, it is necessary to perform a consistency evaluation to ensure uniformity of operating speed on the given road section.

5.2.25.2 Consistency Evaluation

By exceeding the design speed on a specific bend or from a significant difference in operating speed on two successive sections because of a large rate of change of curvature, potential design inconsistencies can result. Krammes et al. [12] showed that, as applied in the US, there are weaknesses in the design speed concept. “The weaknesses make it ineffective at identifying and addressing operating-speed inconsistencies that horizontal bends, whose design speed is less than the driver’s desired operating speed. This inconsistency increases accident potential”.

The process of checking for consistency includes:

- Calculating the change in degree of bend (ΔDC) between successive elements.
- Calculating the change in 85th percentile speed between successive elements.
- Calculate the difference between the 85th percentile speed and the existing or selected design speed Vd for the investigated bend or tangent.

Definition of Good Design:

A design is considered ‘well’, i.e., there is consistency in the horizontal elements, if $\Delta DC \leq 5$ and $\Delta V \leq 6$mph (10km/h) between successive design elements and $|V85\text{-}Vd| \leq 6$mph (10km/h) for the investigated bend or tangent.

A design is considered ‘fair’ if $5 \leq \Delta DC \leq 10$, 6mph$\leq \Delta V85 \leq$12mph (20 km/h) and 6mph$\leq |V85\text{-}Vd| \leq$12mph. These road sections exhibit some minor inconsistencies in geometric design. To correct these defects, traffic warning devices may be used. However, despite traffic warning devices, road sections with $5 \leq \Delta DC \leq 10$ have a double rate of accidents on average, compared with well-designed roads.

A design is considered 'poor', i.e., there are strong inconsistencies in the horizontal elements, if V85>12 mph between successive elements and |(V85>Vd)|>12 mph (or 20km/h) for the investigated bend or tangent. Average accident rates are more than four times higher than on the 'well' designed roads. The next step is to memorise the initial value of the selected design speed and make corrections that minimise differences between successive road features.

5.2.26 Design Speed Selection Process for Various Road Classes

Introduction and definitions

The first step in the design process is to define the function that the facility has to serve [1]. Sections 5.3.1 and 5.3.2 present a list of road classes, which are found in Reference 1 and State DOT manuals. Design speed selection criteria will be discussed for the most common road types. The following is a short explanation of some terms:

The Interstate System

This is a part of the principal arterial system and is a network of routes selected by the state. The highest class of arterial road is a freeway, which is defined as an expressway with fully controlled access, meaning that access to and egress from these facilities are permitted only at controlled locations such as entrance and exit ramps [1]. On the other hand, expressways have partial control of access, i.e. access or egress may be permitted directly from or to abutting property or via a limited number of at-grade junctions [1]. The level of service, which is the quality of traffic flow, provides a rational and cost-effective basis for the selection of design speed [1].

Control of access

This is the degree of through traffic interference with conflicting vehicles and/or pedestrian movement entering, leaving or crossing the road. A facility is ranked higher if its control of access is superior. Table 5.1 presents the level of service that corresponds to the road type and location [1].

Table 5.1 Type of area and appropriate level of Service [1]

Road Type	Rural Level	Rural Rolling	Rural Mountainous	Urban and Suburban
Freeway	B	B	C	C
Arterial	B	B	C	C
Collector	C	C	D	D
Local	D	D	D	D

5.2.26.1 Urban

5.2.26.1.1 Principal arterial definition

The principal urban arterial system serves the major CBDs (Central Business Districts) of urbanised areas, the largest traffic volume corridors and the longest trip distances and carries a high proportion of the total area travel even though it constitutes a relatively small percentage of the total roadway network. The system should be integrated both internally and between major rural connections [1]. The designer must be familiar with the characteristics of both systems to provide adequate transition as the arterial traffic flows between the rural and urban settings [1].

This system is not restricted to controlled access functional routes. The main characteristic of urban arterials is mobility with limited or restricted service to local development. Urban principal arterials provide travel services to major traffic movements. Before designing an urban arterial, it is important to establish its extent and need. If the need is established, steps must be taken to protect the design from such things as roadside development or unplanned locations for major traffic generators.

5.2.26.1.1.1 Principal arterial Design Speed Selection

Generally, the design speed of urban roads should reflect the desired safe operating speed during off-peak hours, but should not be so high that it exceeds the limits of prudent construction, right-of-way and socioeconomic costs. Design speed for urban arterials generally ranges from 60 to 100 km/h, and occasionally may be 50 km/h. The lower speeds (less than 60 km/h) apply to the central CBDs and intermediate areas. Higher speeds are more applicable to the outlying business and developing areas. On urban roads, a design speed of 100 km/h is normally used. The corridor of the main line may also be relatively straight, and the location of junctions may permit high-speed design. Under these conditions design speed of 110 km/h is desirable because a higher design speed means the road has a better overall quality and higher safety. Where terrain is mountainous, a design speed of 80 km/h or 100 km/h may be used, which is consistent with driver expectancy. However, it should be pointed out that the difference between 80 and 100 km/h is large. Generally, urban multilane roads have a 10 km/h lower design speed than corresponding rural multilane roads. For acceptable degrees of congestion, rural and suburban principal arterials should be designed for level of service C. Heavily developed sections of metropolitan areas may necessitate the use of level of service D [1].

5.2.26.1.1.2 Principal arterial—Design Speed Selection—Other References

For urban principal multilane roads in low-density developed areas and for posted speed limits of 80 or 90 km/h, the design speed should not be less than 100 km/h in mountainous terrain. For urban principal multilane arterials in high-density areas and for posted speed limits of 70 or 50 km/h, the lower range of design speed should be between 60 km/h and 80 km/h, and the upper range should be between 70 and 90 km/h [13]. The recommendations for urban (and rural) multi-lane freeway systems are summarised in Table 5.2.

Table 5.2 Urban/rural principal arterial system: multilane roads (interstate and other roads)

Terrain characteristics	Flat/rolling	Mountainous
Extent of access control	Full	Full
One-way DHV 1950-2999	+	+
Recommended LOS	B	C
Posted speed limit	90 (105) km/h	70 (100) km/h
Design speed	100-140 km/h	80-120 km/h

5.2.26.1.1.3 Principal arterial—Other Important Points

Hiersche and Lamm [8] studied the effect of design speed on accident rates and accident costs. Their results, as presented in Table 5.3 (below), show that the accident rate decreases significantly as the design speed increases from 60 to 80 km/h. For higher design speeds, higher operating speeds lead to more severe accidents [13].

Table 5.3 Design speed versus accident rate and accident cost rate [13]

Design Speed, km/h	60	70	80	100
Vehicle kilometres (absolute), 10^6 vehicle kilometres	52.3	12.0	234.5	52.0
Vehicle kilometres travelled, per cent	14.6	5.2	65.7	14.5
Accident rate, accident/10^6 vehicle kilometres	2.12	1.78	1.15	1.11
Accident cost rate, dm/100 vehicle kilometres	3.62	3.85	4.21	5.21

5.2.26.1.1.4 Other Principal Arterials—Multilane Roads

Multi-lane arterials carry heavier traffic volumes, have more frequent junctions, and have more development of adjacent land. The frequency of at-grade crossings has a significant effect on accidents and road capacity.

A divided arterial is an arterial that has separate lanes for traffic in opposite directions. It is recommended that all new arterials requiring four or more lanes be divided. It is not recommended that an arterial be divided, unless it has two full lanes in each direction of travel and the median width is 1.2 m or more. It also must be constructed or marked in a manner to preclude its use by moving vehicles except in emergencies or left/right turns. The major advantages of dividing a multilane arterial are increased safety, comfort and ease of operation.

A divided arterial is generally designed for high-volume and high-speed operation, for which a flowing type of alignment should be provided [1].

A multilane undivided arterial is the narrowest arterial on which each traffic lane is intended to be used (going in one direction). Overtaking is accomplished on lanes that are not in use by oncoming traffic. The ability to pass without travelling in the lane of oncoming traffic results in safer. Undivided arterials with four or more lanes are most applicable in urban and suburban areas where there is considerable development of adjacent land [1].

5.2.26.1.1.5 Other Important Issues

The minimum required sight distance at all points is the stopping sight distance, because overtaking does not require the use of a lane with oncoming traffic.

5.2.26.2 Urban Minor Arterial Definition

A minor arterial street system includes all arterials that are not classified as principal. The system emphasises more land access, and its traffic mobility is more restricted compared with the principal arterial system. It distributes travel to smaller geographic areas and enables urban connections to rural collector roads.

5.2.26.2.1 Urban Minor Arterial -Characteristics

A minor arterial street system interconnects and augments the principal arterial system. It accommodates trips of moderate length at a limited level of travel mobility and is useful for local bus routes.

5.2.26.2.2 Urban Collector Definition

The collector street system provides both land access service and traffic circulation within residential neighbourhoods and commercial and industrial areas.

5.2.26.2.2.1 Urban Collector Characteristics

Urban collector streets provide higher levels of mobility than local streets. The collector street system can penetrate residential neighbourhoods. It distributes trips from the arterial to the local system and vice versa, i.e. from the local streets in residential neighbourhoods to the arterial system. In the CBD and other areas of similar development and traffic density, the collector system can include the entire street grid. It can also have local bus routes as well as pedestrian traffic. It often accommodates public utility facilities within the right-of-way.

In a typical street grid, the closely spaced junctions usually limit vehicular speeds. Therefore, the design speed for local streets is not applicable, and the selection process does not require complicated analysis. Nevertheless, the longer sight distances and bend radii are commensurate with design speeds that are higher than the conventional values that guarantee safety, but sometimes might be more practical and feasible.

5.2.26.2.2.2 Urban Collector—Design Speed Selection—AASHTO

Posted speeds for urban collectors should generally be less than or equal to 50 km/h, although higher design speeds of 60 and 70 km/h are being considered by some states (Illinois, US, for example). For consistency in design elements, a minimum design speed of 50 km/h should be used for urban collector streets, depending on the available right-of-way, terrain, adjacent development and other area controls. To ensure desired operating speeds, road features should be selected appropriately, as presented in Table 5.4 below.

Table 5.4 - Urban collector design characteristics

Terrain Characteristics	Flat/rolling	Mountainous
Extent of access control	none	none
One-way DHV <1300	+	+
One-way DHV 1300-1850	+	+
One-way DHV >1850	+	
Recommended LOS	D	D
Posted speed limit	50 (60) km/h	30 (40) km/h
Design speed	60-80 km/h	40-50 km/h

5.2.26.2.3 Urban Local Definition

A local street is identified as having the most inferior control access. It primarily permits direct access to abutting lands and connections to urban roads, which serve vehicular travel, pedestrian and bicycle traffic. It offers the lowest level of mobility and usually contains no bus routes. Service to through-traffic movement is deliberately discouraged.

5.2.26.2.3.1 Urban Local Characteristics

These distribute local traffic between neighbourhoods. Design speed is not a major factor for local streets or is limited in low-speed environments. To ensure low speeds on local streets, the designer must create a safer and pleasant environment. This can be done using a more restrictive geometry.

5.2.26.2.3.2 Urban Local Design Speed Selection—AASHTO

In design elements, a design speed ranging between 30 and 50 km/h may be used to enable reasonable consistency among different design elements. It depends on available right-of-way, terrain, adjacent development and other area controls. Low operating speeds are desirable for safe accommodation of pedestrians, cyclists, and access along these streets [14].

Design speed exceeding 50 km/h in residential areas may require longer sight distances and increased bend radii. However, such procedures contrast with the basic concept of a local street. Moreover, the limited vehicular speed makes the process of design speed selection simpler and less meaningful [1].

5.2.26.2.3.3 Urban Local Design Speed Selection—Other References

Urban areas usually involve many disruptions to flow. Drivers are more likely to adjust their desired speeds to the imposed speed limits. Consequently, the design speed selection process for a newly redesigned road must consider the likely speed limit regime for the road [3].

Poe and Mason [14] concluded: "If planners and designers desire operating speeds below 40 km/h, then this data indicates that sections should have a maximum radius of 60 metres with respect to horizontal curvature. Similarly, sites with higher mean speeds were associated with longer sight distances, flatter grades, and wider and more open cross sections". They concluded that specific design speed ranges should encompass maximum bend radii, maximum widths for cross-sections and guidance on the effects of sight distance.

5.2.26.2.4 Other Important Points

Posted speeds on urban local roads should be equal to or less than 30 km/h. Designers should consider traffic volume as well as the characteristics of the neighbourhood. For example, in a school zone, where the posted speed limit is 20 mph, design speed should not be higher than 25 mph. An effective balance between non-motorised and motorised users of residential areas is essential.

A good transportation system design directs passing cars along major roads. Local roads must be used by local traffic (especially residential) to preserve lower speeds (ref. Table 5.5) and pedestrian involvement. It is very complicated to reorganise a residential roadway system due to the unexpected modification of the traffic pattern. Thus, the possibility of such changes should be incorporated into the preliminary designs. Fast movement can be discouraged using curvilinear alignments and discontinuities in the street system. This can encourage the balance between vehicles, cycles and pedestrians.

The design of a local transportation system must recognise the following factors:

- Safety for both vehicle and pedestrian traffic.
- Efficiency of service for all users.
- Liveability.
- Economy of land use, construction and maintenance.

Table 5.5 Recommended posted speed limits and design speed for urban local streets

Posted speed limit	20 (30) km/h
Design speed	30–50 km/h

5.2.26.3 Rural

5.2.26.3.1 Principal Arterial Definition

Principal rural arterials include the motorway system and most rural roads [1]. In terms of capacity, multi-lane roads are the largest arterial roads. AASHTO [1] defines an expressway as one with fully controlled access. Roads must provide high levels of safety and efficiency, moving high volumes of traffic at high speeds. They should have a minimum of two through-traffic lanes for each direction [1].

5.2.26.3.1.1 Principal Arterial Characteristics

A principal rural arterial system consists of a network of routes. It is characterised by the following:

- Corridor movement with trip lengths and densities suitable for substantial nationwide travel.
- Movements between all or virtually all urban areas with populations over 50,000 and many with populations over 25,000 [1].

5.2.26.3.1.2 Principal Arterial Design Speed Selection—AASHTO

A freeway is the most difficult class of roadway because of the need to provide a high standard of operation by overcoming the unusual and sometimes unfavourable prevailing conditions. These roads are based on traffic volume and should be constructed to the highest possible standards [1]. Rural roads generally have four through-traffic lanes except for approaches to metropolitan areas, where six or more lanes may be provided [1]. In some cases, fewer lanes may be sufficient to accommodate a given volume of traffic at the desired level of service, which is usually LOS B for rural roads [1]. Design speeds of 100 km/h are normally used for level terrain, and according to Table 5.3, it is the lower bound. Design speed should not be less than the posted speed limit. An upper design speed of 140 km/h is generally acceptable. AASHTO suggests a design speed of 110 km/h for rural roads. In rolling and mountainous terrain, a design speed may drop to 100 km/h or even 80 km/h to meet the driver's expectations. The selection of design speed must consider the recommended ranges (subsection 5.2.26.1.1) and should be based on terrain characteristics, traffic volume and available financial resources.

Roads are designed based on a 20-year traffic forecast, but remain in service for a much longer time. The designer must seriously consider any initial construction cost savings for lower design speeds; otherwise, such a design can result in higher expenses in the future due to environmental damage and inconvenience to traffic that accompanies the reconstruction of major facilities.

5.2.26.3.1.3 Principal Arterial Design Speed Selection—International References

Australian Rural Road Design [13]

A design speed of over 100 km/h is allocated when the accompanying geometrics are compatible with the terrain or when the road is considered to be important enough to justify the additional costs of achieving the added quality service, and a design speed up to 130 km/h could be used. For these types of roads, high uniform speeds are expected. On roads of consistently high geometric standards, speeds on large radius bends and straights are not significantly different [21].

5.2.26.3.1.4 Principal Arterial Design Consistency Issues

Once a design speed has been chosen for different sections of the road, it is necessary to perform a consistency evaluation to ensure the design meets the expectations of drivers. The drop in operating speed from section to section should not be greater than 10 km/h.

5.2.26.3.2 Two-Lane Roads -Definition

Two-lane arterials constitute the bulk of the arterial system. These roadways are adequate when traffic volumes are light and long sight distances are generally available [1]. Two-lane rural roads account for over 85 per cent of all arterial and collector mileage in the US [11].

5.2.26.3.2.1 Two-Lane Roads Characteristics

Rural principal arterials, excluding motorways, are a network of continuous routes that serve and interconnect various regions that are not served by the main roadway system. They have partial free access (ref. Illinois' DOT manual). An important role of major two-lane roads is linking large traffic generators. To maintain smooth traffic flow, it is essential to maintain consistent high speeds and infrequent overtaking delays. High-volume two-lane roads are sometimes used as connectors between two major dual carriageways. They expect to reduce their travel time and improve the journey quality. Two-lane roads in rural areas should provide all-weather accessibility, often for low traffic volumes [3].

5.2.26.3.2.2 Two-Lane Roads Design Speed Selection—AASHTO

A design speed of 110 km/h is suggested for this type of road. The recommended level of service is B for moderate traffic volumes and C for heavier traffic [1].

5.2.26.3.2.3 Two-Lane Roads Design Speed Selection—DelDOT

The Delaware Department of Transportation (DelDOT) manual refers to the minimum value for arterial roads to select the design speed for two-lane roads. It recommends using 100 km/h for new construction and 80 km/h for 4-R improvements. There are also recommended design speed values for collectors and local roads, depending on the ADT (in Figure 5.4 of reference [5]).

5.2.26.3.2.4 Two-Lane Roads Design Consistency Issues

Many experts believe that abrupt changes in operating speeds lead to accidents on two-lane rural roads and that these speed inconsistencies are mostly derived from abrupt changes in road geometry. Therefore, providing longer roadway sections with relatively consistent alignment is an important step toward reducing critical driving manoeuvres and enhancing traffic safety. Even though design speed has been used for several decades to determine sound horizontal alignments, it is still possible to have certain inconsistencies in the road alignment. At low and intermediate design speeds on portions of relatively flat alignment, interspersed between the controlling curvilinear portions, operating speed profiles may exceed the design speed in the access sections by substantial amounts, especially on two-lane rural roads.

Design procedures for rural road alignments for which the design speed is less than the driver's desired operating speed cannot guarantee alignment designs that promote uniform operating speed profiles. The relationship between accident rates and design speed is not consistent. According to one approach, the accident rate decreases with increasing design speed up to 80 km/h. However, according to [13], reducing rural speed limits from 100 km/h to 90 km/h (higher speeds) has been seen to reduce casualties by about 11 per cent. Table 5.6 presents recommended design speed ranges on two-lane roads for several different geometric and traffic conditions [1].

Table 5.6 - Rural principal arterial system: two-lane roads design characteristics

Terrain characteristics	**Flat/rolling**	**Mountainous**
Extent of access control	Partial/Free	Partial/Free
Recommended LOS	B (C)*	B (C)*
Posted speed limit less than:	90 (80) km/h	80 (70) km/h
Design speeds	70–100 km/h	60–90 km/h**

*Los C when higher traffic volumes exist.

** Add a climbing lane if there are more than 10 per cent of heavy vehicles whose speed is likely to fall below passenger vehicles on steeper grades.

5.2.26.3.3 Design Speed Selection Process for Rural Principal Arterial System Summary

Rural principal arterial motorways and other multi-lane roads should provide safe and smooth driving at high speed. Therefore, the roadway design process must derive maximum values for design speed without ignoring the speed limit. In general, design speed values for principal arterial (multi-lane roads) should be between 100 and 140 km/h.

5.2.26.3.4 Minor Arterial

Minor rural arterials are the links from the urban centres to larger towns and are spaced to provide a relatively high level of service to developed areas.

5.2.26.3.5 Rural Collector Definition

Rural collector roads are two-lane roads that form part of the rural road system. These roads must be built to a high standard and be compatible with traffic volume and topography. The basic information which is necessary for the design of collector roads involves safety, traffic volumes, terrain controls and alignment. Major collector roads serve:

(1) Counties not on arterial routes, larger towns not directly served by the higher systems and other traffic generators of equivalent intra-country importance, such as schools, freight traffic and country parks.
(2) Places with nearby large towns or cities, or with routes of higher classification.
(3) The more important intra-county travel corridors.

Minor Collector Roads:

(1) Accumulate traffic from local roads and connect all developed areas within reasonable distances of collector roads.
(2) Provide service to the remaining smaller communities.
(3) Link locally important traffic with the rural hinterland [1].

5.2.26.3.5.1 Rural Collector Characteristics

The design speed selection procedure is not a rigorous one but is flexible, giving them the chance to apply their own experience. Generally, on alignments that encompass a design speed of 90 km/h or less, operating speed varies along the route and is consistently greater than the design speed [1].

5.2.26.3.5.2 Rural Collector Design Speed Selection—AASHTO

Low design speeds (60 mph and below) should generally be applied on roads with curvilinear alignment on rolling or mountainous terrain or where environmental conditions dictate. High design speeds (80–100 mph and above) should generally be applied to roads with level terrain or where environmental conditions are favourable. Intermediate design speeds should be applied on roads "where terrain and other environmental conditions are a combination of those described for low and high design speeds" [1]. According to AASHTO [1], ADT should be a basis for design. Economics (business and development) and environmental aspects affect the future traffic volume that the specific road is supposed to handle. It influences the selection of the number of lanes and the initial and operating costs. The desired level of service is chosen according to the importance of the road in the transportation system. Such information can assist in determining the lower limit (design speed). For level terrain, this should not be less than 50 km/h. For rolling terrain, it should be at least 40 km/h and for mountainous terrain, at least 30 km/h. The upper limit of the design speed is approximately 20 km/h higher than the lower limit. The designer should strive for higher values than the minimum, where conditions of safety dictate and costs allow.

5.2.26.3.5.3 Rural Collector Design Speed Selection—Other References And Proposed Methodology

On roads where drivers are expected to travel at speeds below 100 km/h, the operating speed is influenced by road characteristics and its surroundings. To select a reliable design speed range, an iterative procedure is necessary. In low-speed and intermediate-speed environments, the design speed may vary over the length of the road due to relatively low posted speed limits on specific road sections; it is better to lower the design speed there. This may save money by avoiding undesirable over-design. Another important design criterion is consistency.

Initial values for design speed can be based on speeds found on similar neighbouring roads, or otherwise, they can be obtained from Table 3 of Ref. [21] in accordance with the particular terrain (flat, rolling or mountainous) and the approximate range of horizontal bend radii. Based on this selection, an experimental alignment can be created. Operating speeds that encompass this alignment should be checked against those in the table. If they differ significantly, the initial design must be changed or refined by the selection of a different bend radius, etc. A consistency evaluation should be conducted to ensure that the difference between design speed and operating speed is within acceptable limits. By using the iterative procedure, the engineer might be able to obtain unique design speed values for different road sections that best suit a particular location, environment or community.

5.2.26.3.6 Rural Local Roads Definition

The local rural road system constitutes all rural roads not classified as principal arterials, minor arterials or collector roads [1].

5.2.26.3.6.1 Rural Local Roads Characteristics

The rural local road system primarily provides access to land adjacent to the collector network and serves travel over relatively short distances [1].

5.2.26.3.6.2 Rural Local Roads Design Speed Selection—AASHTO

In general, design speed should be selected as appropriate for the environmental and terrain conditions. In the case of rural local roads, low design speeds are generally applicable on bends on rolling and mountainous terrain or where environmental conditions dictate. Similarly, high design speeds are generally applicable to roads on level terrain or where other environmental conditions are favourable. Intermediate design speeds would be appropriate where terrain and other environmental conditions are a combination of those described for low and high speeds. The influence on design speed from traffic for rural collector roads, as discussed in section 5.9.2.5.2, applies to rural local roads too. Table 5.7 presents recommended design characteristics for rural local roads as a function of terrain condition.

Table 5.7 Recommended design characteristics for rural local roads as a function of terrain condition

Terrain characteristics	Flat/rolling	Mountainous
Extent of access control	Free	Free
Recommended LOS	D	D
Posted speed limit less than:	50 (80) km/h	30 (70)km/h
Design speed	60–100 km/h	40–70 km/h

5.2.27 Speed factors on two-lane rural roads (A case study)

In setting speed limits and designing roadways, the mean free-flow speed and variability of speeds across drivers are considered important safety factors. There is still much to learn about the factors of free-flow speeds, despite a large body of past research on speeds. The existing models estimate a specific speed percentile, and the results are sometimes difficult to interpret because they do not distinguish between the mean speed factors and the speed dispersion factors. It is possible to occur that a road with a high mean speed and low speed variability has the same 85th speed per centile as a road with a much lower mean speed but higher speed variability. Modelling the entire free-flow speed distribution suggested by some authors [1, 2] might rectify this problem. Without any irrefutable link, the relationship between speed and crashes has been studied. There are studies ongoing to establish which factor has an impact on safety—the mean speed or the speed dispersion, either opinion is defendable. The crash severity is increased with an increase in mean speed, while the frequency of interactions between vehicles is increased with an increase in speed variability. An increase in the deviation between a motorist's speed and the average speed of traffic is related to a greater chance of a crash, as indicated by the studies. This was also found that, regardless of the speed limit, the speeds increased with better road geometries. There is low speed dispersion on horizontal bends with radii values of less than 328 ft (100 m), and as the radii increased, the range of speed dispersion also increased, as found by Collins et al. [4]. Where the speed limit exceeded the design speed or where design inconsistencies were present, the speed dispersion is decreased with increasing pavement width in segments (Figueroa and Tarko).

These findings demonstrated the potential importance of road design components on speed distribution and justified developing a model that includes the road geometry factors of both the mean speed and speed dispersion. This section presents free-flow speed models at tangents and horizontal bends on two-lane rural roads and identifies the factors of mean speed and speed dispersion.

Review of Speed Factors

The effects of the geometric and environmental conditions on mean speeds were studied by Yagar and Van Aerde [5]. The roadway grade, the lane width, the land use, the road access and the speed limit were found to be related to mean speeds. The negative effect of the traffic volume on the 10th, 50th, and 90th percentiles of speed was included by Yagar [6]. Although the magnitude of the coefficients varied for each speed, the models shared the same speed factors.

The 85th percentile speeds on tangents depend primarily on their length and the radii of the bends preceding and following the tangent segment as found by Polus et al. [7]. Other elements were identified as less important, such as the spirals, the speed limit, the enforcement level, the cross-section width, the longitudinal slope, the side slope, the general terrain, the driver's attitude and the vehicle's acceleration and deceleration capabilities. For different combinations of tangent length and bend radius, prediction models of the 85th percentile speeds were developed.

As the curvature is widely documented as a primary speed factor, the relationships between the 85th percentile speed at the beginning, middle and end points of horizontal bends and the degree of curvature and its squared term were investigated by Islam and Seneviratne [8]. A model that estimates the 85th percentile maximum speed reduction into horizontal bends as a function of the radius, the approach tangent length, and the 85th percentile speed on the approach tangent was developed by McFadden and Elefteriadou [9]. The 85th percentile speed models for different combinations of horizontal and vertical alignment conditions using the radius or the rate of vertical curvature as explanatory variables were developed by Fitzpatrick et al. [10]. However, with reference to cross-section dimensions and other important road elements, the study failed to identify them as speed factors.

The modelling approach typically used in most studies focused on the effects of isolated or restricted alignment conditions on a specific per centile speed, typically the 85th per centile. Although the 85th percentile speed is widely used to approximate operating speeds, other percentiles have been suggested to represent a high percentage of drivers in road design [11-12]. To develop models with the capability of predicting speeds along a roadway segment, considering the entire free-flow speed distribution and based on multiple roadway factors rather than just a set of horizontal and vertical alignment combinations, further refinement is needed.

Proposed Methodology

The following form is found in most of the existing speed models:

$$V_i = \sum_k b_k X_{ik} + \varepsilon \qquad (1)$$

where at site I, the mean or a specific per centile speed is Vi, and at site i, the value of the k exogenous variable is Xik, bk is the regression parameter associated with variable k, and ε is the normally distributed disturbance term. These models cannot predict any percentile other than the specific one for which they were developed, and this is the limitation of these models. Also, the inability to evaluate the speed variability at a site is another significant limitation of these models.

Assume that some road characteristics, the mean speed and the standard deviation of free-flowing vehicles in a segment are affected. At a site i, with the assumption of normally i th distributed speeds, any percentile speed at the site (Vip) can be calculated by multiplying the corresponding Zp value with the standard deviation of individual speeds (σi) and adding that product to the mean speed value (mi). The Zp value is the standardised normal variable corresponding to a selected percentile; for example, Z50 = 0 and Z85 = 1.036. This function can be represented as a statistical model by adding a normal disturbance term (ε). The percentile speed model has the following form:

$$V_{ip} = m_i + Z_p \cdot \sigma_i + \varepsilon = \sum_j a_j \cdot X_{ij} + \sum_k b_k \cdot (Z_p \cdot X_{ik}) + \varepsilon, \tag{2}$$

Where the effect of the Xj parameter on the mean speed is represented by the coefficient aj, and the effect of the Xk parameter on the standard deviation of individual speeds is represented by the coefficient bk. Medical studies [13-14], to obtain percentile estimates, have used comparable modelling approaches with panel data. In the present study, the proposed approach goes a step further by calibrating the mean and standard deviation terms in a single step.

To emphasise that ordinary least squares regression is applied to panel data, the model in Equation (2) is denoted as OLS-PD.

When multiple observations on individuals are available, the panel data approach is frequently used for econometric applications. For all spots, percentile speeds from the 5th to the 95th percentile, in increments of five, were calculated and by multiplying all explanatory variables by the Zp value corresponding to a respective percentile, the panel was created. The panels have more degrees of freedom by having a higher number of observations than typical cross-sectional data sets, thereby reducing collinearity between the explanatory variables and improving the efficiency of the parameter estimates [15].

By adding site-specific and percentile-specific random effects, the OLS-PD model can be further improved to avoid bias in estimating the model parameters caused by unknown factors not incorporated in the regression model [16]. Poe and Mason, Jr. used a mixed-effects model [17] to account for the random effect from the observation sites while modelling the fixed geometric effects on the 85th percentile speeds. A random-effects model was applied by Tarris et al. [1] by cross-sectioning individual drivers as a group and the location of different speed sensors over time.

Data Collection

Based on the alignment characteristics, road maps were used to identify candidate segments. The free-flow speeds and geometry characteristics of road segments with diverse configurations, cross-section dimensions, roadside clear zones, sight distances, curvature and other features were included in the collected data.

For this study, the data was collected at 158 selected spots on two-lane rural road segments. On most of the segments, more than one spot was selected. The only restriction on speed was imposed in the spot selection, ensuring that it must not be affected by any traffic signal or stop sign. The spots were selected at various distances from and inside horizontal bends. For the selection of spots with respect to vertical bends and junctions, the same approach was used. To observe higher variability in driver behaviour, the distance from the spot to the different road features varied from segment to segment. The selected spots were located at least 91.44 m apart when multiple spots were selected.

For each spot location, the following road characteristics and geometric features were collected:

- General characteristics: terrain type, pavement surface, residential development, and posted speed limit.
- Tangents: grade, sight distance, cross-section dimensions, and roadside obstruction.
- Horizontal bends: radius, maximum superelevation rate, length, and advisory speed.
- Junctions: junction type and presence of channels and auxiliary lanes.

- Distance from the spot to the beginning of horizontal bends and the middle of junctions, if present.

The carriageway and three traversable surface types: paved, gravel and untreated, were considered in dividing the cross-section. The shoulder width represented the obstruction-free clearance distance from the edge of the roadside. By ignoring any local or isolated features, a measuring wheel with one-inch precision was used to obtain all dimensions.

To measure the sight distance at each spot according to the AASHTO standard for stopping sight distance, a ranging laser with a 0.1 ft (0.03 m) precision was used [18]. To measure the grade and the superelevation rate, an electronic level with a 0.1 ft/ft (0.03 m) precision was used.

On weekdays during daylight hours and favourable weather conditions, the free-flow speeds should be measured. To identify free-flow vehicle time, headways of five seconds or more were used. The average number of speed observations made was 360, with at least 100 observations at each spot.

5.2.28 Summary of Observed Road Characteristics and Speeds

The data set that was collected for 158 spots was composed of road characteristics and free-flow speeds. Only those segments were included that had the posted speed limits of 80 or 90 km/h. Some very short segments served as transitions to small towns with speed limits lower than 50 mph, and those were not included. Road segments, containing straights, horizontal bends, vertical bends and junctions were included. Segments showing advisory speed signs from 55 to 80 km/h, containing sharp horizontal bends, were also included.

Even though the selected segments had speed limits of 50 and 55 mph, the mean speed had a range of 37 km/h. A similar range of 35.4 km/h was observed as the 85th percentile speed. An indication of the potential significance of the geometric components as speed factors might be the large variability in speed, compared to the small range in speed limit.

A sight distance of 595.6 m was observed. For safe and efficient driving, the available sight distance is of the utmost importance and is a function of the terrain type and the change in the horizontal and vertical alignments. The spots were mostly located on flat terrain; the other 16 spots were located in semi-mountainous terrain, and 48 spots were located on rolling terrain. The road grade was observed with an absolute range of 13.4 per cent. The grade length was not recorded, and none of the upgrades seemed to have the length necessary to make lorries reduce their speeds.

By counting the number of driveways located a quarter mile before and after each spot, the intensity of the residential development in the segment was recorded. Thirty-four spots, having more than ten residential driveways per mile, were located in segments. For the three traversable shoulder types, there was a large diversity in cross-section dimensions. The lane width was 1.23 m, and the range in clear zone distance, composed of the width of the three shoulders, was 22.7 m. The configurations of shoulder widths and types vary from segments having all three shoulder types and a clear zone of more than 11 m in each direction to segments with a shoulder width of 1.5 m, followed by a guardrail in each direction.

The design of horizontal bends, as suggested by AASHTO (18), should be based on an appropriate relationship between the design speed, curvature, the superelevation rate and side friction. The range in curvature values was very large, although the difference in speed limit was only 5 10 km/h. Equivalent to a range in radii of 1,928.4 m, the degree of curvature had a range of 15.48 degrees. The maximum superelevation rate and bend length were also large, 10.55 per cent and 838.8 m, respectively. The selection of the maximum superelevation rate depends on the climate and terrain conditions, the area type, and the frequency of slow vehicles.

Correlation Between Model Variables

Some correlation, as expected, is shown by the cross-section components. The paved (r = - 0.52) and the untreated (r = - 0.70) shoulder widths were negatively correlated with the gravel shoulder width. In contrast, the road (r = 0.70) and the untreated shoulder (r = 0.48) widths were positively correlated with the paved shoulder width. The untreated shoulder width (r = 0.41) was also positively correlated with the road width. The pavement width variable, the road width and the paved shoulder width were combined, which had a lower correlation with the gravel shoulder width (r < 0.43) and no significant correlation with the untreated shoulder width.

For other variables, the pavement and gravel shoulder widths (r <- 0.38) are negatively correlated to the high residential development variable and positively correlated to the speed limit variable (r = 0.71). In other words, narrower cross-sections and an 80 km/h speed limit were generally associated with tangent segments having high residential development. In addition, a road with a 50 mph speed limit and gravel shoulder widths (r =

0.53 between speed limit and shoulder width) has wider untreated shoulder widths (r = -0.35 between speed limit and untreated shoulder width).

There was no multicollinearity between the variables, in spite of the considerable correlation between model variables, and to enable the model estimation, no variables had to be removed. It must be stressed that an increase in the standard deviation of parameter estimates has a strong correlation between explanatory variables (but not multicollinearity), but does not prevent obtaining a best fit to the data, nor does it affect inferences on mean responses or new observations (19). This means that any systematic bias of estimation is not caused if all the correlated variables are present in the model and the inferences are made within the region of observations, in the presence of strong correlation between explanatory variables. The only consequence is sometimes a need for a larger sample to improve the model's precision in estimating individual impacts of a strong correlation between variables in the model.

Free-Flow Speed Models

Near a horizontal bend, a model of a speed profile was presented. From the speed point of view, the road section is divided into four segment types: tangents, deceleration segments, effective bends and acceleration segments. To change speed between the tangent and bend, the deceleration and acceleration segments are used by drivers.

For each type of segment, separate models were developed. The models were developed for the deceleration and acceleration segments, and the points were determined where an average driver starts and ends changing their speed. To assign the 158 observation spots to the four types of segments, this information was used. This involved an iterative procedure, and the details of the calibration process are presented elsewhere [16]. Only the models for tangent segments and for effective bends are presented here because of space limitations. Horizontal bends were divided into flat and sharp bends to improve the fit of the tangent and bend models to the data.

Bends that have a radius of more than 1700 ft were identified as flat bends. On the cross-section, the speeds on the flat bends are significantly dependent. By using the model for tangent segments with a fixed adjustment for the presence of bends, the speeds are estimated. On pronounced bends, the speeds are strongly dependent on their sharpness, but do not depend on the cross-section.

The models were estimated from observations at 85 spots located on tangents and flat bends, and 14 spots located on pronounced horizontal bends. The speed models were calibrated using SAS software. The calibration results for the tangent model and the calibration results for the bend model are to be presented in a table. By analysing the coefficient of determination, the mean speed estimates and the residuals, the performance of the models was evaluated. The best specification of the OLS-PD model of speeds on tangents, in mph, is presented by the following model:

$$\begin{aligned} V_p &= 57.137 - 0.071 \times TR - 3.082 \times PSL_{50} - 0.131 \times GRA - 1.034 \times RES + 2.38 \times 10^{-3} \times SD \\ &\quad - 1.67 \times 10^{-6} \times SD^2 - 0.422 \times INT + 0.040 \times PAV + 0.394 \times GSW + 0.054 \times USW - 2.233 \times FC \\ &\quad + 5.982 \times Z_p + 1.428 \times (Z_p \times PSL_{50}) + 0.061 \times (Z_p \times GRA) + 0.292 \times (Z_p \times INT) \\ &\quad - 0.038 \times (Z_p \times PAV) - 0.012 \times (Z_p \times CLR) \end{aligned} \tag{3}$$

where:
TR = per cent of lorries.
PSL50 = equal to 1 if the posted speed limit is 50 mph; equal to 0 if the posted speed limit is 55 mph;
GRA = road grade, per cent.
RES = equal to 1 if the segment has 10 or more residential driveways per mile; 0 otherwise; SD = sight distance, feet;
INT = equal to 1 if a junction is located 350 ft before or after the spot; 0 otherwise; PAV = pavement width, includes the road and both paved shoulder widths, feet; GSW = total gravel shoulder width, feet;
USW = total untreated shoulder width, feet.
CLR = clearance distance, includes the width of the total gravel and total untreated shoulders, feet; FC = equal to 1 if the spot is located on a flat bend (radius larger than 1700 ft); 0 otherwise.
Zp = standardised normal variable corresponding to a selected percentile.

This indicates that 84.4 per cent of variability is explained, and the adjusted coefficient of determination (R2) of the tangent model is high. However, the generated panel data along the percentile dimension creates speed data variability, which is principally explained by the Zp factor (the higher the Zp value, the higher the speed). The standard deviation of the model is 3.38 km/h. Sixty-six per cent of the mean speed estimates have residuals lower than 2.11 mph, and only three per cent of the mean speed estimates have residuals higher than 4.22 mph. To conclude that the model provides reasonable estimates, this simple evaluation is very helpful.

The best specification of the OLS-PD model to calculate speeds on horizontal bends, in mph, is the following:

$$V_p = 47.664 + 3.44\times10^{-3}\times SD - 2.639\times RES - 2.541\times DC + 7.954\times SE - 0.624\times SE^2$$
$$+ 4.158\times Z_p + 0.236\times(Z_p\times DC) - 0.199\times(Z_p\times SE) \quad (4)$$

where:
DC = degree of curvature, degrees.
SE = maximum superelevation rate, per cent.

In the bend model, the adjusted R2 value is also high, indicating that 93.2 per cent of the variability is explained. The standard deviation of the model is 1.7 mph (2.8 km/h). Residuals are lower than 1.7 mph in 11 estimates, and residual values are higher than 3.4 mph only in one estimate. The performance of the mean speed estimates for both models may be shown in a table. With no apparent bias, the models seem to provide reasonable mean speed estimates.

Mean Speed and Speed Dispersion Factors

Ten different variables are included in the tangent model in Equation 3, six of them representing mean speed and speed dispersion factors. The first intercept term applied the mean speed along with the following 11 variables, while the standard deviation is applied to the second intercept Zp along with the five variables whose names start with Zp. It is indicated that the variable increases the mean speed by a positive sign of a regression parameter in the first group of variables, while it is indicated that the variable increases the variability of individual speeds by a positive sign of a regression parameter in the second group of variables.

The strongest mean speed factor and speed standard deviation factor are the speed limit binary variable. As the base condition, the speed limit was set as 55 mph. A lower mean speed and a higher variability of the individual speeds, at the same time, are associated with the speed limit of 50 mph. It is therefore indicated as a second finding that some drivers follow the speed limit closer than others, and this difference in compliance further differentiates the individual speeds.

The mean speed up to a specific value is increased by an increase in sight distance in the tangent sections, as bound by the linear and quadratic terms in the equation. The sight does not increase any additional increase in the tangent mean speed distances higher than 712.6 ft.

As expected, the mean speed increases with the increase in any of the lateral dimensions of the cross-section. The gravel shoulder surprisingly had the strongest impact. With strong visual contrast between the gravel and tarmac, the roadway delineation is improved.

The range of individual speeds increases by reducing the distance between the roadside obstructions and the travel lanes. The reason may be that cautious drivers respond to the extra risk (narrow clear zone) by driving at a slower speed than more aggressive drivers.

The mean speed is slightly reduced, and the speed dispersion is slightly increased due to the presence of a junction within 106.7 m of any spot on the tangent. One interpretation for this impact is that cautious drivers respond to the extra risk presented by the vehicles entering and exiting the junction as opposed to fast and aggressive drivers. Due to a high residential driveway density, there is a reduction in mean speeds, and an analogous interpretation also applies for the same.

The effect of the remaining variables on the model on the speeds is easy to explain.

- As expected, an increase in the lorry percentage reduces the mean speeds.
- An upgrade reduces the mean speed and increases the dispersion.
- A downgrade increases the mean speed and reduces the dispersion.

When the spot is located on a flat bend, the model provides an adjustment in mean speed.

Four different variables are included in Equation 4, in the bend model; two of them represent the mean speed and the speed dispersion factors. The mean speed is applied by the first intercept term and the following five variables, while the standard deviation is applied by the second intercept Zp and the two variables whose names start with Zp.

The speed limit could not be included as a factor, as all the bends evaluated for the model have the same speed limit of 55 mph. However, the highest posted speed limit allowed for two-lane rural roads is 55 mph in Indiana, US; therefore, in the sample, any speed reduction forced by adverse curvature can be the highest. Therefore, a segment on a sharp bend with a 55-mph posted speed limit is expected to compel a larger speed reduction to negotiate it than a segment on a comparable bend with a 40-mph posted speed limit.

The bend is a strong factor of both the mean speed and the speed deviation, represented by the degree of curvature and the superelevation rate. As expected, the mean speed is reduced and the speed dispersion increased with an increase in the degree of the bend. The impact mechanism for the superelevation rate is not as obvious. A linear and a quadratic factor for the superelevation rate were included in the model. Superelevation rates higher than 6.4 per cent will reduce the mean speed, although the net impact has to consider the effect of the degree of the bend. The correlation analysis showed that the degree and the superelevation rate were positively correlated.

Therefore, any change in the superelevation rate used to evaluate its impact on bend speeds needs to include the corresponding change in the degree of bend.

An increase in the bend sight distance causes an increase in the mean speed. In this case, like the tangent model, the increase in speed is not restricted. Due to high residential development in the segment, a reduction in mean speed occurs on bends. The impact mechanism is the same as in the tangent model, but stronger.

5.2.29 Safe Spacing (headway) between moving vehicles

Safe spacing (headway) between moving vehicles may be calculated with respect to stopping distances at a particular speed.

i. Considering the driver's reaction time t1 = 2 Seconds,
   From the equation of the law of motion, the distance required by the driver to understand, S1 = u x t1
   Initial speed = u = 80 kph = (80 x 1000/3600) = 22.22 m/sec,
   S1 = u x t1 = 22.22 x 2 = 44.44 metres,

ii. Distance required to stop a vehicle t2 = 4 Seconds,
   From the equation of the law of motion, the distance required to stop a vehicle = S2 = ut – (1/2) x ft^2,
   Initial speed = u = 80 kph = (80 x 1000/3600) = 22.22 m/sec,
   Final speed = v = 0, stopped condition,
   Considering the time, t2 = 4 sec. to stop the vehicle,
   Deceleration = f = (u-v)/t2 = (22.22 -0) / 4 = 5.55 m/sec^2,
   S2 = u x t2 - 0.5 x f x $t2^2$ = 22.22 x 4 - 0.5 x 5.55 x 4 x 4 = 88 – 44.44 = 43.56 m,

Therefore, total distance = S = S1 + S2 = 44.44 + 43.56 = 88 metres = 90 metres (approximately).

Therefore, for a vehicle moving at a speed of 80 km/h, 90 metres are required to stop in a tentative time of t1 + t2 = 2+4 = 6 seconds. Hence, the safe spacing between the vehicles should be a minimum of 90 metres, which is moving at a speed of 80 kph. For estimation by the drivers, the spacing should be a little more than 'X' metres, at the speed 'X' km/h.

5.2.30 Conclusion

The mean free-flow speed and its variability across drivers are important safety factors. Most of the existing models present the factors of the 85th percentile speed and do not distinguish the mean speed from the speed dispersion factors. Furthermore, the existing models cannot estimate the entire range of speed variability at a site. The entire speed distribution has been utilised as an innovative approach to develop speed models instead of focusing on a particular percentile. This has been accomplished by representing the percentile speed as a linear combination of the mean and the standard deviation.

Free-flow speed measurements and two-lane rural road characteristics were used to develop speed models. The models demonstrated their efficiency in identifying relationships between diverse road characteristics and speeds. One important contribution is the fact that the impact of the cross-section dimensions is present in the tangent speed model. In addition, it is equally easy to quantify the impacts of the variables on mean speed and on speed standard deviation. The tangent speed model includes ten different variables, six of them as both mean speed and speed dispersion factors. The bend model includes four different variables, with the degree of bend and the superelevation rate representing the mean speed and speed dispersion factors. The advantages of the models developed in this study, over the traditional OLS models, include predicting any user-specified percentile, involving more design variables than traditional OLS models, and separating the impacts on mean speed from the impacts on speed dispersion.

5.2.31 Summary and Recommendations

The current design speed selection procedure, based on AASHTO guidelines, does not provide enough flexibility in selecting design speed, and the alignment consistency is not considered. Also, the current DOT road design procedure does not incorporate consistency evaluation. Many non-engineering issues have been recognised to influence the process of design speed selection; however, they are still not incorporated in the road design process. There is a need for a logical, flexible and easy-to-use design speed selection procedure, and expert systems are an example of this. Expert systems can be a potential tool for the design speed selection procedure, having successfully overcome some of these problems. In other words, the need for a new and novel approach to the design speed selection process can be justified and successfully realised through expert system software. More attention is required for theoretical and experimental research to continue updating and improving the road design process. In the design process, findings obtained through research should be implemented. Current research in design speed selection for consistency purposes, driver expectancy and how to design wisely, keeping safety in mind, has already been discussed. To make the road design procedure safer and more efficient, attention and time should be given to current findings.

To advance more professional and safer solutions, the design engineer should be open to new ideas as necessary. Furthermore, it is also necessary to stay critical and selective, because to create roads that operate well and are safe, the characteristics of the local environment and traffic patterns play a key role. This section has presented methods to select design speeds in the US and other countries. This section includes a case study on research that contains innovative ideas along with the latest research on design speed. It is intended to provide a comprehensive, complete and logical procedure to select the design speed that will systematically guide design engineers through the design speed selection process for a variety of areas, topography and classifications, for example, urban or rural arterials, collectors, local roads, etc.

References

Application of Speed Limits for Road Safety,
Sandipan Goswami, ACADEMIA,
https://www.academia.edu/71613348/Application_of_Speed_Limit_for_Road_Safety?sm=b&rhid=34073423158

CHAPTER 6

VISIBILITY DURING THE NIGHT, RAIN AND FOG, AND STREET LIGHTING

Abstract: This chapter covers various topics, such as giving a general overview of inadequate visibility during rain, fog and nighttime driving with precautionary measures, measuring street lighting, visibility due to headlight glare, hazards from restricted visibility in adverse weather conditions, highway lighting and design of highway lighting.

6.1 General

Night visibility of a roadway has been the subject of numerous research studies. Many factors influence the night visibility of roadway obstacles. The parameter that seems to be a key to the entire problem is the roadway illumination or brightness as seen by a vehicle driver. The following topics are covered in this chapter:

- streetlighting and illumination
- visibility restricted by a vehicle's headlight glare, and
- hazards of restricted visibility during adverse weather conditions

6.2 Streetlighting Illumination Measurement

A section of the roadway may be selected to measure illumination and to study the illumination patterns. An area of 44.5 metres × 25.5 metres on the roadway may be selected. Here, the distance between two consecutive light posts is 44.5 metres, and it is measured as the length along the travel direction of the road. The width of the roadway intended to be illuminated by the streetlights is 25.5 metres. The height of light posts was initially decided to be 12 metres. The area 44.5 metres × 25.5 metres may be divided by grid lines marked on the road surface longitudinally and transversely at an interval of 0.5 metres.

The illumination from reflected light on the road at every grid point at an interval of 0.5 metres may be measured by an illumination metre and recorded in a raw data file along the length and width of the area. The data in the raw data file is to be processed by triangulation, and illumination contours are to be created by using HEADS Pro or other similar software. HEADS Pro is available from www.techsoftglobal.com along with a tutorial. The intended purpose is to study illumination in lux on various contours and locations, and whether it fulfils the required technical specification.

Study Parameters:
The type of lamp (mercury, sodium, halogen, LED)
The power of the lamp
The height of the lamp
The spacing of the lamp posts

6.2.1 Checking Illumination Patterns

Two basic illumination patterns may be checked and investigated by contour studies, with different input parameters as mentioned above:

Illumination is approximately uniform.
Illumination is considerably non-uniform.

All roadway lighting systems in current use are directional, and practically all roadway obstacles are 3-dimensional, so these are to be considered. Due to the overhead mounting of most light fittings, it is important to consider the downward-directed light and to have some surfaces on the target orientated on non-vertical planes to reveal this characteristic of the lighting system. The road surface brightness is approximately uniform, and the targets are seen against very nearly the same background brightness. The vertical planes of an object have very low brightness, whereas the inclined faces have higher brightness. These differences result in variations in visibility. A non-uniform road surface means that the vertical planes are very dark, and the background brightness is low, so the contrast is very poor. However, the inclined planes of the target object can be considerably bright and thereby result in a good overall contrast. A field study is likely to bring out the following points:

i. A uniform road surface brightness lighting system has been developed to see objects in silhouette (negative contrast), and has moderately high visibility indices, which are nearly the same at all locations along the carriageway centre and ahead of the driver.

ii. The lighting system, which directs light onto the inclined planes, causes stronger brightness than that on the vertical plane and thus reduces the overall contrast with the background.
iii. The lighting system, which creates approximately uniform road surface brightness, is not affected by shapes on the road. A 3-dimensional object on the road provides more information about the visual conditions than does a 2-dimensional object.
iv. For non-uniform roadway brightness, for the same objects at the same locations, the variation in brightness is such that at certain locations the target objects are seen against a very low background brightness and at some locations against a much higher background brightness.
v. The average roadway brightness with a non-uniform pattern is approximately the same as that of the uniform pattern.

The role of roadway brightness patterns, target size and shape for visibility of roadway obstacles at night is demonstrated above for the two conditions, which may be considered extremes of engineering practice.

6.2.2 Night-time Visibility is an Important Component of Road Safety

An important component in the present accident trend is nighttime car accidents. Although nighttime traffic is reduced, nocturnal accidents have very serious consequences. The fatality rate resulting from accidents occurring at night is 38 per cent higher than that of persons injured during the day. In these circumstances, it is necessary to find ways to reduce traffic hazards that occur through elucidation of night visibility conditions and improvement of traffic conditions at night. This discussion presents the visibility conditions that lead to an accident at night, considering the darkness, streetlighting and traffic light action.

An important factor in night visibility research is contrast luminance. This can be described as an objective and a subjective contrast. Objective contrast is a physical quantity that can be measured and calculated. In practice, several models are used, of which the simplest is the ratio of the illumination of two neighbouring areas. In this case, the threshold contrast matters, which shows how great the difference between the light and the object environment is to be noticed by human eyes. For object recognition, the contrast must be greater than the contrast threshold 'Ks'. Since light luminance density depends on the degree of luminance and reflection, illuminated objects in the headlights' field can be seen in different ways. The subjective contrast is a quantity determined by the sensation.

Most importantly for vision is the difference between environment illumination and object illumination. The human eye can only distinguish an object from above the light threshold. This threshold depends on the environment's light luminance and the angle of the seen object. This threshold is determined in the laboratory and is presented as diagrams or text values in spreadsheet tables. The primary criterion to observe an obstacle is to have a 'K' medium contrast (background):

where: Lobject is the object illumination and Lenvironment is the environment illumination.

This contrast can be positive or negative (Fig. 1). The left image shows that

$$K = \frac{L_{object} - L_{environment}}{L_{environment}},$$

the positive contrast is a light-coloured person (illuminated from the front in the third picture) on a dark background. In this case, the size and shape of the obstacle are visible. If the person appears dark and is in front of a light background, the contrast is negative, as shown on the right (illuminated from behind, in the first picture). In the middle picture, when the contrast is very small ('stealth'), the contrast ratio has a numeric value close to zero, as frequently found in nighttime accidents. This contrast may appear small in a poorly illuminated street by provoking so-called physiological blindness. For streetlighting, there are certain parameters, but there are deviations from normal: changing luminance parameters (dust, dirt, ageing, wear), poor location, and shielding by obstacles (buildings, trees, advertisements).

Figure 6.1 Contrasts depending on the relative position of the light source

Figure 6.2 shows the effect of a vehicle coming from the opposite direction; here, the background pedestrian is in contrast due to the increasing light in the background.

Figure 6.2 Model of a typical case encountered at night

6.3 Visibility with Vehicle Headlight Glare

Despite the logical connection between visibility during the nighttime and safety, and the abundant evidence that headlamp glare reduces visibility, there is little direct evidence linking headlamp glare to crash risk (NHTSA, 2007).

6.3.1 Link Between Headlight Glare and Accidents

There is wide acceptance that glare reduces visibility, and reduced visibility, albeit in a context unrelated to glare, appears to be related to the risk of crashes involving pedestrians along rural high-speed roads.

Two driver behaviours are to be noted, first, that drivers respond differently when driving in locations that have an inherently high accident risk than they do when they are driving in locations with an inherently low accident risk. The second hypothesis is that the same drivers will exhibit some of the same behaviours when they are exposed to headlight illumination by oncoming vehicles as when they were driving in high-risk locations. The inference that could be derived if these hypotheses were confirmed is that oncoming headlight illumination is associated with increased accident risk.

6.3.2 Factors Affecting Headlight Glare

The headlight beam patterns of different vehicles and lighting characteristics are mentioned below:

- light source: halogen and high-intensity discharge (HID) headlamps
- mounting height: passenger car and light truck/sports utility vehicle (SUV) headlamps

- optics type: reflector-based headlamps and projector-based headlamps
- beam pattern: visual optical-aim left (vol) and visual optical-aim right (vor) headlamps
- aim: correct aim: downward mis aim occurs when the headlamps are pointed slightly down and upward mis aim occurs when the headlamps are pointed slightly up.

The headlamps were analysed using computer calculation and simulation tools to determine, for an average driver, the detection distance to the targets along a straight two-lane highway with an undivided carriageway under different headlamp types and aim conditions, the brightness of the contrast-reducing scatter in the eye caused by different oncoming headlamp types and the discomfort experienced by a driver exposed to these different conditions.

6.3.3 Headlights

In theory, it is possible to eliminate headlight glare by simply removing the headlights, but such a solution is not realistic. Driving at night is essential for some people, and the purpose of headlights and of low-beam headlamps (since these are designed to be used in the presence of oncoming traffic) is to provide the drivers with enough light to see but not so much light that the safety of oncoming drivers might be compromised.

Present low beam headlamp patterns do not appear to provide sufficient visibility to see and react to potential roadway hazards at driving speeds higher than 48 to 64 km/h, except in areas with high levels of ambient lighting (i.e., urban areas). However, higher intensities will be deemed unacceptable by most drivers facing such oncoming headlamps.

6.3.4 Lighting to Mitigate Glare

On the low beam setting, light from headlamps is directed away from oncoming vehicles, whereas on a full setting, the light is directed at oncoming traffic, which reduces the visibility of the drivers. The recovery time from looking at bright lights is longer for older drivers.

Various potential countermeasures for reducing the negative effects of glare are available. Some of the countermeasures involve comprehensive changes to the requirements for headlamps, and others involve technological developments that might change how headlamps are implemented on vehicles.

A prototype safety-based adaptive forward-lighting system (SAFS) was developed (Bullough et al., 2008). The review of drivers' visual needs suggested that headlamps with higher intensities than low beams would improve visibility.

Various studies and published literature summarise the existing knowledge as below:

- Headlamp glare is quite probably related to increased risks for drivers as it reduces visibility. Research is beginning to establish the role of visibility in safety.
- Drivers' perceptions of glare (i.e., discomfort) are often different from the other negative effects of glare (i.e., recovery), probably explaining in part the negative responses to blue HID headlamps, because they elicit greater discomfort even when they do not diminish visibility more than halogen headlamps, i.e., at the same luminosity.
- The preponderance of misaimed headlamps, even in the small sample sizes used in these studies, suggests that more consistent, correct headlamp aim could improve visibility and reduce glare conditions and create more agreeable conditions for drivers.
- The present low beam headlamp patterns do not appear to provide sufficient visibility at higher driving speeds for which they are commonly used, even when no oncoming vehicle headlamps are present.

Countermeasures for reducing headlamp glare were judged as having high, medium or low potential to reduce glare or improve visibility as follows:

- Mounting height: low headlamp mounting heights might slightly reduce glare, but might also slightly reduce forward visibility.
- Aim: high—it appears that some new vehicles and many in-use vehicles have mis-aimed headlamps that reduce visibility (when pointing downward) and increase glare for other drivers (when pointing upward). Periodic adjustment of the direction should result in more acceptable visual conditions.
- Optical design: low—within the limitations of a headlamp system that produces just a low beam pattern, there is little evidence that this plays a large role in the amount of glare experienced by drivers.

- Low beam distribution: headlamps have a low beam under a broad range of driving conditions with glare control, and much of the evidence suggests that any improvements will be nominal. Most evaluations of low beam distributions, however, have assumed straight, flat roads; improvements to low beams under more complex conditions have been rarely studied.
- Adaptive beam distributions: high—dynamic approaches can significantly change glare and visibility, but will involve investment from the vehicle lighting industry, higher consumer costs and longer times to achieve improvements.
- Colour—blue headlamps increase discomfort, so restricting short-wavelength output could reduce this but would likely have little impact on visibility reductions from headlamps.
- Cleaning and maintenance: medium—ten to 20 per cent of vehicles had dirty or damaged headlamps, or with condensation inside the front lens. In practice, cleaning and maintaining headlamps should not be difficult and should improve visibility and reduce glare for other drivers.

6.3.5 Night Visibility Improvements

6.3.5.1 Nature of the Problem

Headlight design is currently based on a compromise between the need for adequate illumination of the road ahead and the need to avoid dazzling the eyes of the oncoming drivers. During the past half-century, several modifications have been made to control headlight beam configuration. These changes in intensity, beam pattern and aim have improved the visual environment for night driving.

Changes in beam configuration alone, however, cannot possibly provide adequate light to enable the driver to operate their vehicle safely during many common night driving situations because of the nature of the necessary design compromises. This is especially true on two-lane highways when vehicles are opposite each other. Because of the extremely small angular separation between the approaching vehicles and the variation in possible roadway geometry, it is necessary to limit the luminous output in the upper left quadrant of the headlight beam to keep glare out of the eyes of oncoming motorists.

Some luminance must be aimed in this direction, so that the driver can see things approaching from the left/right and can negotiate left-hand/right-hand bends. Also, the design is limited by physical considerations, such as filament size, in determining how sharp this cut-off should be. Therefore, the possibility of developing improved vision is extremely limited.

Improved capacity to see at night with reduced glare can be expected to reduce the frequency of accidents. This is perhaps the major benefit. There are, however, many other aspects of nighttime driving that can be expected to benefit from better visibility. Increased time will be available for driver decision-making because of the increased distances at which road obstacles can be detected. It should be possible to maintain somewhat higher driving speeds, with consequent increased traffic flow and greater utilisation of roads during off-peak nighttime hours.

Although an improved headlight system might produce an increase in night speed, this appears unlikely considering the small and variable shifts in speeds encountered after installing fixed illumination systems. As a result of the elimination of glare, there would be a reduction in stress, a reduced sensation of 'tunnel' driving with the need for more exact control of a lateral position, and an extension of the night driving capability of older drivers whose glare adaptation responses have deteriorated.

6.3.5.2 Beam Use Study

The observations showed that during normal nighttime driving with clear weather conditions, over 75 per cent of drivers were using their headlights improperly, i.e., using low beams when high beams should have been used because they were neither meeting another vehicle nor following closely behind one.

6.3.5.3 Polarised Head Lighting System

A polarised system relies on the principle that two linear polarisers with their planes of polarisation perpendicular to each other permit only a negligible amount of light flux to come out of the second polariser. Polarisers are attached to the headlights with their transmission axes orientated at 45 degrees from the vertical. Another polariser, called the analyser, with the same transmission axes (i.e., parallel to those over the headlights), is installed in front of the driver's eyes in a position similar to that of the sun visor.

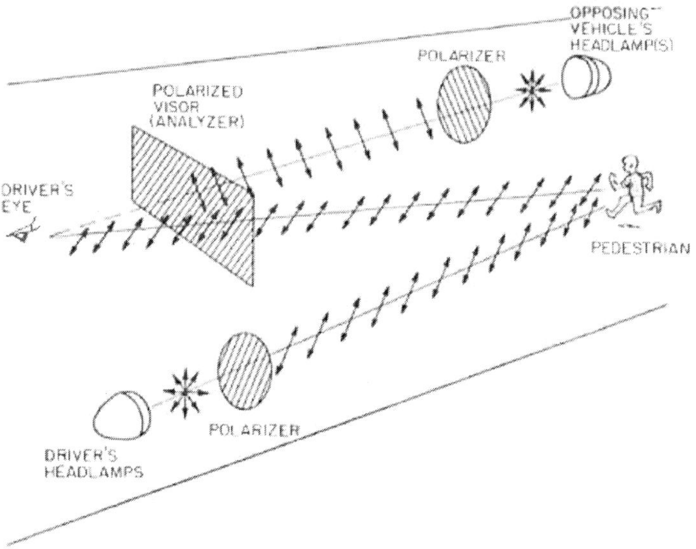

Figure 6.3 Polarised head lighting system: Light from the driver's own head lamps reflected from an object passes through the analyser; light from opposing head lamps does not.

1. Polarisation appears to be the only practicable method by which adequate vehicle head lighting can be provided without causing disturbing levels of glare to motorists going in the opposite direction.
2. Polarisation is technically and economically feasible on modern vehicles.
3. Polarisation is advantageous in terms of improved visibility, and
4. It will enhance vehicular control, safety and comfort, and probably improve traffic flow and safe utilisation of the roads at night.

This analyser is constructed such that it can be put out of the way when no oncoming vehicle is present. Because the transmission axes of the headlight polariser and the analyser are parallel, light from the driver's headlights that is reflected from pedestrians, signs, road surface markings and other parts of the roadside environment will be transmitted through the analyser to the driver's eye. However, when another vehicle equipped in the same manner approaches, the transmission axes of the polarisers over its headlights will be perpendicular to that of the original first vehicle's analyser. Thus, only negligible light is transmitted, and neither driver is blinded by glare from the approaching headlights.

6.3.5.4 Increased Detection Distances with Polarised System

When glare was radically reduced using polarisation, the difference in detection distances between the opposed and unopposed driving modes was greatly reduced. In the opposed mode, the use of polarisation increases detection distances over those obtained in the unpolarised mode, regardless of the target or the intensity of the lamp used with the polarisation. In the unopposed mode, the increased detection distance will also be true if the analyser is not in use and if the polarised illumination is equal to or greater than that from the unpolarised standard headlights.

Visibility without the analyser and without an oncoming vehicle would be equivalent to that of high beam for most types of targets. However, in the absence of an oncoming vehicle, the detection distances shown in Figure 4 for the polarised headlight included the use of the analyser because randomising the presentation of the test factors to the subjects precluded the omission of the analyser for unopposed trials. Also, this constitutes the situation that results if the driver becomes lazy and does not bother to move the analyser out of their line of sight when there are no oncoming vehicles.

6.3.5.5 Substantially Higher 'Safe' Driving Speeds

An important question is how the use of a polarised system affects the safety of the average motorist. Using the data on the visibility of a pedestrian at the road edge for each headlight system, it is possible to calculate a safe speed at which the normal driver would be able to stop before striking the pedestrian. The results of this calculation are given in Table 1. The average speed observed during the beam use study was 100 kph on dry road surfaces and 90 kph on wet road surfaces. The discussion is based on the use of high-intensity polarised lamps. The use of existing high-beam lamps with polarisers would result in 'safe' speed levels about halfway between those of conventional headlights and the high-intensity lights when an oncoming vehicle is present. Without an oncoming vehicle, the 'safe' speed would be less than with a conventional high beam but well above that of a low beam. Overall, the use of existing high-beam lamps with polarisers (no increase in lamp power) would improve night driving over that of conventional headlights except for when there is oncoming traffic. However, the improvement would not be sufficient to raise the 'safe' speed to the level required to see and stop for a pedestrian at current highway speeds.

6.3.5.6 Safe Spaces Between Moving Vehicles

Safe spacing between moving vehicles may be calculated with respect to stopping distances at a particular speed.

i. Considering the driver's reaction time t1 = 2 Seconds,
Using an equation from the law of motion, the distance required by the driver to understand, S1 = u x t1
Initial speed = u = 80 kph = (80 x 1000/3600) = 22.22 m/sec,
S1 = u x t1 = 22.22 × 2 = 44.44 metres,

ii. Distance required to stop a vehicle,
Using an equation from the law of motion, the distance required to stop a vehicle = S2 = ut – (1/2) x ft^2,
Initial speed = u = 80 kph = (80 × 1000/3600) = 22.22 m/sec,
Final speed = v = 0, stopped condition,
Considering the time, t2 = 4 sec. to stop the vehicle,
Deceleration = f = (u-v)/t2 = (22.22 -0) / 4 = 5.55 m/sec^2,
S2 = u x t2 - 0.5 x f x $t2^2$ = 22.22 x 4 - 0.5 x 5.55 x 4 x 4 = 88 – 44.44 = 43.56 m,

Therefore, total distance = S = S1 + S2 = 44.44 + 43.56 = 88 metres = 90 metres (approximately).

Therefore, for a vehicle moving at a speed of 80 kph, 90 metres are required to stop in a tentative time of t1 + t2 = 2+4 = 6 seconds. Hence, safe spacing between vehicles travelling at 80 kph should be a minimum of 90 metres. For approximate estimations by the drivers, the spacing should be more than 'X' metres, at the speed 'X' kph.

Table 6.1 Safe Stopping speeds

SAFE SPEEDS FOR STOPPING (CONVENTIONAL VERSUS POLARIZED HEADLIGHTS)

Condition		Safe Speed (mph)			
		Conventional Headlights		High-Intensity Polarized Headlights	
Opposing Vehicle	Pavement	Low Beam	High Beam	With Analyzer	Without Analyzer
Yes	Dry	40 (64 kph)	40 (64 kph)	63 (101 kph)	— a
Yes	Wet	35 (56 kph)	35 (56 kph)	50 (80 kph)	— a
No	Dry	50 (80 kph)	80 (128 kph)	73 b (117 kph)	80 (128 kph)
No	Wet	40 (64 kph)	65 (104 kph)	58 b (93 kph)	68 (109 kph)

[a]Not applicable--polarized system requires use of analyzer when meeting opposing vehicle; if not in use, it is the same as high beam.

[b]With analyzer in use--if the driver does not use analyzer when there is no opposing vehicle, safe speed would be as indicated in "without analyzer" column.

6.3.5.7 Dirty Windscreens

Due to reduced light transmission and the oncoming headlights illuminating the dirt particles, a dirty windscreen effectively reduces the contrast between the target and its background. No appreciable depolarisation was noted from the dirty windscreen (8). The visibility loss from a dirty windscreen when using the high-intensity polarised lamps was about 6 per cent (620 ft/188 m versus 660 ft) compared to 10 per cent (530 ft versus 590 ft) with conventional headlights. This result does not imply that driving with a dirty windscreen should be ignored. This study only involved the effect of the dirt on visibility, and the problems related to the distraction and confusion caused by dirt on the windscreen were not considered.

6.3.5.8 Polariser Misalignment

Some misalignment of the polarisers between the two approaching vehicles is bound to exist because of road crown, superelevated bends and other movement, including rough road surfaces and potholes. Misalignment will increase the light transmission because of the nature of polarisation.

Dynamic studies of vehicular roll have shown that a maximum of seven degrees of roll occurs in moving vehicles on normal paved highways. As shown in Figure 5, a misalignment of more than twice this value, or 15 degrees, has an insignificant effect on the driver's detection capabilities (14).

It is therefore concluded that superelevated curves and other causes of vehicular roll will not adversely affect a polarised headlighting system.

6.3.5.9 Driver Behaviour

High-glare conditions appeared to make drivers behave more conservatively by judging the brighter sources to be somewhat closer than they were and thereby allowing a somewhat greater margin of safety. This, of course, could have an adverse effect on traffic flow, particularly when a high volume exists on the main road. Fortunately, nighttime traffic is usually not near capacity, particularly for most unlit roads where the glare problem is especially acute.

Gap acceptance studies clearly show that, if a polarised head lighting system is introduced to obtain superior visibility, care should be taken in the design of the analyser to protect drivers during encounters with vehicles from the side. The design of the analyser may be particularly important for older drivers, because the glare tolerance of individuals is increasingly reduced with age. Drivers over 65 years of age may be seriously handicapped at night, and indeed, they voluntarily and substantially reduce their night driving.

6.4 Hazard of Restricted Visibility in Adverse Weather Conditions

The topics are as below:

- overview of driving in rain and fog
- statistics
- pre-trip preparation
- issues with driving in the rain
- issues with driving in fog
- basic tips for driving in the rain
- basic tips for driving in fog
- summary

6.4.1 Overview of Driving in Rain and Fog

Figure 6.4 Driving in the rain

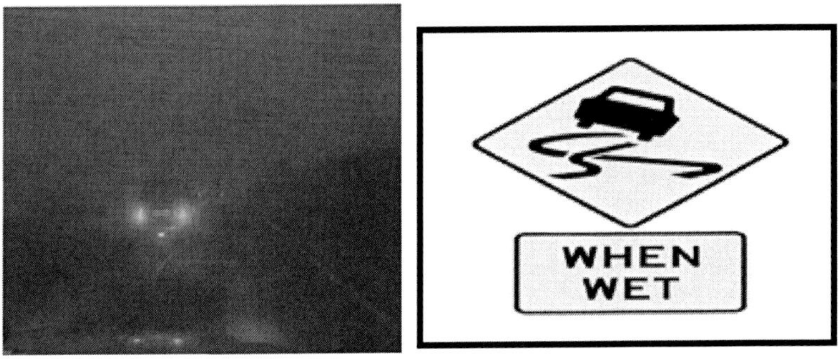

Figure 6.5 Driving in the fog

a) **Driving in the rain:** This reduces driver perception in several ways and is especially debilitating at night. It directly affects perception by seeing through rain, but also produces visibility changes through its action on headlamps, windscreens, the road itself and road markings. We normally see an object when light from a source, the sun, streetlamps, and vehicles' headlamps, reflects from the object back to the eye. Rain interferes with this process in several ways.

b) **Driving in the fog:** Fog is an obvious reason for accidents because drivers cannot see ahead on the roads. Foggy conditions also perpetuate accidents because they affect judgements of speed and distance. The effects are a result of reduced contrast. We see objects, not based on their absolute brightness or darkness, but on their difference between the object's brightness and the background. Fog lowers contrast significantly, causing objects to become fainter and less distinct.

6.4.2 Statistics

1. There are four times as many fatal accidents during rainy conditions as during snow or sleet conditions (Bureau of Transportation Statistics).
2. The most significant weather condition that leads to fatalities, injuries and property damage losses is rainfall, based on *Large Truck Crash Facts* (published by the Federal Motor Carrier Safety Administration).
3. Reduced visibility is a significant factor in 42 per cent of all vehicle crashes and contributes to the danger inherent in any manoeuvre requiring a fast and accurate visual response. Reduced visibility can be caused by lighting and weather conditions, such as glare, dawn, dark, artificial light, rain, sleet, snow and fog (US DOT).
4. As per the Federal Highway Administration, visibility distance is reduced by fog and heavy precipitation, as well as wind-blown snow, dust and smoke.
5. Low visibility conditions cause increased speed variance, which increases crash risk.

6.4.3 Preparation

1. Rain and fog brake lights, and turning signals (indicators) before heading out, as being seen by other drivers may help avoid a collision.
2. Check tyres for correct pressure reduce visibility; drivers are required to inspect headlights, taillights,
3. and tread condition.
4. Tyres which are improperly maintained or have poor treads will generally make handling and stopping more difficult and are potentially more dangerous.
5. As a part of any vehicle's regular maintenance, the windscreen fluid should be filled and the wipers replaced.
6. Taking time to clean the windscreen (inside and outside) thoroughly can help with having the best visibility possible.

6.4.4 Driving in the Rain

1. Rain and fog create several potential problems, such as reduced visibility, creating a veil between the vehicle and the driver, less traction, less control of steering (hydroplaning) and longer braking distances.
2. Often, the first rainfall after a dry spell creates the most difficulties for drivers, as the sun heats the tarmac and keeps oil and grease from passing vehicles on the road surface. Oil becomes very slick with the introduction of rain, especially at road junctions where traffic stops. It is the combination of oil and water which can make it tough to stop or control a vehicle.
3. At higher speeds, tyres can ride on top of the water on the road surface. This is called hydroplaning. Tyres with very shallow treads hydroplane more easily.

6.4.5 Driving in Fog

1. The area in front of a vehicle that is clearly illuminated and visible is the driver's sight zone. High beam headlight settings will produce glare that lights up the fog but does not increase the sight zone ahead of the vehicle. If an object, person or vehicle enters the sight zone in front of a vehicle, the driver may require a sudden stop or evasive manoeuvre.
2. The average person, who is not drowsy, ill or distracted, takes 1.5 seconds to recognise danger and begin to react; this is called reaction time.
3. At 65 kph, a vehicle travels 27.4 metres during the reaction time. This distance is covered at full speed before the vehicle begins to slow down or steer around an obstacle and is called the reaction zone. When a vehicle is driven so fast that the lights do not fully cover the reaction zone, the driving speed is too fast to react safely to the problem.

Figure 6.6 Driver facing an unexpected obstacle

6.4.6 Basic Tips for Driving in the Rain

1. Allow more travel time and drive slower than normal. This will give drivers the freedom to allow greater following distances, cope with heavy traffic that is moving more slowly than normal and enable drivers to navigate any unexpected detours. The driver must turn on the headlights and main beam.
2. Watch out for pedestrians as they may be distracted by an umbrella or other rain gear, and falling rain can distort sounds, so pedestrians may not hear an approaching vehicle.
3. Give large lorries and buses extra room, as the drivers may have a harder time seeing other vehicles due to blind spots, and large tyres displace more water than other vehicles.

Figure 6.7 Drivers must drive more slowly

1. Brake earlier and with less force than normal, increase the stopping distance between the vehicle in front.
2. Use turn indicators earlier than normal so other drivers know your intentions.
3. Watch out for standing water since this can cause hydroplaning, which occurs at higher speeds when tyres ride up on top of the water on the road surface.
4. When hydroplaning, do not turn the wheel suddenly; release the pedal and steer straight until the vehicle regains traction, and avoid making sudden turns or slamming on your brakes.
5. Ventilate the car; rain causes humidity levels to increase, and the vehicle's windows become foggy when the vehicle is operated while it is raining.

Figure 6.8 Standing water and higher speeds cause hydroplaning

6.4.7 Basic Tips for Driving in Fog

1. To bring the vehicle to a stop or to steer for a safe path, more time and space are needed.
2. Use low beam headlights.

3. If the vehicle has daytime running lights, that is not an adequate substitute for turning on the low beam headlights as in the normal settings.
4. Know the appropriate speed for rainy and foggy driving conditions.
5. Start by increasing the stopping distance from 3 to 4 seconds to at least seconds and add to it generously.
6. Vehicle size, braking capabilities, the slipperiness of the road and the driver's awareness are contributing factors to determine an appropriate speed.
7. Remember that there may be drivers coming behind you; make sure that they can spot your vehicle by using appropriate flashers or warning lights.

Figure 6.9 Driving in fog

6.4.8 Summary for Driving in Rain and Fog

1. Rainstorms and fog can occur at almost any time of the year, depending on the territory.
2. Basic inspections and preventive maintenance prior to trips can help assure that visibility is clear while driving.
3. Seeing clearly ahead of and around your vehicle is critical to safe operation.
4. Many factors contribute to poor visibility; these include poor weather, smudgy windows, poorly working lights, weak wiper blades, etc. and can make it a challenge to see clearly ahead of and around your vehicle.
5. Being seen by other motorists also helps to minimise the risk of collisions.
6. Allow extra time to reach your destination and slow down when weather conditions interfere with visibility of your controls.

6.4.9 Effects of Fog on Driving and Accidents on Conventional Roads

An Australian study found that on conventional roads, fog reduces the accident rate by six to ten per cent, with the largest reduction in the more severe categories of accidents. Although comprehensive accident rate data in fog is not available, it appears that fewer fatal accidents occur during fog than during clear weather conditions.

This effect has been attributed to the driver's awareness of fog as a hazard, but the slight reduction of speed observed on conventional highways in fog may also be a contributing factor, particularly with respect to reducing the severity of accidents.

6.4.10 Effects of Fog on Driving and Accidents on Main Roads and Motorways

On main roads and motorways, fog has proved to have the opposite effect. A study by the California Transportation agency showed that the fatality rate for fog accidents on main roads and motorways was almost twice that of non-fog accidents. On main roads and motorways, the probability of multiple-vehicle accidents was greater in fog than in clear weather. It has been observed that fog does not have much effect on the mean number of vehicles per accident because there have also been more single-vehicle accidents in fog. In fog, 98 per cent of all accidents involved four vehicles or fewer. The table below indicates that nearly two-thirds of all accidents that occurred in fog involved nine or more vehicles. These accidents accounted for less than 0.01 per cent of all accidents and about 0.25 per cent of fog accidents.

6.4.11 Drivers' Behaviours

Studies have revealed that traffic behaviour does not change significantly when visual range is reduced by fog, except for a slight reduction in average vehicle speed under some conditions. Because of a reduction in visibility, however, there is an increased probability of overdriving. For free-flowing traffic operating on high-speed main roads and motorways with good geometric design, there appeared to be no consistent change in speed variability, lateral position or collision course time, which is the vehicle headway and speed differential with the front vehicles, and that can be directly related to visibility restrictions caused by fog.

6.4.12 Fog Abatement

Dry ice seeding and other techniques have successfully dispersed cold fog, i.e., fog with liquid water droplets colder than 32° centigrade. However, such fogs comprise only a small fraction of all fog in many countries. For highways, dry-ice seeding might be useful at a few junctions with high traffic volumes in a limited area. Most fogs in northern hemisphere countries occur at temperatures above 0°C and in denser forms usually involve industrial pollution. The presence of pollution particles in the air will not cause fog. However, when cold atmospheric conditions are right for the formation of fog, hygroscopic particles present in most industrial effluents will increase its density. Therefore, reductions in industrial pollution should decrease the occurrence of very dense fogs that are a safety problem for motorists. The clean air standards set forth by the Environmental Protection Agency are leading in that direction. Efforts to achieve a reliable warm fog dispersal capability that might be economically feasible for highway use are far from encouraging. Four general fog dispersal techniques have been attempted:

- Thermal (evaporation of droplets),
- Chemical (evaporation and/or altering drop size),
- Electrical (coalescence of charge droplets), and
- Mechanical.

Fog dispersal techniques require large installations, standby procedures and maintenance expenses, but these often become ineffective in dissipating the fog. These techniques may be feasible in limited areas or cases, such as at busy junctions and roundabouts.

The relatively short persistence of most fogs limits the utility of portable dispersal equipment and installations that require long start-up times. Another limiting factor with portable dispersal equipment is the requirement for repeated application to the same roadway section. For example, with a 5-knot wind at 45 degrees to the centreline and a device that initially clears a 100-ft wide path, fog would begin closing in again within approximately ten seconds. Therefore, if a 60-second repeated-application cycle is required to keep the roadway above some minimum visibility level during periods of dense fog, a two-mile section of the road between interchanges might need as many as eight vehicle-mounted dispersal devices (assuming five miles of travel per vehicle during each dispersal cycle at an average speed of 40 mph).

Natural vegetation or landscaping has been applied in some specific locations to prevent shallow fogs from drifting over the road.

6.4.13 Guidance Systems

A variety of methods designed to assist drivers during fog have been proposed, and some have been tried. These include both active and passive guidance systems. Because of the clear economic infeasibility of justifying active guidance systems, such as 'automated highways', as possible solutions to the fog problem alone, they will not be discussed here. These types of systems might be justifiable for general purposes and, if so, would provide a solution to the fog problem by eliminating the driver from the control loop.

Under certain nighttime conditions, some highly directional types of fixed-lighting systems have proved to be effective in providing additional guidance information in fog. However, as is shown in Figure 6.9, the fog accident problem is particularly severe between the hours of 6 a.m. and 9 a.m., with the peak occurring about 7:30 a.m. Approximately 50 per cent of the fog accidents occur in daylight.

Fixed-lighting systems can be designed to be effective for the 50 per cent of the fog accidents that occur at night. However, the additional expense of such equipment would probably not be recovered from the resulting reduction in accidents. Conventional types of streetlighting may be useful in light and moderate fogs, but are not of much help in dense fog.

Reflective road surface marking results in only a slight increase in visibility during daytime fog and has virtually no effect on traffic behaviour. The use of inset 'pancake' marker lights has been suggested to provide guidance. They are relatively expensive and do not provide any information on stopped vehicles ahead. Perhaps such a display technique, combined with the information on lane occupancy obtained from instrumentation similar to the passing-aid system, would work.

Results of research on rear lighting systems indicate that current rear lights are ineffective in fog, especially during the daytime. A major problem is the possible misuse of a two-level or more tail-lighting system.

The most successful of the proposed plans involves the use of variable message signs that warn drivers of fog conditions ahead and advise them as to the desirable operating speed. Although these signs have little influence on mean operating speeds, they do reduce speed variance when the sign is set at approximately the mean speed of traffic. All the signs are set much below the prevailing speed, a bimodal speed distribution, which increases the likelihood of rear-end collisions. Part of the reason for these results may be the lack of reliability of information provided by such signs in the past. Particularly, where manual changing of speed limits or folding advisory signing was required, the messages were often exposed long after the reduced visibility condition ended. Therefore, many drivers may have assumed that such lighting is meaningless.

A major installation of remotely controlled warning signs is being evaluated. There are three large overhead signs in each direction, with the final sign at the point of maximum accident occurrence. The variable speed message is to be controlled on the recommendation of a state police officer stationed in the signed area and is based on the distance they can see and traffic flow in the area. Research is currently investigating traffic and fog detection equipment and automatic control of the signs.

Preliminary reports indicate that the Oregon installation is having a beneficial effect in terms of accidents and traffic flow. Traffic flow parameters, such as mean speed, speed variance and headway, may be more useful in determining that the required information is displayed rather than developing devices for the detection of fog density.

Use of optical and electronic devices for the detection of fog appears to be too expensive for widespread use and too sensitive to small changes in local conditions within a few metres of the measurement station. Research currently underway is attempting to relate the output of such a device to a meaningful fog index for drivers. The Federal Highway Administration in the USA is currently planning research aimed at developing a speed advisory system that will inform drivers of the status of other vehicles' speeds on the road ahead, but beyond the limit of the sight distances. Such a system might involve a simple fog-no-fog detector for system activation and a speed sensor with a simple roadside sign indicating an advisory speed.

The speed would be set at a location 1/4 to one mile ahead at approximately the mean speed of traffic. The logic devices for controlling the system should probably be interconnected so that no sign calls for a speed that is more than 10 to 20 km/h higher than the following sign, except under exceptionally critical conditions.

The installation of such a system should reduce the speed of traffic gradually and should help to reduce multivehicle secondary accidents. Because the system will be designed to alert drivers to any slowing of traffic ahead, regardless of cause, it is expected that the system will be useful for many situations in addition to those produced by fog.

In summary, there is no technology currently available that will cost-effectively solve the fog problem. Current research on warning advisory signage for fog appears to hold significant potential. The only real safe advice at present is, if the situation gets bad enough, close the road.

Lessons learned during the field tests are mentioned below, many of which have been addressed through design modifications:

- VHF communication entailed a cumbersome Communications Commission permit process.
- Cellular communications were expensive and not reliable enough.
- Power consumption was too high for cellular communications with solar power.
- Fog chamber testing was time-consuming.
- Raindrops and frost formation may have caused false readings.

Table 6.2 Sample Statistics of Accidents in Fog on US Highways

		Accidents in Fog		
Number of Vehicles per Accident	Total Number of Accidents	Number	Percentage of All Accidents	Percentage of Involved Vehicles
1	149,798	3,631	2.4	20.3
2	262,346	4,911	1.9	54.9
3	38,344	786	2.0	13.2
4	7,623	223	2.9	5.0
5	1,708	81	4.7	2.3
6	471	43	9.1	1.4
7	160	23	14.4	0.9
8	47	15	31.9	0.6
9 or more	39	25	64.1	1.4

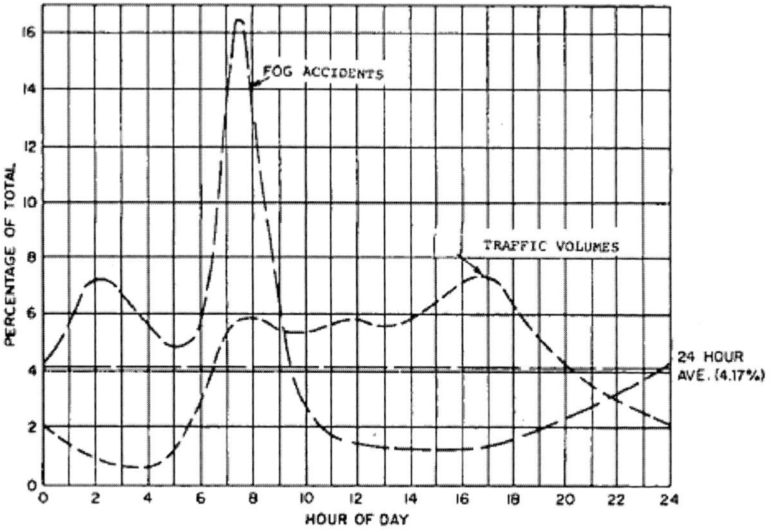

Figure 6.10 Hourly distribution of traffic volumes and accidents in fog

6.5 Highway Lighting

6.5.1 Overview

Today, streetlighting is commonly done using HID lamps. Low-pressure sodium (LPS) lamps became commonplace after World War II due to their low power consumption and long life. Late in the 20th century, high-pressure sodium (HPS) lamps were preferred, taking further the same virtues. Such lamps provide the greatest amount of photopic illumination for the least consumption of electricity. However, white light sources have been shown to double driver peripheral vision and improve driver brake reaction time by at least 25 per cent, enable pedestrians to detect road surface trip hazards better and to facilitate visual appraisals of other people. Studies comparing metal halide and HPS lamps have shown that at equal photopic light levels, a street scene illuminated at night by a metal halide lighting system was seen as brighter and safer than the same scene illuminated by a HPS system.

Two national standards now allow for variation in illuminance when using lamps of different spectra. In Australia, HPS lamp performance needs to be reduced by a minimum value of 75 per cent. In the UK, illuminances are reduced with a higher S/P ratio. New streetlighting technologies, such as LED or induction lights, emit a white light that provides high levels of scotopic lumens, allowing streetlights with lower wattages and lower photopic lumens to replace existing streetlights. However, there have been no formal specifications written around photopic/scotopic adjustments for different types of light sources, causing many municipalities and street departments to hold back on implementation of these new technologies until the standards are updated. Eastbourne in East Sussex, the UK, is currently undergoing a project to see 6000 of its streetlights converted to LED, and will be closely followed by Hastings in early 2014. Many UK councils are undergoing mass-replacement schemes to LED, and though streetlights are being removed along many long stretches of UK motorways (as they are not needed and cause light pollution), LEDs are preferred in areas where lighting installations are necessary.

6.5.2 Benefits of Highway Lighting

The major purpose of highway lighting includes the prevention of accidents and increasing traffic safety. Studies have shown that darkness results in many crashes and fatalities, especially those involving pedestrians; pedestrian fatalities are 3 to 6.75 times more likely in the dark than in daylight. Several decades ago, when automobile crashes were far more common, streetlighting was found to reduce pedestrian crashes by approximately 50 per cent. Furthermore, in the 1970s, lighted junctions and highway interchanges tended to have fewer crashes than unlighted junctions and interchanges. Lighting also reduces crime, with one 2019 study suggesting that appropriate lighting reduces night crime by 36 per cent. Towns, cities and villages use the unique locations provided by lampposts to hang decorative or commemorative banners. Many communities in the US use lampposts as a tool for fundraising via lamppost banner sponsorship programmes, first designed by a US-based lamppost banner manufacturer.

6.5.3 Disadvantages of Lighting

The major criticisms of streetlighting are that it can cause accidents if misused and cause light pollution.

Health and safety

Two optical phenomena need to be recognised in streetlight installations.

- The loss of night vision because of the reflex of drivers' eyes. As drivers emerge from an unlit area into a pool of light from a streetlight, their pupils quickly constrict to adjust to the brighter light, but as they leave the pool of light, the dilation is much slower, so they are driving with impaired vision. As a person gets older, the eye's recovery speed slows, so driving time and distance under impaired vision increase.
- Oncoming headlights are more visible against a black background than a grey one. The contrast creates greater awareness of the oncoming vehicle.
- Stray voltage is also a concern in many cities. Stray voltage can accidentally electrify lampposts and has the potential to injure or kill anyone who touches the post.

There are also physical dangers from the posts of streetlamps, other than children climbing them for recreational purposes. Lampposts pose a collision risk to motorists and pedestrians, particularly those affected by poor eyesight or under the influence of alcohol. This can be reduced by designing them to break away when hit (known as frangible, collapsible, or passively safe supports), protecting them with guardrails or painting the lower portions to increase their visibility. High winds or accumulated metal fatigue also occasionally topple streetlights.

Light Pollution May Affect Astronomical Work

Light pollution can hide the stars and interfere with astronomy. In settings near astronomical telescopes and observatories, LPS lamps may be used. These lamps are advantageous over other lamps, such as mercury and metal halide lamps, because LPS lamps emit lower intensity, monochromatic light. Observatories can filter the sodium wavelength out of their observations and virtually eliminate the interference from nearby urban lighting. Full cutoff streetlights also reduce light pollution by reducing the amount of light that is directed at the sky, which also improves the luminous efficiency of the light.

Light Pollution May Affect Ecosystems

Streetlights can impact biodiversity and ecosystems—for instance, disrupting the migration of some nocturnally migrating bird species. In the Netherlands, Philips found that birds can get disorientated by the red wavelengths in streetlighting, and in response developed alternative lighting that only emits green and blue wavelengths of the visible spectrum. The lamps were installed on Ameland in a small-scale test. If successful, the technology could be used on ships and offshore installations to avoid luring birds towards the open sea at night. Bats can be negatively impacted by streetlights, with evidence showing that red light is the least harmful. As a result, some areas have installed red LED streetlights to minimise disruption to bats, as reported in a study published in *Science Advances*.

Energy Consumption

As of 2017, globally, 70 per cent of all electricity was generated by burning fossil fuels, a source of air pollution and greenhouse gases, and approximately 300 million streetlights are using that electricity.

Cities are exploring more efficient energy use, reducing streetlight power consumption by dimming lights during off-peak hours and switching to high-efficiency LED lamps. Many councils in the USA are using a part-night lighting scheme to turn off lighting at quieter times of night. This is typically midnight to 5:30 am. There have, however, been questions about the impact on crime rates. Typical collector road lighting in New York State costs $6,400/mile/year for HPS at 8.5 kW/mile or $4,000 for light-emitting diode luminaries at kW/mile. Improvements can be made by optimising directionality and shape, however. Transitioning to wide-angle lights enabled a doubling of the distance between streetlights in Flanders from 45 m to 90 m, cutting annual streetlighting electricity expenditures to €9 million for the 2150 km long network that was retrofitted, corresponding to ca. €4186/km.

6.5.4 Streetlight Control Systems

Several streetlight control systems have been developed to control and reduce the energy consumption of a town's public lighting system. These range from controlling a circuit of streetlights and/or individual lights with specific ballasts and network operating protocols. These may include sending and receiving instructions via separate data networks, at high frequency over the top of the low voltage supply or wireless.

Streetlight controls are smarter versions of the mechanical or electronic timers previously used for streetlight on-off operations. They come with energy conservation options like twilight saving, staggering or dimming. Many streetlight controls come with an astronomical clock for a particular location or a global positioning system (GPS) connection to give the best on-off time and save energy.

Some intelligent streetlight controls also come with a global system for mobile (GSM) communications, radio frequency (RF) or general packet radio service (GPRS) communication, and are user-adjustable according to latitude and longitude (low-cost type) for better management and maintenance. Many streetlight controls also come with traffic sensors to manage the lux level of the lamp according to the traffic and to save energy by decreasing the lux when there is no traffic. The United States, Canada, India and many other countries have started introducing streetlight controls to their road lighting for energy conservation, streetlight management and maintenance purposes.

6.5.5 Economics

Streetlight controls can be expensive compared to normal timers, and can cost between $100 and $2,500, but most of them return the investment between six months and two years. As the equipment's lifetime is seven to ten years, it saves energy and cost after the initial investment has been recouped.

6.5.6 Leading Manufacturers

Australia

- Osram Sylvania
- Westinghouse

North America

- American Electric Lighting (formerly ITT, later Thomas & Betts), US
- Cooper Lighting division of Cooper Industries, US
- General Electric, US
- Osram Sylvania, US
- Westinghouse Lighting Corporation (formerly Angelo Brothers) division of the new Westinghouse Electric, US

Europe

- CU Phosco Lighting, United Kingdom
- Trilux, Germany
- Osram, Germany

- Philips, the Netherlands
- Siemens, Germany
- Svetlina AD, Bulgaria
- Thorn Lighting, United Kingdom
- Schréder, Belgium
- DE NOOD, the Netherlands

Asia

- Matsushita Electric Industrial Co., Japan
- Sanyo, Japan
- Bajaj Electricals, India
- K-Lite Industries, India

6.5.7 Measurements of Highway Lighting

Measurement of Nighttime Luminance

Streetlighting may be measured by using a computer-linked meter and a distance-measuring instrument over large sections of the roadway system, but it can pose a data collection challenge. The automatic mobile reading systems and the use of photography methods were developed that enabling quicker data collection. The photographic method was used, and two digital single-lens reflex cameras were calibrated and field tested for measuring nighttime luminance.

Evaluating Roadway Illumination

The performance of roadway illumination can be evaluated by luminance and small target visibility (STV) methods (Zhou et al. 2009).

Luminance is a measure of the amount of light reflected from a surface (Illuminating Engineering Society 2000; Illuminating Engineering Society 2008; International Commission on Illumination 2010), and it is measured in candela per square metre (cd/m^2). It is what is perceived by the human eye as the brightness of the road surface.

STV is a metric used to determine the visibility of an array of targets on the roadway (Illuminating Engineering Society 2000). The recommended method for conflict points, including junctions, is surface or horizontal illuminance (CEN 2008c, Illuminating Engineering Society 2000). Also, vertical illuminance, which helps drivers to see pedestrians and objects on a crossing, should be measured at a height of 1.5 metres above the roadway at the crossing.

Quantity and Quality of Roadway Illumination

Luminance was statistically related to night/day crash frequency ratio as concluded by four different studies (Cobb et al. 1979; Green and Hargroves 1979; Hargroves and Scott 1979; Scott 1980) that evaluated the relationship between illumination parameters (illuminance, luminance, uniformity and glare) on crashes. Within the luminance range of 0.5–2.0 cd/m, an increase in average surface luminance of 1.0 cd/m results in a 35 per cent reduction in nighttime crash frequency ratio, as estimated by Scott (1980). Crashes might increase as uniformity of lighting increases beyond a certain level due to a reduction in contrast between an object and its surrounding visual environment, which is noted by the CIE, in a review of 62 studies (International Commission on Illumination 1992) from 15 nations.

At 18 trunk road junctions, the illuminance was evaluated (Oya et al. 2002), each with at least 10,000 AADT, using data from one year before and four years after. Illuminance levels of 30 lux or more can positively help to reduce nighttime crashes, as shown by the results of calculations with illuminance data at each junction. The results were found to be significant at the one per cent level. It was also found that although the study could not find any statistical significance for this category of lighting level, illuminance levels between 20 to 30 lux can reduce nighttime crashes. Next, an illuminance of ten lux or more is needed for drivers to have good visibility of pedestrians at a junction, as observed by a Japanese study (Minoshima et al. 2006), and an illuminance uniformity ratio of 0.4 will make a junction safer.

The illuminance from three different sets of LEDs and one set of HPS luminaires was measured by Medina et al. (2013), and the measured values were compared with the estimates derived from computer analysis with AGi32® lighting software. On dry days under open skies with no full moon, the measurements were taken, and the results showed that the lighting was in close agreement, but with significant differences between the measured values and software-estimated values.

This may be attributed to the luminaire-specific differences, underscoring the need to perform periodic audits to verify if in-situ lighting levels meet the design specifications. Conducting audits on streetlighting with handheld metres over large sections of the roadway system can pose both a data collection and safety challenge for the data collection personnel.

The efforts have resulted in the development of automatic mobile reading systems to overcome this challenge, and the use of photography that enables quicker data collection from either junctions or road segments. A new measurement system for collecting illuminance data was developed for the Florida DOT by Zhou et al. (2009).

Every 17.5 feet, the data was collected from a vehicle moving at 30 mph with a computer linked to a lighting metre and a distance measuring instrument. To transform the measurements made at the top of the moving vehicle to the equivalent measurements at six inches above the road surface, an inverse square method was used, and a Wilcoxon test was used to compare the measurements. The median of differences between the two was not significantly different from zero, as shown by the results.

The feasibility of LED roadway luminaires was also explored (Schmidt et al. 2014) by analysing eight different LED luminaires produced by different manufacturers and three HPS luminaires with power ratings of 150 W, 250 W, and 400 W, for economic analysis, annualised life cycle costs were used, while the technical feasibility was determined by comparing in-situ measurements to recommended IES standards. Only one LED luminaire conformed to the IES standard for a moderately busy, medium pedestrian conflict road with R3 road surface, as shown by the results of the studies. Also, the existing HPS were economically outperformed by one of the eight studied LED luminaires in terms of life cycle costs. Therefore, although LED luminaires are a promising technology, more technological advancement would be needed to accurately confirm their use in roadway illumination, as concluded by the study.

Existing installation methods for roundabout illumination, that is, luminaires hanging from fixed heights on poles, do not necessarily provide the best visibility for drivers, and they can also be energy/cost intensive (Bullough et al. 2014). A new lighting approach called Eco luminance was consequently evaluated, which relies on both illuminance and luminance using a combination of roadside vegetation to provide visual delineation, lower-level lighting such as landscape lighting to reinforce delineation, pedestrian-level lighting to provide illumination for important safety hazards and retroreflective elements to provide clues about road geometry. In New York, the Eco luminance approach was tested, and the results show comparable approach speeds and initial costs for Eco luminance and conventional lighting. The Eco luminance approach used only a quarter of the energy compared to the conventional illumination method.

For performing illumination audits of junctions, a method was developed using light sensors attached to a handle and a data logger for recording both illumination and position via GPS coordinates (Niaki et al. 2014), making it much less time-consuming. By walking across the exit/entrance line of each junction leg, the measurement can be made, and then the mean junction illuminance can be obtained by averaging. In Montreal, the case studies of 85 junctions indicated that about 59 per cent of junctions had sub-standard lighting levels.

Compared to existing protocols, although this method can simplify the measurements, it increases the safety risk for both personnel and equipment since they must be in the active travel lane to collect data. Additionally, before all the junctions are walked across, the measurements with this method may lack luminance constancy, as there may be voltage fluctuation on-site.

The relationship between road lighting levels and safety using five years of crash data and road lighting measurements from mid-block road sections was studied in New Zealand (Jackett and Frith, 2013). For luminance conversions at specific settings of camera exposure, the lighting levels were obtained by the photographic method, and polynomials of the sixth order were calibrated for pixels.

The results of a study at 152 mid-block road sections showed that the most important performance measure in predicting crashes on road sections is average luminance, and, for predicting crashes on road sections, the uniformity was insignificant. It was noted that a similar result was established in an earlier study.

Next, the lighting data was applied to junctions, but as compared to road sections, the results were not very strong. The photographic method was used in the study, which was fundamentally different from the photographic method. If the pixel-to-luminance conversion approach is not linked to the camera's own calibration constant, then it is only applicable to the specific exposure conditions (shutter speed, F-number, and ISO sensitivity) used in the calibration. However, if the approach is used by linking to the camera's calibration constant, it is applicable for all exposure conditions as long as the same camera is used. This is very important to note as light conditions can vary a lot in the field, and to get the best measurement, the exposure conditions may need to be modified. Also, the photographic method was used on roundabouts. As the roundabouts, unlike conventional junctions, have a visual obstruction at the centre, making it impossible to see the entire travel path in one view, the method requires a different approach.

The effect of lighting quality and quantity on the night/day (ND) crash frequency ratios at rural junctions using negative binomial regression to model illuminance, luminance, and crash data from 99 lighted and unlighted junctions was investigated by Bhagavathula et al. (2015).

It was indicated by the results that a one-lux increase in the average horizontal illuminance results in a seven per cent reduction in the ND crash ratio at all rural junctions. At the lighted junctions, a one lux increase in average horizontal illuminance results in a nine per cent decrease in the ND crash ratio, while for unlighted junctions, a one lux increase in average horizontal illuminance results in a 21 per cent reduction in the ND crash ratio. ND crash ratios are smaller at stop-control junctions than junctions with signals, and ND crash ratios are lower at junctions with a posted speed limit of less than or equal to 40 mph than those with a posted speed limit greater than 40 mph.

The relationship between lighting level and crashes on roadways was investigated in another study by Gibbons et al. (2015). From selected states and the *Highway Safety Information System,* the crash data was obtained, while lighting measurements were collected in situ with a mobile road lighting measurement system. There was no benefit to illumination beyond a certain level on an urban interstate, as observed in the results; the illumination level was about five lux. It was concluded that there is a potential to reduce lighting requirements on main roads and motorways by as much as 50 per cent while maintaining traffic safety. The results indicated that the relationship between lighting level and safety was not as strong as that of lighting presence (lit or unlit) and safety.

6.5.8 Design of Highway Lighting

On motorways or main roads, the lights are normally provided at junctions and at junctions located in non-urban or rural sections outside built-up areas. Lighting is also provided at bus stops, lorry lay-bys, Toll plazas and on the service roads.

Table 6.3 Recommended horizontal luminance for roundabouts

Functional Classification	**Maintained Average Horizontal Illuminance in Lux/FC for different pedestrian area classifications**			**Uniformity Level (E_{avg}/E_{min})**
	High	**Medium**	**Low**	
Major/Major	34.0/3.4	26.0/2.6	18.0/1.8	3:1
Major/Collector	29.0/2.9	22.0/2.2	15.0/1.5	3:1
Major/Local	26.0/2.6	20.0/2.0	13.0/1.3	3:1
Collector/Collector	24.0/2.4	18.0/1.8	12.0/1.2	4:1
Collector/Local	21.0/2.1	16.0/1.6	10.0/1.0	4:1
Local/Local	18.0/1.8	14.0/1.4	8.0/0.8	6:1

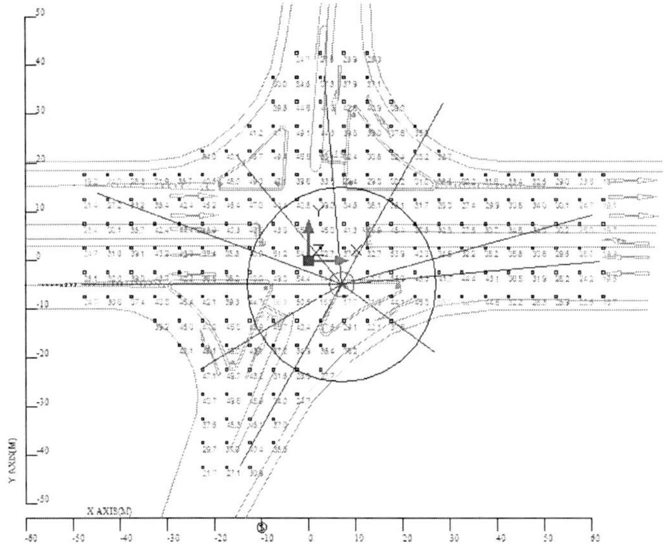

(Source: Illuminating Engineering Society 2008)

HM: 20M HIGH MAST

Luminaire Schedule					
Symbol	Qty	Label	Lumens	LLF	Description
⟶▷	10	A	47000	0.750	BGENF 22R(2X400W HPSV/T)

HORIZONTAL ILLUMINATION LEVEL AT G/L.:

Numeric Summary							
Label	CalcType	Units	Avg	Max	Min	Min/Avg	Min/Max
JUNCTION	Illuminance	Lux	36.48	62.1	14.6	0.40	0.24

Figure 6.11 Illumination design for high mast lighting at a four-road junction

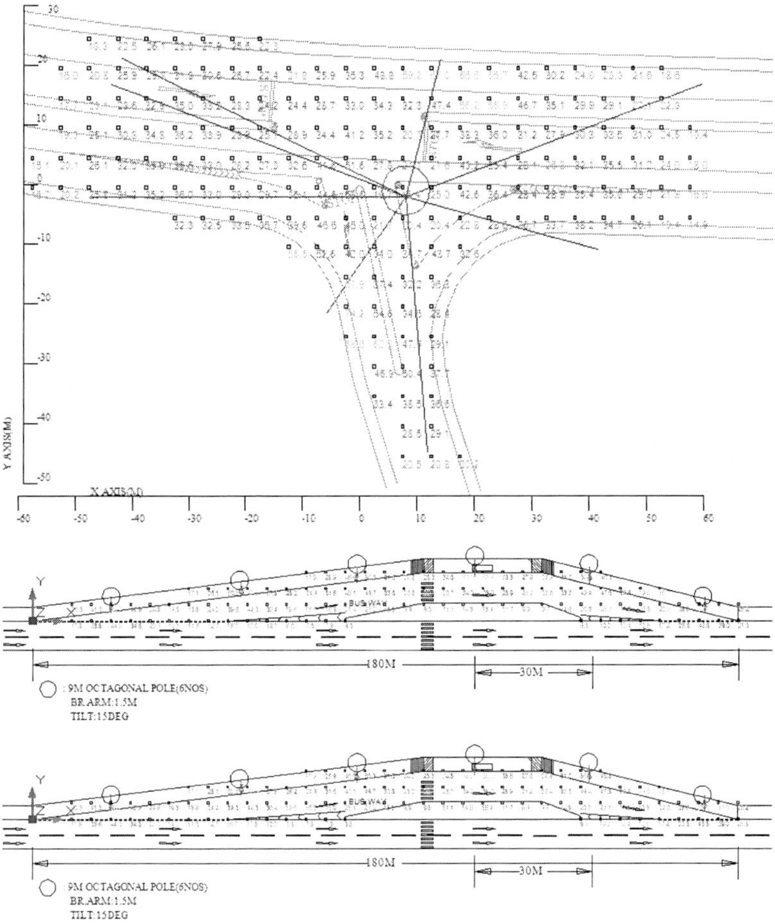

HM: 20M HIGH MAST

Luminaire Schedule					
Symbol	Qty	Label	Lumens	LLF	Description
→▷	8	A	47000	0.750	BGENF 22R(2X400W HPSV/T)

HORIZONTAL ILLUMINATION LEVEL AT G/L.:

Numeric Summary							
Label	CalcType	Units	Avg	Max	Min	Min/Avg	Min/Max
JUNCTION	Illuminance	Lux	32.56	74.2	14.9	0.46	0.20

Figure 6.12 - Illumination design for high mast lighting at a three-road junction

Luminaire Schedule					
Symbol	Qty	Label	Lumens	LLF	Description
	9	A	26500	0.750	JET- BGJRL 250SV/T

HORIZONTAL ILLUMINATION LEVEL:

Numeric Summary							
Label	CalcType	Units	Avg	Max	Min	Min/Avg	Min/Max
TRUCK LAY BYE	Illuminance	Lux	26.83	63.2	6.6	0.25	0.10

Luminaire Schedule					
Symbol	Qty	Label	Lumens	LLF	Description
	6	A	26500	0.750	JET- BGJRL 250SV/T

HORIZONTAL ILLUMINATION LEVEL:

Numeric Summary							
Label	CalcType	Units	Avg	Max	Min	Min/Avg	Min/Max
BUS BAY	Illuminance	Lux	27.07	58.8	7.5	0.28	0.13

Figure 6.13 Typical bus bay (single-sided) lighting with 9 m octagonal pole

HM 1 TO HM 4: 25M HIGH MAST(4NOS)

Luminaire Schedule					
Symbol	Qty	Label	Lumens	LLF	Description
	22	A	47000	0.750	BGENF 22R(2X400W HPSV/T)

HM NO	LUM"A"
HM1	5NOS
HM2	6NOS
HM3	6NOS
HM4	5NOS

TOLL CANOPY CONSIDERED AS FLOATING OBSTRUCTIONS AT AHEIGHT OF 6M FROM G/L.
HORIZONTAL ILLUMINATION LEVEL AT EL.0M FROM G/L.:

Numeric Summary							
Label	CalcType	Units	Avg	Max	Min	Min/Avg	Min/Max
RHS 50 LUX ZONE	Illuminance	Lux	53.01	80.3	25.6	0.48	0.32
RHS 30 LUX ZONE	Illuminance	Lux	32.05	58.7	14.5	0.45	0.25
LHS 30 LUX ZONE	Illuminance	Lux	31.06	57.6	13.8	0.44	0.24
LHS 50 LUX ZONE	Illuminance	Lux	52.92	76.9	26.9	0.51	0.35

Figure 6.14 Typical lorry lay-by (single-sided) lighting with 9 m octagonal pole

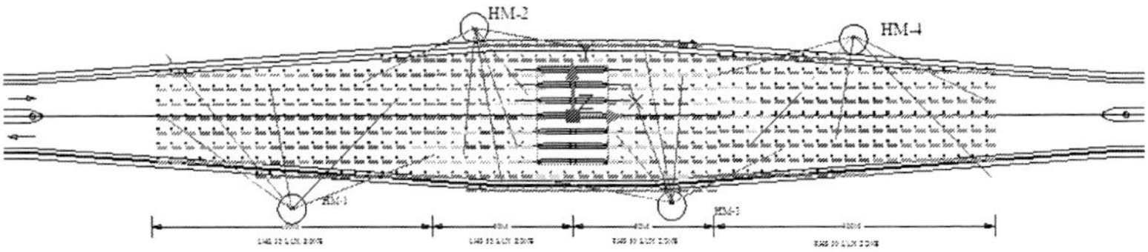

Figure 6.15 Toll plaza lighting with a 25 m high mast

Table 6.4 Sample of bill of quantities of lighting for junctions, a toll plaza, a lorry lay-by & bus bay

Project :
Consultant :

Bill of Quantity for Junction, Toll Plaza , Truck Lay-Bye & Bus Bay Lighting

Sr. No	Junctions	Mast/Poll height(m)	No of Mast/Poll	Fixture Quantity	Bajaj Cat ref	Lamp wattage
1	Proposed Junction 138+800	20m HM	1	6	BGENF 22R	2 X 400w SV (T)
2	Proposed Major Junction at 153+310	20m HM	1	6	BGENF 22R	2 X 400w SV (T)
3	Proposed Major Junction at 1 153+800	20m HM	1	6	BGENF 22R	2 X 400w SV (T)
4	Proposed Junction Plan 163+150	20m HM	1	5	BGENF 22R	2 X 400w SV (T)
5	Proposed Junction Plan 165+750	20m HM	1	5	BGENF 22R	2 X 400w SV (T)
6	Proposed Junction 176+950	20m HM	2	12	BGENF 22R	2 X 400w SV (T)
7	Proposed Junction Plan 179+500	20m HM	1	5	BGENF 22R	2 X 400w SV (T)
8	Proposed Junction Plan 221+365	20m HM	2	12	BGENF 22R	2 X 400w SV (T)
9	Toll Plaza At Km 153+200	25m HM	4	22	BGENF 22R	2 X 400w SV (T)
10	Typical Bus Bay (Single Sided)	9m Pole	6	6	JET - BGJRL 250w HPSV/T	1 X 250w SV (T)
11	Typical Truck Lay Bye (Single Sided)	9m Pole	9	9	JET - BGJRL 250w HPSV/T	1 X 250w SV (T)

6.6 Example for Design of Lighting for the Main Carriageway

Details of Coverage Area

Streetlight Arrangement: Double Arm
Roadway Length: 45.0 m

Roadway Width: 26.0 m

Mounting Height: 13.0 m
Overhang from Edge: 1.00 m
Lantern Tilt Angle: 10.0°

Lantern Spacing: 45.0m
Maintenance Factor: 0.70

Lantern Layout continues Beyond Calculation Area Left: Yes
Lantern Layout Continues Beyond Calculation Area Right: Yes

Types of Luminaries

Catalogue Number: BGEST 400 SV-MOMENTUM (Bajaj Electricals Limited) Lamp Wattage: 400 W

Lamp Quantity: 4 Nos.

Lighting Parameters

Average Horizontal Luminance (E average)	= 34
Lux Maximum Horizontal Luminance (E max)	= 81
Lux Minimum Horizontal Luminance (E min)	= 14
Lux Uniformity Ratio (E min / E average)	= 0.401 (1 : 2.49)
Diversity Ratio (E min / E max)	= 0.170 (1 : 5.88)

Electrical Parameters

Total Wattage	= 1800 W
Total Wattage/Luminary = 450 W LPD (Watt / Sq. Ft)	= 0.14
LPD (Watt / Sq. M)	= 1.54
Watt / Sq. M / 100 Lux	= 4.48

Lighting on the Main Road

Figure 6.16 Layout of the area between two double-arm streetlights installed on the median

Main Road: Installation Data

Luminary Type 1: BGEST 400 SV-MOMENTUM
Type:
Catalogue No: BGEST 400 SV-MOMENTUM
Length: 0.000m.
Width: 0.010m.
Height: 0.000m.

No. of Lamps:		1
Lumens per Lamp:	48000.0 l m	
Light Output Ratio:	86.3 per cent	
Rated Power:	450.0 W	

Main Road: Luminary Layout

* Installation data includes luminaries beyond the edge of the roadway calculation area (shown in blue)

Table 6.5 Coverage Area on the Carriageway of Main Road

S. No.	Type	X (m)	Y (m)	H (m)	O°	T°
1	1	0.00	12.00	13.00	0.00	10.00
2	1	0.00	14.00	13.00	180.00	10.00
3	1	45.00	12.00	13.00	0.00	10.00
4	1	45.00	14.00	13.00	180.00	10.00

Table 6.6 Main road: luminance values in lux Horizontal Luminance on Work Plane

Luminance: Average = 34.3 lx   Maximum = 81.0 lx   Minimum = 13.8 lx
Max / Min = 5.877   Average / Minimum = 2.491

Table 6.6 Main Road: Luminance values in lux in the above table extended horizontally (along the x-axis from 0.5 to 14.5)

Metres (X-Axis)	0.50	1.50	2.50	3.50	4.50	5.50	6.50	7.50	8.50	9.50	10.50	11.50	12.50	13.50	14.50
(Y-Axis)															
0.50	14.9	15.4	16.0	16.7	16.9	17.4	17.6	17.9	18.1	18.2	18.2	17.9	17.5	16.9	16.3
1.50	17.8	18.4	19.5	20.6	21.3	21.9	22.0	21.9	21.8	21.6	21.2	20.8	20.1	19.2	18.3
2.50	21.2	22.2	24.2	25.6	27.0	27.4	27.2	26.7	26.0	25.3	24.5	23.7	22.7	21.6	20.5

3.50	25.627.1	30.1	32.4	34.1	34.1	33.2	32.0	30.7	29.3	28.0	26.825.4	24.0	22.7
4.50	31.033.5	37.0	40.5	42.4	42.1	40.2	38.0	35.8	33.7	31.6	29.828.2	26.5	24.7
5.50	37.641.1	45.5	50.2	51.7	50.3	47.6	44.1	40.8	37.9	35.2	32.630.6	28.5	26.6
6.50	45.249.5	54.8	58.7	59.9	58.0	53.7	49.3	45.3	41.4	38.1	35.132.5	30.2	28.1
7.50	53.859.1	63.9	65.7	64.1	62.4	58.0	52.8	48.1	43.9	40.1	36.834.0	31.5	29.3
8.50	62.568.5	70.4	69.8	67.1	63.7	59.7	54.5	49.5	45.2	41.3	37.934.9	32.4	30.2
9.50	71.571.6	71.3	70.1	67.5	63.9	59.9	55.3	50.1	45.8	41.9	38.535.5	33.0	30.8
10.5	70.571.0	69.9	68.3	66.6	63.2	59.5	55.2	50.3	46.0	42.2	38.935.9	33.4	31.3
11.5	77.174.2	70.7	67.4	64.8	62.1	58.7	55.0	50.2	46.0	42.4	39.236.3	33.8	31.6
12.5	81.077.2	72.1	67.4	64.1	61.3	58.3	54.9	50.0	45.9	42.4	39.336.5	34.0	31.9
13.5	81.077.2	72.1	67.4	64.1	61.3	58.3	54.9	50.0	45.9	42.4	39.336.5	34.0	31.9
14.5	77.174.2	70.7	67.4	64.8	62.1	58.7	55.0	50.2	46.0	42.4	39.236.3	33.8	31.6
15.5	70.571.0	69.9	68.3	66.6	63.2	59.5	55.2	50.3	46.0	42.2	38.935.9	33.4	31.3
16.5	71.571.6	71.3	70.1	67.5	63.9	59.9	55.3	50.1	45.8	41.9	38.535.5	33.0	30.8
17.5	62.568.5	70.4	69.8	67.1	63.7	59.7	54.5	49.5	45.2	41.3	37.934.9	32.4	30.2
18.5	53.859.1	63.9	65.7	64.7	62.4	58.0	52.8	48.1	43.9	40.1	36.834.0	31.5	29.3
19.5	45.249.5	54.8	58.7	59.9	58.0	53.7	49.3	45.3	41.4	38.1	35.132.5	30.2	28.1
20.5	37.641.1	45.5	50.2	51.7	50.3	47.6	44.1	40.8	37.9	35.2	32.630.6	28.5	26.6
21.5	31.033.5	37.0	40.5	42.4	42.1	40.2	38.0	35.8	33.7	31.6	29.828.2	26.5	24.7
22.5	25.627.1	30.1	32.4	34.1	34.1	33.2	32.0	30.7	29.3	28.0	26.825.4	24.0	22.7
23.5	21.222.2	24.2	25.6	27.0	27.4	27.2	26.7	26.0	25.2	24.5	23.722.7	21.6	20.5
24.5	17.818.4	19.5	20.6	21.3	21.9	22.0	21.9	21.8	21.6	21.2	20.820.1	19.2	18.3
25.5	14.915.4	16.0	16.7	16.9	17.4	17.6	17.9	18.1	18.2	18.2	17.917.5	16.9	16.3

Table 6.6 (continued) main road: luminance values in lux in the above table is extended horizontally (along the x-axis from 15.5 to 29.5)

Metres (X-Axis) (Y-Axis)	15.5	16.5	17.5	18.519.5	20.521.5	22.523.5	24.5	25.0	26.50	27.50	28.50	29.50
0.50	15.7	15.2	14.7	14.414.1	13.913.8	13.813.8	13.9	14.1	14.4	14.7	15.2	15.7
1.50	17.5	16.8	16.3	15.815.5	15.215.0	15.015.0	15.2	15.5	15.8	16.3	16.8	17.5
2.50	19.4	18.5	17.8	17.216.7	16.416.2	16.216.2	16.4	16.7	17.2	17.8	18.5	19.4
3.50	21.4	20.2	19.3	18.618.0	17.717.4	17.317.4	17.7	18.0	18.6	19.3	20.2	21.4
4.50	23.2	21.8	20.8	19.919.2	18.818.5	18.418.5	18.8	19.2	19.9	20.8	21.8	23.2
5.50	24.8	23.2	22.0	21.020.3	19.919.6	19.519.6	19.9	20.3	21.0	22.0	23.2	24.8
6.50	26.2	24.5	23.2	22.121.3	20.820.5	20.4 20.5	20.8	21.3	22.1	23.2	24.5	26.2
7.50	27.3	25.6	24.1	23.022.2	21.721.4	21.3 21.4	21.7	22.2	23.0	24.1	25.6	27.3
8.50	28.2	26.5	25.0	23.923.1	22.522.2	22.1 22.2	22.5	23.1	23.9	25.0	26.5	28.2
9.50	28.9	27.3	25.9	24.823.9	23.323.0	22.9 23.0	23.3	23.9	24.8	25.9	27.3	28.9
10.5	29.5	27.9	26.6	25.424.6	23.923.6	23.5 23.6	23.9	24.6	25.4	26.6	27.9	29.5

11.5	29.9	28.4	27.1	26.0	25.1	24.4	24.0	23.9	24.0	24.4	25.1	26.0	27.1	28.4	29.9
12.5	30.1	28.7	27.4	26.2	25.3	24.6	24.2	24.1	24.2	24.6	25.3	26.2	27.4	28.7	30.1
13.5	30.1	28.7	27.4	26.2	25.3	24.6	24.2	24.1	24.2	24.6	25.3	26.2	27.4	28.7	30.1
14.5	29.9	28.4	27.1	26.0	25.1	24.4	24.0	23.9	24.0	24.4	25.1	26.0	27.1	28.4	29.9
15.5	29.5	27.9	26.6	25.4	24.6	23.9	23.6	23.5	23.6	23.9	24.6	25.4	26.6	27.9	29.5
16.5	28.9	27.3	25.9	24.8	23.9	23.3	23.0	22.9	23.0	23.3	23.9	24.8	25.9	27.3	28.9
17.5	28.2	26.5	25.0	23.9	23.1	22.5	22.2	22.1	22.2	22.5	23.1	23.9	25.0	26.5	28.2
18.5	27.3	25.6	24.1	23.0	22.2	21.7	21.4	21.3	21.4	21.7	22.2	23.0	24.1	25.6	27.3
19.5	26.2	24.5	23.2	22.1	21.3	20.8	20.5	20.4	20.5	20.8	21.3	22.1	23.2	24.5	26.2
20.5	24.8	23.2	22.0	21.0	20.3	19.9	19.6	19.5	19.6	19.9	20.3	21.0	22.0	23.2	24.8
21.5	23.2	21.8	20.8	19.9	19.2	18.8	18.5	18.4	18.5	18.8	19.2	19.9	20.8	21.8	23.2
22.5	21.4	20.2	19.3	18.6	18.0	17.7	17.4	17.3	17.4	17.7	18.0	18.6	19.3	20.2	21.4
23.5	19.4	18.5	17.8	17.2	16.7	16.4	16.2	16.2	16.2	16.4	16.7	17.2	17.8	18.5	19.4
24.5	17.5	16.8	16.3	15.8	15.5	15.2	15.0	15.0	15.0	15.2	15.5	15.8	16.3	16.8	17.5
25.5	15.7	15.2	14.7	14.4	14.1	13.9	13.8	13.8	13.8	13.9	14.1	14.4	14.7	15.2	15.7

Table 6.6 (Continued) Main road: luminance values in lux in the above table is extended horizontally (along the x-axis from 30.5 to 44.5)

Metres (X-Axis) (Y-Axis)	30.5	31.5	32.5	33.5	34.5	35.5	36.5	37.5	38.5	39.5	40.0	41.50	42.50	43.50	44.50
0.50	16.3	16.9	17.5	17.9	18.2	18.2	18.1	17.9	17.6	17.4	16.9	16.7	16.0	15.4	14.9
1.50	18.3	19.2	20.1	20.8	21.2	21.6	21.8	21.9	22.0	21.9	21.3	20.6	19.5	18.4	17.8
2.50	20.5	21.6	22.7	23.7	24.5	25.2	26.0	26.7	27.2	27.4	27.0	25.6	24.2	22.2	21.2
3.50	22.7	24.0	25.4	26.8	28.0	29.3	30.7	32.0	33.2	34.1	34.1	32.4	30.1	27.1	25.6
4.50	24.7	26.5	28.2	29.8	31.6	33.7	35.8	38.0	40.2	42.1	42.4	40.5	37.0	33.5	31.0
5.50	26.6	28.5	30.6	32.6	35.2	37.9	40.8	44.1	47.6	50.3	51.7	50.2	45.5	41.1	37.6
6.50	28.1	30.2	32.5	35.1	38.1	41.4	45.3	49.3	53.7	58.0	59.9	58.7	54.8	49.5	45.2
7.50	29.3	31.5	34.0	36.8	40.1	43.9	48.1	52.8	58.0	62.4	64.7	65.7	63.9	59.1	53.8
8.50	30.2	32.4	34.9	37.9	41.3	45.2	49.5	54.5	59.7	63.7	67.1	69.8	70.4	68.5	62.5
9.50	30.8	33.0	35.5	38.5	41.9	45.8	50.1	55.3	59.9	63.9	67.5	70.1	71.3	71.6	71.5
10.5	31.3	33.4	35.9	38.9	42.2	46.0	50.3	55.2	59.5	63.2	66.6	68.3	69.9	71.0	70.5
11.5	31.6	33.8	36.3	39.2	42.4	46.0	50.2	55.0	58.7	62.1	64.8	67.4	70.7	74.2	77.1
12.5	31.9	34.0	36.5	39.3	42.4	45.9	50.0	54.9	58.3	61.3	64.1	67.4	72.1	77.2	81.0
13.5	31.9	34.0	36.5	39.3	42.4	45.9	50.0	54.9	58.3	61.3	64.1	67.4	72.1	77.2	81.0
14.5	31.6	33.8	36.3	39.2	42.4	46.0	50.2	55.0	58.7	62.1	64.8	67.4	70.7	74.2	77.1
15.5	31.3	33.4	35.9	38.9	42.2	46.0	50.3	55.2	59.5	63.2	66.6	68.3	69.9	71.0	70.5
16.5	30.8	33.0	35.5	38.5	41.9	45.8	50.1	55.3	59.9	63.9	67.5	70.1	71.3	71.6	71.5
17.5	30.2	32.4	34.9	37.9	41.3	45.2	49.5	54.5	59.7	63.7	67.1	69.8	70.4	68.5	62.5
18.5	29.3	31.5	34.0	36.8	40.1	43.9	48.1	52.8	58.0	62.4	64.7	65.7	63.9	59.1	53.8

19.5	28.1	30.2	32.5	35.1	38.1	41.4	45.3	49.3	53.7	58.0	59.9	58.7	54.8	49.5	45.2
20.5	26.6	28.5	30.6	32.6	35.2	37.9	40.8	44.1	47.6	50.3	51.7	50.2	45.5	41.1	37.6
21.5	24.7	26.5	28.2	29.8	31.6	33.7	35.8	38.0	40.2	42.1	42.4	40.5	37.0	33.5	31.0
22.5	22.7	24.0	25.4	26.8	28.0	29.3	30.7	32.0	33.2	34.1	34.1	32.4	30.1	27.1	25.6
23.5	20.5	21.6	22.7	23.7	24.5	25.2	26.0	26.7	27.2	27.4	27.0	25.6	24.2	22.2	21.2
24.5	18.3	19.2	20.1	20.8	21.2	21.6	21.8	21.9	22.0	21.9	21.3	20.6	19.5	18.4	17.8
25.5		6.3	16.9	17.5	17.9	18.2	18.2	18.1	17.9	17.6	17.4	16.9	16.7	16.0	15.4
14.9															

Horizontal luminance on the work plane

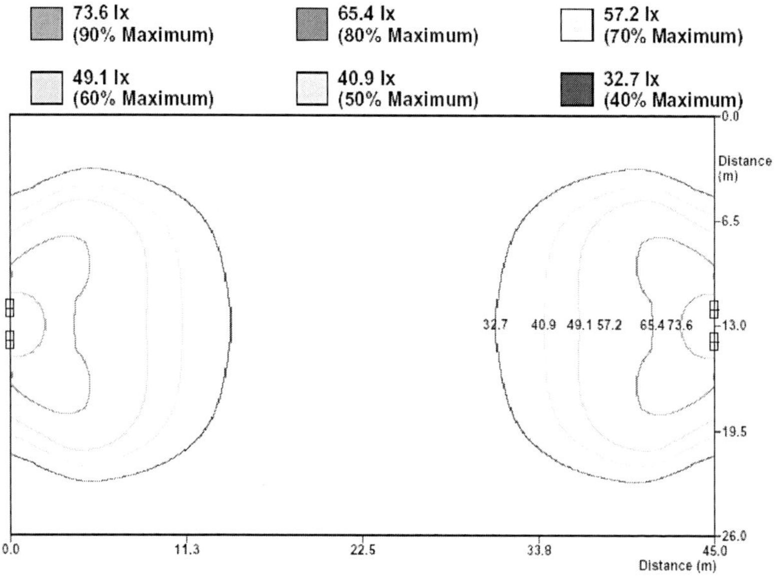

Figure 6.17 Main road: iso-lux contour diagram between two lights Horizontal Luminance on the work plane

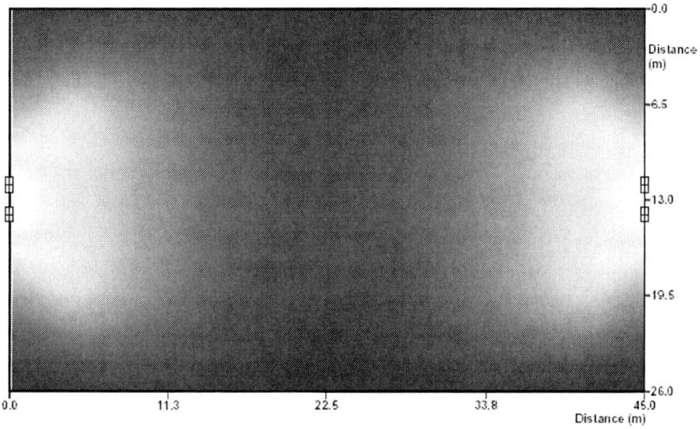

Figure 6.18 Main road: grey-scale diagram between two lights Horizontal luminance on workspace

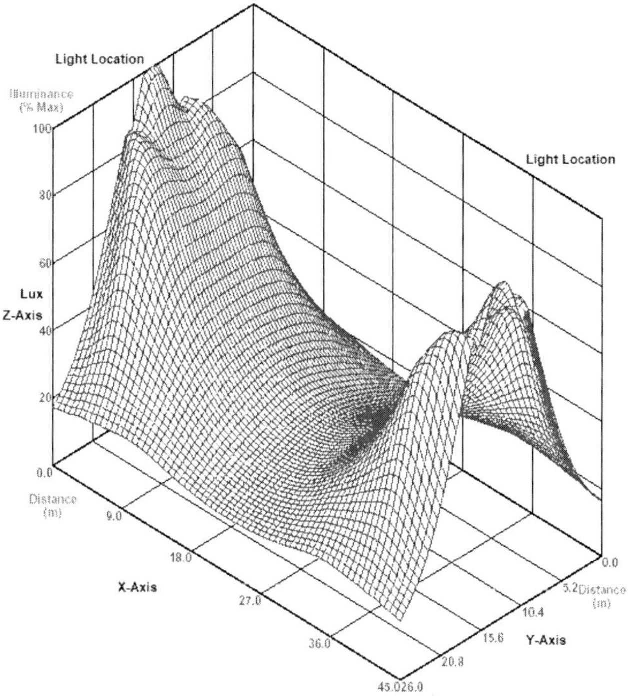

Figure 6.19 Main road: 3D-lux web between two lights, end of example for design lighting for main carriageway

Lighting Design for Each Service Road

Calculation Area Details

Streetlight Arrangement:	Single Sided
Roadway Length:	30.0m
Roadway Width:	7.5m
Mounting Height:	8.0m
Overhang from Edge:	1.50m
Lantern Tilt Angle:	10.0°
Lantern Spacing:	30.0m
Maintenance Factor:	0.70

Lantern Layout continues Beyond Calculation Area Left: Yes
Lantern Layout Continues Beyond Calculation Area Right: Yes

Type of Luminary

Catalogue Number: BGEST 250 SV- MOMENTUM (Bajaj Electricals Limited) Lamp Wattage: 150 W
Lamp Quantity: 2 Nos.

Lighting Parameters

Average Horizontal Illuminance (E average) = 30 Lux
Maximum Horizontal Illuminance (E max) = 581 Lux
Minimum Horizontal Illuminance (E min) = 12 Lux
Uniformity Ratio (E min/E average) = 0.398 (1: 2.51)
Diversity Ratio (E min/E max) = 0.211 (1: 4.74)

Electrical Parameters

Total Wattage = 534 W
Total Wattage / Luminaire = 267 W LPD (Watt / Sq. Ft) = 0.22
LPD (Watt / Sq. M) = 2.37
Watt / Sq. M / 100 Lux = 7.79

6.7 Example for Design of Lighting on the Carriageway of Each Service Road

Figure 6.20 Layout of the area between two single-sided streetlights installed on the side of the service road

Service Road: Installation Data

Luminary Type No. 1: BGEST 250 SV-MOMENTUM
Type:
Catalogue No: BGEST 250 SV-MOMENTUM
Length: 0.750 m. Width: 0.750 m.
Height: 0.000 m.

No. of Lamps: 1
Lumens per Lamp: 27000.0 l m
Light Output Ratio: 78.0 per cent Rated Power: 267.0 W

Service Road: Luminary Layout

* Installation data includes luminaries beyond the edge of the roadway calculation area (shown in blue)

Table 6.7 Coverage area on each service road

S. No.	Type	X (m)	Y (m)	H (m)	O°	T°
1	1	0.00	1.50	8.00	180.00	10.00
2	1	30.00	1.50	8.00	180.00	10.00

Table 6.8 Carriageway of Service Road Luminance Values in Lux

Horizontal Luminance on Work Plane

Luminance:Average = 30.5 lx   Maximum = 57.5 lx   Minimum = 12.1 lx
Max / Min = 4.74   Average / Minimum = 2.511

Metres (X-Axis) (Y-Axis)	0.50	1.50	2.50	3.50	4.50	5.50	6.50	7.50	8.50	9.50	10.50	11.50	12.50	13.50	14.50
0.54	22.6	25.1	30.7	32.3	29.2	26.2	24.0	22.5	20.9	18.6	16.3	14.6	13.4	12.5	12.1
1.61	31.1	37.1	44.0	41.8	36.9	32.3	28.5	25.5	22.9	20.3	17.8	15.9	14.5	13.6	13.1
2.68	41.5	49.5	51.9	49.2	43.4	37.1	31.7	27.5	24.4	21.6	19.0	16.9	15.4	14.5	14.0
3.75	52.3	55.3	54.9	51.8	45.7	39.6	34.1	29.3	25.6	22.6	20.0	17.8	16.2	15.2	14.7
4.82	56.8	55.7	54.9	52.5	47.0	40.9	35.6	30.5	26.5	23.3	20.6	18.3	16.7	15.6	15.1
5.89	57.6	56.4	55.5	55.1	50.7	43.2	36.6	30.9	26.8	23.7	20.9	18.5	16.7	15.6	15.1
6.96	56.2	54.6	53.9	53.5	50.9	43.7	36.5	30.6	26.6	23.6	20.8	18.3	16.5	15.4	14.8

Table 6.8 (continued) Carriageway of each service road luminance values in lux

The above table is extended horizontally (along the x-axis from 15.5 to 29.5)

Metres (X-Axis) (Y-Axis)	15.5	16.5	17.5	18.5	19.5	20.5	21.5	22.5	23.5	24.5	25.50	26.50	27.50	28.50	29.50
0.54	12.1	12.5	13.4	14.6	16.3	18.6	20.9	22.5	24.0	26.2	29.2	32.3	30.7	25.1	22.6
1.61	13.1	13.6	14.5	15.9	17.8	20.3	22.9	25.5	28.5	32.3	36.9	41.8	44.0	37.1	31.1
2.68	14.0	14.5	15.4	16.9	19.0	21.6	24.4	27.5	31.7	37.1	43.4	49.2	51.9	49.5	41.5
3.75	14.7	15.2	16.2	17.8	20.0	22.6	25.6	29.3	34.1	39.6	45.7	51.8	54.9	55.3	52.3
4.82	15.1	15.6	16.7	18.3	20.6	23.3	26.5	30.5	35.6	40.9	47.0	52.5	54.9	55.7	56.8
5.89	15.1	15.6	16.7	18.5	20.9	23.7	26.8	30.9	36.6	43.2	50.7	55.1	55.5	56.4	57.6
6.96	14.8	15.4	16.5	18.3	20.8	23.6	26.6	30.6	36.5	43.7	50.9	53.5	53.9	54.6	56.2

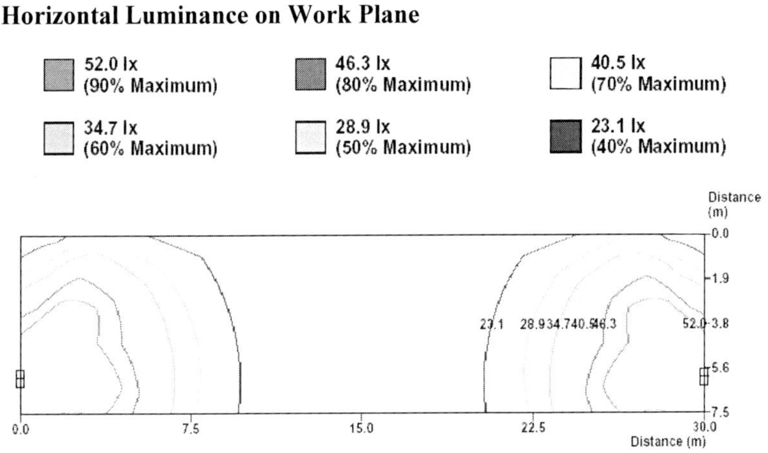

Figure 6.21 Carriageway of service road: ISA-lux contour diagram between two lights

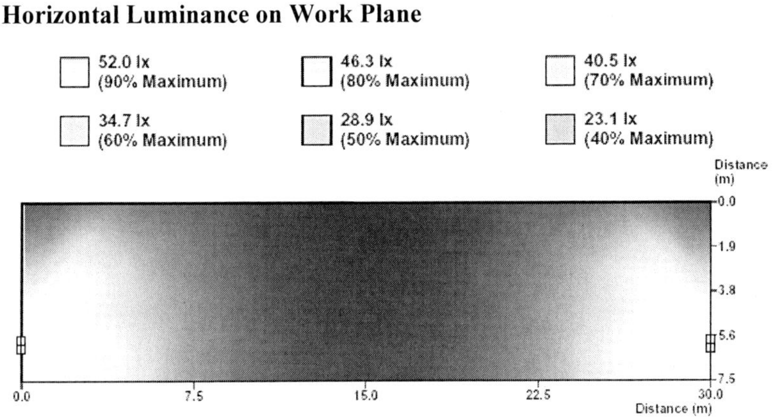

Figure 6.22 Service road: grey-scale diagram

Horizontal Luminance on Workspace

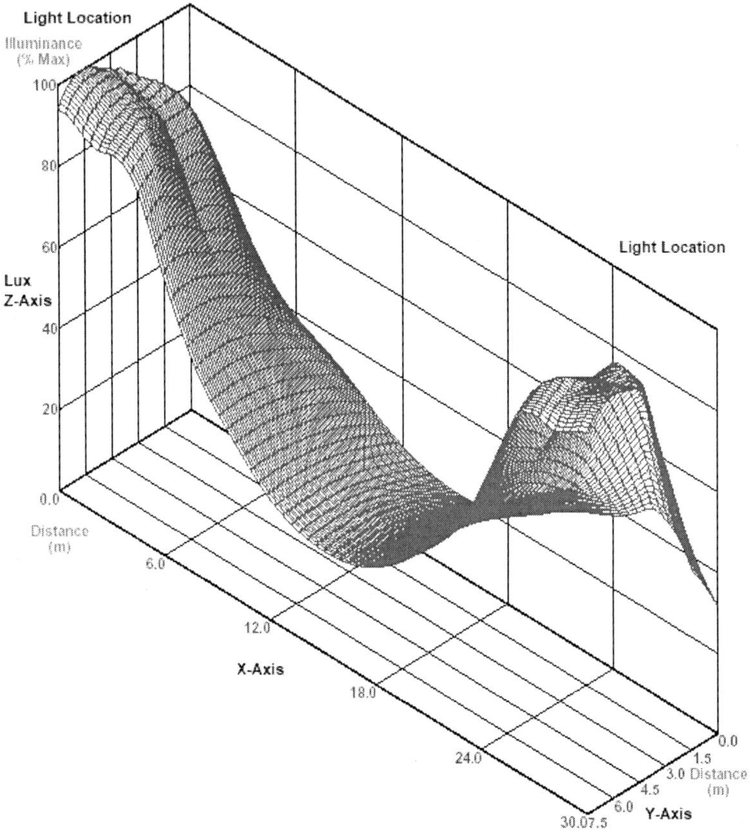

Figure 6.23 Carriageway of service road: 3D-lux web between two lights, end of example for lighting design for service road

6.8 Product Reference

Bajaj Electricals Ltd. 15/17 Sant Savta Marg, Reay Road, Mumbai 400 010, Tel: +91 22 2372 4192
Fax: +91 22 2373 0504, Website: www.bajajelectricals.com

References

Credit for using some figures/text to:

www.Safetyfirst.com
https://www.safetyfirst.com/

Address:
Safety First Systems, LLC,
1055 Parsippany Blvd,
Suite 204,
Parsippany, NJ 07054,
USA,

Telephone:
201-267-8900
888-603-6987

Reference Studies

The dissertation "The Impacts of Illumination Levels on Nighttime Safety at Roundabouts" was presented to the Academic Faculty by Franklin E. Gbologah for his PhD in Civil Engineering at the School of Civil and Environmental Engineering, Georgia Institute of Technology, May 2016. Copyright © 2016 by Franklin E. Gbologah.

CHAPTER 7

CONSTRUCTION AND OPERATIONAL SAFETY

Abstract: This chapter covers a general overview, road construction work zone safety, traffic management plans, proper traffic control, safe and separate work areas, wearing proper safety equipment, ensure proper visibility and avoiding blind spots, work supervision by a competent person, site-specific safety programmes, operational safety, built-up areas, non-built-up areas, major highways, regulation for road users, rating roads for safety and safety tips for road users and road authorities/agencies.

7.1 General

Road safety, being a multidimensional social problem, means focusing on construction, operation and maintenance. Before the road becomes operational, there are various components in its construction. The safety of the workers is critical in the construction of roads, bridges and tunnels, and therefore proper care needs to be taken by the construction agencies as well as the project authorities. The workers may be naïve or ignorant about various safety issues, but the project safety manager must be vigilant about safety in all aspects of the construction activities.

To optimise safety efforts alongside the work whilst maintaining the schedule, representatives of the safety manager may be present. Before starting (such as casting a bridge deck), the various aspects must be checked by the team comprised of design and construction engineers and safety officers to endorse the work and maintain records in a logbook, recording the date, time, activity details and the names of the construction and safety supervisors, for reference by the safety audit team.

7.2 Road Construction Work Zone Safety

In the summer, the roadwork season is in full swing, which means more work zones are popping up on motorways, main roads and streets. Long-distance travel by more people also takes place during this time of year to visit the beach, mountains, and other holiday spots, in addition to people travelling for their regular work.

The Census of Fatal Occupational Injuries (CFOI) by the Bureau of Labour Statistics (BLS) 2019 may be referred to, and the statistics show that 142 construction workers died due to incidents involving motor vehicles. Amongst these, 81 were from collisions with another vehicle, 34 were collisions with objects other than vehicles, and 25 were from non-collision incidents such as overturned or jack-knifed vehicles.

7.2.1 Traffic Management Plan

To maintain the existing traffic flow, proper management should be present at every construction project. The management plan should consist of a temporary traffic diversion and control plan to protect workers by safely controlling traffic around or through the work zone. There should also be a proper traffic control plan inside the work zone to manage the flow of heavy machinery and equipment, construction vehicles and the movements of workers.

7.2.2 Traffic Control

An advanced warning area should be ahead of the work zone with warning signs alerting motorists of upcoming changes in driving conditions There should be a transition area using traffic control devices for lane closures and shifts in traffic pattern, a buffer area, the work zone and a termination area to allow traffic to resume normal flow and a sign indicating the end of the work zone.

Traffic control devices such as cones, barrels, barriers or signs should comply with the Highway Administration's *Manual on Standard or Uniform Traffic Control Devices* (MUTCD in the US) along with any requirements of the state transport agency.

7.2.3 Safe and Separate Work Areas

When several activities take place at the same time, the work zones for construction are usually busy areas. To avoid accidents, devices such as cones, barrels and barriers are used to clearly delineate specific areas of the work zone, such as material storage, areas where heavy equipment is being used, vehicle parking and safe areas for workers on foot to move about in.

7.2.4 Safety Equipment

All personnel should wear proper safety clothing inside the work zone. Management should provide the workers with personal protective equipment (PPE) that includes helmets, steel-toed boots, highly visible clothing and hearing protection depending on the noise levels.

The country's standard, similar to standards developed by the American National Standards Institute (ANSI), should be met or exceeded by all the PPEs. All clothing should meet ANSI Class 2 or 3 standards and should be highly visible, such as in bright fluorescent orange or lime/yellow and contain reflective material, especially if working at night.

7.2.5 Be Alert to Your Surroundings

Everyone should always be mindful of what's going on around them, regardless of what the job duties are. Walking around or behind reversing or manoeuvring vehicles should be avoided.

Any worker inside the work zone should face towards the traffic or have a spotter available when their back is turned towards the traffic, if possible. Spotters should also be used in addition to monitoring traffic to watch the movement of vehicles and heavy equipment inside the work zone and alert workers to any potential dangers.

7.2.6 Ensure Proper Visibility and Avoid Blind Spots

Vehicles, including dumpers, lorries and heavy equipment such as compactors, pavement planers, excavators, pavers and rollers, are constantly moving inside the work zone. All mirrors and visual aid devices should be attached and operating properly, including backup alarms and lights, and must always be checked by the operators.

It should be remembered that the driver has a limited sight line. People present nearby should always stay in visual contact with the driver. The rule of thumb is that if you cannot see them, then they probably cannot see you.

7.2.7 Work Supervision by a Competent Person

Whenever construction work is being executed, a competent person should be on the job site. The person must be capable of identifying existing and predictable hazards in the surroundings, or working conditions which are unsanitary, hazardous to employees, and is authorised to take prompt corrective measures to eliminate them. The competent person needs to conduct hazard assessments and regular inspections of the worksite.

A supervisor needs to select the appropriate measures to be taken by workers and to approve the appropriate types of traffic control devices. The supervisor assigned to the work zone should be informed by the workers about any unsafe hazards or equipment so that they can be corrected immediately.

7.2.8 A Safety Meeting at the Start of Each Workday

To have a quick safety meeting before work begins is a good idea in addition to ensuring that all personnel at the jobsite have the required proper training. Briefing the workers on the work activity scheduled each day and notifying them about all potential hazards is essential, as the work conditions can change greatly from day to day in the work zone. This is also to ensure that all the workers are wearing the proper PPE required.

7.2.9 Safety Program should be Site-Specific

Creating a safe working environment is appropriate for the site, which can go a long way to prevent accidents, because every road, bridge, tunnel or underpass construction project is different, and each work zone has its own unique hazards and challenges. A site-specific safety programme also needs a plan for first aid and emergency medical care in the event of an accident, and safety training schedules for all employees.

7.2.10 Staying Hydrated

Whilst performing construction work, the workers are susceptible to overexertion and heat-related illnesses. Asphalt temperatures can easily rise to 30° F or higher than the surrounding air temperature, and asphalt absorbs 95 per cent of the sun's rays.

Workers need to stay hydrated during road construction by drinking plenty of water or liquids high in electrolytes, like sports drinks or coconut water. To avoid heatstroke, dehydration and heat exhaustion, the workers should also get out of the heat and sun to the maximum extent possible, especially on extremely hot days and wear protective headgear.

7.2.11 Case Study—Vehicular Movement in the Work Zone Incident

A mini dumper was on the way to the workshop at a site within the work zone, and a helper was sitting beside the driver. A tipper came from the opposite direction, crossed the lane and hit the mini dumper on the front right side. The mini dumper came in between the tipper and the side retaining wall and was dragged for approximately 8–10 m. Due to the impact of the collision, the helper of the mini dumper fell on the road and sustained severe abdominal and waist injuries. The driver of the mini dumper also sustained an injury to his right shoulder in the process. The road they were on turned sharply and was sloped.

Basic Causes:

- Loss of control of the mini dumper.
- Brake failure in the tipper.
- Sharp turn with a vertical gradient in the road.
- Insufficient warning signs, speed breaker not painted.
- Travelling with a helper in the mini dumper.
- No self-starter in the mini dumper.
- Old and outdated technology on the mini dumper.

Recommendations/Control Measures/Actions

- Identification of dangerous accident black spots.
- Training drivers.
- Visual warning signs in black spots and painting speed breakers for better visibility.
- Daily checking of the brake, horn and lights of all vehicles.
- No passengers in a mini dumper.
- Medical check-up of all drivers and operators for vision, hearing.
- Ensure that no driver is under the influence of alcohol by checking with a breath analyser.
- Avoid use of non-standard/improvised vehicles.

7.3 Operational Safety

The methods and measures used to protect road users from being killed or seriously injured caused by motorised vehicular traffic are referred to as road traffic safety measures. Pedestrians, cyclists, motorists, vehicle passengers, and passengers on public transport are typical road users. To ensure that in the event of a crash, the impact energies remain below the threshold likely to produce either death or serious injury is the intended outcome of a safe system approach. Depending upon the level of protection offered to the road users involved, this threshold varies in every crash scenario. For example, at speeds higher than 30 km/h, the chances of survival for an unprotected pedestrian diminish rapidly, whereas the critical impact speed is 50 km/h for side impact crashes and 70 km/h for head-on crashes for a properly restrained motor vehicle occupant.

A hierarchy of control is necessary, as sustainable solutions for classes of road safety have not been identified, particularly for rural and remote roads. At the highest level, requirements of all key areas are to be considered for the prevention of serious injury and death. At the second level, there is real-time risk reduction, and the users are provided with a specific warning at severe risk to enable them to take mitigating action. At the third level, reducing the crash risk involves applying the road-design standards and guidelines, such as AASHTO, improving driver behaviour and enforcement.

One of the world's largest public-health and injury-prevention problems is related to road traffic crashes. As the victims were adequately healthy before their crashes, the issue is more acute. More than one million people are killed on the world's roads each year, and 50 million are injured, as reported by the World Health Organization (WHO). Traffic accidents were the leading cause of death among children between 10–19 years of age, as published by the WHO (2004). The problem was most severe in developing countries, and some simple prevention measures could have reduced the number of deaths, was also mentioned in the report. The fatalities and killed or seriously injured (KSI) rates, usually expressed per billion passenger kilometres, are the standard measures used in assessing road safety interventions. KSI rates were replaced by crashes per million vehicle-miles in countries using older road-safety paradigms. As the impact speed affects the severity of injury both to vehicle occupants and to pedestrians, the vehicle speed within the human tolerances for avoiding serious injury and death is a key goal of modern road design. For occupants, the probability of death for drivers in multi-vehicle collisions increased as the fourth power of impact speed, as found by Joksch (1993).

The mathematical term δv or delta-V, meaning change in velocity, is often referred to. Sudden or unexpected severe acceleration or deceleration, which is difficult to measure, is the cause of injuries. However, vehicle speeds before a crash can be estimated by crash reconstruction techniques.

Therefore, the abrupt speed change is used as a surrogate for acceleration. The KSI risk curves used actual crash-reconstruction data, which led to the human tolerances for serious injury and death referenced above, as identified by the Swedish Road Administration. In the modern road-safety paradigm, which focuses on the human tolerances for serious injury and death, interventions are generally much easier to identify. For example, the installation of an appropriate crash barrier is simply required for the elimination of head-on KSI crashes. Roundabouts, which reduce speed on approach, result in a very few KSI crashes.

The old road-safety paradigm of crash risk is a far more complex matter. The driver (such as driver error, illness or fatigue), the vehicle (brake, steering or throttle failures), or the road itself (lack of visibility or sight distance, poor roadside clear zones, etc.) are contributory factors to highway crashes. Interventions may intend to reduce or compensate for these factors, or to reduce the severity of crashes. A comprehensive outline of intervention areas can be observed in management systems for road safety. The fatality risk is increased to maximum when a collision type is either pedestrian or between vehicles, as revealed by a study conducted in Finland.

Interventions related to the design of roadway networks for newly constructed districts are of another class, in addition to management systems for the road networks in built-up areas. The configurations of a road network which inherently reduce the probability of collisions are explored by such interventions. For the prevention of road traffic injuries, interventions are often evaluated. For road-traffic safety, it can be helpful to classify roads into three usages:

- Built-up urban streets with slower speeds, greater densities and more diversity among road users
- Non-built-up rural roads with higher speeds
- Major highways (motorways/interstates/freeways/autobahn, etc.) are reserved for vehicles and are often designed to minimise crashes.

Whilst most injuries occur on urban streets, most fatalities occur on rural roads, making motorways the safest in relation to distance travelled. As reported, in 2013, German autobahns carried 31per cent of motorised road traffic (in travel-kilometres) whilst accounting for 13 per cent of Germany's traffic deaths. The autobahn fatality rate of 1.9 deaths per billion travel kilometres compared favourably with the 4.7 fatality rate on urban streets and 6.6 fatality rate on rural roads.

Table 7.1 Accidents in various Road Classes

Road Class	Injury Crashes	Fatalities	Injury Rate[rate 1]	Fatality Rate[rate 1]	Fatalities per 1000 Injury Crashes
Autobahn	18,452	428	82	1.9	23.2
Rural	73,003	1,934	249	6.6	26.5
Urban	199,650	977	958	4.7	4.9
Total, Average	291,105	3,399	401	4.6	11.6

1. ^ a b per 1,000,000,000 travel-kilometres

7.3.1 Built-up Areas

Most roads are cambered (crowned), having rounded cross sections, to reduce water and snow accumulation. This helps to prevent frost damage as well as increase traction in poor weather. Some roads, particularly on bends, are now surfaced with porous bitumen to enhance drainage. To channel storm water away, grooves are often cut into the surface of concrete surfaces on main roads, and rumble strips at the edges of main roads alert inattentive drivers by creating a loud noise when driven over. In some cases, there are raised markers or studs between lanes to reinforce the lane boundaries, and these are often reflective. In pedestrian areas, to slow down the cars, speed bumps are often placed, preventing them from going too fast near pedestrians. Posts are also installed across the width of pavements to restrict the movements of motorcycles.

Poor and damaged road surfaces can lead to safety problems. If excessive asphalt or bituminous binder is used in asphalt concrete, the binder can 'bleed or flush' to the surface, leaving a very smooth, slippery surface that creates low friction when wet. Certain kinds of stone aggregate become very smooth or polished under the constant wearing action of vehicle tyres, again leading to poor wet-weather traction. All these problems contribute to wet-weather skids or slips and crashes by increasing the braking distances or contributing to loss of control. Road surfaces that are insufficiently sloped result in poor drainage, causing standing water on the surface, leading to hydroplaning.

On some roads, the lanes are delineated with cat's eyes, Botts' dots or reflective raised markers that do not fade like paint. Botts dots are not suitable where the road surface becomes icy in the winter, because frost and snowploughs can break the glue that holds them to the road, despite being embedded in short, shallow trenches carved in the roadway, as is done in the mountainous regions of California.

Road hazards and intersections in some areas are marked at several locations, roughly at travel times of five, twenty and sixty seconds in advance, so that drivers are reminded repeatedly and they are less likely to attempt violent manoeuvres. Road signs and pavement marking materials are mostly reflective, incorporating small glass spheres or prisms to reflect light from vehicle headlights back to the driver's eyes more efficiently.

Figure 7.1 - Pedestrian Crossing, line markings and street furniture
Figure 7.2 A kerb extension at a pedestrian crossing

Figure 7.3 A roundabout used at a four-way junction to improve safety.
Figure 7.4 Utrecht's specially painted bicycle-only lanes

Turning across traffic poses several risks, whether by turning left in right-hand drive countries or turning right in left-hand drive countries. There is a serious risk of a collision with oncoming traffic. Since this could be a head-on collision, injuries are common. In built-up areas, it is the most common cause of fatalities. Another major risk is a rear-end collision whilst waiting for a gap in oncoming traffic.

Countermeasures for these types of collisions include:

- Addition of left/right turn lanes.
- Providing protected turn phasing at junctions with signals.
- Using indirect turn treatments as seen in Michigan, US.
- Converting conventional junctions to roundabouts.

In the absence of these facilities, as a driver is about to turn:

- Keep the wheels straight, so that in the event of a rear-end shunt, the vehicle is not pushed into the oncoming traffic.
- When it seems clear, look at the road you are about to enter. There may be the illusion that, after a time, shows an oncoming vehicle as further away and travelling slower. Looking away breaks this illusion.

There is no presumption of negligence for a collision at a junction, and circumstances may dictate that, for left-hand vehicles, a right turn is safer than turning left.

Left or right turns are to be provided with the same time gap, as recommended in the publication *Geometric Design of Highways and Streets* by AASHTO. Some states have recognised this in statute, and a presumption of negligence is only raised because of the turn if and only if an installed sign prohibited it. For older drivers, turning across traffic has been seen to be problematic.

Designing for pedestrians and cyclists

Pedestrians and cyclists are among the most vulnerable road users and in some countries constitute over half of all road deaths. Interventions aimed at improving the safety of non-motorised users include:

- Footpaths of suitable width for pedestrians.
- Pedestrian crossings should be close to the desire line, allowing pedestrians to cross safely.
- Segregated pedestrian routes and cycle lanes should be away from the main highway.
- Bridges tend to be unpopular with pedestrians and cyclists due to the additional distance and climbing needed in the absence of escalators.
- Underpasses can pose heightened risk from crime if not designed well, but can work for cyclists in some cases.
- Traffic calming and speed humps.

- Low speed limits that are rigorously enforced, possibly by speed cameras.
- Shared space schemes giving ownership of the road space and equal priority to all road users, regardless of mode of use.
- Pedestrian barriers to prevent pedestrians from crossing at dangerous locations.
- Cycling infrastructure.
- Protected junctions.

In the mid-20th century, American ‘passive traffic safety’ measures were adopted. These created roadways, which allowed motorists to travel at high speeds, but these measures de-prioritised cycling and pedestrian facilities. They included excessively wide streets, clear zones adjacent to roadways, wide turning radii and a focus on protecting drivers from the consequences of high speeds. Whilst giving automobiles maximum convenience, the passive traffic safety measures were sought to avoid influencing the behaviour of drivers. Design-orientated traffic safety improvements actively slow drivers down by narrowing roadways, which also helps pedestrians and cyclists, and will hopefully be instated by the recent street policies.

Pedestrian advocates question the effectiveness of schemes for pedestrians to remain safe from vehicles. Successful pedestrian schemes tend to use at-grade crossings, such as crossings close to the intended route and avoid bridges and underpasses. Successful cycling schemes, in contrast, avoid frequent stops even if some additional distance is involved, because cyclists need to expend more energy stopping and starting.

Shared space

Many of the safety measures used to be potentially non-restrictive, such as speed limits, traffic calming, road signs and road markings, traffic lights, Belisha beacons, pedestrian crossings, cycle lanes, etc., and the net amount of danger was increased as these were treated by drivers as an opportunity to drive faster, making this worse than the latter state.

A new approach, known as 'shared space,’ was developed during the 1990s, which removed many of these features in some places, attracting the attention of authorities around the world. Hans Monderman developed this approach by considering that if drivers are treated like idiots, they act as idiots and proposed that trusting drivers to behave was more successful than forcing them to behave. It was also suggested by Professor John Adams, an expert on risk compensation, that traditional traffic engineering measures assumed that motorists were selfish, stupid, obedient automatons who needed to be protected from their own stupidity and non-motorists were treated as vulnerable, stupid, obedient automatons to be protected from cars. The 'shared space' approach leads to a significant reduction in traffic speeds, the virtual elimination of road casualties and a reduction in congestion. Some similarities are shared by ‘living streets’ with shared spaces. Using lower speed limits enforced using special signage, road markings, traffic calming measures and by giving pedestrians priority over motorists, the reduction of traffic speeds in community and housing zones was also achieved.

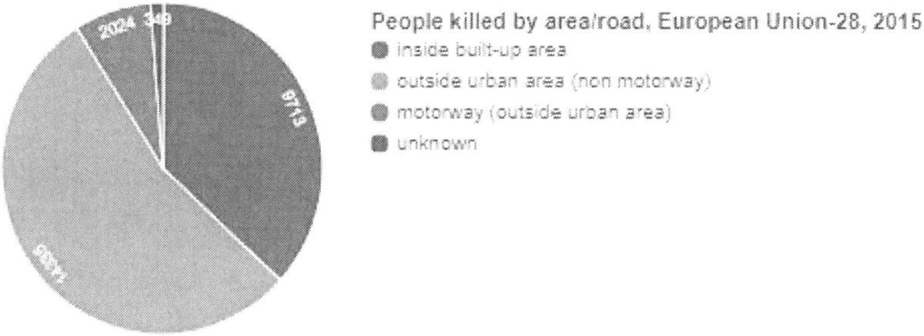

Figure 7.5 Cycling infrastructure in Brighton (UK)
Figure 7.6 A shared space Amsterdam (the Netherlands)

7.3.2 Non-built-up Areas

Rural roads, not including motorways, are outside built-up areas. These roads are not classified as urban roads or motorways. In this category of roads, most people were killed (54.3per cent) in the European Union in 2015, and that is more than in urban areas (36.8 per cent). However, such numbers are likely to vary from country to country.

7.3.2.1 Major Motorways

Major highways are designed for safer high-speed driving and generally have lower levels of injury per vehicle km than other roads. The German autobahn fatality rate of 1.9 deaths per billion-travel-kilometres compared favourably with the 4.7 death rate on urban streets and 6.6 death rate on rural roads, as observed in 2013.

Safety features include:

- Provisions of limited access from the adjoining properties and local roads.
- Provisions of interchanges or grade-separated intersections.
- Provisions of median dividers between traffic lanes to reduce the possibility of head-on collisions.
- Removal of trees and other roadside obstacles obstructing visibility.
- Provision of service roads for more vulnerable road users and slower vehicles.
- Placements of effective energy dissipation devices such as guard rails, wide grassy areas and sand barrels.
- Eliminating road toll booths and using FASTag or similar systems.

The ends of some guardrails on high-speed roads in the US are covered with impact attenuators, designed to gradually absorb the kinetic energy of a vehicle and slow it more gently before it can strike the end of the guardrail head-on, at high speed. To dissipate the kinetic energy, there are several mechanisms. A system of sand-filled barrels, known as Fitch Barriers, uses momentum transfer from the vehicle to the sand. To absorb energy and gradually stop the vehicle, other systems include steel members which tear or deform on impact.

Figure 7.8 Guard Rail on Road Kaluga Oblast (Russia)
Figure 7.9 The Pan-American Highway with a central median and no freestanding obstructions

In some vehicles, tired drivers are awakened by a loud noise as they release the steering wheel and drift off the edge of the road. Major roads have ‘tone bands’, which are impressed or cut into the edges of the roadway. Because of the sound they create, they are known as ‘rumble strips’. The use of ‘raised rib’ markings is an alternative method, which consists of a continuous line marking with ribs across the line at regular intervals. To separate the edge of the hard shoulder from the main carriageway, these were first specially adopted as an edge line marking on motorways. Improved visual delineation of the carriageway edge in wet conditions or at night is the objective of this marking. If the drivers stray from the carriageway and run onto the marking, it provides an audible/vibratory warning to vehicle drivers.

To reduce the need for tyre-traction and increase stability for vehicles with high centres of gravity, motorways are banked on curves. To reduce driver fatigue and increase the carrying capacity of the roadway, the US has developed a prototype automated roadway. Presently, studies have been conducted on roadside units participating in future wireless vehicle safety communications networks.

Compared to ordinary roads, motorways are far more expensive and space-consuming, so motorways are only used as principal arterial routes. Motorways bear a significant portion of the total length of motorised travel in developed nations; for example, in 2003, the UK’s 3533 km of motorways represented less than 1.5 per cent of its total length of roads, but carried 23 per cent of total road traffic.

People die on the hard shoulder more than on the carriageway itself. Without other vehicles passing a parked car, despite having hazard lights, drivers can be unaware that the vehicle is stopped. By putting the cab seat behind their truck, the truck drivers indicate that they are parked. The police in the UK park their vehicles on the hard shoulder at a slight angle so that passing drivers can see down the side of their vehicle and are therefore aware of the stopped vehicles. In the countries that have toll collection booths, 30 per cent of highway crashes occur in the vicinity of toll collection booths; these can be reduced by switching to electronic toll systems.

Vehicle Safety

Buses and Coaches

To reduce the chance of a crash occurring, safety can be improved in various simple ways. The safety of a person travelling by bus or coach is greatly increased by avoiding rushing or standing in unsafe places on the bus or coach and following the rules on the bus or coach itself. To improve safety, including safety bars for people to hold onto, various other safety features can also be implemented into buses and coaches.

The main ways to stay safe when travelling by bus or coach are as follows:

- Arrive early so that one does not have to run to catch the bus or coach.
- Always follow the queue at the bus stop.
- Do not board or alight other than at an official stop.
- Never board or alight at a red light.
- Board the bus only after it has come to a halt without rushing or pushing others.
- Do not stand, sit or travel on the footboard of the bus.
- Do not put any part of your body outside a moving or a stationary bus.
- Whilst in the bus, refrain from making noise as it can distract the driver.
- Always hold onto the handrail if standing in a moving bus, especially on sharp turns.
- Always adhere to the bus safety rules.
- If the driver is using the mobile phone, ask them politely to refrain from using it.

Cars

By reducing the chances of a driver's error or by designing vehicles to reduce the severity of crashes that may occur, safety can be improved. For safety-related vehicle devices, systems, design, and construction, most industrialised countries have comprehensive requirements and specifications. These may include:

- Passenger restraints such as seat belts reinforced by laws requiring their use, and airbags.
- Headlights and reflectors.

- Driver assistance systems such as electronic stability control.
- Fire-retardant interior materials, standards for fuel system integrity and the use of safety glass.
- Sobriety detectors. These interlocks prevent the ignition key from working if the driver breathes into the device and it detects significant quantities of alcohol. These have been used by some commercial transport companies or suggested for use with persistent drink-driving offenders on a voluntary basis.
- The Dutch Reach—use the far hand on the handle when opening the door to avoid hitting cyclists, as this forces you to turn around so you can look at the road.

Motorbikes

Motorcycle riders are nine times more likely to crash and 17 times more likely to die in a crash than car drivers, as shown by road casualty statistics in the UK. Unlike in enclosed vehicles such as cars, the higher fatality risk is due in part to the lack of crash protection, combined with the high speeds at which motorcycles typically travel. The percentage of intoxicated motorcyclists in fatal crashes is higher than other riders on roads, according to US statistics. For the safety of motorcyclists, helmets also play a major role. The National Highway Traffic Safety Administration (NHTSA) in 2008 estimated that helmets are 37 per cent effective in saving the lives of motorcyclists involved in crashes.

Lorries

It has been estimated by the European Commission's Transportation Department that up to 25 per cent of accidents involving lorries can be attributable to inadequate cargo securing. Loss of cargo, loss of lives, loss of vehicles and hazards for the environment are all caused by improperly secured cargo involved in severe crashes. Use of 'Dunnage bags' is one way to stabilise, secure and protect cargo during transportation on the road. Dunnage bags are placed in the voids between the cargo and are designed to prevent the load from moving during transport.

For implementing corporate road safety programs, 'Together for Safer Roads (TSR)' has developed best practices that include data management and analysis, route mapping, investment and upkeep of fleets, safety policies and training for employees and first aid/safety training in case of the occurrence of collisions.

Police

Hundreds of people die each year due to the police chasing suspects driving at high speeds. Such pursuits are allowed by different jurisdictions in different circumstances. If these are restricted to violent felonies only, fewer injuries might occur.

7.3.2.2 Regulation for Road Users

Motor Vehicle Users

Motor vehicle drivers are required to pass a driving test that is dependent on jurisdiction, driver's age, road type and vehicle type. Additional training and licensing, conforming to restrictions on driving after consuming alcohol or various drugs, restrictions on the use of mobile phones, covered by compulsory insurance, wearing seat belts and complying with the speed limits, may be needed for drivers of public transport and goods vehicles. Motorcycle riders may additionally be obliged to wear a motorcycle helmet. Drivers of certain vehicle types may be subject to maximum driving hour regulations.

The jurisdictions of the US states Virginia and Maryland have implemented specific regulations, such as prohibiting the use of mobile phones whilst driving, and limiting the number of passengers accompanying young and inexperienced drivers. In 2017, the National Safety Council released the "State of Safety Report", which ranked states according to these road safety regulations.

Restrictions proposed by insurance companies for highway safety for drivers include a curfew imposed on young drivers to prevent them driving at night, an experienced supervisor to chaperone the less experienced driver, forbidding the carrying of passengers, zero alcohol tolerance, raising the standards required for driving instructors and improving the driving test, vehicle restrictions such as restricting access to high-performance vehicles, a sign placed on the back of the vehicle (an N- or P-plate) to notify other drivers of a novice driver and encouraging good behaviour in the post-test period.

Although providing safe roads is primarily the responsibility of the government, the challenges of development and equity require that all segments of society engage and contribute, including the private sector. To advance the business case for having safer roads, private and public sector coalitions, like Together for Safer Roads (TSR) and the Road to Zero Collision, exist to work alongside governments. They help companies meet their duty of care to employees and the wider community to minimise fleet-related dangers. By improving employee health and safety, protecting assets, reducing losses in productivity and healthcare costs and enhancing the efficiency and effectiveness of supply chains, the campaigns also benefit businesses.

Some countries or states have already implemented some of these ideas through Vision Zero's networks. According to when and where the person drives, the 'pay-as-you-drive' scheme adjusts insurance costs.

Pedal bicycle users

Pedal cyclists conforming to restrictions on driving after consuming alcohol or various drugs and complying with restrictions on the use of mobile phones, wearing a bicycle helmet and following certain speed limits, should be covered by compulsory insurance, depending on jurisdiction, road type and age.

Pedestrians

'Jaywalking' may be prohibited depending on the jurisdiction. Frequent road safety education helps to make children aware of the dangers of the road.

Animals

Collisions with animals are usually fatal to the animals, and occasionally to drivers as well.

7.3.2.3 Rating Roads for Safety

Major roads in Europe have been assessed with a road protection score by the EuroRAP initiative since 1999. Based on how well the road is designed to protect the occupants from being severely injured or killed if a head-on, run-off, or intersection crash occurs, this resulted in a star rating for roads, and an evaluation with four stars represents a road where surviving is most likely. Thousands of road sections across Europe where road-users are routinely maimed and killed for want of safety features have been highlighted by the scheme. The accidents occur sometimes for little more than the cost of safety fencing or the paint required to improve road markings.

To rate the probability of a collision on the road, there are plans to extend the measurements. To inform planning and authorities to make targets, these ratings are being used. In Britain, two-thirds of all road deaths happen on rural roads. This is a bad score when compared to the high-quality motorway network. The single carriageway configuration claims 80 per cent of rural deaths and serious injuries, whilst roads without metal beam crash barriers cause 40 per cent of rural car occupant casualties that hit roadside objects, such as trees. Improvements in safety features for rural roads and driver training are hoped to reduce this statistic.

The number of designated traffic officers in the UK fell from 15 to 20 per cent of police force strength in 1966 to seven per cent in 1998, and between 1999 and 2004 by 21 per cent. Whether the reduction in traffic collisions per 100 million miles driven over this time has been due to robotic enforcement is a question for debate.

Roads in the United States are not safety-rated by the government on their actual safety features, to inform the media and public. However, in 2011, over 800 persons were killed across the US by 'non-fixed objects' that include roadway debris as found in the Traffic Safety Audit conducted by the NHTSA. California had the highest number of total deaths from all crashes. New Mexico had the highest chance for an individual to die from experiencing any vehicle-debris crash.

7.3.2.4 KSI statistics

It was estimated by the WHO in 2010 that 1.24 million people were killed worldwide and an additional 50 million were injured in motor vehicle collisions. Young adults aged between 15 and 44 years account for 59 per cent of global road traffic deaths. Other key facts, according to the WHO report, are:

- Road traffic accidents are the leading cause of death of young people, aged 15–29 years.
- On the roads, 91 per cent of the world's fatalities occur in low and middle-income countries, even though these countries have approximately half of the world's vehicles.
- Half of those dying on the world's roads are vulnerable road users such as pedestrians, cyclists and motorcyclists.
- Without action, road traffic crashes are predicted to result in the deaths of around 1.9 million people annually by 2020.

- Only 28 countries, representing 416 million people, which is seven per cent of the world's population, have adequate laws that address all five risk factors (speed, drink-driving, helmets, seatbelts and child restraints).

It was estimated that around 60 million people died on the roads during the 20th century, which is around the same number as World War II casualties.

At the OECD level, the comparatively poor improvements in pedestrian safety have become a concern. The Joint Transport Research Centre (JTRC) of OECD and the International Transport Forum convened an international expert group and published a report entitled "Pedestrian Safety, Urban Space and Health in 2012". The population growth, vehicle registration, and estimated vehicle kilometres travelled have various impacts on the death trend from OECD countries, as mentioned by the BITRE/Australia agency.

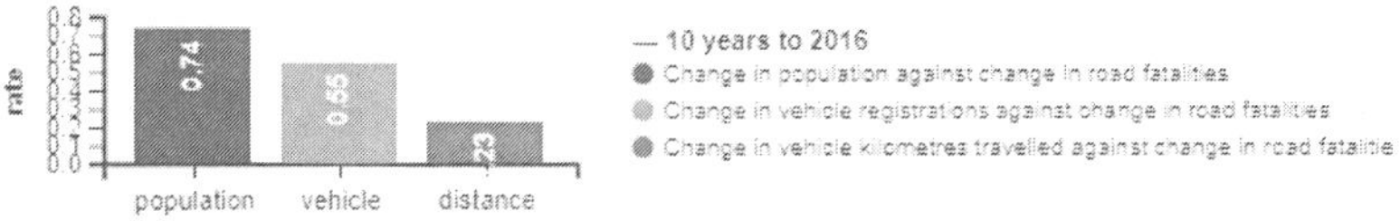

Figure 7.10 Road fatalities per changes in population, vehicle registrations and travelled kilometres

7.3.2.5 Safety Tips for Road Users

- Wear a seat belt.
- Put the mobile phone away.
- Stick to the speed limit.
- Check your blind spot.
- Don't drive into someone else's blind spot.
- Don't drink and drive.
- Don't drive tired.
- Turn on your headlights.
- Follow warning signs/boards.
- Check the tyre treads, brakes and wipers.

7.3.2.6 Road Safety Tips for Road Authorities/Agencies

- Organise funds for improving road safety.
- Establish a dedicated department for improving road safety and data management.
- Conduct safety audits periodically, identify safety deficiencies and accident black spots.
- Implement an intelligent transport system (ITS).
- Install CCTV surveillance, FASTag or similar devices.
- Take measures to discipline traffic.
- Install effective incident management and emergency medical care.
- Driver training and licensing.
- Safety education campaigns.
- Legislation to enforce safety rules.
- High penalties for offences.
- Traffic calming and speed control.
- Planning and design.

References

Some figures/text are from "Road Safety Guidelines" by the Asian Development Bank. "Road Safety Guidelines" was largely written and developed by Ross Silcock Ltd. in association with the Transport Research Laboratory (TRL) Overseas Centre, the United Kingdom (UK). It was produced under the Asian Development Bank (ADB) regional technical assistance (RETA 5620: Regional Initiatives in Road Safety), which has also resulted in the production of a comprehensive set of guidelines for decision makers on road safety policy.

CHAPTER 8

INTELLIGENT TRANSPORT SYSTEMS (ITS) AND FASTAG

Abstract: This chapter will cover various topics, such as a general overview of ITS applications, satellite applications in advanced traffic management, an advanced vehicle detection system, automated message signs for traffic, improvement measures for protecting traffic against accidents in various situations, case studies and resources on appraising ITS technologies and implementation of ITS, and the FASTag system.

8.1 General

ITS is increasingly implemented for improved road safety, enabling discipline-enforced seamless traffic flow by advanced traffic monitoring and traffic guidance, with the application of computers on satellite and Internet technology. ITS systems gather data from the road information system database, update it continuously with the latest information, and then use the data to improve the management of the transport system. The transport administration is provided with the latest information on roads and traffic conditions, based on which various transport decisions are made, and they guide road users about travel details and enhance traffic safety.

8.1.1 ITS Helps Achieve Objectives

ITS can help in achieving policy objectives by the transport planners in several ways. It helps manage undisciplined traffic, unsafe driving, all road users' safety, congestion, pollution, poor accessibility and even social exclusion. It is also effective in reducing journey times and improving reliability, and simply by changing people's perceptions as well. It improves the efficiency of the transport systems' functioning. In supporting the economic and retail in parking guidance systems, it can be useful. To achieve transport policy objectives, it is vitally important to consider ITS as a means. In some circumstances, ITS may not be the best means of achieving transport policy objectives, but in many other circumstances, it will be effective. Applications are therefore to be selected strategically for the latter applications, and not the former.

Examples of ITS

Section 8.4 of this chapter will provide some applications of ITS in action, including costs and evaluations, wherever available. However, in this section, it is worth introducing some of the applications of ITS.

To provide road users with up-to-the-minute information at their current locations, on incidents/delays and how to avoid them, the service outputs real-time information, and is highly useful for the public as well as private road transport. On the roads, such information can ensure improved travel safety.

- The inventories of transport infrastructure in an area (e.g. the condition of the road network) are maintained in a relational database management system (RDBMS). This is used to inform the road users by using geographical information systems (GIS) for better management and prioritising the maintenance work.
- By using ‘Smartcard’ or ‘FASTag’ on public transport, the road users may be provided with the best deal for the cluster of trips that they might be making in a particular period. It also provides the operator(s) with detailed information about the passengers’ travel habits. The second application helps apportion revenue between operators, as well as for service planning.
- Real-time information for route planning is also possible, for both public transport and car users.
- Displaying parking guidance systems helps drivers by reducing search time for parking.
- Public transport information in various formats may be made available (including visible messages for people who are deaf or hard of hearing).

8.1.2 Learning Summary

Up to this point, we can:

- Understand the main applications of ITS.
- Set out the principles of possible ITS applications.
- Key ITS applications in actual locations around a region.

8.1.3 Structure of the Rest of This Chapter

The other sections in this chapter describe in more detail the different applications—what they are, and how they work in the ITS. Then it looks at how ITS applications can be appraised to ensure that they meet policy and other objectives. The final, and longest section, provides information on a range of actual examples of ITS applications that have been implemented, and their effects.

8.2 ITS Applications

Different ITS applications are discussed in this section, their ways of working, and their contribution to reducing transport energy use

8.2.1 Satellite-Based Communications

Various ITS applications depend on Internet-based satellite communications; these shall be discussed more in the subsequent sections of this chapter. Although these have been typically provided by the USA, the EU has also embarked on one of its biggest research, development and delivery projects. The EU intends to put in place its own satellite communications system, GALILEO.

8.2.2 Incident Detection

As soon as ITS detects an incident on any transport system, it communicates the information to a control centre. In response to certain types of incidents, traffic management strategies can be put into effect. This reduces their impact on the traffic system, and further control of the system is done using ITS. In case of an accident that occurred on a motorway, this is detected by overhead or roadside CCTV cameras and is received by the control room, which activates the variable message signing (VMS) to:

a) Stop the traffic that is too close to the accident on that route by lane closures, lane control and temporary speed limits.
b) The VMS is used to advise the traffic further away from the accident to take another route.

Hence, roadside incident detection by ITS can save considerable energy by redirecting traffic away from the area that is congested due to the incident, and by managing the speeds of traffic moving towards the incident, to reduce congestion.

8.2.3 Traffic Signage for Variable Speed Limits

At peak hours of the day, it can be effective to lower speed limits to maximise road capacity and reduce congestion. The congestion is caused by the impatient drivers, lane-changing and the 'wave propagation' effect on all following vehicles. To control this, traffic signing is required for variable speed limits together with, if possible, some form of automatic enforcement (e.g. average or point speed cameras). The reduced speed and less congestion will also benefit energy consumption.

8.2.4 Ramp Control

At the junctions and grade-separated interchanges on a motorway, it is necessary to regulate the flow of traffic along a slip road at peak hours by using the ITS' ramp control signage. Sensors on the main road detect traffic density, the optimum level as well as spacing of joining traffic is calculated, and their access onto the main road is regulated by traffic signals. This minimises the congestion on the main road caused by joining traffic.

Figure 8.1 Ramp Metering (Source DfT 2004)

8.2.5 Traffic Signal Control

In relation to actual demand for the road network in real-time, the linked and isolated signalised traffic junctions are managed by ITS. The traffic levels, speeds and queue lengths can be detected by inductive loop detectors embedded in the road surface. A local signal control computer receives this information, and if necessary, this local centre communicates with the zonal cell, with a computer controlling the signals for a whole area of a town or city. The communications are maintained as locally as possible to minimise the communication times and costs. The most effective cycle times and green times are computed for the signals by the signal controllers. If, for example, the maximum cycle time is 120 seconds, the signal controller cannot override this. This real-time signal control typically increases the capacity of a group of linked signals. In the short term, such an increase in capacity should reduce congestion and, in turn, energy consumption. If the traffic levels increase because of the reduced congestion, such benefits may not be available.

These signal control systems can give priority to certain vehicles, most typically, trams and buses. Detecting a priority vehicle is on approach and the current signal phase is red, the signal control system can bring a green signal forward, or if the current signal phase is green, it may delay a red signal, so that the priority vehicle does not have to stop and wait at the junction. However, user-defined limits are adapted to the delay of a red signal.

Ambulances, fire engines, rescue cranes, police and VIP vehicles are also considered priority vehicles. The detection of an approaching priority vehicle is by satellite through the GPS linked to the public transport operator's control room. With inductive loop detection embedded in the road surface, all public transport vehicles, along with emergency vehicles, are given equal priority; therefore, it is less efficient.

VMS with real-time information can also be linked to signal control systems for vehicles. In case bad congestion is detected by the detectors at one junction or in one street, the VMS receive this information from a human controller, and subsequently the drivers can be advised of alternative routes and/or modes (e.g. 'congestion in outer ring road around the auditorium, use Central Avenue').

Smart signal control systems are also adaptable as a form of access control, entering certain restricted areas. For example, if a road is particularly environmentally sensitive and is used by many public transport vehicles, therefore, queuing may be undesirable in that location. To 'move the queue' from the sensitive area to a less sensitive one and permit the optimum amount of traffic into the sensitive area, signal timings can be regulated.

This technique can be further linked to pollution monitoring. Signals reacting in real time can help move forward the queuing traffic around, such that the build-up of pollution hotspots does not take place in certain parts of town.

8.2.6 Parking Management

Parking management may need the following provisions:

- Real-time parking availability.
- Distribution and management of limited parking spaces.
- Pricing.
- Controlling enforcement.

ITS has various applications in relation to these provisions. Parking guidance systems have traditionally linked microwave communication, inductive loops or infrared at the entrances and exits of car parks. These also display occupancy and queuing by VMS on their key links into and around the town or city centres and advise drivers where they are most likely to find a space.

To date, there are no effective parking guidance systems that inform drivers about where to park on the street. The parking guidance highly merits the investment, whereas the distribution of parked cars between car parks is skewed. Therefore, a few car parks are very popular, whilst the others remain only partially full. The parking guidance system may be of less use in case the vehicles are evenly distributed at most times of day, and there is low demand for parking at all car parks.

Parking guidance systems in Southampton, in the UK, have reduced parking search time by 50 per cent.

Figure 8.2 Parking Guidance using VMS (Source DfT 2004)

ITS can be very useful in enforcing parking space management. Paper permits for parking are relatively easy to copy, therefore prone to fraud. It is a much better option to adopt the FASTag system. Such electronic checking equipment can also help with surveillance in case of any investigation. ITS can record the data conveyed to the local control room. ITS is increasingly used in parking payment systems. It is very common in towns and cities of Europe to make payments by mobile phone apps for on-street parking in real time. The users register, and when they wish to park, send a message to an on-street ticket machine, and pay the bill through their bank account later.

8.2.7 Management of Demand Responsive Transport (DRT)

DRT is a form of public transport with diversion/flexibility that takes the users to the place where they want, when they want, instead of operating on fixed routes at fixed times. From the user perspective, the most favoured public transport is the taxi, which comes with a matching price tag. DRT comes with a functional facility between a taxi and a conventional bus. To maximise the flexibility and the efficiency of the service, a sophisticated booking/scheduling system is frequently employed. In case some passengers have different destinations, this system may cause a delay to other passengers who have their destinations off the route. Studying the bookings of passengers, ITS may decide the grouping of passengers and also the route that can minimise the delay to the passengers. This has several objectives as follows:

- To ensure the highest possible utilisation of the vehicle and the driver(s).
- Minimised travel time saves passengers' time and fuel energy.
- ITS can provide convenience for the user. If someone wants to divert off the most direct route to pick up someone else, he may have to put up with some inconvenience in the normal route to get alternative routes. ITS makes it easy to modify the scheduling system, such as defining a tentative route, maximum diversion, and maximum journey time, for passengers already on the vehicle.
- ITS can provide users with booking facilities in a variety of ways by phone, SMS and by accessing the website.
- ITS can combine routing using software to link between the driver and the control centre, the router software can indicate to the driver where they must go next. Otherwise, trips may be forgotten. There is often a risk that a driver, by mistake, may miss out a pick-up or drop people off in the wrong order.
- ITS can incorporate additional bookings made at short notice, whilst the vehicle is out on the road. If someone needs a trip, the scheduling software can quickly check whether any vehicle is nearby to pick them up and, if possible, then ITS sends a message to the driver whilst they are en route.
- ITS can send an alert to users on their mobiles when a vehicle is reaching them and is close by, so that they can get ready to be picked up, reducing dwell times and increasing vehicle utilisation.
- ITS can store user details based on their ID card numbers, i.e., their address, mobile phone number, most common trips, disabilities, etc., in a database, to simplify and speed up the booking procedure.

8.2.8 Management of Vehicle Fleet and Freight

ITS can manage fleets to ensure it is utilised to maximum efficiency. For any travel agency that has even a small fleet of vehicles, fleet management is an immensely important activity. It is therefore relevant to even small local agencies and authorities, as well as to bus operators, which may or may not be owned by the public sector. Each vehicle should be able to communicate its location, journey purpose and its state, e.g., whether running normally or malfunctioning, directly to a central control room by using a mobile phone. Certain buses with automatic vehicle location (AVL) systems use roadside beacons. However, these are suitable only for locating vehicles within a limited geographical area, most likely, in a single city.

ITS can monitor how vehicles are used:

- Schedule and reschedule vehicles for travel trips more efficiently.
- Assess the need for more or fewer vehicles to carry out a set of assignments, for example, deliveries. It checks the average time taken and how far the drivers deviate from the average.
- Assess individual driver performance with respect to the time taken to carry out a delivery, fuel consumption related to driving style.
- Manage services in real time. If a bus operation control system finds, from AVL, that all the buses on a particular route in one direction are running late, it can stop one or more of the buses, shift the on-board passengers to other buses, and put them back into return service in the opposite direction, to maintain the return service without any delay. Thus, by implementing AVL for fleet management reasons, bus operators can realise savings from fleet efficiency.

8.2.9 Speed Detection and Improvement of Road Safety

Speeding aggravates both the risk of an accident occurring and its severity. All EU countries have serious concerns about reducing the number of road accidents. Sweden has the most radical target for accident reduction, with its 'Vision Zero', that there should not be any road deaths. Many other countries also have quantified targets for reduction. ITS can make a major contribution to achieving such targets, as follows:

- Monitoring using point speed cameras; these cameras detect the speed of a vehicle at a short point on the road, such as at an accident blackspot, using radar detection, and conventional camera film. The cameras are not always installed, so it is not always effective. The vehicles detected exceeding the posted speed limit are charged a penalty by receiving a fine, and in some states, a driver's licence may also be suspended for some duration.
- Studies have found that, at various sites, both casualty and death rates have come down. At camera sites, there was also a reduction of fatalities per annum. The statistics for the reduction in accidents are different in different states.

- Average speed cameras are installed over a strategically selected stretch of road and are linked to number plate recognition systems. These cameras calculate the average speed of a car over that stretch. Similar enforcement is done with point cameras, but these cameras use digital technology, so they are 'on' at all times.

Figure 8.3 Speed camera (source DfT 2004 and author)

- Alert signs for drivers about their speed are also done by ITS, but without any enforcement. For example, on entering a town, vehicles exceeding the urban speed limit will be detected and the signage will flash a message "Slow Down—Speed Limit is 50 kph". These have been found to reduce the speeds of vehicles and motorcycles by 2 to 20 kp/h at various sites
- Intelligent Speed Adaptation (ISA) uses satellite-based GPS technology to indicate the location of a vehicle and the speed limits. 'Active' ISA then exercises automatic control over the vehicle's engine and braking system so that the driver cannot exceed the speed limit.
- ISA is being trialled in the UK, the Netherlands and Sweden. Evaluation of the UK trial indicates that mandatory active ISA could achieve significant annual fuel savings.

The effect of speed detection and its control results in two benefits: first, less energy is consumed by reducing the speeds of vehicles travelling at higher speeds; and second, there will be a reduction in accidents and their severity. This has been observed in many cities where 'mode shift' is observed and rates of bicycle use are higher. Cyclists prefer safer roads, and so more cyclists feel tempted to try cycling a try.

8.2.10 Vulnerable Road User systems (e.g. Puffin crossings)

Children, the elderly and disabled people are identified as vulnerable road users. They are all disproportionately represented in road accidents, especially as pedestrians. As they have reduced mobility and/or sensory perception, a conventional transport system may not fully meet their needs. ITS can be used to provide them with roads that are easier and safer to use. For example:

- Cameras may sense at pedestrian crossings how long people are taking to cross the road and can give them more time if they need it.
- Audible public transport information for blind and partially sighted people. In some cases, this can be provided only when it is needed—a pocket-sized device that can be activated at a bus stop or other information point will give them the audible information that they need. Audible location finders guide blind and partially sighted people to travel and reach their destination.
- Visible public information guides for hearing disabled and partially hearing disabled people to travel and reach their destination.

Real-time indicators for the next stop, in visual and audible form, are provided on buses and trains.

8.2.11 Planning for Multimodal Trip

A mode shift from cars to public transport can achieve many broad objectives, whereas transport authorities are interested in providing the best transport information for the public. ITS can do multimodal trip planning services by communicating details to the public on the phone, Internet, WAP and/or SMS. The general details may be made available on the website.

By selecting their origin and desired destination in the system by ITS, the users will get a variety of trip options on a variety of modes of transport. The timings and costs for each mode should also be available. For such services, ITS require an effective, technical back-up, which is considerable.

Regular updating of the details in ITS is the most difficult part, to ensure that all the information on the system or systems into which it links is up to date in real time. Accurate information (e.g. shortest route) can be communicated to the public only when such updated details are maintained.

Effective functioning of multimodal travel information systems was carried out to a certain extent, but was never evaluated. Various other existing studies are based on limited data. It is not easy to rely upon the conclusions. No evaluation has been planned so far for pilot multimodal travel information systems in four test cities in Germany.

- It was observed that the behavioural changes were marginal in various cases. There is some limited evidence to show that business travellers are more willing to change mode, travel time or route in response to travel information; and therefore, the more unknown the journey, the more likely that the travel information will be heeded.
- Literature studies may reveal that the information and data are not in themselves leading to their use. This may be linked to finding by another study. Most systems are provided publicly, and private companies appear unwilling to invest at present.
- A number of multimodal portals are available for studies, and interested users may visit those sites to learnabout the studies and conclusions.

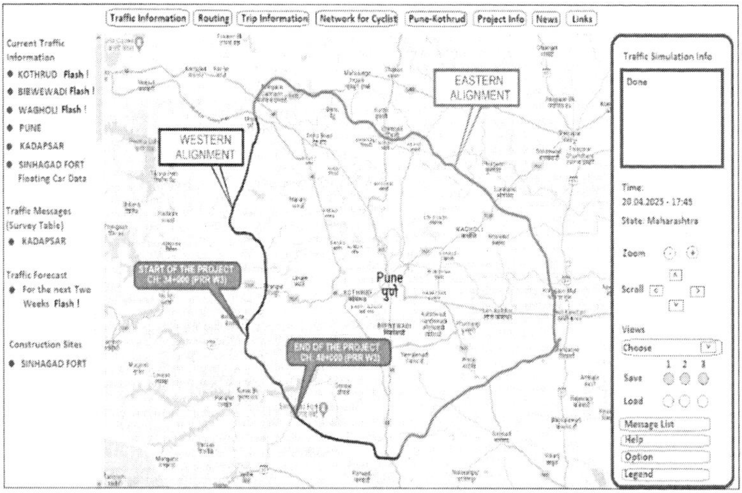

Figure 8.4 Bayern info and real-time travel and traffic information

8.2.12 Passenger Information Systems

ITS can provide real-time information to public transport passengers (RTPI) through a variety of media such as at-stop displays, SMS messaging and the Internet. The AVL link is used for such services. A typical system is schematically described in Figure 8.5. These systems theoretically result in some energy savings. Such a system also promotes a modal shift from car to bus, but an example of such a shift is very rare.

8.2.13 Route Guidance and Navigation

ITS can be effective in fleet management and AVL to provide drivers with very detailed route guidance information using GPS. This is a standard device fitted to cars and has a beneficial effect as it can reduce the number of kilometres driven and, in turn, reduce energy consumption. This provision can be used in conjunction with congestion information advising drivers to choose other routes to avoid traffic jams.

8.2.14 Variable Message Signs (VMS)

ITS can provide information to travellers by VMS. Based on the real-time site situation, the message on a VMS can change. A control centre remote from the site creates a message and communicates it to road users by means of a sign. The advantage of VMS is that it reflects the current conditions of the route and changes the sign display from time to time. These are highly effective to advise drivers on motorways or city streets about difficult driving conditions (e.g. "bridge is closed temporarily to high vehicles for maintenance work"), to give notice of incidents (e.g. "accident between kilometres 14 and 15") and to advise about various alternative routes that can be taken.

8.2.15 Road User Charging (RUC), Tolling and Access Control

ITS can be a very effective means for RUC, tolling and access control. To pay back the loans from banks or other financial institutions that were used for constructing a road, bridge or tunnel, tolls are collected from road users. The charge is applied for providing the road users with a congestion-free, high-standard drive, raising money for maintenance of the road, building new infrastructure and for service and safety improvements. For all these purposes, this means of collecting the funds is essential.

Some examples are given as follows:

- RUC can be enforced by automatic number plate recognition (ANPR) cameras, by taking a photograph of the number plates, and a computer compares the number plate with those stored in a database that have paid their charge that day. ITS can recognise the number plate and register the number on the database when they pay. The payment can be done either in cash at any pay-station (e.g. a fuel station), or by pre-registering and then paying on the Internet, by SMS or by phone. All these systems must be interlinked to provide an effective service from such a complex ITS application. The cost of implementing such a system setup was around €300 million, and the operating cost is around €3.20 per charge payer.
- The 'FASTag' system is used for collecting revenue at many toll bridges, motorways and all congestion charging schemes in some cities. Vehicle owners register for the 'FASTag' that is 'read' by the beacon when the car passes through a charging point. They are then deducted from the pre-paid amount of the FASTag, and ANPR is used for enforcement.
- The congestion charging scheme in Singapore uses an in-car unit (IU)—see Figure 8.6, which must be fitted in every vehicle. A card similar to a phone card is to be bought and inserted into the IU before driving into a charging area. Charges may vary by location and by time of the day, but a beacon activates the IU when the vehicle drives past a charging point, and the payable amount of money is deducted from the card in the IU. In this case, ANPR is also used for enforcement.

8.2.16 Public Transport Payment

ITS can also function for payment of fares for public transport in several important ways:

- Electronic ticket-issuing machines on buses and trains provide more specific ticket types, which benefits the passenger. These machines also provide the operator with some management information, which is useful to the operator (e.g. certain ticket machines detect the speed of a bus along a route).
- Ticket issuing machines at stations and stops reduce the time of the overall public transport journey by reducing the time spent waiting for tickets and boarding.

- Smartcards retain information about the individual user as well as facilitating a wider range of fare options (e.g. discount, concession, etc.), which is not possible with paper-based tickets. They can also be used to reduce the chances of ticketing fraud.

ITS thus benefits the passengers as well as the operators.

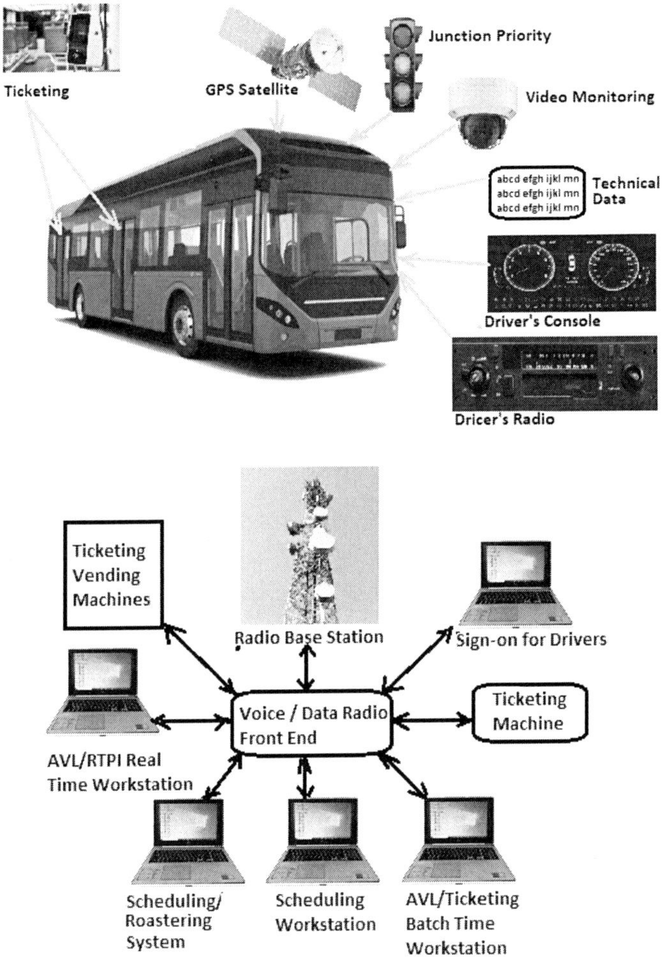

Figure 8.5 Schematic representation of RTPI system (source—City of Edinburgh Council)

Figure 8.6 Electronic Road Pricing In-vehicle Unit (IU) in Singapore

8.2.17 UTMC Systems

Urban traffic management and control (UTMC) systems combine many functions of ITS. A typical UTMC system can be made up of:

- Links with external systems.
- A database.
- Communication links between the different parts of the UTMC system.
- Communicating information links from the UTMC system to users.

The UTMC approach is for the exchange of information within a system and between systems. This is illustrated schematically in Figure 8.7.

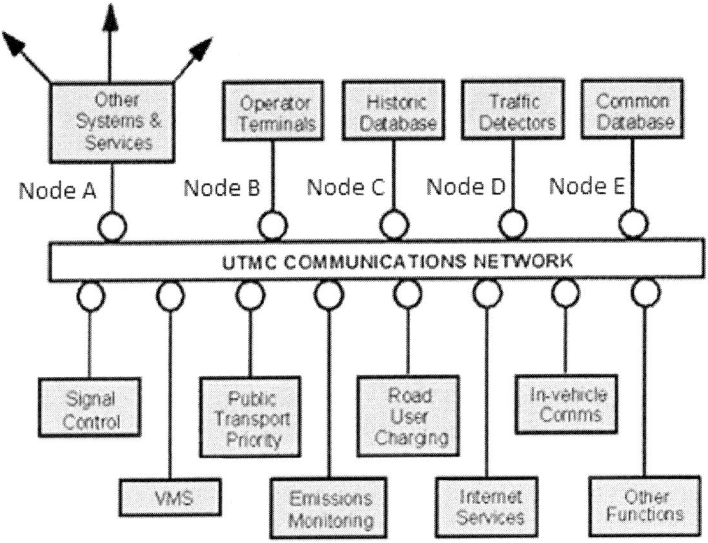

Figure 8.7 Schematic representation of the UTMC system

- Node A: Other systems from which information is received.
- Node B: UTMC traffic management centre—there should be one in the system, although it may be distributed around several locations.
- Node C: "intelligent outstations" used to control elements of the system that interact with users, the "controlled units".
- Node D: "controlled units", such as VMS, traffic signals, parking signals, road pricing, enforcement cameras and so on.
- Node E: Mobile units—for example: the Singapore IU.

8.3 Case Studies and Resources for Appraising ITS

8.3.1 A Useful Tool for Appraising ITS

The "ITS Planning Guide (2004)", by the UK Department for Transport, is a very useful resource for transport planning professionals who are looking for schemes with an ITS element. It provides details of ITS applications as a means of resolving transport-related issues to achieve transport objectives. It contains a directory of ITS tools with a user's guide, case studies of ITS applications, and some reviews of the evaluation of ITS tools around Europe.

The guide is available free from:
DfT Publications,
PO Box 236, Wetherby,
West Yorkshire, LS23 7NB, UK
E-mail: dft@twoten.press.net
Product code: **45RVD02310.**

8.3.2 ITS in Action for Public Transport Payment—the Oyster Card

The public transport smartcard 'Oyster card' is used in London. Some Oyster cards are used like season tickets (abonnements), whilst they can also be used as stored value tickets. Users can top them up with money, even if they also use their Oyster as a season ticket on some other part of the transport network. The card communicates with the card reader by using radio frequencies and is contactless. It is also the world's first public transport ticket in its category to feature price-capping. Travellers can use it to pay single fares, but if the cash value of their single fares in a day exceeds the price of a one-day ticket, then they are only charged for that one-day ticket. The Oyster card is particularly important on buses in London, where traditionally, passengers used to pay the driver a cash fare, which used to keep the buses longer at bus stops, causing considerable delay.

The use of a smartcard also gives operators much better data about how many passengers travelled on which trains and buses. This can make the distribution of revenue to different operators much more accurate than if there were only paper tickets. However, it depends on the nature of the financial contract between the operator and the public transport authority.

To implement the Oyster card, all ticket barriers at all underground stations are equipped with card readers, and ticket machines are installed on all buses and trams. The machines for selling tickets also sell or top up the Oyster card whilst keeping the existing magnetic strip ticketing provision active.

8.3.3 ITS Application for Parking Guidance System—Aalborg, Denmark

This section gathered details from the Jupiter project website (http://www.euroweb.net/jupiter/). The transport project 'Jupiter', was developed under the European Commission's 'THERMIE' programme.

Real-time information on the number of unoccupied parking places in each parking facility is covered by Aalborg's parking guidance system in Denmark. On the main roads leading to the city centre, as well as on the outer peripheral roads around the city centre, parking information is displayed on a number of VMS. The circular roads can effectively distribute traffic to the parking spaces on the periphery of the city centre.

The parking guidance system serves nine parking facilities (seven overground car parks and two multi-storey car parks) with provision for around 3000 spaces around the city centre. A total of 39 VMS and eight conventional signs are linked to the system. Figure 8.8 shows the main VMS of the parking guidance system and the location of the nine parking facilities.

8.3.4 ITS Implemented for Access Control and Road Pricing

In 1994, Rome adopted limited access to the LTZ of the city east of the Tiber, an area of 4.6 km^2. In 1998, access to the area was limited to specific users who were required to pay for an annual permit.

Using a vehicle-ground beacon and without user intervention, the automation of identification and/or the applicable tariff for vehicle entrance into the restricted area was implemented. The access control system is accomplished using a series of gates. The enforcement is active during the weekdays from 6.30 am to 6.00 pm and on Saturday from 2.00 pm to 6.00 pm.

Based on the technology used for the 'TELEPASS' system, the following types of technology infrastructure were put into effect:

- TV camera and infra-red illuminators.
- Microwave transponder.
- On-board unit with smart card.

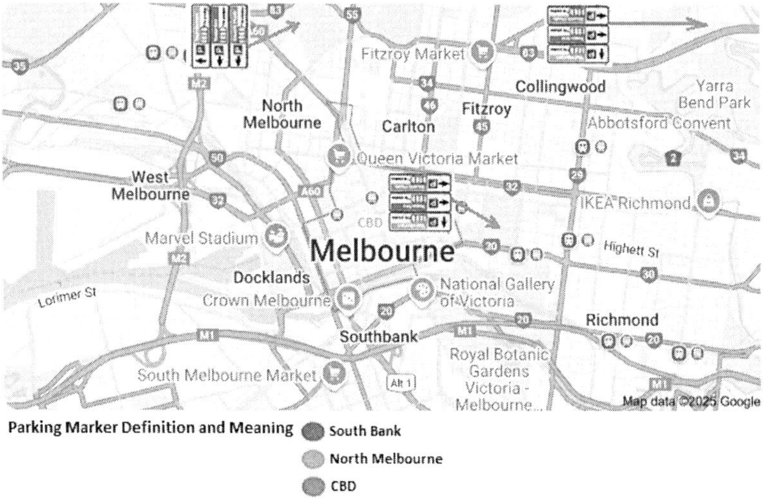

Figure 8.8 Parking Guidance System

Provisions:

- Full real pricing scheme—real charging, real users and real revenues.
- Area covered by system: 4.6 km^2.
- Number of charging points: 22+1 entrance gate.
- Number of users: 30,000 resident vehicles, 30,000 service vehicles (free access), 50,000 plates for disabled people (free access), 29,000 authorised individuals and 8000 freight delivery vehicles (have to pay for access).
- Number of trips per day: about 70,000.

The scheme resulted in:

- A ten per cent decrease in traffic during the day.
- A 20 per cent decrease in traffic during the restriction period.
- A 15 per cent decrease in the morning peak hour (8.30 am–9.30 am).
- A ten per cent increase in two-wheeled vehicles.
- A six per cent increase in public transport use.

Various details of the scheme, apart from the cost of its implementation, are available at http://www.progressproject.org/. Select 'Project Reports' and download 'Deliverable 5.2'.

In Stockholm, vehicles passing any one of 18 points at the boundary of the city centre (inbound or outbound) have their number plates photographed during the scheme's operating hours (06:30 am–0630 pm, Monday to Friday). The charge varies by time of day between €1 and €2 per pass (maximum payment €6 per vehicle per day). As enforcement, if the driver does not pay the relevant charge within five days, they receive a reminder and an administrative charge of around €7. If another 4 weeks have elapsed, they are fined €50.

ITS at present may operate with provisions like FASTag, fitted to the car, which communicates with the scanners on gantries at the charging points. The vehicle owner then pays the charges by direct debit. The scheme can reduce traffic levels within the cordon considerably.

8.3.5 ITS in Action for Multimodal Trip Planning

The EU IST carried out the project TRANS-3 and the earlier CAPITALS PLUS project in France; a multimodal trip planning website, http://www.transbale.com/index.php, was set up and run for the duration of the project. The website is still active except for the journey planning element, as the funding was not continued.

To objective is to achieve a shift from the individual car to the alternative modes. Transbasel is a solution proposed centred on a multi- and inter-modal journey planner. The journey planner proposes a choice of trips with different modes in response to each request. An evaluation involved a sample of only 88 users who completed a pop-up questionnaire on the website. It was found that 83 per cent did not change their behaviour as a result of the information, seven per cent changed route, six per cent changed travel time and four per cent changed mode. However, due to the small sample size, the results are indicative only.

8.3.6 ITS in Action for VMS and Route Guidance

Across Europe, the EU's 'TEMPO' programme funds ITS projects. In the current tranche, Southern European Road Telematic Implementations (SERTI) (www.serti-mip.com) covers long-distance road operations across Spain, Italy, Germany, Switzerland and some parts of France. Around 25 partners are involved in the project, including seven public authorities and 14 motorway companies. Some examples of their work are available directly from their website. No costs or quantified benefits are included on the website.

8.4 Technologies and Implementation of ITS

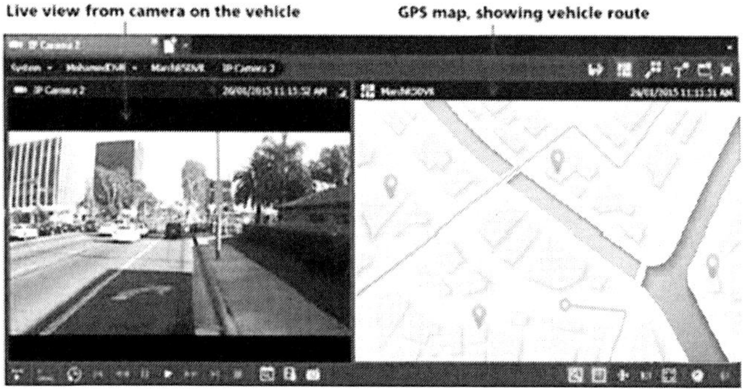

Figure 8.9 ITS uses a graphical user interface to display the road network system and ITSs' GIS-enabled data points

ITS is a hi-tech system for various applications which aims to provide road users, authorities and operators with innovative services relating to different modes of transport and traffic management. ITS can enable the above-mentioned users to have 'smarter' use of transport networks, making them better informed and making travel safer and more coordinated.

A GPS-enabled viewer displays the route and location of the vehicle only when GPS is available in combination with a camera system and an open street map. Some of these technologies include calling for emergency services in case of an accident, cameras used to enforce traffic laws and displaying speed limits. Although ITS may be applicable to all modes of transport, it is defined as a technical environment for a regional network in a state or country. It also provides interfaces for other modes of transport [1]. Thus, ITS improves the road safety and efficiency of road transport, traffic management, mobility, etc. [2].

Recently, governments of developed countries have shown increasing interest in ITS with a focus on national security. The proposed ITS systems involve surveillance of the roadways. Funding for various projects is made either directly through the relevant department or with their approval. In emergencies when large casualty numbers arise because of a natural disaster or terrorism, rapid mass evacuation of people is required in urban areas.

In the developing countries, the shift of people from the countryside to towns and cities created suburbs. Private vehicles are affordable only to a limited part of the population, and the rest of the population demands multimodal transportation systems of motorcycles, buses and trains. This produces considerable hazards from air pollution, creates significant safety risks, and generates feelings of inequity in society. Developing countries, such as China, India and Brazil, although largely rural, are undergoing rapid urbanisation and industrialisation, and a motorised infrastructure is being developed simultaneously with private motorisation. The countries with populations having significant wealth need a highly dense multimodal transportation system for the poor, and that is thwarted by the highly motorised transportation system for the rich.

The various applications of ITS technologies vary in respect of basic management systems such as car navigation; traffic signal control systems, container management systems, VMS, ANPR or speed cameras to monitor the traffic systems, such as CCTV systems, and including more advanced applications such as integrating live real time data and feedback from various sources, such as parking guidance and information systems; weather information; de-icing of bridges; etc. Accurate predictive techniques are now being developed to allow advanced modelling by comparing historical baseline data. Some of these technologies are discussed in the following sections [4].

Figure 8.10 Traffic monitoring gantry & wireless communication dish antenna (Source: Wikipedia)

8.4.1 Wireless Communications

In ITS, radio modem communications are widely used for short and long-range communication over UHF and VHF frequencies. These are various forms of wireless communication technologies.

The Intelligent Transportation Society of America and the United States Department of Transportation are promoting WAVE and the dedicated short-range communications (DSRC) standard. Short-range communications up to 350 m can be accomplished with IEEE 802.11 protocols. The range of these protocols can be extended by using mobile ad hoc networks or mesh networking.

Longer range communications by using infrastructure networks such as WiMAX (IEEE 802.16), Global System for Mobile Communications (GSM), or 3G are well-established methods and are recommended. However, extensive and very expensive infrastructure is required for implementing these methods. Planning and determining an appropriate business model are required to support this infrastructure. Auto insurance companies have utilised Telematics 2.0 as ad hoc solutions to support eCall and behavioural functionalities.

8.4.2 Computational Technologies

There are some recent advancements in vehicle electronics that have led to a move towards fewer, more capable and expensive computer processors with hardware memory management and real-time operating systems on a vehicle. Sophisticated software applications are implemented, which include model-based process control, artificial intelligence, and ubiquitous computing, by the new embedded system technology. Amongst these the most important is artificial intelligence ITS.

Figure 8.11 - RFID E-Z Pass reader attached to the pole. Its antenna is used in traffic monitoring in New York City by using a vehicle re-identification method (Source: Wikipedia)

The 'floating car' or 'probe' data is collected from various transport routes. The following four methods have been used to obtain the raw data:

- Triangulation method: mobile phones were used to communicate messages in developed countries in the mid-2000s. The cars used to contain one or more mobile phones, and they periodically transmit their presence information to the mobile phone network. As a vehicle moves, the signal of any mobile phones inside the vehicle also moves. By measuring and analysing network data using triangulation, the data was converted into traffic flow information, pattern matching or cell-sector statistics in an anonymous format. With more congestion, there are more cars, more phones, and thus, more communications.
- In metropolitan areas, the distance between antennas and towers is shorter, and that results in increased accuracy. The advantage is that only the mobile phone network is leveraged; no new infrastructure is required along the road. However, there is a disadvantage where the triangulation method may fail, especially in areas where the same mobile phone towers serve more than one parallel route, such as a motorway with a service road, a motorway and a railway line, or a street that also has a bus lane. Therefore, the popularity of the triangulation method started declining in the last decade.
- Vehicle re-identification: this system uses FASTag technology. A set of detectors is mounted along the road for vehicle re-identification methods. A unique identification number is detected for the vehicle at a location, and further down the road, it is re-detected. By detecting and comparing the times between pairs of sensors of a specific device, the travel times and speed are calculated. By using the radio frequency identification (RFID) serial numbers from electronic toll collection (ETC) transponders (FASTag), enforcement, if applicable, can be done.
- GPS-based methods: vehicles with satellite navigation systems that have two-way communication with a traffic data provider are increasing in number, providing for vehicle speed detection. Instead of using dedicated hardware, modern methods use smartphone-based solutions using Telematics 2.0 approaches.
- Smartphone-based rich monitoring: smartphones can use various sensors to track traffic speed and density. To detect the traffic speed and road quality, the accelerometer data is monitored by collecting data from smartphones used by car drivers. The real-time traffic density and possibility of traffic jams may be identified by collecting

audio data and GPS tagging provisions of smartphones. In Bangalore, India, this was implemented as a part of research for an experimental system, Nericell.

8.4.3 Floating Car Data/Floating Cellular Data

Floating car data technology has superiority over various other methods of traffic measurement:

- Implementation and functioning are less expensive compared to sensors or cameras.
- The coverage includes all locations and streets.
- Set up is fast and maintenance is low.
- Operates during heavy rain and other adverse weather conditions.

8.4.4 Sensing

With the advancements in telecommunications and information technology, coupled with ultramodern state-of-the-art microchips, RFID, and inexpensive intelligent beacon sensing technologies, the technical capabilities of ITS have been enhanced. The intelligent transportation systems provide road users across the world with enhanced safety benefits. Intelligent vehicle technologies are infrastructure-based networked systems, and they are the sensing systems for the ITS. Infrastructure sensors (for example, in-road reflectors) are indestructible devices that are embedded in the road or installed in the surroundings of the road (e.g., on buildings, posts, and signs) as required. These may be dismantled during road maintenance manually or by sensor injection machinery for rapid deployment. Vehicle-sensing systems for infrastructure-to-vehicle and vehicle-to-infrastructure, electronic beacons for identification communications, video ANPR and vehicle magnetic signature detection technologies of ITS are deployed at desired intervals to increase continuous monitoring of vehicles operating in critical regions.

Saw-cut loop detectors are buried in the road to detect vehicles at junctions. These are shown as the rectangular shapes of loop detector sealant at the bottom part of Figure 8.12.

8.4.5 Inductive Loop Detection

Figure 8.12 – Saw-cut loop detectors for vehicle detection buried in the road (Source: Wikipedia)

8.4.5.1 Video Detection

ITS implementation uses video cameras as another form of vehicle detection by traffic-flow measurement and automatic incident detection as a non-intrusive method of traffic detection. The video detection systems that are used in ANPR do not involve installing any components directly into the road surface or roadbed.

Cameras take the details of the passing vehicles, which are fed into processors that analyse the changing characteristics of the video image. Typically, the cameras are mounted on poles or structures above or at the side of the roadway. Some initial configuration is required to input the baseline background image to the processor. The initial configuration data includes known measurements such as the distance between lane lines and the height of the camera above the roadway. The detection of traffic can be done by a single video detection processor receiving data simultaneously from one to eight cameras, depending on the brand and model. The video detection system generates output about the lane-by-lane vehicle speeds, vehicle counts and lane occupancy readings. The system may also generate outputs including gaps, headway, stopped-vehicle detection, and wrong-way vehicle alarms.

8.4.5.2 Bluetooth Detection

Bluetooth is an accurate and inexpensive way to transmit the position of a moving vehicle. The sensing devices of ITS along the road detect Bluetooth devices installed in passing vehicles. Interconnected sensors may calculate travel time and provide data for origin and destination matrices. Bluetooth measurement has the following advantages compared to other traffic measurement technologies:

- It is non-intrusive and cheaper.
- Setting up Bluetooth systems is generally quick with little or no calibration.
- The devices are increasingly prevalent in vehicles, and the amount of data collected over time becomes more accurate and valuable for travel time and estimation purposes.

By using audio signals, it is also possible to measure traffic density on a road from the cumulative sound of the tyres, engine, honks and air turbulence. Roadside-installed microphones pick up the various vehicle noises, and then the audio signal processing techniques estimate the traffic state. The accuracy of such an audio data processing system compares well with the methods used by other systems.

8.4.5.3 Radar Detection

Radars are mounted on the side of the road to measure traffic flow, for stopped and stranded vehicle detection purposes. Radar, like the video systems, also reads the initial environment during the setup, and can distinguish between vehicles and other objects. Traffic flow radars use the 'side-fire' technique to count the number of passing vehicles and estimate traffic density by looking across all traffic lanes in a narrow band. A 360-degree radar system may be used for stopped vehicle detection (SVD) and automatic incident detection. They scan all lanes along large stretches of road. Radar is more efficient and provides better performance over longer ranges than other technologies [9]. SVD radars are recommended on all smart motorways [10].

8.4.5.4 Information Fusion from Multiple Traffic Sensing Modalities

The data from the different sensing technologies can be combined in intelligent ways to determine the accurate state of traffic. Data fusion may combine the advantages of the different individual methods, based on an approach that utilises sound, imagery and sensory data [11].

8.4.6 Emergency Vehicle Notification Systems

The eCall is a European initiative that can assist motorists in the case of a collision. Automobile manufacturers must equip all new cars with an eCall provision [12]. In case of an accident, the eCall is generated either manually by the driver or passengers of the vehicle or automatically via activation of in-vehicle sensors [13]. When activated, the device will establish an emergency call carrying both voice and data directly to the nearest emergency point or public safety answering point (PSAP) [13]. This enables the vehicle occupants to send a minimum data set to the trained eCall operator. The minimum data set contains information about the incident, including time, precise location, the direction the vehicle was travelling and vehicle identification. Provisions of the eCall system depend on the manufacturer; commonly, it could be mobile phone-based with a Bluetooth connection to an in-vehicle interface. Other systems include an integrated eCall device, or a functionality of a broader system like navigation, a telematics device, or a tolling device. The standardisation of eCall should be done by the standards institute of respective countries, and commitment from various states.

The EU used S-band satellite communication from the EC-funded project 'SafeTRIP' and developed an open ITS system that will improve road safety and provide resilient communication. Within the EU, such a system will also provide greater coverage of the emergency services.

Figure 8.13 - Congestion pricing gantry at North Bridge Road, Singapore (Source: Wikipedia)

Figure 8.14 Automatic speed enforcement gantry (Source: Wikipedia)

8.4.7 Automatic Road enforcement and Traffic enforcement camera

A traffic enforcement camera system, consisting of a camera and a vehicle-monitoring device, is used to detect and identify vehicles exceeding the speed limit or some other violation of traffic rules, including automatic ticketing of offenders based on the registration plate.

The tickets for the violation penalty are sent by mail, and the applications include:

- Speed cameras are used to detect vehicles exceeding the legal speed limit. To detect a vehicle's speed, many such devices use radar or electromagnetic loops buried in individual lanes of the road.
- Red light cameras that detect vehicles ignoring red traffic signals.
- Bus lane cameras that identify vehicles travelling in lanes reserved for buses.
- Level crossing cameras detect vehicles crossing an at-grade railway line illegally.
- Cameras detecting vehicles crossing double yellow lines.
- High-occupancy vehicle lane cameras detect vehicles violating HOV requirements.

8.4.8 Variable Speed Limits

Figure 8.15 - Variable speed limit sign in the United States (Source: Wikipedia)

Variable speed limit signs change speed limits during congestion and other situations, usually to slow the traffic down.

The M25 motorway around London is Britain's most heavily travelled section, between junctions 10 and 16. Variable speed limits combined with automated enforcement have been in force since 1995. Initial results indicated savings in journey times, smoother-flowing traffic, and a fall in the number of accidents, and subsequently the implementation was made permanent in 1997 [14].

8.4.9 Cooperative systems on the road

ITS can enable communication on the road car-to-car or car-to-infrastructure, and vice versa. The goal of cooperative systems is to increase road safety by using sensor infrastructure and necessary communication. For central processing, the data available from vehicles is acquired and transmitted to a server. Active windscreen wipers indicate rain, and frequent braking suggests congestion, and these activities are detected and collected by regional control. The driving recommendations are processed by the server and transmitted by electronic messages to vehicles.

The systems ensure the most efficient, safe, secure and comfortable journey for drivers.

The ITS World Congress is an annual trade show that promotes ITS technologies. The event is organised each year in a different region, for example, Europe, the USA or Asia-Pacific [18]. In 1994, the first ITS World Congress was held in Paris.

8.4.10 Smart Transportation—New Business Models

New mobility and smart transportation models are emerging globally.

- Schemes such as bike sharing, car sharing and scooter sharing are gaining popularity.
- In many cities, electric vehicles and charging schemes are being implemented.
- The connected car scheme has an emerging market.
- Commuters and operators all over the world are interested in new, smart parking solutions.
- The various new models and schemes provide opportunities for solving issues in urban areas.

8.4.11 ITS in the Connected World

Service providers of mobile phones are not limited to just connectivity; they are becoming significant players in enabling communications for ITS. Apps are being continuously launched to take mobile payments, provide data insights and navigation tools, offer incentives and discounts, from calling taxis to shopping.

8.4.12 Payments and Billing Flexibility

Earning revenue from an asset, business, etc., is created by these new mobility models. A flexible settlement and billing platform enables quick and easy sharing of revenues and an overall better customer experience. Discounts, loyalty points and rewards and direct marketing are made available to the users, for example.

8.4.13 ITS Scenario in Europe

On 7th October 2004 in London, a network of national ITS associations was launched, with the secretariat at ERTICO-ITS Europe [19]. For the implementation of ITS, various bodies are also connected, such as public authorities, industry players, infrastructure operators, users, national ITS associations and other organisations. To improve transport safety, security and network efficiency, ERTICO focuses on work programmes and initiatives alongside taking measures to reduce environmental impact.

8.4.14 ITS Scenario in the United States

The United States have an ITS chapter that holds a yearly conference to promote and showcase ITS technologies and ideas. This conference is attended by representatives from the Department of Transportation of each state, city, town, and county within the state.

8.5 FASTag

The next generation of traffic management will be the electronic toll collection (ETC) system, for various advantageous features; it is therefore obvious to opt for the ETC system at tolls. Various highway authorities have rolled out a programme for ETCs at tolls on national highways: FASTag.

Figure 8.16 The FASTag trademark

FASTag is a device that employs radio frequency identification (RFID) technology for making toll payments directly from the prepaid account linked to it. It is affixed on the windscreen of the vehicle and enables the driver to drive through toll plazas and other cameras. FASTag is valid for five years, and once purchased, it can be topped up.

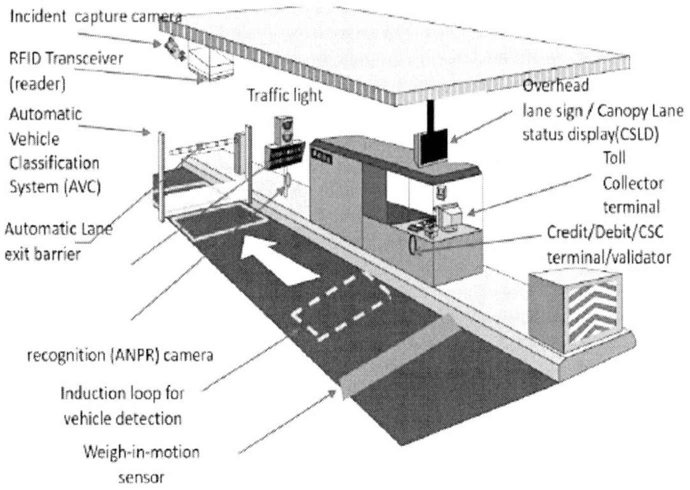

Figure 8.17 Tolling mechanism

FASTag offers non-stop movement of vehicles through toll plazas and the convenience of cashless payments nationwide.

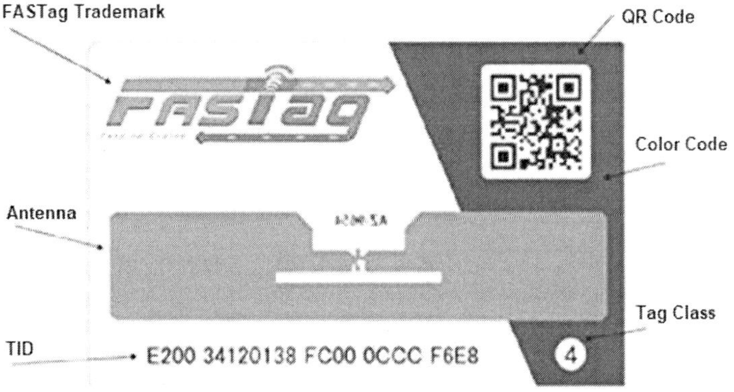

Figure 8.18 Standard arrangements of embedded details in the windscreen tag

8.5.1 Benefits of Using FASTag

1. Ease of payment—no need to carry cash for toll transactions, so it saves time.
2. Non-stop movement of vehicles leads to lower fuel costs.

3. Online recharge—FASTag can be recharged online using a credit/debit card/NEFT/ RTGS or Internet banking.
4. SMS alerts for toll transactions, low balance, etc.
5. Online portal for customers.
6. Valid for five years.
7. Incentive: cashback of ten per cent on all toll payments using FASTag in a specified year.

Reduced air pollution.
Reduced use of paper.
Analytics for better highway management.
Reduction in the management of the toll plaza.

8.5.2 Provision for an Incentive/Discount on Toll Payments Using FASTag

The cashback amount for a particular month will be credited to the FASTag account within a week of the subsequent month.

8.5.3 Flexibility for FASTag Option, Applicable to Any Particular Kind/Category of Vehicle

Adopting FASTag is voluntary at this stage of the programme. However, it may be made mandatory in future at some locations. FASTag is applicable for all categories, kinds, makes and types of vehicles.

8.5.4 Implementing the 'FASTag' Programme

A highways management company is a company incorporated by a national highways agency in association with a national payment company, and is implementing this programme with the help of toll plaza concessionaires, FASTag issuer agencies and toll transaction acquirer (selected banks).

8.5.5 Roads Using FASTag

FASTag is presently operational at toll plazas in various developing and developed countries. More toll plazas will be brought under the FASTag programme in the future.

RFID technology is at the core of FASTag. For the improvement of road safety, it plays an important role in identifying vehicles violating traffic rules and collecting details of drivers. RFID technology uses an electronic product code (EPC) through which every vehicle can be uniquely identified. This code is different from the vehicle's registration number and exclusive to it on a global scale. Each EPC code is a 13-digit number. FASTag is issued by GS1, a standards body, which ensures that each code is unique and in sync with the global standards put in place for correct product identification. The code needs to be standardised to ensure that the data coded inside is not read differently at different levels.

Figure 8.19 Manual toll collection

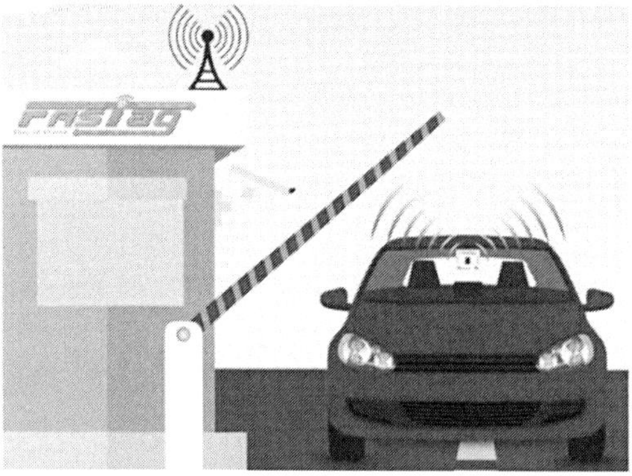

Figure 8.20 – Electronic toll collection using FASTag

8.5.6 Charges for FASTag

FASTag has a one-time fee, say USD 100. This refundable security deposit depends on the vehicle type. Users have to check other details on the website of issuing agencies.

8.5.7 Buying of FASTag and the Required Documents

Customers may visit any of the point of sale (PoS) locations at toll plazas/issuer agencies to get the FASTag account created. The customer needs to submit a copy of the following documents along with the application for FASTag:

1. Registration certificate (RC) of the vehicle.
2. Passport-size photograph of the vehicle owner.
3. Details as per the category of the vehicle owner (viz. individual/corporate)

8.5.7.1 For Individuals

The following documents need to be submitted,

Copy of ID and proof of address
One passport-size photograph.
Copy of driving licence
Copy of PAN Card
Copy of passport

For instance, a valid driving licence would be sufficient for address & ID proof. The originals of the above documents may also be required for verification.

8.5.7.2 Corporate Customers

- The issue of FASTag is subject to verification of the original documents mentioned above by the issuing agency.
- Customers are advised to bring originals of the above documents whilst visiting the PoS for the purchase of a FASTag.
- The FASTag will be fixed on the vehicle's windshield by the representative of the issuing agency only.

- Any representative of the vehicle owner can purchase the FASTag by providing the above documentation and information.

8.5.8 Information to the Customer on How Much Toll has Been Debited

The customer will receive an SMS with the requisite details to their registered mobile number as soon as the toll transaction takes place. Periodic statements of account may also be obtained on the website of the issing agency after registration by the customer.

8.5.8.1 Specific Lanes at the Toll Plaza for FASTag

For making payment at toll plazas by the FASTag, the vehicles must go in the lanes demarcated for FASTag. However, a vehicle, although having a FASTag, may use any lane at toll plazas for making a cash payment. Penalty for a vehicle entering the FASTag lane without a valid FASTag will be charged double the applicable toll amount in cash.

8.5.8.2 Monthly Pass or Local Pass

The monthly pass facility is specific to each toll plaza. The customer may call the customer care of the participating bank/issuer for further details.

8.5.8.3 Where to Finally Pay the Toll

The issuing agency will electronically deduct the applicable toll fee from the customer's account linked to the FASTag. This deduction will be done post the toll transaction. The customer needs to fund the account which is linked to FASTag adequately. This recharge process may be termed a top-up.

8.5.8.4 Recharging by the Customer for a FASTag Account

You may recharge your FASTag account by making a payment by cheque or online using a credit/debit card/NEFT/RTGS, or through Internet banking. A FASTag account can be recharged up to the maximum limit.

8.5.8.5 the Minimum Value for Top-up

The minimum recharge amount is specified.

8.5.8.6 Crossing a Toll Plaza, If the FASTag Equipment is not Working

The standard operating procedure (SoP) for the FASTag user may be referred to.

8.5.8.7 One FASTag for Two Vehicles is not Permitted

The customer will need to buy two separate FASTags for each vehicle.

8.5.8.8 FASTag Bought for One Vehicle Cannot Be Used for Another Vehicle

FASTag is issued to each vehicle based on the KYC documents submitted. If a FASTag, for instance, bought for a car is used for a lorry, then the FASTag shall get blacklisted, resulting in the denial of vehicle passage from FASTag lanes on toll plazas. FASTag users are advised to desist from such practice.

8.5.8.9 Loss of the FASTag

The customer will need to call the customer care number of the issuer agency to block their FASTag. Once the customer opens a new account, the issuer agency will transfer the balance to the new account.

8.5.8.10 has the Correct Fee Been Deducted from My FASTag Account

The customer receives an SMS alert at their registered phone number each time an amount is deducted from their FASTag account. Fees payable at each toll plaza may be seen on the authority's website in addition to those displayed at the toll plaza.

8.5.8.11 Report an Incorrect Deduction and Refunds

The customer can report an incorrect deduction by contacting the issuer agency. The agency will review the customer's request and take the necessary action.

8.5.8.12 Blocking FASTag Account If the Vehicle is Lost

The customer can contact the agency's customer care department.

8.5.8.13 When a New Vehicle is Fitted with an RFID Tag That is to Be Activated

The customer can contact the agency's customer care department.

8.5.8.14 Use of the Tags on FASTag Lanes Purchased Earlier

Only the RFID tags purchased earlier may be used on FASTag lanes. In case of any problems, please contact the PoS of the issuer agency. Any other type of tag will not work on FASTag lanes.

8.5.8.15 FASTag Can Be Used for Passage Through Toll Plazas on the Roads

Customers can check the list of tolls on the relevant website.

8.5.8.16 How to Get the FASTag Checked/Rectified

If the customer has a FASTag and also a balance in their linked account, but they could not go through the ETC (electronic toll collection) lane, then to get the FASTag checked/ rectified, whether it is active or not, refer to the steps under the SoP for FASTag users. The customer can contact the agency's customer care department.

8.5.8.17 Concessions Available for Local Vehicles

If the customer resides within 10 4 km of a particular toll plaza, they can have access to local concessionary fees via FASTag to get the concessions available for local vehicles.

8.5.8.18 What If the Customer Relocates to Another City?

FASTag will work on all enabled toll plazas. In case of a change in city or address, the customer can contact the agency's customer care department.

8.5.8.19 Use of More Than One FASTag by One Particular Vehicle

It is strictly prohibited to purchase two FASTag for one vehicle. In case of multiple FASTags, the customer will be responsible for operational issues.

8.5.8.20 in Case the Customer Sells/Transfers Their Car

If the car is sold or transferred, only the issuing agency needs to be informed.

8.5.8.21 In case the FASTag gets damaged

The customer can contact the agency's customer care department to arrange for a duplicate.

8.5.8.22 Use of FASTag at places other than Federal/National highways

Usage of FASTag may be extended to car parks or paying for other amenities.

8.5.8.23 Addressing any grievance of the vehicle owner/user

The customer can contact the agency's customer care department. The highway agency can be approached as the next escalated level.

8.6 Conclusion

This chapter has discussed how ITS can bring real benefits in various contexts in various countries, and the implementation of ITS. ITS can be very useful but projects are expensive. It is to be ascertained that the ITS solutions will meet the required transport objectives before investing in them.

References

Credit for using some figures/text to:
Wikipedia, Web page: Intelligent Transport System webpage:
https://en.wikipedia.org/wiki/Intelligent_transportation_system

CHAPTER 9

GREEN HIGHWAYS

Abstract: This chapter contains a general overview of the use of electric vehicles and compressed natural gas (CNG) as fuel on green highways, eliminating pollution significantly, vehicles integrated with renewable energy sources for sustainable mobility on green highways, green highways trees, utilising recycled materials for eco-friendly construction and conclusion.

9.1 General

Green Highway projects enhance road safety by creating a natural environment using trees and plants. Electric vehicles (EVs) can be monitored by ITS systems and devices, even when stationary and without discharging batteries. Solar panels keep the cells charged. The upgrade, rehabilitation and new construction work for road infrastructure will have green cover and access to reusable materials.

9.1.1 Top Facts About Green Highways Projects

Under a green highway project, two-lane roads with paved shoulders or a four/six-lane configuration will undergo major upgrades or rehabilitation.

- The projects may require loans. The World Bank provides loans to promote green highway schemes. This is a very noble motive to save the planet.
- Once the green highways project is completed, road users will have improved user comfort, a pleasing travel experience and enhanced safety. This will attract more road developers and will help generate high revenue for the governments as well as boost income for the people of those areas. The smooth and drivable roads will fulfil the socio-economic requirements of the region.

- Improved roads will cut down on travel time and, therefore, save valuable working hours and energy. Due to smooth driving, wear and tear on the vehicles will be minimised.
- Maintenance will also be carried out whilst it is operational. Maintenance will be done every five years on flexible surfaces and every ten years on rigid surfaces.
- The essential components are sustainable development and maintenance of highways, institutional capacity enhancement, road safety and research and development.
- Construction may require a Highway or Bridge Engineering Laboratory with high-standard material testing to assess the quality of works and the impact of interventions on climate resilience.
- The projects may help achieve enhanced road safety by adopting advanced measures such as safety audits at the design, implementation, operation and maintenance stages. Road users as well as local people will have a pollution-free, safe environment. ITS shall impose strict surveillance and control towards illegal actions by road users, which may be more effective on electric-operated vehicles.
- Research and development studies may be done to research construction using fly ash, RAP and other demolition waste on embankments, using waste plastic, modifiers in bituminous works, stabilising soil and pavement layers, planting vegetation and slope protection using bio-engineering solutions.

9.1.2 Key Points to be Addressed

The primary purpose of a green highway is to reduce pollution and hazards and make the highway eco-friendly. The following measures have been identified:

- Planting trees will provide a pollution-free and eco-friendly corridor, as discussed in Section 9.2.
- Vehicles may be using CNG as fuel, as discussed in Section 9.3.
- Vehicles may be using electrical energy integrated with rechargeable energy sources for sustainable mobility and enhanced road safety, as discussed in Section 9.4.
- Recycled materials may be used in construction, as discussed in Section 9.5.

9.2 Green Highways with Trees

Vision of the Planting Policy

A green highways scheme has the following objective: to develop eco-friendly motorways/freeways with the participation of the community, farmers, NGOs, the private sector, institutions, government agencies and the forest department for economic growth and development of the region in a sustainable manner.

Air and water pollution are important issues and need to be resolved as a matter of urgency. The scheme adheres to a policy framework that includes the planting of trees to help reduce the impact of air pollution and dust, provide shade on hot roads during summer, reduce the impact of noise pollution and soil erosion, prevent glare from the headlights of oncoming vehicles and generate employment.

Planting trees can also achieve any country's forest cover target. If the mandate is for a forest cover of 33 per cent of the land, but much less is covered by forests, then the difference can be met by planting trees outside the forest areas. This is hardly achievable on existing highways because of inadequate right-of-way, but green highways have wider rights-of-way for planting an adequate number of trees.

The policy should have a strict auditing system whereby payment to agencies will be released by the government only if 90 per cent of the trees planted in the previous year have survived. It is only about how many trees survived and how useful this is for the local community, and not about how many trees were planted. The policy will focus on planting new trees only.

9.2.1 Policy Objectives

To facilitate planting trees, the following objectives are proposed:

- Establish a policy framework for planting along the green highways.
- Reduce the impact of air pollution and dust by planting trees and shrubs.
- Provide essential shade on glaring hot highways during summer.
- Reduce the impact of high levels of noise pollution caused by the increasing number of vehicles, where the number of trees is inadequate.
- Protect embankments from soil erosion.

- Prevent headlight glare from oncoming vehicles.
- Reduce the snow accumulation and fog intensity.
- Reduce the effect of wind during storms.
- Create employment opportunities for local people in planting and maintenance.

9.2.2 Purposes of Planting

In highway development, loss of vegetation is an inevitable consequence. It is therefore the responsibility of the highway authorities to offset this loss by eco-friendly corridor development and management. The highway authorities must have an agenda to enhance the aesthetics of the highway corridor at all possible locations. Highways should not be considered just as a means of transportation, but rather as an integral part of the physical environment and socio-economic development of the region.

Due to budgetary limitations, the land needed for planting and landscape improvement is not considered during the DPR stage. The acquisition plans for the highway projects do not have adequate rights-of-way (RoW) for tree planting. As a result, after construction, when the planting starts, it is only done wherever space is available, and at some places, only three to four rows of trees can be planted. The width of the remaining RoW is not sufficient to accommodate even a single row of plants. To ensure sufficient width of the RoW along the green highways, it should be made mandatory to consider avenue planting. The DPR consultants should essentially consider land for tree planting, and this should be cross-verified in the plans. The highway authorities must accept the project cost for such provisions that are made mandatory by the government.

9.2.3 Institutional Arrangements and Financing Patterns

In the PPP (public-private-participation) mode of highway development projects, it is the responsibility of highway authorities to develop a green corridor along the highways for aesthetic enhancement of the project by planting ornamental trees and landscaping and turfing with grasses and ornamental shrubs.

Planting may be done by outsourcing the work to experts and experienced agencies, and this can be done by the nodal office of the highway authority. The nodal agency will be the forest division for highways passing through forest areas. The nodal agency officer will be the engineer. In case of a government-funded highway, the concessionaire will be the nodal agency in cases where roadside planting is included in the concession agreement.

Nodal agencies will take up the planting work if the contractor/ concessionaire fails to implement the planting programme within the stipulated period, as per their agreement, or assign the work to another agency. As per standard specifications, the responsibility of ensuring compliance will remain with the nodal agency.

The selection of species will be site-specific as per the guidelines or as per the recommendation of the adjoining forest department. A memorandum of understanding (MoU) shall be signed by the project authority with the planting agency for strict compliance with the technical specifications and on the legal rights of the land, as well as forest products.

One per cent of the total project cost (TPC) should be exclusively reserved for the planting fund in a separate account by the highway authority. In case of BOT projects, assigning an agency for planting and maintenance should be done by a committee consisting of the nodal agency officer, a representative of the monitoring agency and the concessionaire. In case of publicly funded projects, a representative of the highway authority will also be a member of the committee.

The district-level advisory committee should meet once every three months and report to the monitoring cell. The chairperson of the committee should be the nodal agency officer.

9.2.4 Land Requirement

An action plan will be prepared based on the inventory, with respect to additional landscaping measures and road users' amenities. If the available land width is not sufficient to implement this plan, acquisition of additional land should be seriously considered, keeping in mind the following requirements during feasibility/DPR studies.

- To provide flatter side slopes in cuts and fills to match the adjacent land.
- To provide enough space for planting suitable trees and plants in a specified number of rows.
- To provide sufficient area for parking, look-out spots and other aesthetic features.
- A space of 3 to 5 metres should be kept reserved for utilities adjacent to the RoW.

9.2.5 Planting Agency

The entire highway section may be divided into two categories based on the legal status of the existing roadside plantings. The first category includes where the highway alignment passes through forest areas, where the RoW has been identified as protected forests, and permission for tree cutting is required from the forest department under the Forest Conservation Act. On granting permission, the forest department stipulates the conditions not only for compensatory afforestation (CA) by tree planting but also for avenue planting, which is reserved for the forest department. Avenue planting will therefore be taken up by the forest department apart from CA.

In the second category, the RoW for highway corridors is not declared as a protected forest, and the roadside plantings may be done by a specialised contractor as a BoQ (bill of quantities) item or by the concessionaire, if it is included in the concession agreement.

9.2.6 Planting via a Bidding Process

For the RoW in a highway section not declared as a protected forest, the planting and its maintenance may be taken up through outsourcing. This may be done by following a bidding process as per the standard procurement protocol.

9.2.7 Panelled Agencies

The highway authority may appoint an authorised agency and a separate group of agencies specialising in transplanting trees should also be prepared. The bid documents for planting work on federal and state highways should be issued to enlisted agencies only.

9.2.8 Role of Local Government

For roadside plantings, nodal agencies should involve the local institutions, like local state governments, for plantings, watering, protection and sharing of benefits. In selecting the planting agencies, the local governments may be helpful in their areas.

Except for those carried out by the forest department, all other plantings should comply with any statutory conditions. This should be carried out under a MoU between the nodal agency and the planting agency. Wherever feasible, the schemes of other government departments and agencies should be matched and encouraged.

9.2.9 Planting Scheme

The planting scheme could be broadly classified into two categories, which are as follows:

- Tree planting along the highway with grass turf and shrubs.
- Planting on medians or special landscaping on embankments.

9.2.10 Selection of Tree Species for Roadside Plantings

Planting is one of the most important components of soft landscaping. Trees, shrubs and climbers have been used to enhance the soft natural ambience and protect against harsh elements. The planting species are chosen based on their physical growth characteristics, like form and shape, foliage pattern, growth rate, branching pattern, soil characteristics and conditions of the area to be planted, like waterlogged areas, etc. If a pure avenue of a single species is planted for a considerable length of the road, it gives a harmonious and pleasing look. However, mixed planting is better for the environment and creates more robust plants.

The selection of plant types and planting arrangements shall be based on the following considerations:

- Aims and objectives of planting.
- Shape and size of the canopy.
- Texture and colour of the foliage, flowers or fruits in different seasons and stages of growth.
- Adaptability and suitability to agro-climatic regions.
- Growth rate, average age of maturity and replacement cycle.
- After care and maintenance required for sustenance and growth.
- Economic and other social benefits.
- Drawbacks, if any, like prone to insects/pest diseases, animal grazing and human interference.

9.2.11 Planting Patterns

A holistic approach should be planned for the road landscape that is applicable to the entire length. The concept is to maintain visual characteristics and uniformity in the landscape along the highway corridor. In case of non-availability of uniform land for the plantings, different schemes may be developed with local variations in the design. Based on the similarity of available width, soil conditions, temperature, rainfall and topography, the project corridor can be divided into several homogenous landscape sections. To choose the suitable species for a particular section, field surveys may be carried out by studying the local flora and vegetation native to the section. Planting pattern depends on available RoW, and may be decided as follows:

Small to medium-sized ornamental trees should be in the first row along the highways.

- Planting in subsequent rows depends upon the width available for shade-giving varieties, higher than that in the first row. In rural areas, tall shade-giving trees will always be in the last row.
- Planting shrubs on the median.
- Planting ground covering herbaceous species on the median, embankments and landscaped areas.
- Turfing the median, special landscapes and embankments may be done with grass.

Table 9.1 Plants Recommended for the First Row of Planting

S. No.	Soil	Botanical Name
1.	Loamy	Delonix Regia
2.		Cassia fistula
3.		Bauhinia sps.
4.		Cassia nodusa
5.		Jacaranda mimo saefolia
6.		Peltophorum ferrugineum
7.	Waterlogged	Terminalia arjuna
8.		Syzygiumcuminii
9.		Cordia dictoma
10.	Alkaline Soil	Terminalia arjuna
11.		Pongamia pinnata
12.		Albizzia lebbek

Table 9.2 Species recommended for the second and subsequent rows except the last row

S. No.	Soil	Botanical Name
1.	Loamy	Melia azadiracta
2.		Pongamia pinnata
3.		Gravillea robusta
4.		Albizzia lebbek
5.		Dalbergia sissoo
6.		Terminalia arjuna

Table 9.3 Shade-giving trees recommended for the last (or only) row in roadside avenues

S. No.	Soil	Botanical Name
1.	Loamy	Ficus religiosa
2.		Ficus infectoria
3.		Madhuca indica
4.		Mangifera indica
5.		Azadirachta indica
6.		Tamaridus indica
7.		Syzynium cuminii
8.		Dalberjia sissoo
9.		Dalbergia sissoo
10.	Sandy	Azadirachta indica (at pH up to 8.5)
11.	Alkaline Soil (usar)	Pongamia pinnata (pH upto 9.0)
12.		Terminelia arjuna
13.		Syzyniumcuminii
14.	Waterlogged Area	Terminalia arjuna

Tables 9.1, 9.2 and 9.3 list a few species that can generally be planted throughout the country if the soil condition matches. Region-wise choices can be made with the help of local experts from the forest department and the horticulture department. As far as possible, fruit-bearing trees like mango, guava, berries and tartaric are ideal for areas near habitations

throughout the country. Other useful fruit-bearing species like jackfruit, bel, etc., may also be preferred.

The shrubs to be planted in the median shall be of low or medium height to reduce headlight glare. One or two rows of flowering shrubs may be provided according to the width of the median in different sections.

In sections where the median width is less than 1.5 m, only grass turf is appropriate. Some herbaceous species may also be planted as ground cover, not only on the medians but also on special landscapes and embankment slopes.

The species proposed for turfing or ground cover are: cyndon dactylon, cythocline perpurea, solanum nigrum, alternanthera, chlorophytum, eupatorium, wedelia, duranta, portulacca, lpomea, pelia cadrii, beleprone oblongata, tradescantia, wedelia, duranta, portulacca, lpmea, pelia cadrii, beleprone oblongata, tradescantia, asparagus, opheopogon grass, etc. depending on climatic conditions, requirement for water and future management. However, other species suggested by the forest department or the horticulture department can also be used.

9.2.12 Technical Specification

Table 9.4 Ornamental plants (except in the last row)

S. No.	Subject	Data
1.	Distance from embankment	1.0 metres away from the toe of the embankment
2.	Spacing between plants	3 metres
3.	Spacing between rows	3 metres
4.	Size of the pits (normal soil)	60 × 60 × 60 cms
5.	For alkaline soil (Usar)	By Auger
6.	Waterlogged areas	Mounds with varying heights depending on the water level
7.	No. of plants per km.	333
8.	Activity and schedule	As per the agreement
9.	Height of the plant	1.5m to 2m

Table 9.5 Shade-giving plants (in the last row)

S. No.	Subject	Data
1.	Distance from embankment	3.0 metres
2.	Spacing between plants	12 metres
3.	Size of the pits (normal soil)	60 × 60 × 60 cms
4.	For alkaline soil (Usar)	By Auger
5.	Waterlogged areas	Mounds
6.	Species recommended	As per the agreement
7.	No. of plants per km.	84
8.	Height of the plant	More than 2m.

Unsuitable soil may need to be replaced with good-quality soil. The unsuitable soil is to be removed by digging pits and replacing it with gypsum (1 to 3 kg), depending on the pH, along with 2 kg of composite and sand. This lowers the pH and will help the plants survive for longer.

9.2.13 Specifications for Median Planting

If the width of the median varies, use one or two rows of flowering shrubs. In sections where the median width is less than 1.5 metres, only grass turf is recommended. Planting one row of shrubs on a median 2.5/3.0 m wide, and two rows of flowering shrubs on a median 4.5/5.0 m is proposed. On medians 4.5/5.0 m wide, these plants should be 1.5 m away from the edge.

Plants should be 3 m apart, and the planting pits should be 0.6 m in diameter and depth. Therefore, the total number of plants per km in a single row will be 333 and in two rows, 666.

The surface should be well prepared. Loose debris on the median should be hand-compacted and the surface smoothed. Add good-quality soil to an adequate depth to ensure better growth and survival of the grasses and shrubs.

9.2.14 Tree Transplanting

To save the trees along existing RoWs, transplanting should be done instead of cutting the trees down as much as possible. Tree transplanting is a very complex job and requires a high level of expertise and skill, besides special equipment. Tree transplanting depends on the soil and climatic conditions of the area. The location of transplanting should be within 5 km of the existing tree location. At least three months before transplanting, identify the location for transplanting and undertake preliminary treatments.

Winter is a favourable season for transplanting as the tree is dormant or less active. The tree should be pruned or trimmed heavily, leaving a bare framework of large branches. A rectangular trench 40 to 50 cm wide and 1 to 2 m deep should be dug all around the stem, enough distance away from it. Depending upon the size of the specimen, all the roots are cut.

9.2.15 Protection Measures

The fencing for single row plantings may be done using either iron fencing or brick/concrete tree guards. Barbed wire fencing can be used for multiple rows. A five-strand barbed wire fence, with cross strands, stretched on angled iron poles, is recommended. The angled iron poles should be fixed four metres apart. Bamboo/thorn fencing may also be used if these are suitable. Barbed wire fencing and the iron guards should be as per the relevant specifications.

The planting must not be a hazard to road traffic by restricting visibility, and therefore, the first row of planting should be a suitable distance from the road. The outer side of curves, medians, junction corners and cut slopes are the most vulnerable locations in this regard. Trees should not be planted within 14 m of the centre line of the traffic lane to provide recovery areas for vehicles that run off the road.

The second and subsequent rows of planting should be 3 m away from the first row, as per the availability of land. Growth of vegetation close to a bend may lead to obstructions and cause accidents. Tall and overgrown plants on and near the bend must be removed. On the innermost lane of the carriageway, a stopping sight distance of 170 m at 100 kph must be ensured. Planting must not be done in highway medians up to 1.5 m wide, as this leaves little room for maintenance.

9.3 Use of Compressed Natural Gas (CNG) As Fuel

CNG has many advantages compared to conventional petroleum products. For fleet vehicles, waste disposal lorries and delivery vehicles, CNG is the smart and affordable choice. There are savings to be had in terms of fuel consumption cost, reduced emission levels and extended life of the vehicle.

Reduced Fuel Cost

CNG offers almost 50 per cent savings over conventional products such as petrol and diesel fuel. In the recent past, the average cost per litre of petrol in many countries has risen approximately 10 to 30 per cent. In the present economic climate, there is no better alternative to CNG to fuel vehicles.

Environment Friendly

CNG is one of the cleanest transportation fuels in the market today. Because of its lower carbon content, CNG burns cleaner than any other petroleum-based product. CNG produces the least emissions and pollutants than conventional liquid fuels. Compared to other petroleum products, CNG produces 20–30 per cent fewer greenhouse gas emissions and 95 per cent fewer tailpipe emissions. The CNG fuel systems are completely sealed, and therefore, there are no emissions.

Reduced Maintenance Cost

CNG has no lead content, so there is no fouling, and thus the spark plug life is extended. The crankcase oil is neither diluted nor contaminated by CNG, resulting in extended intervals between oil changes and tune-ups. CNG does not react to the metals, so pipes and mufflers last longer. Maintenance costs are reduced, and the overall life expectancy of the engine is extended.

Performance Advantages

Natural gas has an octane rating of about 127; therefore, CNG is superior to petroleum-based products. Since natural gas is already in a gaseous state, CNG vehicles experience less vapour locking and start well even during severe cold or hot weather conditions.

Safety Benefits

Fuel storage tanks for CNG reduce the likelihood of accidental release because they are stronger and safer than gasoline or diesel tanks. The risk of fire and ground contamination is reduced, as with any accidental release, CNG disperses quickly into the air instead of onto the ground. Additionally, CNG produces very little to no emissions during refuelling.

CNG has a high auto-ignition temperature (540°C) and a narrow range (5–15 per cent) of inflammability, making it less likely to auto-ignite on hot surfaces. This means it will not burn if its concentration in the air is below 5 per cent or above 15 per cent. Such a high ignition temperature and a limited range of flammability ensure that accidental ignition and combustion are very unlikely. CNG is compared to conventional fuels in Table 9.6.

Table 9.6 Comparisons between CNG and other Conventional Fuels

Properties	**Unit**	**Petrol**	**Diesel**	**LPG**	**CNG**
Relative Density	Water = 1	0.74	0.84	0.55	-
Relative Density	Air=1	-	-	1.285	0.64
Auto-ignition					
Temperature	Degree C	360	280	374	540
Flammability Range	per cent in air	1-8	0.6-5.5	2.2-9.0	5-15
Flame Temperature	Degree C	2030	1780	1983	1900
Octane Number	-	87	-	93	127

9.3.1 Background

Liquefied natural gas (LNG) and CNG-based applications were considered due to the growing concern with air pollution from diesel vehicles in the European transportation sector, despite the obstacles and the range of alternative fuels being significant. This transition in the road transport sector [1] seeks to meet the following objectives:

- To include an assessment of the potential impact of bio-methane.
- To assess the present and possible future market for natural gas as a road transport fuel in global, regional and national markets.
- To assess the role of government initiatives in the use of CNG/LNG-fuelled vehicles.

Natural gas vehicles (NGVs) have a significant presence, most notably in China, Iran, India and Italy.

9.3.2 Analysis of global levels of natural gas-fuelled vehicles and fuel usage

All land-based motor vehicles, including original equipment manufacturers' (OEM) vehicles, factory-approved conversions and post-sale conversions, are called NGVs by NGV Global (formerly the International Association for Natural Gas Vehicles). CNG, LNG and bio-methane or renewable natural gas (RNG), which can be gaseous or liquid, are used as fuels. NGV Global provided comprehensive statistics on gas-fuelled vehicles, showing that, in 2018 [2], there were around 26 million NGVs operating worldwide.

The global NGV population has grown from 2000 to 2018, as shown in Figure 9.1. Between 2005 to 2017, the growth of NGVs in the total vehicle population has increased tentatively from 0.5 per cent to 1.5 per cent [3]. The growth is significant, and it is most noticeable in the Asia-Pacific region, followed by Latin America.

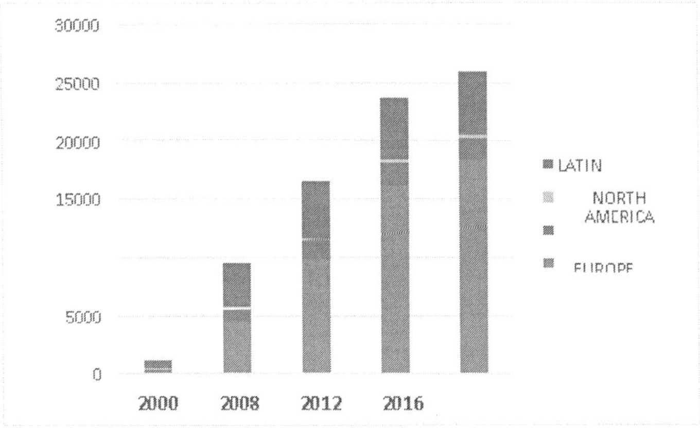

Figure 9.1 Global NGV numbers by region, 2000 to 2018
Source: NGV Global http://www.iangv.org/current-ngv-stats/

The country-wise distribution of NGVs in 2018 is shown in Figure 9.2 and Table 9.7. It is worth noting that the six countries account for about 80 per cent of the total vehicle population, with China, Iran and India accounting for over 50 per cent.

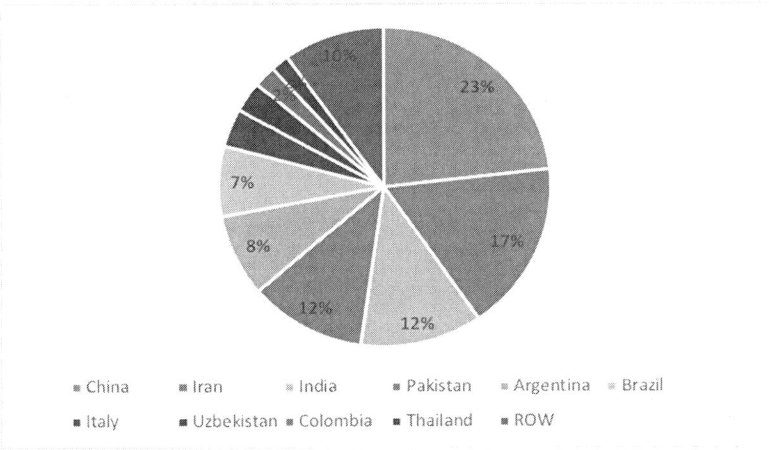

Figure 9.2 NGVs by Country, 2018. Source: IANGV

Table 9.7 Main NGV Countries, 2018

Country	**Number of NGVs**	**NGVs as a percentage of total vehicle population**	**NGVs as a percentage of total NGV population.**
China	6,080,000	3.7 per cent	23.2 per cent
Iran	4,502,000	31.9 per cent	17.2 per cent
India	3,090,139	1.5 per cent	11.8 per cent
Pakistan	3,000,000	14.0 per cent	11.5 per cent
Argentina	2,185,000	9.5 per cent	8.4 per cent
Brazil	1,859,300	2.3 per cent	7.1 per cent
Italy	1,004,982	2.0 per cent	3.8 per cent
Uzbekistan	815,000	40.8 per cent	3.1 per cent
Colombia	571,668	4.7 per cent	2.2 per cent
Thailand	474,486	1.5 per cent	1.8 per cent
Total	23,582,575	n/a	90.1 per cent
ROW	2,580,989	-	9.9 per cent
World total	26,163,564	1.52 per cent	100.00 per cent

Source: IANGV

These numbers should, however, be treated with some caution.

- Methods of data collection, data consistency and the variations in reporting yearly data vary by country.
- Definitions vary from country to country as to what constitutes an NGV. Some include all small engines, such as agricultural machines, whilst others consider only motorbikes.
- Dual-fuelled vehicles, in practice, may use relatively small volumes of natural gas.
- Some countries do not identify vehicles separately, with LPG and natural gas, and the former has a larger proportion.

Completely reliable data is not available on the types of natural gas vehicles, but it may be ascertained that most of the NGVs are small vehicles. According to the International Energy Agency (IEA) (2017), large lorries and other HGVs that used CNG or LNG accounted for about one per cent of the total stock of NGVs in 2015. Most of these vehicles were to be found in China, the USA and the European Union. The importance of gas-fuelled HGVs is likely to grow as they represent a more prospective market for NGVs than cars in terms of fuel consumption.

The estimated global demand for gas in transport was 130 Bcm in 2017, according to IEA (2018). However, this estimation included energy dissipated in the delivery through pipelines. Analysis of the sector for the highest consuming countries, showing the split between gas used for pipeline transport and road transport, is shown in Table 9.8.

Table 9.8 Gas consumption for transport (IEA definition).
Top ten countries, 2016 (Bcm)

Country	Road transport	Pipelines	Total
Russia	1.1	34.2	35.3
USA	1.8	19.1	20.9
China	20.8	-	20.8
Iran	8.1	-	8.1
Canada	0.1	4.2	4.3
Argentina	2.9	1.2	4.1
India	3.0	-	3.0
Thailand	2.9	-	2.9
Brazil	1.8	1.1	2.9
Turkmenistan	2.0		2.0

Source: IEA

The demand for gas since 2000 is shown in Table 9.9, illustrating the increasing role of gas in road transport.

Table 9.9 IEA data on Gas consumption for transport, from 2000 to 2016 (Bcm)

Vehicle type	Average consumption in GGE
Class 8 Truck	12,889
Delivery Truck	1,974
Light Truck	683
Taxi	2,813
Car	480

	2000	2005	2010	2015	2016
Road	3.6	11.9	31.8	47.7	48.8
Pipeline transport*	63.3	73.6	71.2	65.4	69.4
Total	66.8	85.5	103.0	113.0	118.2

Source: IEA,
*i.e. gas consumed during delivery through pipelines and compressors, etc.

- Table 9.4 considered the countries listed in both Tables 9.8 and 9.9, showing the annual average consumption per vehicle, by calculating the diesel litres equivalent (DLE) [7] and gallons of gasoline equivalent (GGE) [8]. This is on the assumption that all the gas is used by road vehicles. It indicates a wide range of values for average consumption. Table 9.11 shows the average consumption for all vehicles in the USA in 2017. Table 9.11 shows that when the numbers fall within the expected range of vehicle consumption (except Thailand), the levels of NGV consumption are higher than expected, given the proportion of small vehicles in the total existing NGVs. The possible reasons are anomalies and inaccuracies in the data, high levels of utilisation for smaller NGVs, for example, many NGVs in India are taxis, and so will have a higher usage rate than privately owned vehicles and lower fuel efficiency for NGVs, for example, in China, many NGVs are low-cost conversions of gasoline-fuelled cars and, therefore, relatively inefficient, as well as having higher system losses.

Table 9.10 Gas consumption for transport (IEA definition excluding pipelines).
top ten countries

Country	Number of NGVs	Gas consumption 2016 (Bcm)	Average consumption	
			in DLE	in GGE
China[9]	6,080,000	20.81	3,533	11,743
Iran	4,502,000	8.08	1,853	6,158
India	3,090,139	2.97	993	3,300
Pakistan	3,000,000	1.96	676	2,246
Argentina	2,185,000	2.86	1,351	4,490
Brazil	1,859,300	1.79	994	3,303
Italy	1,004,982	1.35	1,386	4,608
Uzbekistan	815,000	0.10	127	421
Colombia	571,668	0.71	1,285	4,272
Thailand	474,486	2.88	6,260	20,809

Source: IANGV, IEA and author's calculations

Table 9.11 Annual average fuel consumption by vehicle type in the USA 2017
Source: www.afdc.energy.gov/data/

In summary, although presently the uses are limited in a small number of countries, the number of NGVs has been growing globally. It is also noted that to date, most users of NGVs are only cars and other light vehicles, and just one per cent of HGVs are NGV users. However, these past users and the consumption levels may not remain the same soon, and the HGV sector is expected to be the most promising for future natural gas users.

9.3.3 the Main Advantages and Challenges of NG As a Transport Fuel

The advantages and challenges of natural gas over other transportation fuels are:

- Environmental.
- Financial.
- Availability of the vehicle and fuelling.

The role of bio-methane as a transport fuel has evolved significantly and may be considered separately.

9.3.4 The Environmental Advantages of Natural Gas as a Transport Fuel

The reduction of the adverse impact on the environment by road traffic has been a significant policy feature of governments of various countries for many years. There is a wide range of measures in place around the world to establish legislation to improve fuel efficiency, reduce CO_2 and limit atmospheric pollution, and these have been introduced by a few countries and/or cities. Fuel efficiency measures for smaller vehicles are the main emphasis; however, HGVs are also increasingly under consideration. Vehicle manufacturers also undertake necessary actions through improved engine and vehicle design. Reducing sulphur emissions is clear-cut, but using natural gas vs conventional fuels, the advantages are still not clear.

Based on emissions at the two stages, well-to-tank (WTT) and tank-to-wheel (TTW), combined as well-to-wheel (WTW), the environmental impacts of vehicle fuels are considered.

Natural gas generally has better environmental performance than diesel and petrol, but this is an area of continuing debate and is not agreed upon by all studies. Some studies have suggested that methane leakage might increase the overall level of emissions and have a much higher global warming potential than CO_2 (Le Fevre 2017). Additional emissions arise both from losses in the upstream and at the combustion stage, escaping into the atmosphere, referred to as 'methane slip'.

9.3.5 Cars

Environmental performance in terms of emissions of CO_2 equivalent per unit of distance travelled is the most usual measure. Research done by the European Commission's Joint Research Centre (2014) and a more recent study by Thinkstep (2017) are summarised in Table 9.12. The Natural Gas Vehicles Association (NGVA) funded the latter study and provided detailed estimates of the impact of methane emissions. Methane emissions are estimated to be 27 per cent of the WTT emissions and six to eight per cent of total WTW greenhouse gas (GHG) emissions for a CNG vehicle [11].

The designs of engines are adapted specifically for natural gas as fuel, but dual-fuel engines are still likely to be susceptible to some degree of methane slip. It is therefore expected that the overall amount of methane emissions should be reduced. The estimated methane emission is 0.05–0.2 per cent during fuel dispensing, as found by the same study.

Table 9.12: Well-to-wheel GHG emissions for cars using different fuels in CO_2 eq/km

Mode	**Study**	**Petrol**	**Diesel**	**LPG**	**CNG**	**LNG**
Car	JEC	125-185	115-145	154-160	137-185	-
Car	Thinkstep	169	140	-	131	-
VW Golf	T&E	130	132	-	123-141	-
Car 2020	JEC	83-132	79-106	106-110	94-122	-
HGV	Thinkstep	-	1074		908	912

Source: Thinkstep (2017). Note figures for LNG HGVs are based on HPDI engines

9.3.6 HGVs

Greater prospects for natural gas, both in terms of growth rates and volumes consumed, were likely to be with HVGs, where LNG was likely to feature as an option; this was noted in the paper published by the OIES (Le Fevre 2014). Although several recent studies have started to emerge, it has been hard to measure the environmental impact of different fuels in the HGV sector under real-world conditions. In Table 9.13, the Thinkstep (2017) analysis of HGV emissions is shown, which included a blend of natural gas with bio-methane.

In Figure 9.3, a summary is given of a more recent study by the Sustainable Gas Institute at Imperial College (SGI) (Piers et al, 2019). Both studies agree that emissions from natural gas-fuelled lorries could be around 15 per cent lower than diesel lorries. However, it is noted in the SGI report that, in some circumstances, such as dual-fuel lorries used for urban deliveries, GHG emissions could be greater.

From an atmospheric pollution standpoint, the natural gas in HGVs is more attractive, though new diesel engines are being enforced and designed to meet the Euro VI fuel standard [12], and these are narrowing the gap between the two options. The new engines are achieving much lower particulate emissions and nitrogen oxides (NOx) emissions than previously, as stated by Grigoratos et al (2019).

Table 9.13 Well-to-wheel GHG emissions for HGVs using different fuels in CO_2 eq/km

Mode	**Diesel**	**CNG**	**LNG**	**80 per cent CNG + 20 per cent SNG**	**80 per cent LNG + 20 per cent SLNG**
HGV	1074	908	912	738	749

Source: Thinkstep (2017). Note figures for LNG are based on HPDI engines

Overall, natural gas has some environmental advantages over other fossil fuels, and this has led to some preferential support from governments and attracted subsidies for infrastructure, vehicle purchases and lower duty on fuel. Although it has been questioned whether these moves represent good value for money [15], and because natural gas does not provide a clear path to full decarbonisation and existing fuels have improved in terms of both lower CO_2 and reduced atmospheric emissions, the environmental case for conventional (i.e. fossil) natural gas in transportation is not overwhelming.

To date, due to the costs of the batteries and their size, which have limited the effective payload, electric vehicle applications in the HGV sector have been limited. This has enhanced the longer-term prospects for gas as a fuel. Although mostly for local use due to range restrictions, there is growing interest in electric freight vehicles [14]. In the past ten years, lithium battery costs have fallen by 80 per cent. Gas use in the HGV sector and improvements in battery technology could be a significant game-changer.

9.3.7 the Economic Case for Natural Gas As Transport Fuel

The trade-off between the slightly higher capital cost of a CNG or LNG vehicle and the price differential with petrol or diesel may make natural gas more economically viable.

For HGVs, CNG provides up to 1600 km range compared with 500–1000 km for CNG [16]. Vehicle configuration [17] determines the choice between LNG and CNG. The density and configuration of gas pipeline networks and the extent of LNG availability determine the difference in refuelling costs. In China and the USA, the preferred fuels are LNG and CNG, respectively.

9.3.8 Vehicle and Fuelling Availability

The relatively small number of gas-fuelled vehicles and the limited number of refuelling stations imposed constraints on the rate of uptake of NGVs in the past.

In some countries, NGV cars are still not widely available, but in the HGV sector, good progress has been made. There were operational problems together with significant methane slip in early versions of dual-fuelled lorries, and they were not a great success. Manufacturers such as Volvo, Scania and Iveco, which have all introduced new gas HGVs in 2017/18, are now much more widely available from other manufacturers in Europe and smaller commercial vehicles are also becoming available. CNG minibus vehicles powered by IVECO's 3.0-litre, 140 hp engine [21] are operated by Blue Bus Innovations in the UK.

To summarise, cheap gas in relation to other fuels is required for gas in road transport—this can be achieved either by an oversupplied market, as in the USA, or state subsidies (including higher duties on competing fuels), as is the case in most other countries. The environmental case is most compelling for bio-methane.

9.4 EVs (Electrical Vehicles) Integrated with Renewable Energy Sources for Sustainable Mobility in Green Highways Vision

Governments in various countries across the world have been tackling the problem of air-polluting emissions by committing significant resources to improving air quality. Achieving the goal of reducing air pollution will require both the private and public sectors to invest in clean energy technology. Transitions from conventional houses to smart houses and from conventional vehicles to EVs are also needed. Integrating renewable energy sources (RESs) such as solar photovoltaics, wind energy systems and diverse varieties of bio-energies will be necessary. Additionally, the scope for decarbonisation within the transportation sector itself shall be there. Paradoxically, it seems to happen that the same transportation sector might also present an opportunity for speedy decarbonisation. Statistics show that the transportation sector is responsible for 14 per cent of global GHG emissions. However, plug-in electric vehicles (PEVs) are a viable clean technology, amongst the numerous options for public transportation, vehicle light-weighting, start-stop trains, improved engine technology, fuel substitution and production improvement, hydrogen, power-to-gas, and natural gas heavy fleets to reduce transportation emissions. This section examines the feasibility of significantly reducing GHG emissions by integrating EVs with RESs for sustainable mobility, with studies available from the European Union (EU).

Mainly, the transport sector is responsible for air pollution in European cities, as it produces almost a quarter of all GHG emissions. Since 2007, there has been a decrease in emissions, but they remain higher than in 1990. Road transport was considered responsible for more than 70 per cent of GHG emissions in 2014, as can be observed in Figure 9.3.

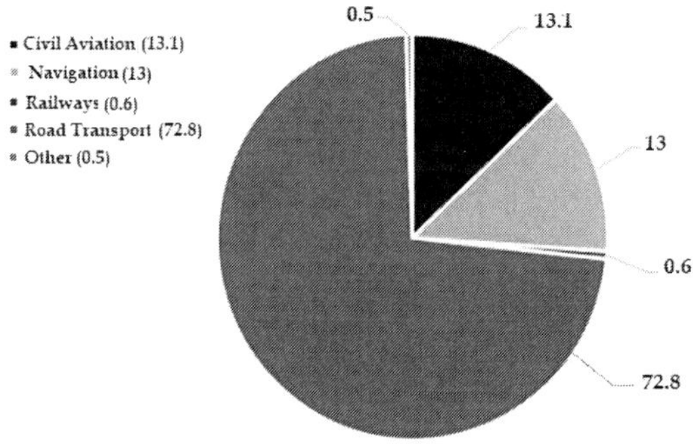

Figure 9.3 GHG emissions from transport by mode in 2014 [4].

The Paris Agreement was adopted by 196 countries on 12 December 2015, at the UN climate change conference in Paris. This was an additional international measure to increase efforts to address climate change through a reduction of global GHG emissions and restriction of the global average temperature rise below 2°C. Based on the understanding that 2°C is the maximum allowable emissions threshold, after which irreversible climate harm would have occurred, this was agreed by the Intergovernmental Panel on Climate Change (IPCC). To resolutely instigate lower GHG emissions to prevent unsustainable climate conditions from occurring [1, 2, 3], all efforts must consequently be on deck. Despite each country having different sectors contributing to its respective GHG emission profiles, each is anticipated to accomplish its obligation by exerting the relevant efforts.

The global shift towards a low-carbon and circular economy has been considered by the European Commission since 2016, which has adopted a low-emission mobility strategy. The European strategy for the transport sector includes an irreversible shift to low-emission mobility, as it is mandatory to reduce air pollutants, which damage health.

GHG from transport was expected to be at least 60 per cent lower than in 1990 and be firmly on the path towards zero. The strategy integrated a broader set of measures to support Europe's transition to a low-carbon economy and to jobs, growth, investment and innovation. The quality of the air, reducing the levels of noise, lowering the levels of congestion and ameliorating the safety [4] will be improved by the strategy.

The measures considered to accomplish air purification have to include:

- Technology to create affordable vehicles thus promoting a decrease in emissions.
- A revolutionary shift from internal combustion engines to alternative sources of energy for transportation.
- Lower emission levels and using fuels such as hydrogen.
- Using innovative biofuels and renewable synthetic fuels, and electricity should be encouraged and accelerated.
- Increased use of low- and zero-emission vehicles.

Consistent support from local authorities is a critical determinant of the success of these strategies. To use vehicles with low emissions from alternative energy sources, these authorities may offer incentives to people and may also encourage the use of other methods of travel, including cycling and walking, public transport and car-sharing/pooling arrangements, which effectually decrease polluting emissions.

However, with respect to the interaction and integration of EVs with renewable RESs, the power grid, the charging infrastructure and the strategies to decrease air pollution, there is a lack of systematic studies. This section of the chapter seeks to investigate the integration of EVs with RESs for sustainable mobility in greatly reducing GHG emissions.

9.4.1 Road Transport and the Menace of Harmful Emissions

About one-fifth of the EU's overall emissions of CO_2, the principal component of GHG, is generated by road transportation. Although these emissions were reduced by 3.3 per cent in 2012, the emissions are still about 20.5 per cent greater than those generated in 1990. Transport is the main sector in the EU where GHG emissions are still increasing. If Europe transforms its road transport sector, it can achieve its long-term transition to a low-carbon economy. Future air pollutant emissions, including GHGs from road transport, can be diminished if EVs are powered with electricity from RESs.

In road transport in the EU, lorries and buses account for 25 per cent of CO_2 emissions and for approximately six per cent of total EU emissions. Despite some improvement in fuel consumption efficiency in recent years, these emissions are continually growing due to the increasing movement of road cargo. The European Commission is currently developing an elaborate plan to decrease CO_2 emissions from heavy-duty vehicles to mitigate these issues.

In EVs, there is no tailpipe CO_2 emission and other pollutants like nitrogen oxides (NOx), non-methane hydrocarbons (NMHC) and particulate matter (PM); they are nearly silent and do not produce any vibration. Technological optimisation and market development are targets for future optimisation of EVs. The main area for technology is the reliability and durability of batteries and super capacitors, reducing battery weight and volume, improving their safety and reducing their cost. Improving hybrid electric power trains, charging infrastructure and plug-in solutions are the other technological challenges.

It is debatable whether the higher number of EVs will result in additional power production to meet the increased electricity demand, but any extra electrical energy needs must be incorporated into the infrastructural system.

Currently, it is impossible to tell how much electricity will be required and what type of production can be sufficient to cover the demand. However, once certainty has been attained, the demand for electricity generation is likely to have an enormous impact on the overall power system. It is estimated that the electricity consumption needed by an 80 per cent share of EVs in 2050 will differ between 3 and 25 per cent of total electricity demand across the member states of the EU.

States with high solar energy production capacity have a charging peak during the day, and will have to adopt different power management strategies from countries which only depend on wind, or combined solar and wind electricity production.

Related to increasing the number of EVs, the main benefit is the significant reduction in direct emissions of CO_2 as well as air pollutants from road transport. Nonetheless, these positive impacts could be partially offset due to additional emissions caused by increased amounts of electricity production as well as continued use of fossil fuels in the power industry. This means that lower emissions of CO_2 and air pollutants, determined by a substantial increase in EVs, could cause higher emissions by electricity generation when it is based on fossil fuel (i.e., coal in thermal power plants) combustion and when there is no reduction in electricity demand in other sectors.

9.4.2 EV Integration with Renewable Energy Sources (RESs)

High standards for power balance control and quality in several regions in Europe have been initiated by the growing wind power and solar photovoltaic (PV) installations [Ref. 13, 14, 15].

Large offshore wind farms tend to have high-power capacity at a single location. Notably, the magnitude of the power fluctuation can be high due to wind speed variations. For instance, the 10-minute average wind power profile at the 160 MW wind farm, Horns Rev, Denmark, is depicted. Within an 11-day period, the wind power generation differs from zero to approximately 100 per cent production. Wind power fluctuations are visible at different times, such as short (intra-hour) and long (several hours) [Ref. 6, 16, 17, 18, 19, 20, 21, 22, 23, 24, 25, 26, 27, 28]. Furthermore, wind power and PV are known as non-dispatchable energy sources since the active power production is variable over time [Ref. 6, 17, 18, 19, 20, 21, 22, 23, 24, 25, 26, 27, 28, 29]. The transmission system operator (TSO) plays a significant role in ensuring the balance between consumption and production, including during the intra-hour time.

9.4.3 EV Integration with Wind Energy

The provision of ancillary services using EVs in power systems together with wind power works well [18]. For high-value services, including ancillary services that often reduce the operating cost to EV owners in the short-term period, EVs should be incorporated, according to Kempton et al. [21, 22]. Pillai and Bak-Jensen [23] examined the benefits of ancillary services provisions by EVs in the western Danish power system. The load frequency control (LFC), at the disposition of secondary reserves, which is assessed through simulation models, was mainly checked. It needs to be explained how EVs can efficiently control power mismatch resulting from the variability of wind power, therefore eliminating the use of conventional power plants.

A similar idea was developed where large amounts of clustered EVs, as well as household apparatus, can be used to offer secondary reserves such as LFC in the power system (Galus et al.) [24].

9.4.4 EV Integration with Solar Energy

Research on the use of solar power through EVs is significantly diversified compared to various studies focusing on wind power and EVs. It is substantially possible to generate electricity from PV at both medium and low voltage levels within the power systems. The utilisation of EVs in the process of distributing grids using PV is considered as an alternative to storing energy, instead of controllable loads, vis-à-vis the suggestions made by various scholars, and is evident from the analysis conducted by Bessa and Matos [29].

Research has revealed that solar power can be easily stored in car batteries for future use at night, by charging during the daytime when solar radiation is at its peak. People understand the significance of maximising the cost of EVs throughout the irradiation period through the idea of a green charge [29]. Additionally, EVs can be charged during the day at parking areas situated, for instance, at work, which is a concept introduced by Birnie [30].

Despite having high penetration capacity from PV, the grids may also have lower voltages. In such situations, the primary constraint is linked to variations in voltage magnitudes along the feeders. These events are likely to occur regularly, but mainly in residential areas that have a high concentration of rooftop PVs. Conversely, various alternatives to mitigate voltage capacity were examined by many studies, for instance,

- Grid reinforcement [32],
- Approaches to reactive power control [34, 35, 36, 37, 38],
- Harmonised active power curtailment [33], and
- Permanent storage of energy [39].

Research has shown that the application of coordinated EV loads within the feeders using high PV penetration has not been satisfactorily analysed.

9.4.5 Electrical Network Integration with the Management of Distribution Grid

Significant changes in the energy, building and transport sectors led by a more sustainable energy system are needed. The development of distributed (or diffused) generation (DG) is the greatest change in the energy sector. Power plants with nominal power lower than 10 MW are connected to the distribution network. DG plants exploit primary energy sources, which are renewable in most cases. These are distributed on the territory and are hence called distributed generation, as they could not otherwise be exploited in a traditional centralised plant. They supply local loads, and they can be operated in a cogenerative mode. In an urban district, PV panels and solar collectors mounted on top of buildings are DGs.

High investment costs arise since, being medium or small plants, economies of scale cannot be applied, and this is one of the main drawbacks of DG. The real problem is the difficulty in predicting and controlling the power produced and sent to the distribution network. So, in the management of the grid to shift to a new paradigm and to ensure passage to a smart grid, the main driver is the distributed generation, together with other distributed energy resources such as EVs and energy storage.

A smart grid is a modern electric power grid infrastructure that:

- Guarantees the reliability of the system and the security of supply.
- Resolves problems related to the power generation from RESs and controls the load.
- Promotes energy efficiency.
- Involves the passive end users.

To implement that, integration with information and communication technology (ICT) is needed.

To ensure that, despite any possible disturbance, the balance is maintained, from load fluctuations to faults determining the unavailability of some grid elements, the electrical network is designed. In the paragraph below, the structure of the electrical grid is investigated.

Power generation occurs mainly in big, centralised power plants, within a traditional network configuration. The energy produced is:

- Put on the transmission network at 'high voltage' (HV network, 132–220–400 kV).
- Then it is transformed into 'medium voltage' (MV network, 15–20–23 kV) through primary substations.
- Eventually, it is converted in secondary substations into 'low voltage' (LV network, 230–400 V) and distributed to the end users.

Whilst most industrial users are connected to the MV network, domestic and commercial users are generally connected to the LV network.

To transfer bulk power from major generators to areas of demand, the HV grid is utilised, which has a network structure to ensure different alternative paths for the power flow in case some of its elements are unavailable owing to a fault. For example, in Italy, the transmission network has reached high automation levels, leading to good reliability and security of supply. The transmission grid is ruled by the TSO, which is 'Terna' in Italy.

The MV and LV grids, on the other hand, have a radial structure. Even if in MV grids, different possible paths are available for the power flow, these grids are operated in a radial way. The distribution system operator (DSO) manages the distribution networks for MV and LV grids.

Originally, the development of the distribution networks was to transfer power just in one direction: from the substations to the final customer. This model is alright if only loads, except for a very few generators, are connected to distribution networks so that they can be regarded as passive. This model needs to be revisited due to the recent large spreading of distributed generation, mainly from renewable sources. If the installed DG grows, reverse power flow can occur. It means that the power does not flow from the substations to the users but vice versa. Thus, the updating of protection and regulation systems is necessary.

A transition to a new management of the electrical network and of the entire power system, as a new model, is needed, which is known as the smart grid. The aim is to move from a system in which power production is centralised and controllable. At the same time, consumption is completely random, so the responsibility for the balance between generation and consumption is entirely on the production side, in a system in which part of the generation is non-programmable. Still, a controllable portion of consumption can counterbalance this. Therefore, the system is moved from a passive grid to an active one in which there is a bi-directional exchange between producers and users.

MVs can send energy to the grid as 'mobile energy' storage units, whilst EVs expend power. Greater adaptability for the framework to react to supply as well as demand may be meant by the expansion in electric vehicle adoption. The 'vehicle-to-grid' administration to purchasers of the EVs could be offered by the energy providers. For this situation, the yearly expense of charging an electric vehicle is taken care of by reserve funds from electric vehicle-to-grid administrations. A unique charger at home could be introduced by the proprietor of the electric vehicle, and the provider will deal with the electric vehicle's battery. Amid peak times when costs are higher, the energy providers will naturally exchange power from the electric vehicle's battery to the grid, possibly by giving money back to the owner. During off-peak times when costs are lower, the battery can be charged.

To incompletely control charging EVs, the utilities plan is to utilise distributed assets, for example, sustainable power source generation, storage and demand response. Grid innovations could deal with the charging of EVs, such as progressed metering. The gadgets support off-peak charging, which permits charging stations to be incorporated with time-based rates. To help with choices, they enable utilities to examine charging station utilisation and practices.

To adjust requests on the grid and increment accommodation, more EV charging stations are important.

Power utilisation is expanded, and convenience is diminished, conceivably if the nearest charging zone may not be along the driver's way. Unwavering quality whilst charging EVs in an area or work zone could be supported by Microgrids. Solar power and wind, as well as storage, would decrease the strain on our power grid. Additionally, power versatility may be improved by the circling capacity into nearby microgrids.

9.4.6 Electric Vehicle Coordination

In case the process involves a large fleet of vehicles, providing supplementary services from EVs is a very plausible option [40]. An aggregator technology is therefore suggested, which may intermediate between the vehicles, the utility organisations, as well as the energy market. The aggregator framework, alongside other forms of the framework, is demonstrated in Figure 9.4 [41, 42].

The EV aggregator can produce essential signals for coordinating the EV fleet based on the data shared between the energy market, TSO and the DSO.

In an LV grid, as PV penetration increases, controlling the EV load can lead to an improvement in the feeder operation and a decrease in the need to invest in infrastructure upgrades. There are constraints to keep in mind for the grids with high penetration. In respect of dependability, quality and price, customers can evaluate improvements. Aiming to reduce long-term variations in voltage magnitude that arise in a decentralised RES generation context, we expect a major change soon in the voltage quality improvement. EV load coordination can facilitate a balance between production and consumption, which can reduce under- and over-voltages.

To allow for greater RES penetration by energy suppliers, the planning rules may be revised. The various options include [43]: (i) decrease circuit distance by redirecting the power route; (ii) improvement of the upstream transformer volume; (iii) improvement of MV/LV transformers with automatic voltage control; (iv) improvement of the grid circuit thermal capacity, (v) improvement of the grid's rated voltage level; (vi) setting up of additional reactive power compensation; (vii) erection of new substations; and (viii) modifications to the grid architecture.

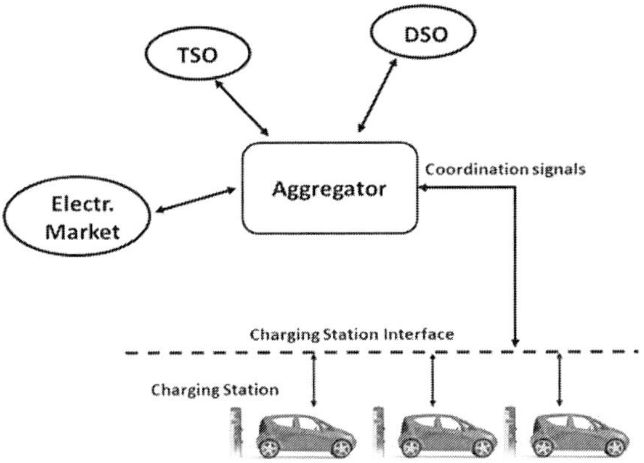

Figure 9.4 Simplified EV coordination framework: Transmission System Operator (TSO), distribution system operator (DSO), electric market and charging stations for EVs.

Automatic power curtailment is recently being looked at within DSOs, if RES-induced voltage variations are to exceed the allowable amounts. To do so, it is required to link all units in a communication infrastructure network, which may not be convenient for owners and might lead to a need for compensation. Extensive research has been done on reactive power methods for LV grids and showed some limitations in the effectiveness of reducing voltage rise [46, 48, 49, 50].

Figure 9.5 An electric vehicle is charging

In view of PV, most PV units previously connected to LV grids work with unit power factor. Additionally, the high R/X ratio (imaginary part divided by the real part of impedance) of LV cables challenges the efficiency of these techniques. The voltage rise reduction in a feeder shall be visible [50] by providing reactive power contribution from all PV units. Besides, active power solutions can also be utilised as an alternative.

9.4.7 Synergies Between EV Needs and PV Distributed Generation

In this subsection, the concept of synergies between dispatchable generation sources and controllable loads, utilising in particular PV distributed generation and EV batteries, is discussed. In the smart grid environment, increasing PV generation would involve storage systems to modulate power injections. In this situation, within smart grids, EVs (V2G) are more advantageous than conventional energy storage (batteries).

With a considerable decarbonisation effect and enhanced resource efficiency, the reduction of pollution can be supported by using RESs to power EVs, as an alternative to conventional fossil fuels. Alternatively, considering sustainability targets, more RES volume is expected, such as a 45 per cent share of renewable energy in the electricity sector by 2030, which is currently just 21 per cent. In Figure 9.6, per unit (p.u.) based graphs of the load demand profile (maximum peak load demand) and PV generation (Watt-peak installed PV capacity) against vehicle mobility pattern profiles (maximum shares of vehicles in motion) are shown. Under unrestricted charging, no interdependence can be drawn between the penetration impact of both EV and PV. In fact, it is anticipated that during the evening rush hour, EV-generated load peaks will occur.

A greater penetration level of DER, along with a mitigation of the intermittent effects of high penetration of PV generation, can be facilitated by more flexibility in EV demand. Transferring the EV demand peak is necessary to prevent bottlenecks and high electricity prices during peak demand, as well as to relieve the excess of PV power generation. Instead of minimising operational expenditures, the objective is principally focused on risk mitigation. The failure to alleviate risk could be costly (e.g. forfeit of the regulator, energy not delivered or damaging the reputation of the utility company).

The plan could encourage effective and harmonious communication among the diverse players within the decentralised energy management system (DEMS) domains. For the smart grid and the deregulated energy markets, it is anticipated that a variety of players will exploit different equipment suppliers. These performers are required to be effectively interconnected and synchronised, resulting in significant execution costs due to the sheer number of procedures and equipment, functions or applications to be integrated.

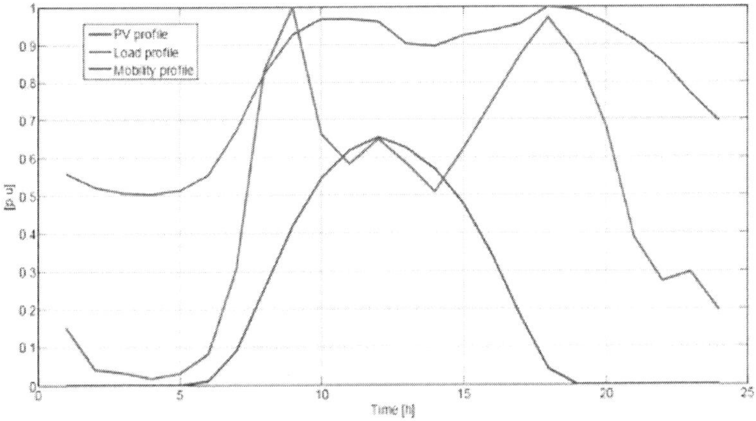

Figure 9.6 Load demand profile against typical EV load and PV generation profiles as a function of time of day.

9.4.8 Conclusion

For green and sustainable transportation systems, EVs represent one of the most promising technologies. A high penetration of EVs will have positive effects and benefits, such as:

- Less reliance on fossil fuels.
- Significant reduction of GHG and toxic pollutant emissions.
- Integration of renewable energy into existing electric grids.

The latest advances related to the interaction and integration of EVs with RESs, such as wind energy, solar photovoltaic panels and EVs, will significantly reduce air pollution, and are reviewed in this section with a detailed discussion on some key concerns and possible solutions. The full contribution by governments, power utilities, EV and aggregators manufacturers, policymakers and owners is required for the successful implementation of the coupled EV-RES technology.

9.5 Using Recycled Materials in Eco-friendly Green Highway Construction

New uses for materials like blast furnace and steel slags, carpet fibres, glass, coal ash byproducts, municipal solid waste, combustion ash, plastic, roofing shingle waste and rubber tyres are coming out of the research on waste byproducts. The use of many of these materials by the highway construction industry can relieve some of the burden of disposal and provide inexpensive construction products. This section provides a summary of their known uses as waste applications by describing the research being conducted on the materials. The research has been mostly conducted in the laboratories and must now be systematically applied in highway construction to determine viability and long-term performance.

Increasing amounts and types of waste are being generated as the world population grows. Various waste will remain in the environment for hundreds, perhaps thousands of years, of those that are produced today. A waste disposal crisis has resulted from non-degradable waste materials, combined with growing consumerism. Recycling waste into useful products is one solution to this crisis.

Research is continually advancing into new and innovative uses of waste materials. Concerning the feasibility, environmental suitability and performance of recycled products in highway construction, a wide variety of studies and research projects are being completed or are in the process of being completed by many highway agencies, private organisations and individuals [1]. Society's need for safe and economic disposal of waste materials is verified in respect of the highway industry's need for better and more cost-effective construction materials by these studies.

Current research on those waste materials that have shown promise as a substitute for conventional materials is summarised in this section (See Table 9.14). Rather than on more commonly followed practices, it primarily focuses on new and innovative uses for waste materials and byproducts by the highway industry.

Table 9.14 Summary of known uses of waste

Waste Material	**Annual Rates in Millions of Metric Tonnes**		**Current and Past Highway Uses**				
	Produced	**Recycled Reused**	**Asphalt Pavement**	**Concrete Pavement**	**Base Course**	**Embankment**	**Other**
Blast Furnace Slag	Not known	14.1	Accepted use as an aggregate in base and surface (friction) coarse, research indicates good performance	Accepted use as a cement additive in granulated form; research is ongoing	Accepted use, good, hard, durable aggregate	Limited but accepted use,	Research in Roller-Compacted concrete, accepted as ice control abrasive
Carpet Fibre Waste	2	Not known	Experimental stages in hot mix asphalt (HMA) and SMA have no field data	Experimental Stages, no field data	No known use	No known use	No known use

Coal Combustion Byproducts							
Coal Fly Ash	45	11	Past use as a mineral filler, research ongoing	Accepted use research ongoing	Used in soil stabilisation	Used in Flowable fills, embankment	Used in all types of PCC
Coal Bottom Ash or Bottom Slag	16	5	Combined ash as a fine aggregate, performance data limited	Use unknown	Use unknown	Used as a sub-base material, embankment	Lightweight concrete, abrasives
Flue Gas Desulfurisation Waste	18	-	Use unknown	Used as a set retarder	Used with cement in soil stabilisation	Used as an embankment material	No known use
Glass	12	2.4	Accepted use, long-term performance research under way	Past research indicated performance problems	Used in dense and open-graded bases	Some research projects under way	Limited use as a paint bead, pipe backfill
Mill Tailings	432	<1 per cent	Accepted use, research indicates good performance	Limited but accepted use	Use unknown	Accepted use	Use unknown

Waste Material	**Annual Rates in Millions of Metric Tonnes**		**Current and Past Highway Uses**				
	Produced	**Recycled Reused**	**Asphalt Pavement**	**Concrete Pavement**	**Base Course**	**Embankment**	**Other**
Coal Combustion Byproducts							
Municipal Waste Combustion Ash	7.3	<10 per cent	Past research indicated good performance - Environmental	No known use	Used in cement-stabilised bases	Used in soil stabilisation	No known use

			questions				
Plastic	14.7	0.3	Used as a binder additive	Experimental stage	No known use	No known use	Used as fence or delineator posts, guardrail block-outs
Reclaimed Concrete Pavement	3	-	Limited use, long-term performance research under way	Limited use, research under way	Accepted use	Accepted use	Used as rip-rap
Reclaimed Asphalt Pavement	91	73	Variety of accepted uses	Experimental stages	Accepted use	Accepted use	Used as shoulder material
Roofing Shingle Waste Industry-Produced Re-roofing Waste	0.4 7.7	<1 per cent	Limite use, d research under way	No known use	No known use	No known use	Used as a pothole patching material
Scrap Tyres	2.3	0.4	Accepted use, extensive research being conducted	Experimental Stages	Used as an insulator	Used with some success research continuing	Being marketed for noise-reduction or retaining wall, moulded posts, many minor uses
Steel Slag	7.5	6.9	Past research indicates good performance	Extensive research, poor performance	Limited use	Accepted use	Ice control
Waste Rock	954	<1 per cent	Accepted use,	Limited but	Use	Use unknown	No known

			research indicates good performance	accepted use	unknown		use

9.5.1 Blast Furnace Slag

Blast furnace slag is an industrial byproduct of iron. Silicates and alumina-silicates of lime and other bases are the primary contents of the slag [2]. In the United States in 1992, with a value of $99.5 million, the quantity of 14 million metric tonnes (t) of blast furnace slag was sold, as reported by the Bureau of Mines [3]. About 90 per cent of this was air-cooled slag, which is primarily used in road construction. Table 9.15 summarises current uses for blast furnace slag in the United States.

Table 9.15 Air-cooled blast furnace slag sold or used in the United States

Use	**1991**		**1992**	
	tity kt	**Value3 ($ 1,000)**	**tity kt**	**Value3 ($ 1,000)**
Asphaltic Concrete Aggregate	1,634	9,577	2,212	13,270
Concrete Aggregate	1,333	9,683	1,325	9,193
Concrete Products	358	2,333	469	2,863
Fill	855	3,950	1,229	5,328
Glass Manufacture	W	W	W	W
Mineral Wool	449	3,056	745	4,872
Railroad Ballast	221	1,288	177	829
Road Base	5,339	30,282	5,894	31,183
Roofing, built-up & shingles	68	771	70	844
Sewage Treatment	W	W	W	W
Soil Conditioning	W	W	W	W
Other[b]	633	5,455	665	6,238
Total[c]	**10,889**	**66,393**	**12,697**	**74,62**

Source: Cheryl Solomon. Slag-Iron and Steel, annual report, U.S. Department of the Interior, Bureau of Mines, Washington, D.C., 1992.

[a]Value based on the selling price at the plant.
[b]Includes ice control and miscellaneous uses.
[c]Data may not add to totals because of rounding.

W = withheld to avoid disclosing company proprietary data; data are included with "other".

In some situations, blast furnace cement can show equal or improved performance over conventional Portland cement concrete [4]. The characteristics of low heat hydration, good long-term strength gain and high chemical resistance are observed in slag cement [5]. The time needed to attain design strength is increased by using slag cement containing more than 80 per cent granulated blast furnace slag [4].

A recent innovative use study examined the effects of slag on roller-compacted, no-slump, lean concrete mixes [6]. Many road agencies also use air-cooled, non-granulated blast furnace slag as an aggregate in HMA surfaces, as a base or sub-base material, as an embankment material, and in snow and ice control [7] and as an aggregate in open-graded friction courses in asphalt pavements [1, 2, 7, 8].

Figure 9.7 photo showing the ground and graded mixed cullet pile located at an asphalt plant.

9.5.2 Steel Slag

Fused mixtures of oxides and silicates—primarily calcium, iron, unslaked lime and magnesium are contained in steel slag. In the steel-making process, it is a byproduct [1]. Steel slag contains a significant quantity of iron, and the highly compressed void structure of steel slag results in a very dense, hard material. In the United States in 1992, 6.9 million tonnes of steel slag at a total value of $21.9 million were sold (the average selling price for steel slag at the plant was $3.02 per metric tonne) [1].

The use of steel slag as an aggregate in HMA in highway construction is the focus of some current research. In a study by the Pennsylvania Department of Transportation, the bituminous mixtures containing steel slag exhibited high stability, high skid resistance and longer heat retention, resulting in easier compaction [9]. However, its use as an aggregate is not cost-effective due to its high absorption rate.

Research has also been conducted on the use of steel slag blended cement [3]. As indicated by the results, the steel slag has a mineral composition similar to that of ordinary Portland cement clinker; therefore, steel slag could become unstable because of its free calcium oxide.

9.5.3 Plastics

Figure 9.8 (a) Sign board and (b) Noise-reduction walls made of recycled plastics filled with rubber and plastics
(Photo source: Oregon Department of Transportation)

The content of plastics is more than eight per cent of the total weight and about 12 to 20 per cent of the volume of the municipal waste stream in the United States [1, 10, 11], and in 1992, approximately 14.7 million tonnes of plastic were discarded; only 0.3 million tonnes or 2.2 per cent were recycled.

In highway construction, the use of virgin polyethene as an additive to asphaltic concrete is not new; however, two new processes, NOVOPHALTR and PolyphaltR, also use recycled plastic as an asphalt cement additive [1, 10, 11].

The recycled low-density polyethene (LDPE) resin, which is generally obtained from plastic trash and sandwich bags, is used by both latter two processes. In pellet form, the recycled plastic is made and added to asphalt cement at a rate of four to seven per cent by weight of binder (0.25 per cent to 0.50 per cent by weight of total mix) [11,12].

Michigan State University is studying the use of recycled plastic in Portland cement. In this study, to replace 20 to 40 per cent of fine aggregate by volume (7.5 to 15 per cent by total volume) in a lightweight concrete mix, recycled high-density polyethene (HDPE) was used. By using either level of HDPE, compressive strengths were reduced. Overall flexural strengths remained constant, and the impact resistance of concrete, which can be related to flexural toughness, is increased.

Many agencies and private companies have been experimenting with the use of recycled plastic for items such as guardrail posts and block-outs, delineator posts, fence posts, noise barriers, signposts and snow poles in highway construction.

- The Federal Highway Administration has approved the use of a guardrail offset block made of 100 per cent recycled wood and plastic [13]. Although the product's initial cost is currently higher than for conventional block material, it is believed that the post will resist damage and deterioration better than conventional materials, thereby resulting in reduced overall life-cycle cost.
- A Carson City, Nevada, company is marketing a noise-reduction wall that contains recycled rubber tyres and recycled plastics [14]. The wall's shell is made of a pultruded thermosetting composite of polyester and glass, and the fill section is made of ground, recycled plastics and rubber tyres.
- In May 1992, Alberta Transportation and Utilities initiated a research project on the use of recycled plastic fence and guardrail posts [15]. The cost of these plastic fences and guardrail posts was somewhat higher than corresponding wood posts.

9.5.4 Glass

Glass discarded annually makes up approximately seven per cent, or approximately 12 million tonnes, of the total weight of US municipal solid waste. Approximately 20 per cent of this glass is being recycled, primarily for cullet in glass manufacturing [10]. Depending on the types of collection methods used, costs, and public factors, the glass may be used in highway construction. In general, in major metropolitan areas only, the large quantities of waste glass needed for such applications are found [16].

(a) (b)

Figure 9.9 (a) Glass for use in asphalt pavement is generally crushed so that 100 per cent passes through the 9.5 mm sieve with not more than eight per cent through the 75 mic. Sieve.

(b) The standard for crushed glass results in uniform gradation, as can be seen here. Most specifications will allow a limited amount of contaminants. Some plastic pieces can be seen among the glass.

In addition, several agencies are routinely using recycled glass in the manufacturing of glass beads to add to paint used for road marking [6].

9.5.5 Municipal Waste Combustion Ash

Two types of ash, namely fly ash and bottom ash, are produced by controlled combustion of municipal solid waste. Most municipal waste combustion (MWC) ash (80 to 99 per cent) is bottom ash, which typically meets the environmental standards for the toxicity characteristic leaching procedure (TCLP) [2]. However, usually a high percentage of heavy metals exists in the fly ash (e.g., lead and cadmium), and the leachate may not meet some environmental standards [18].

Concerns over the environmental acceptability of MWC ash have severely curtailed the initiation of research on the beneficial uses of MWC ash. The Environmental Protection Agency (EPA) estimates that less than 10 per cent of the MWC ash produced in the United States is being used in a limited number of beneficial projects, using incinerator residue as a partial aggregate substitute in an asphaltic concrete base course and using combined MWC ash as an aggregate in stabilised and unstabilised bases and sub-bases.

9.5.6 Scrap Tyres

Considerable research on crumb-rubber-modified asphalt has been conducted since the 1991 passage of the Intermodal Surface Transportation Efficiency Act. This research has addressed both performance and environmental issues; additional research is examining the use of scrap tyre rubber in other highway-related applications.

- The Carson City, Nev., company that is marketing a noise wall that contains recycled rubber tyres and recycled plastics is also researching the use of rubber tyres in lightweight fill, subgrade insulation, and channel slope protection, as well as an additive to Portland cement concrete pavement.
- The North Carolina Department of Transportation recently conducted a laboratory study on the use of ground scrap tyres in Portland cement concrete [21]. After the scrap tyres were processed to remove loose steel and fibres, they were finely ground. The ground rubber was then substituted for fine aggregate in the mix at increments of 10, 20, and 30 per cent by volume of fine aggregate. Tests conducted to determine compressive and flexural strengths showed that these decreased with increasing amounts of rubber.
- A 1992 project in Richmond, Maine, assessed the effectiveness of using tyre chips as an insulating layer in order to limit frost penetration beneath a gravel-surfaced road that experienced severe deterioration during spring thawing [22]. Thermocouples, resistivity gauges, groundwater monitoring wells, and a weather station were installed to monitor the project. After a year, results indicated that a 152-mm-thick tyre chip layer can reduce frost penetration by up to 40 per cent.
- A Mankato, Minn., company is marketing blocks made from recycled tyres for a variety of uses, including landscaping and retaining walls.
- A company in Pittsburgh, Pa., has developed a process that can convert scrap tyres into a form that can be used as poles or stakes. The process, which requires only that the tyres be split and flattened, rolls the tyres in a spiral fashion to form a nearly solid "log" of reinforced rubber material.

9.5.7 Carpet Fibre Waste

In the United States, the carpet industry produces about one billion square metres of carpet per year. Of this, to replace existing carpet, approximately 70 per cent is used; this results in about 1.2 million tonnes of carpet waste produced annually [23]. This increases the total amount of waste fibres produced by the carpet-making industry to an estimated two million tonnes.

- Research on utilising these waste fibres in both asphalt pavements and Portland concrete is being undertaken: fibres were mixed with concrete in a standard drum mixer at a rate of two per cent by volume to determine the effectiveness of using recycled fibres from old carpet for concrete reinforcement [23]. The results indicated that using two per cent carpet waste in the mix had no appreciable effect on flexural strengths but did markedly decrease 28-day compressive strengths. (See Table 9.16) However, toughness indices (I5 and I20) show that the addition of carpet waste can be effective in increasing the concrete's energy-absorption abilities, which was calculated as recommended in ASTM C 1018. Other research focused on the use of waste nylon fibres to reduce plastic shrinkage cracking [24]. Laboratory tests on a limited number of samples indicated no significant difference in compressive or flexural strength from conventional concrete. Results from field testing indicated that the addition of waste nylon fibres into Portland cement concrete panels can reduce plastic shrinkage cracking by approximately 90 per cent. On the use of waste or recycled fibres in a dense-grade asphalt mix, research has also been conducted [25]. To a polymer-modified, dense-graded mix at a rate of 0.3 per cent by weight of total mix, two waste fibres and two commercially available fibres were added. There is no appreciable difference between commercially available fibres and recycled or waste fibres in the asphalt mixture. The results indicated an increase in asphalt content, which was not due to an increase in film thickness from the addition of the fibres, as anticipated; it was due to displacement of the aggregate matrix. This finding implies that additional asphalt was primarily needed to coat the surface area of the fibres.

Table 9.16 Compressive and Flexural Test Results of Fibre-Reinforced Concrete

Fibre in FRC Mix	per cent	Compressive Strength				Flexure Test at 28 days			
		1 day		28 days		Strength		Toughness	
		MPa	CV per cent	MPa	CV per cent	MPa	CV per cent	I5	I20
Concrete Control	0	20.9	4.9	52.6	3.1	4.6	4.0	1.0	1.0
Fibre Mesh PP	0.5	24.2	3.7	52.2	1.4	4.6	3.1	2.6	6.9
Type I waste fibre	2.0	20.9	5.7	39.70	4.4	4.7	2.3	3.3	7.5
II waste fibre	2.0	18.6	7.6	40.7	7.4	4.4	4.3	3.5	9.8

Source: Youjiang Wang, et al., "Fiber Reinforced Concrete Using Recycled Carpet Industrial Waste and Its Potential Use in Highway Construction," in Symposium Proceedings - Recovery and Effective Reuse of Discarded Materials and Byproducts for Construction of Highway Facilities, October 1993.
CV = coefficient of variance

9.5.8 Roofing Shingle Waste

It is estimated that in the United States, between eight and 12 million tonnes of roofing shingles are manufactured each year. Approximately 65 per cent of these shingles are used for re-roofing, and from five to eight million tonnes of old waste shingles are produced annually [26, 27]. In addition, between 400,000 and 900,000 tonnes of waste are produced annually from the manufacture of roofing shingles.

Several studies have focused on the use of roofing shingle waste as an asphalt pavement material.

- Several projects were conducted on the use of roofing shingles in HMA pavements [27]. Their use in dense-graded mixes indicated that the addition of roofing shingle waste can result in a reduction in optimum neat binder content, enhance the ability to densify under compaction, and increase the plastic strain component in permanent deformation measurements. Cold tensile strengths were also reduced, but the impact on the corresponding strains was dependent on the type of shingle waste and the grade of asphalt cement. This finding could indicate that HMA's potential for thermal cracking could be reduced by adding roofing shingle waste.

- The use of roofing shingle waste in stone matrix asphalt mixes was also studied by adding 10 per cent manufactured roofing shingle waste to the mix, resulting in a 25 to 40 per cent reduction in the required neat binder content.
- Five to 7 per cent asphalt shingles by weight of the mix, ground to a uniform consistency, were added to a drum mix plant [28]. No construction problems and no problems regarding pavement performance were reported.
- Experimenting with asphalt cold-patch material made from old roofing material, the resulting patch material showed only minor signs of distress after 22 months of service. In comparison, conventional cold-patch material generally lasts only three to six months.

9.5.9 Coal Combustion Byproducts

When coal is burned in power plants, two types of ash are produced: coal fly ash and bottom ash. Coal fly ash is the very fine ash carried in the flue gas; bottom ash (or slag) is the larger, heavier particles that fall to the bottom of the hopper after combustion.

The physical and chemical characteristics of these ashes vary depending on the type of coal burned. An additional byproduct of the coal combustion process is produced from coal containing sulphur. When this coal is burned, sulphur dioxide is produced; scrubbers are used to limit the amount of sulphur dioxide released into the atmosphere. The resulting waste of this process is flue gas desulfurisation (FGD) waste. An estimated 18 million t of this waste is produced annually in the United States; 136 million t of it is currently stockpiled [29].

Coal fly ash

In coal fly ash, the primary components are silicon dioxide, aluminium oxide, iron oxide and calcium oxide. Extensive research has been conducted on the use of coal fly ash as a highway construction material. Though most of this research has looked at its use as a mineral admixture to Portland cement concrete, research has also been conducted on a variety of other uses, including in soil stabilisation, roller-compacted concrete, and road base stabilisation.

Laboratory investigations determined that the optimum mix was a composite of 84 per cent dense-graded aggregate, 11 per cent ponded fly ash, and five per cent hydrated lime. A test section 230 m long and 20 cm thick was constructed and overlaid with an asphalt base, binder and surface course. After three years of service, the experimental section outperformed the conventional section, and the amount of rutting was significantly lower in the experimental section than in the control section. Except for minor reflective cracking associated with base shrinkage, the experimental section performed well.

Bottom ash

A similar chemical makeup to fly ash exists in the bottom ash has but it has a much coarser gradation. On its use as sub-base material, studies showed that it had sufficient properties to perform adequately [32]. Bottom ash has also been used as an aggregate for lightweight concrete; coal boiler slag has been used as an abrasive in pavement deicing products and as a sandblasting abrasive.

Combined ash

Fly ash and bottom ash, when placed in landfills, are generally combined. Consequently, most current research has focused on the use of combined ash. The physical properties of combined ash, including gradation, specific gravity, and loss on ignition, can vary considerably depending on the type of plant and source of coal. Chemical properties, however, are similar to those found in typical fly ash. The study examined the effects of the ash on indirect tensile strengths and tensile strength ratios of asphalt concrete mixes and indicated that the addition of coal ash at six and eight per cent by weight of aggregate decreased the 24-hour tensile strengths of Marshall specimens compared to the control mix. Several agencies are also conducting research into the use of combined ash as an embankment material.

9.5.10 Flue gas desulfurisation (FGD) waste

On the use of FGD waste, research focused on stabilised road bases and as an embankment material. The research was done by placing two 91.4 m experimental sections containing FGD waste stabilised with seven per cent by dry weight of high early strength, high sulfate-resistant Portland cement. To date, no distress related to the FGD waste has been identified in either pavement section. It was also found that the strength of the cement-stabilised FGD increased when mixed with coal bottom ash. Additionally, surface water and soil leachate were analysed for both sections; the material constituents were compared with EPA drinking water standards and TCLP concentrations. The results showed that none of the EPA heavy metal concentrations were exceeded. However, the drinking water standards were exceeded for sulfates; TCLP standards do not contain values for sulfate levels.

9.5.11 Conclusion

The problems of environmentally safe and efficient disposal of waste continue to grow. The existing landfills in many areas are getting filled up, and have made the establishment of new landfills very difficult. As the waste accepted at municipal solid waste landfills is becoming more restricted, the cost of disposal continues to increase. One answer to all these problems lies in the ability of society to develop beneficial recycling and use for these waste products.

The highway construction industry can effectively use large quantities of this waste. The use of waste byproducts in lieu of conventional materials, for instance, would relieve some of the burden associated with disposal and may provide an inexpensive and advantageous construction product. Current research on the beneficial use of waste byproducts, such as highway construction materials, has identified several promising uses for these materials. Some of these materials are discussed above, which include:

- Blast furnace and steel slag.
- Recycled plastic.
- Rubber tyres.
- Glass.
- Municipal solid combustion ash.
- Carpet fibres.
- Roofing shingle wastes.
- Coal ash byproducts, including fly ash, bottom ash, and FGD waste.

These studies have been conducted primarily in the laboratory. In future, these ideas may be put into action by initiating a systematic programme to determine the viability and long-term performance of these materials in actual highway construction projects.

Reference

Credit for using some figures/text goes to: Ministry of Road Transport and Highways on "Green Highways", Government of India.

Webpage:
https://morth.nic.in/en/green-highways

Appendices and References

Appendix-A

This lists some of the important design software, websites and useful documents, guidelines, and manuals that exist in industrialised countries that would be worth acquiring by road safety professionals in developing countries of the Asian and Pacific region. The Road Engineering Association of Asia and Australasia (REAAA) guide (Reference 7) provides an excellent overview of other road safety resources available from around the world.

Refer to the software for designing and producing sophisticated detail design reports and construction CAD drawings to build safe roads.

Download the software containing tutorials/videos from the website www.techsoftglobal.com,

HEADS Pro is the largest software system for highway engineering and design with all standard and special features for uniform and multiple section highway widening and strengthening, hill roads, low cost rural roads, tunnels geometric design, pavement design/rehabilitation, profile correction, grade separated interchanges, at-grade intersections with type selection and design, swept path analysis, car racing track design, bituminous/asphalt mix design, highway and hill road drainage design, highway lighting design, cad construction drawings, bill of quantities etc. to design and build safe roads. HEADS Pro has CAD-based highway engineering and design. HEADS Pro for the design of roads is available with a very high technological standard and at a very reasonable price for carrying out various work related to highway detail engineering projects. Most of the current projects across the world are widening highways from 2/4 lane to 4/6/8 lane expressways, where the existing road is strengthened with pavement overlay and widened with new construction on the side. In PPP-BOT/EPC mode projects, there are requirements for route realignments, bypasses, vehicular/pedestrian underpasses, elevated flyovers, service roads, bus bays, truck lay-bys, railway bridges, box pushed rail/road underpasses, toll plazas, etc. These make the design quite complex, and with various software products, these are done in parts, while Heads Pro does everything at once, ensuring the highest level of road safety.

Accuracy in construction plays a key role in road users' safety. HEADS Pro provides the most advanced technology solution in satellite-based construction. HEADS Pro enables the GPS-based motor graders, dozers and pavers to perform faster by eliminating a series of survey checks for level verifications (Refer to page 32 of HEADS Pro Technical Specifications, downloadable from the website mentioned above). HEADS Pro Highway Design is presented by 3D Drive Through by UC-win/Road to check visibility in cut areas, horizontal and vertical curves at various speeds at https://youtu.be/2mAyBM8MYB8

ASTRA Pro bridge design software includes design of RCC 'T' girder bridges, composite bridges with steel plate/box girder and RCC deck slab in straight or curved layout with single and multi-span, PSC 'I' girder bridges, PSC box girder bridges, continuous PSC box girder bridges, steel truss bridges, arch suspension bridges, rope cable suspension bridge, cable stayed & extradosed bridges for river bridges with non-linear stage analysis, flyovers, rail/road over bridges, vehicular/pedestrian rail/road underpasses as RCC Box, RCC box culverts (single cell & multi cell), RCC slab culverts, NP3/NP4 pipe culverts, hydrology, RCC pier, wall & box types abutments with counter-fort earth retaining walls, and analyzing span lodged and dislodged conditions, seismic analysis, well, pile and open foundations, bridge bearings, reinforced earth walls (BS 8006), tunnel support structure and RCC lining based on RMR geological mapping, estimation etc.

TransPlan is a powerful software system for traffic analysis and transport planning. TransPlan features signalised intersection capacity analysis as per the *Highway Capacity Manual* (HCM) by the Transportation Research Board, traffic analysis to obtain peak hour factor, seasonal variation factor, traffic homogeneous sections, tollable traffic, ADT, AADT, transport demand elasticity, traffic projections, capacity and level of service (LOS), equivalent standard axle load (ESAL/MSA), economic analysis and financial analysis for a highway project.

Cloud-based process/storage/sharing facilities are available with this software as optional. On procuring multiple licenses, say, four licenses, and installing at four different locations anywhere in the country/world, the project name, username and password may be shared with those four license locations. They can access and share the cloud server to view the design and drawings, modify, re-process, download, and upload the work to the cloud storage 'Techsoft Drive'.

Important documents

1. Guidelines for Accident Reduction and Prevention (International Edition). Institution of Highways and Transportation. 1990. Available (in English) from Institution of Highways and Transportation, 6 Endsleigh Street, London WC1H 0DZ, United Kingdom Tel: (44171) 3872525,Fax:(44171)3872808
2. Guidelines for the Safety Audit of Highways. Institution of Highways and Transportation. 1996. Available (in English) from 6 Endsleigh Street, London WC1H 0DZ, United Kingdom Tel: (44171)3872525, Fax:(44171)3872808
3. Guide to Traffic Engineering Practice: Part 4 Road Crashes. 1988.
4. Available (in English) from National Association of Australian State Road Authorities (NAASRA), Austroads, P.O. Box 659, Haymarket, New South Wales 2000, Australia Tel: (612) 2186218 Fax: (612) 2817458
5. Organization for Economic Co-operation and Development (OECD) Road Research Reports. Various reports and publications, including Targeted Road Safety Programmes (1994), Road Infrastructure Rehabilitation and Safety Strategies in Central and East Europe (1995), Improving Road Safety by Attitude Modification (1994), and Marketing of Traffic Safety (1993). These are available (in English) from OECD Road Transport Research Programme, 2, rue André-Pascal, 75775, Paris Cedex 16, France. Tel:(331) 45248200, Fax:(331)4524 8176
6. Road Safety Engineering Manual. 1992. Available (in English) from the Royal Society for the Prevention of Accidents (RoSPA), Cannon House, The Priory Queensway, Birmingham B6 6BS, United Kingdom, Tel:(44) 2002461, Fax:(44)200 1254
7. Road Safety Handbook. Institute of Transport Economics. 1982. Available (in Norwegian and Russian) from the Institute of Transport Economics, P.O. Box 6110 Etterstad, M-0602, Oslo, Norway Tel: (472) 2573800, Fax: (472) 2570290
8. Road safety signage and Road marking by "3M",
9. https://www.3m.com/3M/en_US/p/c/signage-marking/i/safety/road-safety/ Road Safety Devices by "Dark Eye", https://www.darkeye.in /assets/frontend/imgs/Dark%20Eye%20Catalogue%202020-5%20(2).pdf

Appendix B

This appendix lists some of the important organisations involved and active in road safety, which may be of interest to decision makers in the developing world.

1. International and funding organisations active in supporting road safety

United Nations Economic and Social Commission for Asia and the Pacific (ESCAP) (Transport, Communications and Tourism Division) RajadamnernAvenueBangkok10200THAILAND Tel: (66 2) 288 1234
Fax: (662) 288 1000

World Health Organization (WHO) (Injury Prevention Programme) Avenue Appia 20 United Nations Building, ch-1211 Geneva 27, Switzerland, Tel: (4122) 7912111
Fax: (4122) 7910746

The World Bank 1818 H Street NW Washington, DC 20433 UNITED STATES Tel: (1 202) 477 1234
Fax: (1202) 4776391

Asian Development Bank (ADB)
P.O.Box789, 0980 Manila, PHILIPPINES
Tel: (632) 6326803/6326463
Fax: (632) 6362423

2. Bilateral aid agencies that are the most active in the Asian and Pacific region on road safety

Australian Agency for International Development (AusAID)

G.P.O. Box 887 Canberra, ACT 2601, Australia Tel: (61 6) 278 4000

Fax: (616) 2764880

Overseas Development Administration (ODA) 94 Victoria Street London SW1 E5JL United Kingdom Tel: (44 171) 917 0588
Fax: (44 171) 9170016

Japan International Cooperation Agency (JICA)

P.O. Box No .216, 48/F, Shinjuki Building, 1-1, Nishi-Shinjuku 2-Chome
Shinjuku-ku Tokyo, 163-04 JAPAN
Tel: (813) 33465197
Fax: (813) 33465094

3. International technical organisations with safety interests

Institute of Transportation Engineers 525 School Street, SW Suite 410
Washington, DC 20024-2797 UNITED STATES Tel: (1 202) 554 8050
Fax: (1202) 8635486

Permanent International Association of Road Congress (PIARC) 27 Rue Gueneguad
75006 Paris, FRANCE
Tel: (331) 46337190
Fax: (331) 46338460

International Road Federation (IRF) 525 School Street, SW Washington, DC 20024-2797 UNITED STATES
Tel: (1202) 5442106
Fax: (1202) 4790828

Prevention Routier International,
75 Ruede Mamer L-8081 LUXEMBOURG,
Tel: (35 2) 31 8341
Fax: (352) 31 1460

Organization for Economic Co-operation and Development (OECD) 2 Rue Andre Pascal, F-75775 Paris, Cedex 16
FRANCE
Tel: (331) 45249594
Fax: (331) 45247960

Road Engineering Association of Asia and Australasia (REAAA) Chairman REAAA Technical Committee, c/o Burwood Highway, Vermont South, Victoria 3133, AUSTRALIA
Tel: (613) 98811555
Fax: (613) 98878104

4. Research institutes and organisations undertaking road safety research of relevance to developing countries

ARRB Transport Research Ltd. 500 Burwood Highway Vermont South Victoria 3133 AUSTRALIA
Tel: (613) 98811555
Fax: (613)98878104

International Association of Traffic and Safety Sciences (IATSS) 6-20-2 Yaesu Chuo-ku, Tokyo 104, JAPAN
Tel: (813) 32737884
Fax: (813) 32727054

Federal Highway Administration, Office of International Programs,
US Department of Transportation 400 Seventh Street SW Washington, DC 20590
UNITED STATES
Tel: (1 202) 3660111
Fax: (1 202) 3669626

Institute for Road Safety Research
P.O. Box 170
2260 AD Leidschendam, THE NETHERLANDS Duindoorn 32
Tel: (3170) 3209323
Fax: (3170) 3201261

Swedish National Road & Transport Research Institute (VTI) S-581, 95 Linkoping SWEDEN
Tel: (4613) 203000
Fax: (4613) 141436

Technical Research Centre of Finland (SVOV)
Road, Traffic, and Geotechnical Laboratory Traffic Section/Safety Team
Sahkomiehentie 3 FIN-02150 ESPOO, FINLAND
Tel: (358) 04564591
Fax: (358) 0464850

Institute of Transport Economics
P.O. Box 6110 Etterstad N-0602 Oslo NORWAY Tel: (47 22) 573 800
Fax: (4722) 570290

Institut National de Recherch (INRETS) Surles Transport setleur Sécurité
2 Avenue du Général Malleret-Joinville 94114 Arcueil Cedex
FRANCE
Tel: (3347) 407163
Fax: (3345)475606

Transport Research Laboratory (TRL) Overseas Centre Old Wokingham Road Crowthorne
Berkshire RG116AU UNITED KINGDOM
Tel: (441344) 773131
Fax: (441344) 7703056

5. Research organisations in the region that are active on road safety issues

Central Road Research Institute (CRRI)
P.O. CRRI, Mathura Road, New Delhi 110020, INDIA
Tel: (91 11) 691 2268
Fax: (91 11)6845943

National Transport Research Centre (NTRC), Ministry of Communications, Sector H-8/3, Islamabad, Pakistan
Tel: (9251) 253676
Fax: (9251) 253651

Indian Institute of Technology (IIT-Delhi)
New Delhi 110016, INDIA
Tel: (91 11) 6858703
Fax: (91 11) 6862037

Institute of Road Engineering (IRE) Jl Raya Timur 264
P.O. Box 2UJBB andung 40294 INDONESIA Tel: (62 22) 780 2251
Fax: (62 22)7802253

Traffic Management Research Institute (TMRI)
Public Security Ministry 1 Qianxiang Lane1, Qianrong Road, Wuxi, Jiangsu
People's Republic of China
Tel: (86 510) 5516123
Fax: (86 510) 5515315

Transport Engineering Design Incorporation (TEDI)
Ministry of Transport and Communications 278T on DucThang, Hanoi
VIET NAM
Tel: (844) 8518366

6. Others/individuals/NGOs

European Transport Safety Committee (ETSC) Rue du Cornet 34 B-1040
Brussels, BELGIUM
Tel: (322) 2304106,4004
Fax: (322) 2304215

Global Traffic Safety Trust International (GTST)
c/o Royal Australasian College of Surgeons, Spring Street, Melbourne,
Victoria 3000, Australia
Tel: (613) 6221033
Fax: (613) 6634075

7. Centre for Advancement of Road Traffic Safety (CARTS), Sarojini House, Bagwan Das Lane, Mandi House, New Delhi 110001, India. They work with the Central Road Research Institute (CRRI), as mentioned above, as (5).

Appendix C

References

Chapter 1 ISO and Road Safety Audit

References:

1. ISO 39001, ISO 39002 and ISO 39003
2. Safety check for new and improved road projects and schemes, U.K.

Chapter 2 Road Safety and Administrative Setup

2.1 General Introduction and Overview

Road users killed in various modes of transport as a proportion of all road traffic deaths, Source: D. Mohan, Journal of Transport and Infrastructure, 2002, 9:79 94.

Ref: https://www.who.int/violence_injury_prevention/publications/road_traffic/world_report/chapter2.pdf

2.2 Trends in Road Safety

References:

For more information on ADB, visit: http://www.adb.org

Figure 2.58 Countries with Traffic Related Accident Deaths

Ref.https://en.wikipedia.org/wiki/List_of_countries_by_traffic-related_death_rate

2.3 Road Safety Improvement Plans

References:

1. Gerondeau, C. 1993. *Road Safety Financing and the Use of Motor Insurance Levies*: Proceedings of First Conference on Asian Road Safety (CARS1).Kuala Lumpur, Malaysia: JKR, Jalon Salahuddin.
2. Lundebye, S., and T. E. Wetteland. 1997. Financing of Road Safety Actions: Proceedings of Third African Road Safety Congress. Pretoria, South Africa: Economic Commission for Africa and Organisation for

Economic Co-operation and Development (OECD).
3. FASTag Pro: Recharge any FASTag online without logging in for two minutes
4. https://www.fastagpro.com/recharge
5. RFID technology and what role does it play in FASTag
6. https://www.financialexpress.com/auto/
7. Theory and Applications of Economics in Highway and Transport Planning, by Dr Vinay Maitri, Dr P. K. Sarkar, Standard Publishers Distributors.

2.4 Road Safety Management

References and Key Documents:

1. World Bank Road Safety Policy Seminar for Central and Eastern Europe. 1994. Budapest, Hungary.
2. New Zealand and the Official Committee on Road Safety. 1991. New Zealand National Road Safety Plan. New Zealand.
3. Ross, A., and M. Mwiraria. 1991.Review of World Bank Experience in Road Safety. Technical Paper INU 93. Washington, DC: World Bank.
4. Organization for Economic Co-operation and Development (OECD) 1996. Targeted Road Safety Programs Road Research Report. Paris, France: OECD.
5. ADB/United Nations Economic and Social Commission for Asia and the Pacific (UN/ESCAP) Road Safety Seminar / Workshop for the Asian and Pacific Region. Bangkok, Thailand,1996.
6. UN/ESCAP.1997.Guidelines on Action Plans and Programs. Bangkok, Thailand: UN/ESCAP.

2.5 Road Accident Data Maintenance

2.6 Road Safety Funding by the insurance industry

References:

1. Lundebye, S., and T. E. Wetteland. 1997. *Financing of Road Safety Actions*. Proceedings of the Third African Road Safety Congress. Pretoria, South Africa: Economic Commission for Africa and Organization for Economic Co-operation and Development (OECD).
2. Lowe, P. 1994. *Financing Road Safety Activities*. Paper presented at World Bank Policy Seminar on Road Safety in Central and Eastern Europe. Budapest, Hungary.

3. United Nations Economic and Social Commission for Asia and the Pacific (UN/ESCAP). 1997. *Guidelines on Action Plans and Programmes*. Bangkok, Thailand: UN/ESCAP.
4. Insurance Institute of Highway Safety, 1005 North Glebe Road, Arlington, VA 22201, US.
5. Loss Prevention Association of India, Warden House, Sir Pheroze Shah Mehta Road, Mumbai, India 400001.

2.7 Planning and Design of Safe Roads

References:

1. Transport Research Laboratory (TRL). 1994. Towards Safer Roads in Developing Countries: A Guide for Planners and Engineers. Overseas Centre, TRL, Old Wokingham Road, Crowthorne, Berkshire RG45 6AU, UK.
2. AASHTO. 1990. A Policy in Geometric Design of Highways and Streets.
3. AASHTO, Institute for Transportation Studies, Berkeley, US.
4. NAASRA. 1988. Guide to Traffic Engineering Practice. P.O. Box 489, Milsons Point, Sydney, NSW 2061, Australia.
5. Department of Transport. 1996. Departmental Standards and Advice Notes Department of Transport, The Stationery Office, Publications Centre, P.O. Box 276, London SW8 5DT, UK.
6. ADB. 1998. Vulnerable Road Users: ADBRETA 5620 Project; 1998. Asian Development Bank, P.O. Box 789, 0980 Manila, Philippines.
7. PADECO Co., Ltd. 1996. Non-motorised Vehicles. Report produced for the World Bank, Washington, DC, US.
8. Institution of Highways and Transportation. 1996. Guidelines for The Safety Audit of Highways. Institution of Highways and Transportation, 6 Endsleigh Street, London W1H0DZ, UK.
9. AusRoads 1994. Road Safety Audit. AusRoads, Level 9, Robell House, 287 Elizabeth Street, Sydney, NSW 2000, Australia.

2.8 Improvements of Hazardous Locations

References:

1. Institution of Highways and Transportation. 1990. Highway Safety
2. Guidelines Accident Reduction and Prevention (International Edition). UK.

3. Transport Research Laboratory (TRL). 1991. Towards Safer Roads in Developing Countries: A Guide for Planners and Engineers. Overseas Centre, TRL, Old Wokingham Road, Crowthorne, Berkshire RG456AU, UK.
4. Royal Society for the Prevention of Accidents (RoSPA). 1992. Road Safety Engineering Manual. RoSPA, Cannon House, The Priority, Queensway, Birmingham B46BS, UK.
5. National Association of Australian State Road Authorities (NAASRA). 1988. Guide to Traffic Engineering Practice: Part 4, Road Crashes. NAASRA, P.O. Box 489, Milson Point, NSW 2061, Australia.
6. Environmental and Transport Planning. 1992. Civilised Streets. A Guide to Traffic Calming. 10 Clermont Terrace, Brighton, UK.
7. Bangalore Road Safety Project. 1994. Frazer Town subproject area. Bangalore City Traffic Police/Bangalore University, Bangalore, India.
8. Baguley, C. J. 1995. Progress Towards the National Casualty Reduction Target Since 1991. Proceedings of the Seminar on Road Safety. Damansara. JKR, Government of Malaysia, Jalan Salahuddin, Kuala Lumpur, Malaysia.

2.9 Road Safety Education for Children

References:

1. Department of Transport (DoT). 1995. Road Safety Education in Schools: Good Practice Guidelines. UK: DoT.
2. DoT. 1996. Child Development and the Aims of Road Safety Education: A Review and Analysis. Road Safety Research Report No.1.UK: DoT.
3. More information can be obtained on Road Safety Officers (RSOs) from the Institute of Road Safety Officers, 16 Southlands Avenue, Louth, Lincolnshire, LN11 8EW, UK. Tel. (44 1507) 327028.
4. Child-to-Child Trust, Institute of Education, 20 Bedford Way, London WC1H 0AL, UK. Tel: (44 171) 612 6650; Fax: (44 171)6126645.
5. Information about resources and teaching aids can be obtained from: the Royal Society for the Prevention of Accidents (RoSPA), Cannon House, The Priory Queensway, Birmingham B46BS, UK. Tel: (44 121) 200 2461; Fax: (44 121)2001254.
6. British Institute of Traffic Education Research (BITER), Kent House, Kent Street, Birmingham B56QF, UK. Tel: (44 121) 622 2402; Fax: (44 121) 622 3450.
7. National Road Safety Committee, P.O. Box 7125, Nasinu, Suva, Fiji. Tel: (679) 394886; Fax: (679) 393253.

8. Ram Badan Joshi, Private Boarding School Organization, Kathmandu, Nepal. Fax: (977) 1412572.

2.10 Training and Testing of Drivers

References:

1. Information on driving tests or driver training in the UK can be obtained from: DSA Head Office, Stanley House, 56 Talbot Street, Nottingham NG1 5GU, UK. Fax: (44 115) 9557735.
2. Information on examiner training can be obtained from: The Training Manager, DSA Training Establishment, Paul Waller Avenue, Cardington, Bedford MK453ST, UK. Fax: (44 1234)742728.
3. DSA.1992. *Driving Skills: The Driving Manual*. UK: The Stationery Office (ISBN 0115510540).
4. DSA.1995. Driving Skills: Your Driving Test. UK: The Stationery Office (ISBN 011551158X). Information on 3 and 4 can be obtained from: The Stationery Office, P.O. Box No. 276, London SW8 5DT, UK, or Overseas Sales, The Stationery Office, Fax: (44 171) 873 8203

2.11 Publicity Campaigns for Road Safety

References:

1. Hastings, G., and A. Haywood. 1991. Social Marketing and Communication in Health Promotion: Health Promotion International. UK: Oxford University Press.
2. Glanz, K. et al. (Ed.). 1990. Health Behavior and Health Education. California, US: Jossey-Bass.
3. Organization for Economic Co-operation and Development (OECD). 1971. Road Safety Campaigns: Design and Evaluation—The Use of Mass Communication for the Modification of Road User Behavior. Paris, France: Road Transport Research, OECD (out of print).
4. OECD. 1975. Manual on Road Safety Campaigns. Paris, France: Road Trans- port Research, OECD.
5. OECD. 1993. Marketing of Traffic Safety. Paris, France: Road Transport Research, OECD.
6. OECD. 1994. Improving Road Safety by Attitude Modification. Paris, France: Road Transport Research, OECD.
7. Preusser, D. F., and R. D. Blomberg (Ed. J. A. Rothengatter). 1987. Development and Validation of a Road Safety Public Education Process: Road Users and Traffic Safety. The Netherlands: Van Gorcum.

2.12 Standards for Vehicle Safety

References:

1. United Kingdom Vehicle Inspectorate. Cars and Light Commercial Vehicle Testing (ISBN0115510532).
2. United Kingdom Vehicle Inspectorate. Motorcycle Testing (ISBN0115510052).
3. United Kingdom Vehicle Inspectorate. Heavy Goods Vehicle Inspection Manual ISBN 011551063X).
4. United Kingdom Vehicle Inspectorate. Public Service Vehicle Inspection Manual (ISBN0115510702).
5. United Kingdom Vehicle Inspectorate. Ministry of Transport Testing Guide (ISBN 0 11 551056 7). The Stationery Office.
6. Ing Guido. CITA Working Group, Promotion of the Vehicle Inspection.
7. Paper: Technical Inspection. CITA absi, Rue de la Technologie 21/25B 1082, Brussels, Belgium. Tel: (32 24)69070; Fax: (32 24) 690 795.

Items 2-6 can be obtained from Stationery Office Bookshop, P.O. Box 276, London SW8 5DT, UK. Tel: (44 171) 873 0011; Fax: (44 171)873 8200.

2.13 Traffic Legislation

References:

1. Road Safety (Traffic) Regulations 1988: Road Safety Act 1986. Victoria, Australia.
2. National Committee on Uniform Traffic Laws and Ordinances: Uniform Vehicle Code and Model Traffic Ordinance. 1992. Evanston, Illinois, US.
3. Southern Africa Transport and Communications Commission (SATCC). 1990. *The SATCC Road Traffic Signs Manua*l. Norway: Institute of Transport Economics.
4. SATCC. *SATCC Highway Code*. Norway: Institute of Transport Economics.
5. Zaal, D.1994. *Traffic Law Enforcement*: A Review of Literature. Australia: Federal Office of Road *Safety*.
6. Axup, D. 1993. *Enforcement*: *Traffic Safety Toolbox*. Washington, DC, US: Institute of Transportation Engineers.
7. Legal Research Board. 1992.*Road Transport Rules.* Laws of Malaysia. Malaysia: International Law Book Services.

2.14 Traffic Police and Law Enforcement

References:

1. No. 6 (SW) Region Police Driving School, Wiltshire Police Headquarters, London Road, Devizes, Wiltshire SN102DN, UK. Fax:(44 1380) 734196.
2. Northwestern Traffic Institute, P.O. Box 1409, Evanston, Illinois 60204, US.
3. Axup, D. 1993. Enforcement: Traffic Safety Toolbox. Washington, DC, US: Institute of Traffic Engineers.
4. Baker, J., and L. Frickle. 1986. Traffic Accident Investigation Manual (9th edition). Evanston, US: Northwestern Traffic Institute.
5. Frickle, L. Traffic Accident Reconstruction.1986.US: Northwestern Institute, Evanston (ISBNUS09126420706).
6. Neades, J., and R. Ward. 1996.Principles of Accident Reconstruction: Vols. 1 (Theory) and 2 (Practice). UK: Wiltshire6 Regional Driving School, Devizes (ISBN1900452006).
7. Neades, J., and R. Ward. 1996. Surveying for Accident Investigation. UK: Wiltshire 6Regional Driving School, Devizes (ISBN09126420706).
8. Traffic Patrol Officers Manual. 1997. UK: Wiltshire 6 Regional Driving School, Devizes.

2.15 Emergency Assistance to Road Accident Victims

References:

1. Schwartz, G.R., P. Safer, J.H. Stonc, P.B. Storey, and D.K. Wagner, (Eds.). 1986. *Principles and Practice of Emergency Medicine* (seconded.). Philadelphia, US: W.B. Saunders. (Vol. II, Part IV, Emergency Medical Systems, pp.1,371-1,491.)
2. WHO.1984. Consultation on Emergency Medical Organisation: Report on a WHO meeting, Washington, DC, September 1983. Geneva, Switzerland: WHO.

2.16 Road Safety Research

References:

1. Central Road Research Institute. 1995. Annual Report 1994-95. New Delhi, India.

2. IIT. 1995. The Work Done 1991-1995. New Delhi, India: IIT.
3. TRL. 1994. "Microcomputer Accident Analysis Package, Version 5 (MAAP5) User Manual." UK: TRL.
4. Hills, B. L., and C. J. Baguley. 1993. "Accident Data Collection and Analysis: the Use of the Microcomputer Package MAAP in Five Asian Countries": Proceedings of the Asian Road Safety Conference, Kuala Lumpur, Malaysia, October 1993.
5. REAAA. 1992. Road Safety Resources for Developing Countries. Victoria, Australia: Australian Road Research Board, Ltd.
6. TRL. 1991. Towards Safer Roads in Developing Countries. UK: TRL.

2.17 Costing Road Accidents

References:

1. Chand, M. 1995. "Cost of Road Accidents in India—with Special Reference to Kerala. Pune, India: *Indian Journal of Transport Management*, July 1995.
2. Central Road Research Institute. (CRRI). 1982. *Road User Costs*. New Delhi, India: CRRI.
3. Ghee C. E., et al. 1997. *Socio-Economic Aspects of Road Accidents in Developing Countries*.UK: TRL Report TRL247.
4. Alfaro, J.L., M. Chapuis and F. Fabre (eds.). 1994. *Cost 313 Socio-Economic Costs of Road Accidents*. Final Report. European Commission, 200 Rue de la Loi, BU315/33, Brussels, Belgium.
5. TRL. 1995. *Costing Road Accidents in Developing Countries*. Overseas Road Note 10. UK: TRL.
6. Andreassan, D. 1993. *Road Accident Costs and Their Use*. Victoria, Australia: Australian Road Research Board Special Briefing.
7. Forkenbrock, D., N. Foster, and T. Pogue. 1994. Safety and Highway Investment: University of Iowa.
8. Department of Transport. 1994. "Valuation of Road Accidents." Highways Economics Note No. 1 1994. UK: Department of Transport.
9. Elvik, R. 1995. "An Analysis of Official Economic Valuations of Traffic Accident Fatalities in 20 Motorized Countries." Accident Analysis and Prevention, Vol. 27, No. 2, 1995. US.
10. Department of Transport. 1995. Road Accidents Great Britain 1994: The Casualty Report. UK: The Stationery Office (see also 1968-1991, and 1993).

The following are web links to important sites on this crucial social issue:

1. United Nations on Traffic Accidents Take Toll on Social, Economic Progress, General Assembly Agrees, Adopting Resolution Improving Road Safety
https://www.un.org/press/en/2014/ga11498.doc.htm
2. World Bank on Road crashes are the biggest safety challenge for youths
https://blogs.worldbank.org/endpovertyinsouthasia/road-crashes-are-biggest-safety-challenge-youths
3. Asian Development Bank (ADB) on Technical Assistance for Socio-economic Impact of Road Crashes (Financed by the Poverty Reduction Cooperation Fund)
https://www.adb.org/sites/default/files/project-document/69005/tar-stu-38081.pdf
4. ResearchGate on Social Justice Approach to Road Safety in Kenya: Addressing the Uneven Distribution of Road Traffic Injuries and Deaths across Population Groups
https://www.researchgate.net/publication/45389690_Social_Justice_Approach_to_Road_Safety_in_Kenya_Addressing_the_Uneven_Distribution_of_Road_Traffic_Injuries_and_Deaths_across_Population_Groupss
5. Clinical Epidemiology and Global Health: Despite flows in Transport Policy, it is noticed that many more proven wide social factors are increasing the number of road traffic accidents, including injuries and deaths. A wider social determinant among low socioeconomic groups, including identity crises, largely contributes towards an increased rate of road accidents in India.
https://cegh.net/article/S2213-3984(20)30159-7/fulltext

Chapter 3 Traffic Safety Studies and Measures

References:

1. Alphand, F., U. Noelle, and B. Guichet. 1991. Roundabouts and road safety: State of the art in France. *Intersections without traffic signals, II*, ed. W. Brilon, 107-25. Berlin: Springer-Verlag.
2. American Association of State Highway and Transportation Officials (AASHTO). 2004a. *A policy on geometric design of highways and streets,* 4th ed. Washington, DC: American Association of State Highway and Transportation Officials- 2004b. *A guide for achieving flexibility in highway design.* Washington, DC: American Association of State Highway and Transportation Officials.

3. Balkin, S., and J.K. Ord. 2001. Assessing the impact of speed limit increases on fatal interstate crashes. *Journal of Transportation and Statistics* 4(1):1-26.
4. Bath, S. 1993. *Influence of socioeconomic factors on the exposure to crash risk for young drivers.* Vancouver: Traffic Safety Research Dept., Insurance Corporation of British Columbia.
5. Bicycle Federation of America. 1998. Creating walkable communities: A guide for local governments. Mid-America Regional Council, Kansas City, MO. http://www.bikewalk.org/pdfs/ncbwpubwalkablecomm.pdf.
6. Bissell, H. H., and L. G. Neudorff. 1980. Criteria for removing traffic signals. ITE 1980 Washington, DC: Compendium of Technical Papers, Institute of Transportation Engineers.
7. Bowman, B. L., and R. L. Vecellio. 1994. Effects of urban and suburban median types on both vehicular and pedestrian safety. *Transportation Research Record* 1445, TRB. Washington, DC: National Research Council.
8. Box, P. C. 2000. Curb parking findings revisited. Transportation Research E-Circular. Washington, DC: Transportation Research Board. 2002. Angle parking issues revisited. *ITE Journal* 72 (3): 36-47. 2004.
9. Curb- parking problems: Overview. *Journal of Transportation Engineering* 130 (1): 1-5.
10. Brown, M. 1995. The design of roundabouts: State of the art review. London: HMSO.
11. Brude, U., and J. Larsson. 2000. What roundabout design provides the highest possible safety? Nordic Road Transport Research 12:17-21.
12. Burden, D., and P. Lagerwey. 1999. Road diets: Fixing the big roads. High Springs, FL: Walkable Communities.
13. Butler, P. 1996. Automobile insurance pricing. National Organisation for Women (www.now.org), presented at the National Conference on Women's Travel Issues.
14. Claessen, J. G., and D. R. Jones. 1994. The road safety effectiveness of raised wide medians. Proceedings of the 17th Australian Road Research Board Conference 17 (5): 269-87.
15. Clark, J. E. 1985. High speeds and volumes on residential streets: An analysis of physical street characteristics as causes in Sacramento, California. ITE 1985 Compendium of Technical Papers. Washington, DC: Institute of Transportation
16. Engineers.
17. Committee on Access Management. 2003. Access management manual. Washington, DC: Transportation Research Board.

18. Committee on Geometric Design. 2004. Context-sensitive design around the country: Some examples.
Washington, DC: Transportation Research Board.
19. Congress for the New Urbanism. 2002. Civilizing downtown highways: Putting New Urbanism to work on California's highways. San Francisco, CA.
20. County Surveyors Society. 1994. Traffic calming in practice. London: Landor Publishing.
21. Daisa, J. 2006. Proposed recommended practice: Context-sensitive solutions in designing major urban thoroughfares for walkable communities. Washington, DC: Institute of Transportation Engineers.
22. Duany, A., and E. Talen. 2002. Transect planning. Journal of the American Planning Association 68 (3): 245- 66.
23. Dumbaugh, E. 2005a. Safe streets, livable streets. Journal of the American Planning Association 71 (3): 283- 98-2005b. Safe streets, livable streets: A positive approach to urban roadside design. PhD diss., Georgia Institute of Technology-2006a. Enhancing the Safety and Operational Performance of Arterial Roadways in the Atlanta Metropolitan Region. Report Submitted to the Atlanta Regional Commission-2006b. The design of safe urban roadsides: An empirical analysis. Transportation Research Record 1961:74-82.
24. Ebbecke, J. M., and J. J. Schuster. 1977. Area-wide impact of traffic control devices. Transportation Research Record 644:54-57.
25. Environmental Protection Agency (EPA). 2004. Characteristics and performance of regional transportation systems. Washington, DC.
26. Ewing, R. 1999. Traffic Calming State-of-the-Practice. Washington, DC: Institute of Transportation Engineers- 2001. Impacts of traffic calming. Transportation Quarterly55(1):33-45-2002. Impediments to context-Sensitive Main Street design. Transportation Quarterly 56(4):51-64.
27. Ewing, R., and R. Cervero. 2001. Travel and the built environment. Transportation Research Record 1780: 87- 114.
28. Ewing, R., R. Pendall, and D. Chen. 2002. Measuring sprawl and its impact. Washington, DC: Smart Growth America/U.S. Environmental Protection Agency-2003. Measuring sprawl and its transportation impacts. Transportation Research Record 1832:175-83.
29. Ewing, R., R. Schieber, and C. V. Zegeer. 2003. Urban sprawl as a risk factor in motor vehicle occupant and pedestrian fatalities, American Journal of Public Health 93:1541-45.

30. Farouki, O. T., and W. J. Nixon. 1976. The effect of the width of suburban roads on the mean free speed of cars. Traffic Engineering and Control 17:518-19.
31. Federal Highway Administration. 2001. Geometric design practices for European roads. Publication No. FHWA-PL-01-026. Washington, DC: U.S. Department of Transportation.
32. Fitzpatrick, K., P. Carlson, M. Brewer, and M. Wooldridge. 2001. Design factors that affect driver speed on suburban streets. Transportation Research Record 1751:18-25.
33. Fitzpatrick, K., S. Turner, M. Brewer, P. Carlson, N. Lalani, B. Ullman, N. Trout, E. S. Park, D. Lord, and J. Whitacre. 2006. Improving pedestrian safety at an unsignalized crossing. TCRP Report 112/NCHRP Report 562, Washington, DC: Transportation Research Board.
34. Florida Department of Transportation. 2006. Median handbook—Interim version. Tallahassee: Florida Department of Transportation.
35. Garder, P.E.2004.The impact of speed and other variables on pedestrian safety in Main Accident Analysis and Prevention 36 (4): 533-42.
36. Gattis, J. L. 2000. Urban street cross-section and speed issues. Transportation Research E-Circular. Washington, DC: Transportation Research Board.
37. Gattis, J. L., and A. Watts. 1999. Urban street speed related to width and functional class. Journal of Transportation Engineering 125 (3): 193-200.
38. Geddes, E. 1996. Safety benefits of traffic calming. Vancouver: Insurance Corporation of British Columbia.
39. Geoplan. 1994. Evaluation of pedestrian road safety facilities. Report prepared for the Road Safety Bureau, Roads and Traffic Authority (New South Wales) by Geoplan Urban and Traffic Planning.
40. Gluck, J., H.S. Levinson, and V. Stover. 1999. Impacts of access management techniques. NCHRP Report 420. Washington, DC: Transportation Research Board.
41. Hadayeghi, A., A. S. Shalaby, and B. N. Persaud. 2003. Macrolevel accident prediction models for evaluating safety of urban transportation systems. Transportation Research Record 1840:87-95
42. Hadayeghi, A., A. S. Shalaby, B. N. Persaud, and C. Cheung. 2006. Temporal transferability and updating of zonal-level accident prediction models. Accident Analysis and Prevention 38 (3): 579-89.
43. Hakkert, A. S., V. Gitelman, and E. Ben-Shabat. 2002. An evaluation of crosswalk warning systems: Effects on pedestrian and vehicle behaviour. Transportation Research Part F. 5(4): 275-292.

44. Hall, J. W., C. J. Burton, D. G. Coppage, and L. V. Dickinson. 1976. Roadside hazards on non-freeway facilities. Transportation Research Record 601:56-58.
45. Harwood, D. W. 1986. Multilane design alternatives for improving suburban highways. NCHRP Report 282. Washington, DC: Transportation Research Board, National Research Council. -- 1990. Effective utilization of street width on urban arterials. National Cooperative Highway Research Program Report 330. Washington, DC: Transportation Research Board, National Research Council.
46. Hauer, E. 1999. Safety in geometric design standards. http://members.rogers.com/hauer/Pubs/09Safety%20in%20Geometric %20Design.pdf(accessedJanuary11, 2005).
47. Hauer, E., and P. Garder. 1986. Research into the validity of the traffic conflicts technique. Accident Analysis and Prevention 18:471-81.
48. Heimbach, C. L., P. D. Cribbins, and M. S. Chang. 1983. Some partial consequences of reduced traffic lane widths on urban arterials. Transportation Research Record 923:69-72.
49. Hess, P. M., A. V. Moudon, and J. M. Matlick. 2004. Pedestrian safety and transit corridors. Journal of Public Transportation 7(2): 73-93.
50. Huang, H. F., J. R. Stewart, and C. V. Zegeer. 2002. Evaluation of lane reduction "road diet" measures on crashes and injuries. Transportation Research Record 1784:80-90.
51. Hummer, J. E., and C. F. Lewis. 2000. Operational capacity of three-lane cross-sections. Report FHWA/NC/2000-003. Raleigh: Center for Transportation Engineering Studies, North Carolina State University.
52. Humphreys, J.B. 1978. Safety aspects of curb parking. Washington, DC: Federal Highway Administration. Institute of Transportation Engineers. 2001. A study of angle parking accidents. Traffic Safety Committee, Arizona Section. Washington, DC: Institute of Transportation Engineers.
53. Jacobs, J. 1961. The death and life of great American cities. New York: Random House.
54. Jacobsen, P. L. 2003. Safety in numbers: More walkers and bicyclists, safer walking and bicycling. Injury Prevention 9: 205-09
55. Jacquemart, G. 1998. Synthesis of highway practice 264: Modern roundabout practice in the United States. National Cooperative Highway Research Program. Washington, DC: National Academy Press.
56. Johnston, R. E. 1962. Experience with narrow medians. Proceedings of the 1st Australian Road Research Board Conference 1 (1): 489-99.
57. Kmet, L., P. Brasher, and C. Macarthur. 2003. A small area study of motor vehicle crash fatalities in Alberta, Canada. Accident Analysis and Prevention 35 (2): 177-82.

58. Knaap, K. K., and K. Giese. 2001. Guidelines for the conversion of urban four-lane undivided roadways to three-lane two-way left-turn facilities. Ames: Center for Transportation Research And Education, Iowa State University.
59. Knoblauch, R. L., M. Nitzburg und R. F. Seifert. 2001. Pedestrian crosswalk case studies: Sacramento, California; Richmond, Virginia; Buffalo, New York; Stillwater, Minnesota. Report No. FHWA-RD-00-103.McLean, VA: Federal Highway Administration.
60. Knoblauch, R. L., and P. D. Raymond. 2000. The effect of crosswalk markings on vehicle speeds in Maryland, Virginia, and Arizona. Report No. FHWA-RD-00-101. McLean, VA: Federal Highway Administration.
61. Knoblauch, R. L., B. H. Tustin, S. A. Smith, and M. T. Pietrucha. 1988. Investigation of exposure-based pedestrian accident areas: Crosswalks, sidewalks, local streets, and major arterials. Washington, DC: Federal Highway Administration.
62. Ladrón de Guevara, F. S. P. Washington, and J. Oh. 2004. Forecasting travel crashes at the planning levels: Simultaneous negative binomial crash model applied in Tucson, Arizona. Transportation Research Record 1897:191-99.
63. Lamm, R., B. Psarianos, and T. Mailender. 1999. Highway design and traffic safety engineering handbook. New York: McGraw-Hill.
64. Lee, J., and F. Mannering. 1999. Analysis of roadside accident frequency and severity and roadside safety management. Olympia: Washington State Department of Transportation.
65. Leong, H.J.W. 1970. The effect of kerbed median strips on accidents on urban roads. Proceedings of the 5th Australian Road Research Board Conference 5 (3): 338-64.
66. Levine, N., K. E. Kim, and L. H. Nitz. 1995a. Spatial analysis of Honolulu motor vehicle crashes: I. Spatial analysis. Accident Analysis and Prevention 27 (5): 663-74-1995b. Spatial analysis of Honolulu motor vehicle crashes: II. Zonal generators. Accident Analysis and Prevention 27 (5): 675-85.
67. Litman, T., and S. Fitzroy. 2005. Safe travels: Evaluating mobility management traffic safety impacts. Victoria, BC: Victoria Transport Policy Institute.
68. Lovegrove, G. R., and T. Sayed. 2006. Macro-level collision prediction models for evaluating neighbourhood traffic safety. Canadian Journal of Civil Engineering 33 (5): 609-21.

69. Lucy, W. 2003. Mortality risk associated with leaving home: Recognising the relevance of the built environment. American Journal of Public Health 93 (9): 1564-69.
70. Lucy, W., and R. Rabalais. 2002. Traffic fatalities and homicides by strangers: Danger of leaving home in cities, inner suburbs, and outer suburbs. Charlottesville: University of Virginia.
71. Mahalel, D., and Z. Szternfeld. 1986. Safety improvements and driver perception. Accident Analysis and Prevention 18 (1): 37-42.
72. Marks, H.1957. Subdividing for traffic safety. Traffic Quarterly (July): 308-25.
73. Maycock, G., and R.D. Hall. 1984. Crashes at four–arm roundabouts. TRRL Laboratory Report LR1120. Crowthorne, UK: Transport and Road Research Laboratory.
74. McCoy, P. T., M. Ramanujam, M.Moussavi, and J. L. Ballard. 1990. Safety comparison of types of parking on urban streets in Nebraska. Transportation Research Record 1270:28-41.
75. McCoy, T.A., P. T. McCoy, R.J. Hadan, and V.A. Singh. 1991.Safety evaluation of converting on-street parking from parallel to angle. Transportation Research Record 1327: 36-41.
76. Milton, J. and F. Mannering. 1998.The relationship among highway geometrics, traffic-related elements and motor-vehicle accident frequencies. Transportation 25:395-413.
77. Minnesota Department of Transportation. 2002. Access operations study: Analysis of traffic signal spacing on four-lane arterials. Saint Paul, MN: Minnesota Department of Transportation.
78. Mokdad, A. H., J. S. Mark, D. F. Stroup, and J. L. Gerberding. 2004. Actual causes of death in the United States, 2000. Journal of the American Planning Association 291:1238-45.
79. Naderi, J. R. 2003. Landscape design in the clear zone: The effects of landscape variables on pedestrian health and driver safety. In Transportation Research Board, 82nd Annual Conference Proceedings. Washington, DC: Transportation Research Board. CD-ROM.
80. Nevada Department of Transportation. 1999. Access management system and standards. Carson City: Nevada Department of Transportation.
81. Noland, R. B. 2001. Traffic fatalities and injuries: Are reductions the result of "improvements" in highway design standards? Transportation Research Board 80th Annual Conference Proceedings. Washington, DC: Transportation Research Board. CD-ROM
82. Noland, R. B., and L. Oh. 2004. The effect of infrastructure and demographic change on traffic-related fatalities and crashes: A case

study of Illinois county-level data. Accident Analysis and Prevention 36:525-32.
83. Organisation for Economic Co-operation and Development (OECD). 1998. Safety of vulnerable road users. Publication No. DSTI/DOT/RTR/RS7(98)1/FINAL. Paris: OECD.
84. Persaud, B., E. Hauer, R. A. Retting, R. Vallurupalli, and K. Mucsi. 1997. Crash reductions related to traffic signal removal in Philadelphia. Accident Analysis and Prevention 29:803-10.
85. Persaud, B. N., R. A. Retting, P. E. Garder, and D. Lord. 2002. Crash reductions following installation of roundabouts in the United States. Arlington, VA: Insurance Institute for Highway Safety.
86. Prevedouros, P. D. 2001. Evaluation of in-pavement flashing lights on a six-lane arterial pedestrian crossing. Proceedings of the 71st Annual Meeting of the Institute of Transportation Engineers, Institute of Transportation Engineers. Washington, DC. CD-ROM.
87. Retting, R. A., S. A. Ferguson, and A. T. McCartt. 2003. A review of evidence-based traffic engineering measures designed to reduce pedestrian-motor vehicle crashes. American Journal of Public Health 93 (9): 1456-63.
88. Roberts, I., R. Norton, R. Jackson, R. Dunn, and I. Hassal. 1995. Effect of environmental factors on risk of injury of child pedestrians by motor vehicles: A case-control study. British Medical Journal 310:91-94
89. Robinson, B. W. 2000. Roundabouts: An information guide. Washington, DC: Federal Highway Administration.
90. Robinson, D. 2005. Safety in numbers: More walkers and bicyclists, safer walking and bicycling. Bicycle Federation Australia. http://www.bfa.asn.au/bfanew/pdf/publications/safety_in_numbers.pdf (accessed June 12, 2006).
91. S&K Transportation Consultants. 2000. Access management, location and design. Participant notebook for NHI course 133078. Washington, DC: National Highway Institute, Federal Highway Administration.
92. Sawalha, Z., and T. Sayed. 2001. Evaluating safety on urban arterial roadways. Journal of Transportation Engineering 127 (2): 151-58.
93. Schoon, C. C., and J. van Minnen. 1993. Accidents on roundabouts: II. Second study into the road hazard presented by roundabouts, particularly with regard to cyclists and moped riders. R-93-16. The Netherlands: SWOV Institute for Road Safety Research - 1994. The safety of roundabouts in the Netherlands. Traffic Engineering and Control 35 (3): 142-48.

94. Scriven, R. W. 1986. Raised median strips—A highly effective road safety measure. Proceedings of the 13th Australian Road Research Board Conference 13 (9): 46-53.
95. Seburn, T J. 1967.Relationship between curb uses and traffic accidents. Traffic Engineering 37 (18):42-47.
96. Smart Growth Network. N.d. Getting to smart growth. International City/County Management Association. Chicago. http://www.smartgrowth.org/pdf/gettosg.pdf.
97. Study Tour Team. 2001. Geometric design practices for European roads. Washington, DC: Federal Highway Administration.
98. Surface Transportation Policy Project (STPP). 2000.Meanstreets. http://www.transact.org/PDFs/ms2000/ms2000.pdf, 2002.Meanstreets. http://www.transact.org/PDFs/ms2002/MeanStreets2002.pdf, 2004.Meanstreets. http://www.transact.org/library/reports_html/ms2004/pdf/Final_Mean_Streets_2004_4.pdf,
99. Swift, P., D. Painter, and M. Goldstein. 2008. Residential street typology and injury accident frequency. http://www.newurbanengineering.com.
100. Syrek. D. 1955. Accident rates at intersections. Traffic Engineering 25:312-16.
101. Tester, J.M., G.W. Rutherford, Z. Wald, and M.W. Rutherford. 2004. Matched case-control study evaluating the effectiveness of speed humps in reducing child pedestrian injuries. American Journal of Public Health 94:646-50.
102. Texas Transportation Institute (TTI). 1982. On-street parking: synthesis of safety research related to traffic control and roadway elements. Washington, DC: Federal Highway Administration.
103. Transportation Research Board. 2003. A guide for addressing run-off-road collisions. NCHRP 500, vol. 6. Washington, DC: National Research Council - 2005. TRB special report 282: Does the built environment influence physical activity/ Examining the evidence. Washington, DC: Transportation Research Board.
104. Transportation Research Institute. 1997. Stopping sight distance and decision sight distance. Discussion Paper No. 8.A. Salem, OR: Oregon Department of Transportation.
105. Trowbridge, M.J., and N.C. McDonald, 2008. Urban sprawl and miles driven daily by teenagers in the United States. American Journal of Preventive Medicine 34 (3): 202-06.

106. Turner, D. S., and E. S. Manfield. 1990. Urban trees and roadside safety. Journal of Transportation Engineering 16 (1): 90-103.
107. UK Department for Transport. 1997. Killing speed and saving lives. London: UK Department of Transport - 2007. Manual for streets. London: Thomas Telford.
108. Untermann, R. K. 1990. Street design—Reassessing the function, safety, and comfort of streets for pedestrians. The road less traveled: Getting there by other means, 11th International Pedestrian Conference. Boulder, CO: City of Boulder.
109. Van Houten, R., and J. E. L. Malenfant. 1999. Canadian research on pedestrian safety. McLean, VA: Federal Highway Administration.
110. Van Houten, R., R. A. Retting, C. Farmer, and J. Van Houten. 2000. Field evaluation of a leading pedestrian interval signal phase at three urban intersections. Transportation Research Record 1734:86-92.
111. Vitaliano, D. F., and J. Held. 1991. Road accident external effects: An empirical assessment. Applied Econometrics23:373-78.
112. World Health Organization. 2004. World report on road traffic injury prevention. Geneva: World Health Organization.
113. Zegeer, C. V., and F. M. Council. 1995. Safety relationships associated with cross-sectional roadway elements. Transportation Research Record 1512:29-36.
114. Zegeer, C. V., C. Seiderman, P. Lagerwey, M. Cynecki, M. Ronkin, and R. Schnieder. 2002a. Pedestrian facilities users guide—Providing safety and mobility. Report No. FHWA-RD-102-01. Washington, DC: Federal Highway Administration.
115. Zegeer, C. V., J. R. Stewart, H. H. Huang, and P. A. Lagerwey. 2002b. Safety effects of marked vs. unmarked crosswalks at uncontrolled locations: Executive summary and recommended guidelines. Report No. FHWA-RD-01-075. Washington, DC: Federal Highway Administration.
116. Zeigler, A. J. 1986. Guide to the management of roadside trees. Report No. FHWAIP-86-17. Lansing: Michigan Department of Transportation.

Chapter 4 Highway Safety at Intersections and Roundabouts

References for Intersection Safety (Source of excerpts and illustrations)

1. Savolainen, Andrzej and P. Tarko, Safety of Intersections on High-Speed Road Segments with Superelevation, Joint Transportation Research Program, Civil Engineering, Purdue University, Purdue e-Pubs

2. Safer Intersections for Rural Highways, NCHRP Report 650, NATIONAL COOPERATIVE HIGHWAY RESEARCH PROGRAM, TRB of the National Academies
3. Rural Expressway Intersection Safety, Centre for Transportation Research and Education, Iowa State University

References:

1. American Association of State Highway and Transportation Officials, A Policy on Geometric Design of Highways and Intersections, Washington, D.C., 2001.
2. Bared, J. G. and H. Lum, "Safety Evaluation of Intersection Design Elements (Pilot Study)," Transportation Research Board Conference Presentation, Washington, D.C., 1992.
3. Bauer, K.M. and D. Harwood, Statistical Models of At-Grade Intersection Accidents, Report No. FHWA-RD- 96-125, Federal Highway Administration, McLean, Va., 1996.
4. Blower, D., K.L. Campbell, and P.E. Green, "Accident Rates for Heavy Truck-Tractors in Michigan," Accident Analysis and Prevention 25 (3): 307-321, 1993.
5. Bonneson, J. A. and P.T. McCoy, "Estimation of Safety at Two-Way Stop-Controlled Intersections on Rural Highways," Transportation Research Record 1401: 83-89, 1993.
6. Cameron, A. and P. Trivedi, "Regression Based Tests for Over-dispersion in the Poisson Model," Journal of Econometrics 31, 255-274, 1990.
7. Chang, L.-Y. and F. Mannering, "Analysis of Injury Severity and Vehicle Occupancy in Truck-and Non-truck-Involved Accidents", Accident Analysis and Prevention, Vol. 31, No. 5, 579-592, 1999.
8. David, N.A. and J.R. Norman, Motor Vehicle Accidents in Relation to Geometric and Traffic Features of Highway Intersections, Volume II – Research Report, Report No. FHWA-RD-76-129, Federal Highway Administration and National Highway Traffic Safety Administration, Washington, D. C., 1975.
9. G. D. Hamilton Associates Consulting Ltd., Traffic Conflict Procedures, Manual, Second Edition, Insurance Corporation of British Columbia.
10. Greene, W.H., LIMDEP, Version 7.0, Econometric Software,1998.
11. Greene, W. H., Econometric Analysis, 5th Edition, Prentice Hall, New York, 2002.
12. Hakkert, A. S. and D. Mahalel, "Estimating the Number of Accidents at Intersections from a Knowledge of the Traffic Flows on the Approaches," Accident Analysis and Prevention 10: 69-79, 1978.

13. Hanna, J.T., T.E. Flynn, and W.L. Tyler, "Characteristics of Intersection Accidents in Rural Municipalities," Transportation Research Record 601: 79-82, 1976.
14. Harwood, D.W., M.T. Pietrucha, M.D. Woolridge, R.E. Brydia, and K. Fitzpatrick, Median Intersection Design, National Cooperative Highway Research Program Report 375, Transportation Research Board, National Research Council, National Academy Press, Washington, D.C., 1995.
15. Hauer, E., "The Harm Done by Tests of Significance," Accident Analysis and Prevention, Vol. 36, No.3: 495- 500, 2004.
16. Hauer, E., J.C.N. Ng, and J. Lovell, "Estimation of Safety at Signalized Intersections," Transportation Research Record 1185: 48-61, 1988.
17. King, G.F. and R.B. Goldblatt, "Relationship of Accident Patterns to Type of Intersection Control," Transportation Research Record 540: 1-12, 1975
18. McCoy, P.T., E.J. Tripi, and J.A. Bonneson, Guidelines for Realignment of Skewed Intersections, Nebraska Department of Roads Research Project Number RESI (0099) P471, 1994.
19. McGee, H.W., W.E. Hughes, and K. Daily, Effect of Highway Standards on Safety, National Cooperative Highway Research Program Report 374, Transportation Research Board, National Research Council, National Academy Press, Washington, D.C., 1995. Miaou, S.P., "The Relationship Between Truck Accidents and Geometric Design of Road Sections: Poisson Versus Negative Binomial Regressions," Accident Analysis and Prevention 26 (4), 471-482, 1994.
20. Missouri Highway and Transportation Department, Manual on Identification, Analysis and Correction of High-Accident Locations, Technology Transfer Assistance Program, Missouri Highway and Transportation Department, Second Edition, 1990.
21. Pickering, D., R.D. Hall, and M. Grimmer, Accidents at Rural T-Junctions, Research Report 65, Transport and Road Research Laboratory, Department of Transport, Crowthorne, Berkshire, United Kingdom, 1986.
22. Pietrucha, M.T., T.R. Pieples, and P.M. Garvey, "An Evaluation of the Pennsylvania Road Safety Audit Pilot Program, Transportation Research Record, Transportation Research Board, National Research Council, National Academy Press, Washington, D.C., 2000.
23. Poch, M. and F. Mannering, "Negative Binomial Analysis of Intersection-Accident Frequencies," Journal of Transportation Engineering 122 (2): 105-113, 1996.

24. Shankar, V., F. Mannering, and W. Barfield, "Effect of Roadway Geometrics and Environmental Factors on Rural Freeway Accident Frequencies," Accident Analysis and Prevention 27 (3): 371- 389, 1995.
25. Tarko, A.P., K. C. Sinha, S. Eranky, H. Brown, E. Roberts, R. Scinteie, and S. Islam, Crash Reduction Factors for Improvement Activities in Indiana, Joint Highway Research Project, final report, June 2000.
26. Tarko, A.P. and M. Kanodia, Hazard Elimination Program-Manual on Improving Safety of Indiana Road Intersection and Sections, Report No. FHWA/IN/JTRP-2003/19, Joint Transportation Research Program, West Lafayette, IN, 2004.
27. U.S. Department of Transportation, Manual on Uniform Traffic Control Devices: Millennium Edition, 2001.
28. Vogt, A. and J.G. Bared, Accident Models for Two-Lane Rural Roads: Segments and Intersections, Report No. FHWA-RD-98-133, Federal Highway Administration, McLean, Va., 1998.
29. Vogt, A. Crash Models for Rural Intersections: Four-Lane by Two-Lane Stop-Controlled and Two-Lane by Two-Lane Signalized, Report No. FHWA-RD-99-128, McLean, Va., 1999. 81
30. Vuong, Q., "Likelihood Ratio Tests for Model Selection and Non-nested Hypotheses, Econometrica 57,307-334, 1989.
31. Washington, S.P., M.G. Karlaftis, and F. Mannering, Statistical and Econometric Methods for Transportation Data Analysis, CRC Press, Boca Raton, FL, 2003.
32. Zegeer, C.V., J. Hummer, L. Herf, D. Reinhurf, and W. Hunter, Safety Cost-Effectiveness of Incremental Changes in Cross-Section Design – Informational Guide, Report No. FHWA-RD-87-094, McLean, Va., 1987.
33. Zegeer, C. V., J. R. Stewart, F.M. Council, D. W. Reinfurt, and E. Hamilton, "Safety Effects of Geometric Improvements on Horizontal Curves", Transportation Research Record 1356, 1992.

References for Safety at Roundabouts

Department of Transport of Northrhine-Westfalia, Germany. Empfehlungen zum Einsatz und zur Gestaltung von Mini-Kreisverkehrsplaetzen (Guidelines for the Use and Design of Mini-Roundabouts). Düsseldorf, Germany, 1999.

Queensland Department of Main Roads (QDMR). Relationships between Roundabout Geometry and Accident Rates. Queensland, Australia: Infrastructure Design of the Technology Division of QDMR, April 1998.

Department of Transport (United Kingdom). Geometric Design of Roundabouts. TD 16/93. September 1993.
American Association of State Highway and Transportation Officials (AASHTO). A Policy on Geometric Design of Highways and Streets. Washington, D.C.: AASHTO, 1994.
Pein, W.E. Trail Intersection Design Guidelines. Prepared for State Bicycle/Pedestrian Program, State Safety Office, Florida Department of Transportation. Highway Safety Research Center, University of North Carolina, September 1996.
Service d'Etudes Techniques des Routes et Autoroutes (SETRA—Center for Technical Studies of Roads and Highways). Aménagement des Carrefours Interurbains sur les Routes Principales (Design of Rural Intersections on Major Roads). Ministry of Transport and Housing, December 1998.
Americans with Disabilities Act Accessibility Guidelines for Buildings and Facilities (ADAAG). 36 CFR Part 1191. As amended through January 1998.

Fambro, D.B., et al. NCHRP Report 400: Determination of Stopping Sight Distances. National Cooperative Highway Research Program, Transportation Research Board, National Research Council. Washington, D.C.: National Academy Press, 1997.
Harwood, D. W., et al. NCHRP Report 383: Intersection Sight Distances. National Cooperative Highway Research Program, Transportation Research Board, National Research Council. Washington, D.C.: National Academy Press, 1996.
Austroads. Guide to Traffic Engineering Practice, Part 6—Roundabouts. Sydney, Australia: Austroads, 1993.
Florida Department of Transportation. Florida Roundabout Guide. Florida Department of Transportation, March 1996.
American Association of State Highway and Transportation Officials (AASHTO). Guide for Development of Bicycle Facilities. Washington, D.C.: AASHTO, 1991.

Krammes, R., et al. Horizontal Alignment Design Consistency for Rural Two-Lane Highways. Publication No. FHWA-RD-94-034. Washington, D.C.: Federal Highway
Administration, January 1995.

Sawers, C. Mini roundabouts: Getting them right! Canterbury, Kent, United Kingdom: Euro-Marketing Communications, 1996.

Brilon, W., and L. Bondzio. Untersuchung von Mini-Kreisverkehrsplaetzen

(Investigation of Mini-Roundabouts). Ruhr-University Bochum, Germany, 1999

Chapter 5 Speed Limits

References:

1. Fitzpatrick, K., Carlson, P. J., Brewer, M.A., Wooldridge, M. D., and Miaou, S. P. "Design Speed, Operating Speed, and Posted Speed Practices." National Cooperative Highway Research Program Project 15-18 draft final report, July 2002.
2. American Association of State Highway and Transportation Officials, A Policy on Geometric Design of Highways and Streets (2001).
3. Manual on Uniform Traffic Control Devices for Streets and Highways -- Millennium Edition. Federal Highway Administration, U.S. Department of Transportation, Washington, DC (2000).
4. R. Lamm and E. M. Choueiri. "Recommendations for Evaluating Horizontal Design Consistency Based on Investigations in the State of New York." Transportation Research Record 1122, 1987. Pp. 68-78.
5. Oppenlander. Variables Influencing Spot-Speed Characteristics—Review of the Literature. Special Report 89, Highway Research Board, Washington, D.C., 1966.
6. D.B. Fambro, R. L. Nowlin, S. P. Warren, K.A. Lienau, J.M. Mounce, R. P. Bligh, K. K. Mak, and
7. H.E. Ross. Geometric Guidelines for Suburban High-Speed Curb and Gutter Roadways. FHWA/TX-95/1347-1F, College Station, TX, May 1995.
8. Fitzpatrick, K., Elefteriadou, L., Harwood, D., Collins, J., McFadden, J., Anderson, I. B., Krammes, R. A., Irizarry, N., Parma, K., Bauer, K., and Passetti, K., "Speed Prediction for Two-Lane Rural Highways." Report FHWA-RD-98-171, Federal Highway Administration, Washington, DC (2000).
9. Fitzpatrick, K., Carlson, P. J., Wooldridge, M. D., and Brewer, M. A., "Design Factors That Affect Driver Speed on Suburban Arterials." Report FHWA/TX-00/1769-3. Texas Transportation Institute, College Station, TX (August 1999).
10. Poe, C. and Mason Jr., M., "Analyzing Influence of Geometric Design on Operating Speeds Along Low-Speed Urban Streets: Mixed Model Approach." Transportation Research Record 1737. Transportation Research Board, Washington, DC (2000) pp 18-25.
11. Parker, Jr., M. R., "Effects of Raising and Lowering Speed Limits on Selected Roadway Sections." Final Report, Publication No. FHWA-RD-

92-084, Federal Highway Administration, Washington, DC (January 1997).
12. Fambro, D. B., Fitzpatrick, K., and Koppa, R. J., "Determination of Stopping Sight Distances." NCHRP Report 400, Transportation Research Board, Washington, DC (1997).
13. Krammes, R. A., Brackett, R. Q., Shafer, M. A., Ottesen, J. L., Anderson, I. B., Fink, K. L., Collins, K. M., Pendleton, O. J., and Messer, C. J., "Horizontal Alignment Design Consistency for Rural Two-Lane Highways." Research Report FHWA-RD-94-034. Federal Highway Administration, Washington, DC (1994).
14. Polus, A., Fitzpatrick, K., and Fambro, D., "Predicting Operating Speeds on Tangent Sections of Two-Lane Rural Highways." Transportation Research Record 1737. Transportation Research Board, Washington, DC (2000) pp. 50-57.
15. Chowdhury, M. A., Warren, D. L., and Bissell, H., "Analysis of Advisory Speed Setting Criteria." Public Roads, Vol. 55, No. 3 (December 1991) pp. 65–71.
16. McLean, J.R., "Speeds, Friction Factors, and Alignment Design Standards. "Research Report ARR 154." Australian Road Research Board, Victoria, Australia (1988).
17. McLean, J., "An Alternative to the Design Speed Concept for Low Speed Alignment Design." Transportation Research Record 702. Transportation Research Board, Washington, DC, (1979) pp. 55–63.

Further References:

1. AASHTO, A policy on geometric design of highways and streets,1994
2. Blaschke, Joseph D., John M. Mason; Impact of the AASHTO green book on highway to reliability, TRR-100
3. C.A.O. Flaherty, Editor, Transportation Planning and Traffic Engineering
4. Context-Sensitive Highway Design(http://www.scenic.org/roads.htm)
5. DelDOT Road Design Manual, Chapter 3-DesignStandards2/1992.
6. Elmberg C.M., Effects of speed zoning in urban areas, M.S. Thesis, Purdue University, 5 / 1960.
7. Hassa, Gibreel, and S. M. Easa, Evaluation of Highway Consistency and Safety; Practical Application
8. Hierche E-U, R. Lamn, K. Dieterle and A. Nikpour, "Effects of highway improvements designed in conformity with the RAL-L on traffic safety of two-lane rural highway"; Research Road Construction and Road Traffic Technique, Vol. 431, Minister of Transportation, Bonn, Germany, 1984.

9. Horizontal Alignment Design Consistency for Two-Lane Highways, FHWA-RD-94-034,1995
10. Insurance Institute for Highway Safety, Highway Loss Data Institute, Speed and Speed Limits (http://www.hwysafety.org/safety_facts/qanda/speed_limits.htm)
11. Kalivoda, Eric, The effect of Design Speed on two-lane rural highway safety, PhD. thesis, 1995.
12. Krammes R., Fitzpatrick K., Blaschke J., Fambro D., Speed: Understanding design, operating and posted speed, Report No. 1465-1
13. Lamm R., Psarioanos B., Mailaender T., Highway design and traffic safety engineering handbook, 1999
14. Mason, John M., Poe, Christopher M., Geometric Design Guidelines to Achieve Desired Operating Speed on Urban Streets, Institute of Transportation Engineers 65th Annual Meeting
15. McKean J., An Alternative to the design speed concept for low-speed alignment design, Australia
16. Road Research Board, Low Volume Roads: Second International Conference, Transportation Research Record 702, 1979
17. Messer C.J., J.M. Mounce and R.Q. Braquet. Highway geometric design consistency related to driver expectancy. FHWA, USDOT 1981
18. Mihailo Maletim, VajoAnjus; Standard speeds and flows in the road design process, 1985.
19. Recommended guidelines for subdivision streets; ITE publication.
20. Reudiger Lamn, Elias M., Choueiri, Theodor Mailaender; Comparison of 1280 Rural Road Design, Australia,1999
21. Transafety Reporter, Survey investigates the Relationships among design speeds, operating speeds, and posted speed limits, 1997
22. Washington DOT Manual. Additional References are used in Section 8 – Additional Literature Review
23. "Design Speed Selection Recommendations" by Ardeshir Faghri, PhD, David Capparuccini and Tatiana Cvetek, Department of Civil and Environmental Engineering, University of Delaware, 355 Du Pont Hall, Newark, Delaware 19716, Tel. (302) 831-1446
24. Department of Transportation. Texas TxDOT http://onlinemanuals.txdot.gov/txdotmanuals/rdw/traffic_characteristics.htm

Speed factors on two-lane rural highways

References:

1. Tarris, J., Poe, C., Mason, Jr., J., and K. Goulias. Predicting Operating Speeds on Low-Speed Urban Streets: Regression and Panel Analysis Approaches. In Transportation Research Record 1523, TRB, National Research Council, Washington, D.C.,1996, pp.46-54.
2. Fitzpatrick, K., Carlson, P., Brewer, M., Wooldrige, M., and S. Miaou. Design Speed, Operating Speed and Posted Speed Practices. NCHRP Report 504. National Research Council, Washington, D.C., 2003.
3. Garber, N. and R. Gadiraju. Factors affecting speed variance and its influence on accidents. In Transportation Research Record 1213, TRB, National Research Council, Washington, D.C., 1989, pp. 64-71. 4. Collins, J., Fitzpatrick, K., Bauer, K. and Hardwood, D. Speed variability on rural two-lane highways. In Transportation Research Record 1658, TRB, National Research Council, Washington, D.C., 1999, pp. 60-69.
4. Yagar, S. and M. Van Aerde. Geometric and environmental effects on speeds of two-lane highways. Transportation Research Part A: Policy and Practice, Vol. 17A, No. 4, 1983, pp. 315-325.
5. Yagar, S. Predicting speeds for rural two-lane highways. Transportation Research Part A: Policy and Practice, Vol. 18A, No. 1, 1984, pp. 61-70.
6. Polus, A., Fitzpatrick, K., and D. Fambro. Predicting Operating Speeds on Tangent Sections of Two-Lane Rural Highways. In Transportation Research Record 1737, TRB, National Research Council, Washington, D.C., 2000, pp. 50-57.
7. Islam, N. and P. Seneviratne. Evaluation of design consistency of two-lane rural highways. ITE Journal, Vol. 64, No. 2, 1994, pp. 28-31.
8. McFadden, J., and L. Elefteriadou. Evaluating horizontal alignment design consistency of two-lane rural highways. In Transportation Research Record 1737, TRB, National Research Council, Washington, D.C., 2000, pp. 9-17.
9. Fitzpatrick, K., Elefteriadou, L., Harwood, D., Collins, J., McFadden, J., Anderson, I., Krammes, R., Irizarry, N., Parma, K., Bauer, K. and K. Passetti. Speed Prediction for Two-Lane Rural Highways. Report FHWA-RD-99-171, FHWA, U.S. Department of Transportation, 1999.
10. Bonneson, J. (2001). Controls for horizontal curve design. Transportation Research Record 1751, pp. 82-89.
11. Polus, A., Poe, C. and Mason Jr, J. (1995) Review of international design speed practices in roadway geometric design. International Symposium

on Highway Geometric Design Practices, Conference Proceedings, Boston, Massachusetts, Chapter 5, pp. 1–8.
12. Morgenstern, B., Mahoney, D. and B.Warady. Estimating total body weight in children on the basis of height and weight: Are evaluation of the formulas of Mellits and Cheek. Journal of the American Society of Nephrology. Vol. 13, 2002, pp. 1884-1888.
13. Morrell, C., Pearson, J., Brant, L. and S. Gordon-Salant. Construction of hearing percentiles in women with non-constant variance from the linear mixed-effects model. Statistics in Medicine. Vol. 16, 1997, pp. 2475-2488.
14. C.Hsiao. Analysis of Panel Data. Cambridge University Press, New York, U.S.A., 1986.
15. Figueroa, A. and A. Tarko. Reconciling Speed Limits with Design Speeds. Report FHWA/IN/JTRP-2004/07, Joint Transportation Research Program, Purdue University, 2004.
16. Poe, C. and M. Mason. Analysing the influence of geometric design on operating speeds along low-speed urban streets. In Transportation Research Record 1737, TRB, National Research Council, Washington, D.C., 2000, pp. 18-25.
17. AASHTO. A Policy on Geometric Design of Highways and Streets. 2001.
18. Washington, S., Karlaftis, M., and F. Mannering. Statistical and Econometric Methods for Transportation Data Analysis. Chapman & Hall / CRC Press LLC, Boca Raton, Florida, 2003.

Chapter 6 Visibility during Night, Rain, Fog, and Highway Lighting

Highway Lighting

References:

The Impacts of Illumination Levels on Nighttime Safety at Roundabouts, A Dissertation Presented to The Academic Faculty by Franklin E. Gbologah, in partial fulfilment of the requirements for the degree of Doctor of Philosophy in Civil Engineering, in the School of Civil and Environmental Engineering, Georgia Institute of Technology

1. AASHTO (2010). Highway Safety Manual, 1st Ed. Vol 1. American Association of State Highway and Transportation Officials, Washington, DC.

2. AASHTO (2010). Highway Safety Manual, 1st Ed. Vol 3. American Association of State Highway and Transportation Officials.
3. Abdel-Aty, M., Wang, X., and Santos, J. (2009). "Identifying Intersection-Related Traffic Crashes for Accurate Safety Representation." In the Institute of Transportation Engineers. ITE Journal, No. 79(12), pp. 38-44.
4. Al-Ghamdi, A.S. (2003). "Analysis of Traffic Accidents at Urban Intersections in Riyadh." In Accident Analysis & Prevention, No. 35(5), pp. 717-724.
5. Alisoglu, S. (2010). "Roundabout Vs Signalized Intersections: A Comprehensive Analysis." in Kansas Government Journal. Kansas Department of Transportation: Kansas.
6. Azhar, A. Svante, B. (2011). "Signal Control of Roundabouts." In Procedia-Social and Behavioural Sciences, No. 16, pp. 729-738.
7. Bassani, M., Mutani, G. (2012). "Effects of Environmental Lighting Conditions on Operating Speeds on Urban Arterials." In Transportation Research Record: Journal of the Transportation Research Board, No. 2298(-1), pp. 78-87.
8. Bertolotti, L., Salmon, M. (2005). "Do Embedded Roadway Lights Protect Sea Turtles?" In Environmental Management, No. 36(5), pp. 702-710.
9. Bhagavathula, R., Gibbons, R., and Edwards, C. (2015). "Relationship between Roadway Illumination Level and Nighttime Rural Intersection Safety." In Transportation Research Record: Journal of the Transportation Research Board, No. 2485, pp. 8.
10. Bo, L., Bhagwant, P., Craig, L., and Ravi, B. (2009). "Validation of a Full Bayes Methodology for Observational before-after Road Safety Studies and Application to Evaluation of Rural Signal Conversions." In Accident Analysis & Prevention, No. 41, pp. 574-580.
11. Box, P. (1970). "Intersections" in Traffic Control and Roadway Elements – Their Relationship to Highway Safety, Automotive Safety Foundation, Highway Users Federation for Safety and Mobility, Washington, DC.
12. Box, P. (1970). "Relationship between Illumination and Freeway Accidents." Illuminating Research Institute, New York.
13. Brons, J., Bullough, J., and Rea, M. (2007). "Light Pollution: Thinking inside the Box." In Lighting Journal (Rugby, the UK), No. 72(5), pp. 27-34.
14. Brons, J.A., Bullough, J.D., and Rea, M.S. (2008). "Outdoor Site-Lighting Performance: Comprehensive and Quantitative Framework for

Assessing Light Pollution." In Lighting Research and Technology, No.40(3), pp. 201-224.

15. Bruneau, J.F., Morin, D. (2005). "Standard and Nonstandard Roadway Lighting Compared with Darkness at Rural Intersections." In Transportation Research Record: Journal of the Transportation Research Board, No. 1918, pp. 116-122.
16. Bryer, T. (2011). "Stop-Controlled Intersection Safety: Through Route Activated Warning Systems." FHWA-SA-11-015, Federal Highway Administration.
17. Bullough, J. (2002). "Interpreting Outdoor Luminaire Cutoff Classification." In Lighting Design and Application: LD and A, No. 32(7), pp. 44.
18. Bullough, J., Snyder, J., Skinner, N., Capo, R., Rizzo, P., and Besenecker, U. (2014). "Demonstration and Evaluation of New Approaches for Roundabout Lighting." Transportation Research Board 93rd Annual Meeting. Washington, DC.
19. California Energy Commission. (2005). "Outdoor Lighting Zones: General Information." Sacramento, CA.
20. Casol, M., Fiorentin, P., and Scroccaro, A. (2008). "On Road Measurements of the Luminance Coefficient of Paving." 16th IMEKO TC4 Symposium Exploring New
21. Frontiers of Instrumentation and Methods for Electrical and Electronic Measurements. Florence. CEN. (2008c). "Road Lighting - Part 3: Calculation of Performance." In.
22. Chin, H.C. Quddus, M.A. (2003). "Applying the Random Effect Negative Binomial Model to Examine Traffic Accident Occurrence at Signalized Intersections." In Accident Analysis & Prevention, No. 35(2),
23. pp. 253-259.
24. Cobb, J., Hargroves, R.A., Marsden, A.M., and Scott, P. (1979). "Road Lighting and Accidents." Meeting of the International Commission on Illumination (CIE). Kyoto, Japan.
25. Cottrell, D.Mu, S. (2005). "Utah Intersection Safety: Recurrent Crash Sites: Identification, Issues and Factors." University of Utah, Mountain Plains Consortium.
26. Daniels, S., Brijs, T., Nuyts, E., and Wets, G. (2010a). "Explaining Variation in Safety Performance of Roundabouts." In Accident Analysis & Prevention, No. 42(2), pp. 393-402.
27. Daniels, S., Nuyts, E., and Wets, G. (2008). "The Effects of Roundabouts on Traffic Safety for Bicyclists: An Observational Study." In Accident Analysis & Prevention, No. 40(2), pp. 518-526.

28. De Brabander, B. Vereeck, L. (2007). "Safety Effects of Roundabouts in Flanders: Signal Type, Speed Limits and Vulnerable Road Users." In Accident Analysis & Prevention, No. 39(3), pp. 591-599.
29. Donnell, E. (2015). "Statistical Issues in Relating Roadway Lighting to Safety." Transportation Research Board 94th Annual Meeting. Washington, DC.
30. Donnell, E., Porter, R.J., Shankar, V.N., Gross, F., Hu, W., Gemar, M., and Oh, J. (2011). "Estimating the Safety Effects of Roadway Lighting at Intersections: A Cross-Sectional Study Approach." 90th Annual Meeting of the Transportation Research Board. Washington, D.C.: Transportation Research Board.
31. Donnell, E.T., Porter, R.J., and Shankar, V.N. (2010). "A Framework for Estimating the Safety Effects of Roadway Lighting at Intersections." In SAFETY SCIENCE, No. 48(10), pp. 1436-1444.
32. Elvik, R. (1995). "Meta-Analysis of Evaluations of Public Lighting as Accident Countermeasure." In Transportation Research Record, No. (1485), pp. 112-123.
33. FHWA. (1994). "The 1994 Annual Report on Highway Safety Improvement Programs." O.O.S. U.S. Department of Transportation Federal Highway Administration Washington, DC.
34. FHWA. (1996). "The 1996 Annual Report on Highway Safety Improvement Programs." U.S. Department of Transportation Federal Highway Administration Office of Safety Washington, DC.
35. FHWA. (2000)."Roundabouts: An Informational Guide."FHWA-RD-00-067,
36. FHWA. (2009)."Intersection Safety Issue Briefs: The National Intersection Problem."
37. FHWA-SA-10-005, Federal Highway Administration, Washington, DC.
38. FHWA. (2012)."2012 Guidance Memorandum on Promoting the Implementation of Proven Safety Countermeasures." FHWA (2012). http://safety.fhwa.dot.gov/provencountermeasures/fhwa_sa_12_005.pdf.(12/23/2013).
39. Flannery, A. (2001)."Geometric Design and Safety Aspects of Roundabouts. "In Transportation Research Record, No. 1751(1), pp. 76.
40. Forrest, M. Yussuf, M. (2007)."Highway Safety Information System Guidebooks for the Minnesota State Data Files." Washington, DC.
41. Fotios, S., Boyce, P., and Ellis, C. (2005)."The Effect of Pavement Material on Road Lighting Performance." PPAD 9/100/77 UK.
42. Garder, P. (2012). "Fuel Consumption at a Modern Roundabout Vs. A Signalized Intersection: A Case Study Comparing Two Similar

Intersections in Bangor, Maine." 91st Annual Meeting of the Transportation Research Board. Washington DC.
43. Gbologah, F., Samoylov, A., Guin, A., Hunter, M., Pass, D., and Rodgers, M. (2015). "Evaluation of Roundabout Illumination and Safety." In Submitted to Journal of Accident Prevention and Analysis, No.
44. Georgia Governor's Office of Highway Safety "Crash Injuries. "http://www.gahighwaysafety.org/research/crash-injuries/.(7/14/2014).
45. Gibbons, R. Influence of Pavement Reflectivity on Target Visibility. University of Waterloo 1997
46. Gibbons, R., Guo, F., Du, J., Medina, A., Lutkevich, P., Terry, T., and Li, Q. (2015). "Linking Road Lighting and Crash Safety." Transportation Research Board 94th Annual Meeting. Washington, DC.
47. Green, E.R., Agent, K.R., Barret, M.L., and Pigman, J.G. (2003). "Roadway Lighting and Driver Safety." KTC-03-12/SPR247-02-IF,
48. Green, J. Hargroves, R.A. (1979). "A Mobile Laboratory for Dynamic Road Lighting Measurement." In Lighting Research & Technology, No. (11), pp. 7.
49. Green, W.H. (2003). Econometric Analysis, Fifth Ed. Prentice Hall, Upper Saddle River, NJ.
50. Griffith, M. (1994). "Comparison of the Safety of Lighting Options for Urban Freeways," in Public Roads. Federal Highway Administration: Washington, DC.
51. Gross, F., Lyon, C., Persaud, B., and Srinivasan, R. (2013). "Safety Effectiveness of Converting Signalized Intersections to Roundabouts." In Accident Analysis and Prevention, No. 50, pp. 234-241.
52. Hallmark, S., Hawkins, N., Omar, S., Hans, Z., Kinsenbaw, C., Orellano, M., and Isebrands, H. (2008). "Strategies to Address Nighttime Crashes at Rural, Unsignalized Intersections." Center for Transportation and Education, Iowa State University.
53. Hammond, S., Hunter, C., and Chang, K. (2014). "The Effect of Additional Lane Length on Roundabout Delay". Transportation Research Board 93rd Annual Meeting. Washington, DC: Transportation Research Board.
54. Hardwood, D., Bauer, K., Potts, I., Torbic, D., Richard, K., Rabbani, E., Hauer, E., E., Elefteriadou, L., and Griffith, M. (2003). "Safety Effectiveness of Intersection Left and Right Turn Lanes." In Transportation Research Record: Journal of the Transportation Research Board, No. 1840.
55. Hargroves, R. A. Scott, P. (1979). "Measurements of Road Lighting and Accident Results." In Public Lighting, No. (44), pp. 8.

56. Hauer, E. (1996). "Statistical Test of Difference between Expected Accident Frequencies." In TRANSPORTATION RESEARCH RECORD, No. (1542), pp. 24- 29.
57. Hauer, E. (1997). Observational before-and-after Studies in Road Safety. Pergamon, Oxford, OX, U.K.; Tarrytown, N.Y., U.S.A.
58. Hauer, E. (2005). "Cause and Effect in Observational Cross-Sectional Studies in Road Safety." Transportation Research Board Annual Meeting.
59. Highways Agency. (2007). "Geometric Design of Roundabouts. Design Manual of Roads and Bridges, Vol 6, Section 2, Part 3, Td 16/07." London.
60. Hiscocks, P. Eng, P. (2011)."Measuring Luminance with a Digital Camera." In. Hydén, C.Várhelyi, A. (2000). "The Effects on Safety, Time Consumption and Environment of Large-Scale Use of Roundabouts in an Urban Area: A Case Study." In Accident Analysis & Prevention, No. 32(1), pp. 11-23.
61. IIHS. (2000)."Roundabouts Status Report. "Illuminating Engineering Society. (1999). In Lm 5099 - IESNA Guide for Photometric Measurement of Roadway Lighting Installations.
62. Illuminating Engineering Society. (2000). "Light Trespass: Research, Results, and Recommendations." New York.
63. Illuminating Engineering Society, "Roadway Lighting ", ANSI/IESNA RP-8-00 Reaffirmed 2005. Illumination Engineering Society of North America, New York.
64. Illuminating Engineering Society, "Design Guide for Roundabout Lighting", IES DG-19-08. Illuminating Engineering Society of North America, New York.
65. Illuminating Engineering Society.(2014)."RoadwayLightingAnsi/IesRp8-14.RP8-14, I .E. Society
66. Inanici, M.N. (2006). "Evaluation of High Dynamic Range Photography as a Luminance Data Acquisition System." In Lighting Research and Technology, No. 38(2), pp. 123-136.
67. Institute of Lighting Engineers. (2000)."Guidance Notes for the Reduction of Lighting Pollution." Warwickshire, UK.
68. International Commission on Illumination. (1976). "Calculation and Measurement of Luminance and Illuminance in Road Lighting." CIE Publication No. 30,
69. International Commission on Illumination. (1982)."Calculation and Measurement of Luminance and Illuminance in Road Lighting."CIENo.30-2(TC-4-6), Paris.

70. International Commission on Illumination.(1984)."RoadSurfacesandLighting."CIENo.66,
71. International Commission on Illumination. (1992). "Road Lighting as an Accident Countermeasure." Publication No. 93, Central Bureau of the CIE Vienna. International Commission on Illumination. (2000). "Road Lighting Calculations." CIE 140 - 2000, Vienna, Austria.
72. International Commission on Illumination. (2003). "Guide on the Limitation of the Effects of Obtrusive Light from Outdoor Lighting Installations." In PUBLICATIONS- COMMISSION INTERNATIONALE DEL ECLAIRAGE CIE, No. 150, pp. ALL.
73. International Commission on Illumination, "CIE 115:2010 Lighting of Roads for Motor and Pedestrian Traffic 2nd Edition", CIE Central Bureau, Austria.
74. Isebrands, H. (2009). "Crash Analysis of Roundabouts at Highspeed Rural Intersections." In Transportation Research Record: Journal of the Transportation Research Board, No. 2096, pp. 7.
75. Isebrands, Hallmark, S. (2012). "Statistical Analysis and Development of Crash Prediction Model for Roundabouts on High-Speed Rural Roadways." In Transportation Research Record: Journal of the Transportation Research Board, No. 2312(-1), pp. 3-13.
76. Isebrands, H., Hallmark, S., Hans, Z., McDonald, T., Preston, H., and Storm, R. (2004). "Safety Impacts of Street Lighting at Isolated Rural Intersections." Minnesota Department of Transportation
77. Isebrands, H., Hallmark, S., Hans, Z., McDonald, T., Preston, H., and Storm, R. (2006). "Safety Impacts of Street Lighting at Isolated Rural Intersection - Part 2." MN/RC-2006-35, Minnesota Department of Transportation
78. Isebrands, H., Hallmark, S., and Hawkins, N. (2014). "Effects of Approach Speed at Rural High-Speed Intersections." In Transportation Research Record: Journal of the Transportation Research Board, No. 2402(-1), pp. 67-77.
79. Isebrands, H., Hallmark, S., Li, W., McDonald, T., Storm, R., and Preston, H. (2010). "Roadway Lighting Shows Safety Benefits at Rural Intersections." In Journal of Transportation Engineering, No. 136(11), pp. 949-955.
80. Jackett, M. Frith, W. (2012). "How Does the Level of Road Lighting Affect Crashes in New Zealand - a Pilot Study." New Zealand Transport Agency.
81. Jackett, M. Frith, W. (2013). "Quantifying the Impact of Road Lighting on Road Safety — a New Zealand Study." In IATSS Research, No. 36(2), pp. 139-145.

82. Jensen, S. (2013). "Safety Effects of Converting Intersections to Roundabouts." 92nd Annual Meeting of the Transportation Research Board. Washington, DC.
83. Joksch, H. Kostyniuk, L. (1998). "Modeling Intersection Crash Counts and Traffic Volume." FHWA-RD- 98-096, Washington, DC.
84. Kansas Department of Transportation. (2003)."Kansas Roundabout Guide."
85. Keith, D.M. (2000). "Roadway Lighting Design for Optimization of Upd, Stv and Uplight." In Journal of the Illuminating Engineering Society, No. 29(2), pp. 15-21.
86. Keith, D.M. (2003). "Correlations of Roadway Uud Values to Upd, Uplight and Classifications." In Journal Of the Illuminating Engineering Society, No. 32(1), pp. 29-+.
87. Modern Roundabout-Inventory Database. Downloadedon12/22/2013. http://roundabout.kittelson.com/Roundabouts/Search.
88. Kusuma, A., Koutsopoulos, H. (2011). "Critical Gap Analysis of Dual Lane Roundabouts." In Procedia - Social and Behavioural Sciences, No. 16, pp. 709-717.
89. Lenters, M. (2005)."Safety Auditing of Roundabouts. "National Roundabout Conference. Colorado.
90. Lipinski, M. E. Wortman, R.H. (1978). "Effect of Illumination on Rural at-Grade Intersection Accidents. "In Transportation Research Record, No. (611), p. 3.
91. Lord, D. Persaud, B. (2000). "Accident Prediction Models with and without Trend: Application of the Generalized Estimating Equations Procedure." In Transportation Research Record: Journal of the Transportation Research Board, No. 1717(-1), pp. 102-108.
92. Lyon, C., Haq, A., Persaud, B., and Kodama, S. (2005). "Safety Performance Functions for Signalized
93. Intersections in Large Urban Areas: Development and Applications to Evaluation Left-Turn Priority
94. Treatment." In Transportation Research Record: Journal of the Transportation Research Board, No.1908.
95. Manco, J. (2004)."The Hub of the Circus: A History of the Streetscape of the Circus, Bath."
96. Marks, V. (1977). "Effects of Reduced Intersection Lighting on Nighttime Accident Frequency." Project HR-1003A, Iowa Department of Transportation.
97. McGee, H., Taori, S., and Persaud, B. (2003). "Nchrp 491: Crash Experience Warrant for Traffic Signals." NCHRP 491, T.R. Board, Washington, DC.

98. Medina, J., Avrenli, K., and Benekohal, R. (2013). "Field and Software Evaluation of Illuminance from LED Luminaires for Roadway Applications." In Transportation Research Record: Journal of the Transportation Research Board, No. 2384(-1), pp. 55-64.
99. Minoshima, O., Oka, K., Ikehara, K., and Inukai, N. (2006). "Research on Accident Reduction by Intersection Lighting." Road Engineering Association of Asia and Australia (REAAA). Philippines.
100. Miska, E., Leur, P., and Sayed, T. (1998). "Road Safety Performance Associated with Improved Traffic Signal Design and Increased Signal Conspicuity "http://mutcd.fhwa.dot.gov/pdfs/miska_02.pdf.(12/23/2013).
101. Mittra, S., Washington, S., and Van Schalkwyk, I. (2007). "Important Omitted Spatial Variables in Safety Models: Understanding Contributing Crash Causes at Intersections." TRB 86th Annual Meeting Compendium of Papers. Washington, DC.
102. Møller, M. Hels, T. (2008). "Cyclists' Perception of Risk in Roundabouts." In Accident Analysis & Prevention, No. 40(3), pp. 1055-1062.
103. Monsere, C. M. Fischer, E.L. (2008). "Safety Effects of Reducing Freeway Illumination for Energy Conservation." In Accident Analysis & Prevention, No. 40(5), pp. 1773-1780.
104. Montella, A. (2011). "Identifying Crash Contributory Factors at Urban Roundabouts and Using Association Rules to Explore Their Relationships to Different Crash Types." In Accident Analysis and Prevention, No. 43, pp. 1451-1463.
105. Niaki, M., Saunier, N., Miranda-Moreno, L., Amador, L., and Bruneau, J.-F. (2014). "Method for Road Lighting Audit and Safety Screening at Urban Intersections." In Transportation Research Record: Journal of the Transportation Research Board, No. 2458, pp. 27-36.
106. Nicholas, J. (1991)."A Road Lighting Survey Method for Accident Sites. "Report No.4, North Carolina Department of Transportation. (1999). "North Carolina Crash Report Instruction Manual."
107. O'Flaherty (1997). Transport Planning and Traffic Engineering. John Wiley and Sons, Inc., New York.
108. Ourston Roundabout Engineering (2010)."Leif Ourston Receives Frank Blackmore Award." http://www.ourston.com/updates-archive/114.html.
109. Oya, H., Ando, K., and Kanoshima, H. (2002). "Research on the Interrelation between Illuminance at Intersections and Reduction in Traffic Accidents." In Journal of Light & Visual Environment, No. 26(1), pp._29-34.

110. Painter, Farrington, D. (2001). "The Financial Benefits of Improved Street Light, Based on Crime Reduction." In Lighting Research & Technology, No. 33(1), pp. 3-12.
111. Per Ole, W. (2009). "Effects of Road Lighting: An Analysis Based on Dutch Accident Statistics 1987– 2006." In Accident Analysis and Prevention, No. 41, pp. 123-128.
112. Persaud, B., Council, F., Lyon, C., Ecles, K., and Griffith, M. (2005). "Multijurisdictional Safety Evaluation of Red Light Cameras." In Transportation Research Record: Journal of the Transportation Research Board, No. 1922.
113. Persaud, B.Lyon, C. (2007). "Empirical Bayes before–after Safety Studies: Lessons Learned from Two Decades of Experience and Future Directions." In Accident Analysis & Prevention, No. 39(3), pp. 546- 555.
114. Persaud, B., Retting, R., Garder, P., and Lord, D. (2001). "Safety Effect of Roundabout Conversions in the United States: Empirical Bayes Observational before-and-after Study." In Transportation Research Record: Journal of the Transportation Research Board, No. 1751(-1), pp. 1-8.
115. Pochowski, A.L. (2011). "Review of Statewide Roundabout Policies." 90th Annual Meeting of the Transportation Research Board. Washington DC.
116. Preston, H. Schoenecker, T. (1999). "Safety Impacts of Street Lighting at Isolated Rural Intersections." 1999-17, Minnesota Department of Transportation. Qin, X., Bill, A.,
117. Chitturin, M., and Noyce, D. (2013). "Evaluation of Roundabout Safety." 92nd Annual Meeting of the Transportation Research Board. Washington, DC.
118. Rea, M., Bullough, J., Fay, C., Brons, J., Derlofske, J., and Donnell, E. (2009). "Review of the Safety Benefits and Other Effects of Roadway Lighting." NCHRP 5-19, NCHRP, Washington, DC.
119. Retting, R., Persaud, B., Garder, P., and Lord, D. (2001). "Crash and Injury Reduction Following Installation of Roundabouts in the United States." In American Journal of Public Health, No. 91(4), pp. 628 - 631.
120. Robinson, B., Rodgers, W., Scarborough, W., Kittleson, W., Troutbeck, R., Brillon, Bondzio, L., Courage, K., Kyte, M., Mason, J., Flannery, A., Myers, E., Bunker, J., and Jacquemart, G. (2000)."Roundabouts: An Informational Guide."FHWARD-00-067,
121. Rodegerdts, L. (2008). "Status of Roundabouts in North America."

122. Rodegerdts, L., Bansen, J., Tiesler, C., Knudsen, J., Myers, E., Johnson, M., Moule, M.,
123. Persaud, B., Lyon, C., Hallmark, S., Isebrands, H., Crown, R.B., Guichet, B., and O'Brin, A. (2010). "Roundabouts: An Informational Guide, 2nd Ed." NCHRP 672, Transportation Research Board.
124. Rodegerdts, L., Blogg, M., Wemple, E., Myers, E., Kyte, M., Dixon, K., List, G.,
125. Flannery, A., Troutbeck, R., Brilon, W., Wu, N., Persaud, B., Lyon, C., Harkey, D., and Carter, D. (2007a). "Roundabouts in the United States." NCHRP 572, Transportation Research Board.
126. Rodegerdts, L., Blogg, M., Wemple, E., Myers, E., Kyte, M., Dixon, M., List, G., Flannery, A., Troutbeck, R., Brilon, W., Wu, N., Persaud, B., Lyon, C., Harkey, D., and Carter, D. (2007b). "NCHRP Web-Only Document 94 - Appendixes to NCHRP 572: Roundabouts in the United States.": Transportation Research Board, Washington, DC.
127. Rodgers, M., Hunter, M., Samoylov, A., and Gbologah, F. (2014). "Evaluation of Current Practice for Illumination at Roundabouts: Safety and Illumination of Roundabouts - Phase 1. Estimation of the Safety Impact of Illumination at Roundabouts." G.D. o. Transportation.
128. Rodgers, M., Hunter, M., Samoylov, A., Gbologah, F., and Berrebi, S. (2014). "Evaluation of Current Practice for Illumination at Roundabouts: Safety and Illumination of Roundabouts - Phase 1. A Review of International Roundabout Lighting Practices, Polices, and Standards." GDO. Transportation.
129. Schmidt, S., Long, S., and Harper, J. (2014). "Economic and Technical Feasibility of LED Lighting on State Routes in Missouri." Transportation Research Board 93rd Annual Meeting. Washington, DC: Transportation Research Board.
130. Schmitt, L. Calibration of the HCM 2010 Single-Lane Roundabout Capacity Equations for Georgia Conditions (Phase 2). Georgia Institute of Technology. 2013
131. Schneider, C., Rasband, W., and Eliceiri, K. (2012). "NIH Image to Imagej: 25 Years of Image Analysis." In Nature Methods, No. 9(7), pp. 4.
132. Schoon, C. van Minnen, J. (1994). "The Safety of Roundabouts in the Netherlands." In Traffic Engineering Control No., pp. 142-148.
133. Scott, P. (1980). "The Relationship between Road Lighting Quality and Accident Frequency." LR 929,
134. SETRA. (1998)."The Design of Interurban Intersections on Major Roads: At-Grade Intersections." Bagneux, Cedex.

135. Shankar, V., Albin, R., Milton, J., and Mannering, F. (1998). "Evaluating Median Crossover Likelihoods with Clustered Accident Counts: An Empirical Inquiry Using the Random Effects Negative Binomial Model." In Transportation Research Record: Journal of the Transportation Research Board, No. 1635(- 1), pp. 44-48.
136. Spacek, P. (2004). "Basis of the Swiss Design Standard for Roundabout." In Transportation Research Record, No. (1881), pp. 27-35.
137. Troutbeck, R. (1993). "Capacity and Design of Traffic Circles in Australia." In Transportation Research Record, No. 1368, pp. 68-74.
138. Uddin, W., Headrick, J., and Sullivan, J. (2012). "Performance Evaluation of Roundabouts for Traffic Flow Improvements and Crash Reductions at a Highway Interchange in Oxford, MS." 91st Annual Meeting of the Transportation Research Board. Washington, DC.
139. Ulfarsson, G. Shankar, V. (2003). "Accident Count Model Based on Multiyear Cross-Sectional Roadway Data with Serial Correlation." In Transportation Research Record: Journal of the Transportation Research Board, No. 1840(-1), pp. 193-197.
140. Uncu, I. Kayaku, M. (2010). "Analysis of Visibility Level in Road Lighting Using Image Processing Techniques." In Scientific Research and Essays, No. 5(18), pp. 6.
141. Walker, F. W., Roberts, S.E. (1976). "Influence of Lighting on Accident Frequency at Highway Intersections." In Transportation Research Record, No. (562), pp. 73-78.
142. Wang, X., Abdel-Aty, M., and Brady, P. (2006). "Crash Estimation at Signalized Intersections: Significant Factors and Temporal Effect." In Transportation Research Record: Journal of the Transportation Research Board, No. 1953(-1), pp. 10-20.
143. Wisconsin Department of Transportation. (2001). "Roundabout Guide." 149.Wortman, R. H. Lipinski, M.E. (1974)."Rural at-Grade Intersection Illumination."
144. Wuller, D., Gabele, H. (2007). "The Usage of Digital Cameras as Luminance Meters." Place of Publication: San Jose, CA, USA. Country of Publication: USA. SPIE -The International Society for Optical Engineering.
145. Zhou, H.G., Pirinccioglu, F., and Hsu, P. (2009). "A New Roadway Lighting Measurement System." In Transportation Research Part C-Emerging Technologies, No. 17(3), pp. 274-284.
146. 152 Zirkel, B., Park, S., McFadden, J., Angelastro, M., and McCarthy, L. (2013). "Influence of Sight Distance on Crash Rates and Operating Speeds at Low-Volume Single-Lane Roundabouts in the United

States. "In Transportation Research Record: Journal of the Transportation Research Board, No. 2389(-1), pp. 42-50.

Chapter 7 Construction and Operational Safety

References:

By: Kendall Jones, Editor, ConstructConnect, on 3rd of June 2022
https://www.constructconnect.com/blog/10-road-construction-work-zone-safety-tips

Road traffic safety. From Wikipedia, the free encyclopedia
https://en.wikipedia.org/wiki/Road_traffic_safety

Chapter 8 Intelligent Transport Systems (ITS)

References:

Blokland, M., Mouris, R. (2005). Behavioural Effects of Multimodal Travel Information. Report to Dutch Ministry of Transport, Advies Dienst Verkeer en Vervoer (Travel and Transport Advisory Service), Rotterdam.
Carsten O. M. J. and Tate F. N. (2005). Intelligent speed adaptation: accident savings and cost–benefit analysis Accident Analysis & Prevention, Volume 37, Issue 3, May 2005, pp. 407-416
Winnett, MA & Wheeler, A. H. (2002). Vehicle-activated signs–a large-scale evaluation, TRL Report 548. Please note that websites may be closed after a certain period.
http://www.utmc.gov.uk/research/index.htmhttp://www.utmc.gov.uk
http://www.its-assist.org.uk/links.htm
http://www.trafficlinq.com/its.htm
http://www.eu-spirit.com/http://www.itsnetwork.org/
http://www.cordis.lu/telematics/tap_transport/home.html

1. Directive2010/40/EuoftheEuropeanParliamentandoftheCouncilof7July 2010.eur- lex.europa.eu (http://eur-lex.europa.eu/LexUriServ/LexUriServ.do?uri=OJ:L:2010:207:0001:0013:EN:PDF)
2. "Reducing delay due to traffic congestion. [Social Impact]. ITS. The Intelligent Transportation Systems Centre and Test bed". SIOR, Social Impact Open Repository. Archived from the original on 2017-09-05. Retrieved 2017-09-05. (https://web.archive.org/web/2017090

5144020/http://sior.ub.edu/jspui/cris/socialimpact/socialimpact00438) (http://sior.ub.edu/jspui/cris/socialimpact/s ocialimpact00438)

3. "Smart tech to end fwy congestion". PerthNow. 2020-07-07. Retrieved 2020-10- 07. (https://www.perthnow.com.au/news/transport/way-of-the-future-perths-47m-smart- freeway-to-be-switc hed-on-in-august-ng-b881601834z)
4. Monahan, Torin (2007). "War Rooms" of the Street: Surveillance Practices in Transportation Control Centres" (PDF). The Communication Review.10 (4): 367–389. CiteSeerX 10.1.1.691.8788.doi:10.1080/10714420701715456.S2CID44101831.(http://publicsurveillanc e.com/papers/war_rooms.pdf) (https://citeseerx.ist.ps u.edu/viewdoc/summary?doi=10.1.1.691.8788) (https://doi.org/10.1080%2F10714 420701715456) (https://api.semanticscholar.org/C orpusID:44101831)
5. "Frequently Asked Questions". Intelligent Transportation Systems Joint Program Office. United States Department of Transportation. Retrieved 10 November 2016. (http://www.its.dot.gov/faqs.htm)
6. Tarnoff, Philip John, Bullock, Darcy M, Young, Stanley E, et al. "Continuing Evolution of Travel Time Data Information Collection and Processing", Transportation Research Board Annual Meeting 2009 Paper #09-2030. TRB 88th Annual Meeting Compendium of Papers DVD
7. Mohan, Prashanth, Venkata N. Padmanabhan, and Ramachandran Ramjee. *Nericell: rich monitoring of road and traffic conditions using mobile smart phones.* Proceedings of the 6th ACM Conference on Embedded Network Sensor Systems. ACM, 2008.
8. Ahmed, Hazem; EL-Darieby, Mohamed; Abdulhai, Baher; Morgan, Yasser (2008-01-13). "Bluetooth-and Wi-Fi-Based Mesh Network Platform for Traffic Monitoring". *Transportation Research Board 87th Annual Meeting*. (https://trid.trb.org/view.aspx?id=8482 40)
9. Tyagi, V., Kalyanaraman, S., Krishnapuram, R. (2012). "Vehicular Traffic Density State Estimation Based on Cumulative Road Acoustics". *IEEE Transactions on Intelligent Transportation Systems* 13
10. (3):1156–1166.doi:10.1109/TITS.2012. 2190509.S2CID14434273.(https://doi.org/1
11. 0.1109%2FTITS.2012.2190509) (https://api.semanticscholar.org/CorpusID:14434273)
12. "Comparison of Stopped Vehicle Detection (SVD) Technologies for Smart Motorway Applications."
13. *Ogier Electronics*. Retrieved 4 May 2020. (http://ogierelectronics.com/comparison-of-svd- technology.php)

14. "Smart motorway evidence stock take and action plan". *GOV.UK*. Retrieved 2020-04- (https://www.gov.uk/government/ publications/smart-motorway-evidence-stocktake-andaction- plan)
15. Joshi, V., Rajamani, N., Takayuki, K., Prathapaneni, N., Subramaniam, L.V. (2013). *Information Fusion-Based Learning for Frugal Traffic State Sensing*. Proceedings of the Twenty-Third International Joint Conference on Artificial Intelligence.
16. Keating, Dave. "All Cars In Europe Can Now Call The Police Themselves".*Forbes*. Retrieved 2019- 09-25. (https://www.forbes.com/sites/davekeating/2018/04/01/ starting-today-all-cars-in- europe-can-call-the-police-themselves/)
17. Gleeson, Colin. "New cars to automatically inform authorities of crashes". The Irish Times. Retrieved2019-09-25. (https://www.irishtimes.com/news/ireland/irish-news/new-cars-to-automatically-inform-authorities-of-crashes-1.3447079)
18. *Report (HC15,2004, 05): Tackling congestion by making better use of England's motorways and trunk roads (Full Report)* (PDF), National Audit Office, 26 November 2004, retrieved 2009-09- 17 (http://www.nao.org.uk/publications/nao_reports/04-05/040515.pdf)
19. TrendinRoadAccidents,JapanArchived2009-05-21attheWaybackMachine.nilim.go.jp (http://www.nilim.go.jp/english/conference/asia2006/8-6.pdf) (https://web.archive.org/web/20090 521071120/http://www.nilim.go.jp/english/conference/asia2006/8-6.pdf)
20. 3rd Safety Forum, 25 March 2004
21. European Commission, Directorate-General “Information Society”, Directorate C “Miniaturisation, Embedded Systems and Societal Applications”, Unit C.5“ ICT for Transport and the Environment”, "Towards Cooperative Systems for Road Transport", Transport Clustering Meeting, 8 Nov. 2004. (http://cordis.europa.eu/pub/ist/docs/dir_c/trans/towards_coop_sys_road_trans_en.pdf)
22. "ITS World Congress". *Promotional website*. Retrieved 10 November 2016. (http://itsworldcongress.com/)
23. "Introducing the Network of National ITS Associations!". *Promotional website*. Retrieved 10 November 2016. (http://itsnetw ork.org/)

External Web Links

- Intelligent transportation system at Curlie (https://curlie.org/Science/Technology/Transportation/Intelligent_Systems/)
- ITS Handbook available for free download
- U. S. Department of Transportation–Intelligent Transportation Systems Joint Program Office website
- Guide to Federal ITS Research – U.S. Department of Transportation (http://road-network-operations.piarc.org/)(http://www.its.dot.gov)(http://its.dot.gov/factsheets/pdf/ITS_MiniFactBooklet_V11.pdf)
- ITS Safety Applications Factsheet–U. S. Department of Transportation
- SMART Highways Magazine
- ISO standardization of intelligent transport systems
- CEN standardization of intelligent transport systems (http://its.dot.gov/factsheets/pdf/JPO_SafetySolutions_v3.pdf)(http://smarthighways.net/)(https://www.iso.org/committee/54706.html)(https://www.itsstandards.eu/)
- Traffic Signage for Variable Speed Limits. Readers should have a look at the web page by the following link: http://europa.eu.int/comm/dgs/energy_transport/galileo/index_en.htm,
- Parking Management (CONVERGE project– see the web link as mentioned below), http://www.cordis.lu/telematics/tap_transport/research/projectsum/converge.html). Speed Detection and Improvement of Road Safety Point speed cameras (http://www.dft.gov.uk/stellent/groups/dft_rdsafety/documents/downloadable/dft_rdsafety_610816.pdf - page 4.)
  Intelligent Speed Adaptation (ISA)
  http://www.rws-avv.nl/servlet/page?_pageid=121&_dad=portal30&_schema=PORTAL30 &p_folder_id=7737.
- Multimodal Trip Planning
  www.transportdirect.info,http://www.travel-andtransport.comorwww.ns.nl,and
  http://www.utmc.gov.uk/research/index.htm

The study mentions a number of multimodal portals including http://www.utrecht.nl/smartsite.dws?id=13776&mw=1003&w=18&p=3352&parFrom=3352&infFrom=3352 (in Dutch), http://www.bayerninfo.de/index_e.html(in English), http://www.smarttrek.org/index.html(in English, Seattle area) and

- UTMC Systems (Sourced from http://www.utmc.gov.uk/guidance/index.htm(UKDfT))
- ITS implemented for Access Control and Road Pricing http://www.stockholmsforsoket.se/templates/page.aspx?id=2453
- ITSinactionformultimodaltrip planning

FASTag Systems

References:

1. FASTagPro:Recharge any FASTag online without login in 2minutes https://fastagpro.com/recharge
2. RFID technology and what role does it play in FASTag https://www.financialexpress.com/auto/
3. Theory and Applications of Economics in Highway and Transport Planning, by Dr Vinay Maitri, Dr P. K. Sarkar, Standard Publishers Distributors

Chapter 9 Green Highways

References:

Green Highways (Plantation, Transplantation, Beautification & Maintenance) Policy–2015

Use of Compressed Natural Gas (CNG) as fuel by Vehicles in Green Highways

1. There are some limited prospects in the rail sector, although the volumes are not likely to be material.
2. The currency of data for each country varies; in most cases, numbers are for the years 2016 to 2018.
3. This is an approximation based on global fleet numbers provided by the International Organization of Motor Vehicle Manufacturers, OICA. http://www.oica.net/category/vehicles-in-use/ and NGV Global.

4. Armenia has an even higher level of saturation at 69 per cent, though the total number of NGVs is only 300,000.
5. Energy Aspects (forthcoming).
6. The data is from the IEA's World Energy Balance Master File, which contains detailed statistics underpinning the WEO.
7. 1 m^3 of gas=1.032 litres of diesel.
8. 1 litre of diesel=3.32 GGE
9. Analysis by Energy Insights (forthcoming) suggests that gas consumption in China in 2017 was around 33 Bcm. Whilst this number includes gas consumption in the shipping sector, it suggests an even higher average consumption figure for road vehicles.
10. See, for example, https://www.transportenvironment.org/publications/natural-gas-vehicles-%E2%80%93-road-nowhere
11. 45 per cent arising from production, processing and liquefaction, 32 per cent from transmission, distribution and storage, 15 per cent from LNG feedstock transportation, and eight per cent from dispensing. The WTT element can vary significantly with different gas sources.
12. https://www.dieselnet.com/standards/eu/hd.php.
13. https://www.ngvglobal.com/blog/vw-utilise-100-scania-lng-trucks-greener-logistics-0920.
14. See for example https://www.drax.com/technology/can-electricity-power-heavy-duty-vehicles/and https://www.ukhaulier.co.uk/news/road-transport/trucks/fully-battery-electric-driven-truck-for-heavy-duty-distribution-mercedes-benz-eactros-starts-in-the-murg-valley-emission-free-and-quiet-transportation/./
15. Spiersetal 2019 and Transport & Environment 2018.
16. In 2017 astral NP4×2 tractor unit pulling a tri-axle box van trailer and running at a gross vehicle weight of 30 metric tons has just completed a 1,728 km road journey without needing to refuel.
17. CNG tanks cannot fit on a 6x2 tractor unit–see https://www.gasrec.co.uk/cng-lng/
18. https://www.ngvglobal.com/blog/uk-fuel-differential-freeze-a-boost-for-ngv-hgvs-1110#more-56245.
19. https://www.edie.net/news/6/John-Lewis-and-Waitrose-to-power-delivery-fleet-on-bio-methane-by-28-green-gb-week/.
20. This is based on a $27,000/year for regular diesel and $11,500/year for LNG at about $7.60/MMBtu. (Article in Natural Gas World, 1.10.18)

21. http://www.ngvglobal.com/blog/uk-bus-operator-first-to-run-cng-minibuses-1107#more-56161.
22. https://ec.europa.eu/inea/en/connecting-europe-facility/cef-transport.
23. https://www.lngworldnews.com/eu-funds-german-lng-fueling-project/./
24. https://www.naturalgasworld.com/gasum-to-develop-swedish-lbg-unit-65004.
25. http://www.ngvglobal.com/blog/natural-gas-for-trucks-promoted-by-german-transport-ministry- 0617#more-54069.

Electrical Vehicles Integrated with Renewable Energy Sources for Sustainable Mobility in Green Highways

https://www.intechopen.com/chapters/60938

Author details

Michel aLongo1*, Federica Foiadelli 1 and WahibaYaïci 2

*Address all correspondence to:michela.longo@polimi.it

1. Department of Energy, Politecnico di Milano, Milan, Italy (ii) Canmet ENERGY Research Centre, Natural Resources Canada, Ottawa, Ontario, Canada
2. Environment and Climate Change Canada. Canadian Environmental Sustainability Indicators, Greenhouse Gas Emissions. 2016. Available from: www.ec.gc.ca/indicateursindicators/default.asp?lang=en&n=FBF8455 E [Accessed: 2018-02-20]
3. Government of Canada. Canada's Way Forward on Climate Change: The Paris Agreement. 2016. Available from: http://climatechange.gc.ca/default.asp?lang=En&n=24700154-1 [Accessed: 2018-02-20]
4. Office of the Parliamentary Budget Officer. Canada's Greenhouse Gas Emissions: Developments, Prospects and Reductions. 2016. Available from:http://www.pbo-pb.gc.ca/web/default/files/Documents/Reports/2016/ClimateChange/PBO_Climate_Change_EN.pdf[Accessed: 2018-02-20]
5. European Commission. A European Strategy for Low-Mission Mobility. Brussels. 2016.Available from: https://ec.europa.eu/clima/policies/transport_en

[Accessed: 2018-03-02]
6. Thomas CES. Transportation options in a carbon-constrained world: Hybrids, plug-in hybrids, biofuels, fuel cell electric vehicles, and battery electric vehicles. International Journal of Hydrogen Energy. 2009; 34:9279-9296. DOI: 10.1016/j.ijhydene.2009.09.058
7. Richardson DB. Electric vehicles and the electric grid: A review of modeling approaches, impacts, and renewable energy integration. Renewable and Sustainable Energy Reviews. 2013; 19:247-254. DOI: 10.1016/j.rser.2012.11.042
8. Mwasilu F, Justo JJ, Kim E-K, DoTD, Jung J-W. Electric vehicles and smart grid interaction: A review on vehicle-to-grid and renewable energy sources integration. Renewable and Sustainable Energy Reviews. 2014; 34:501-516. DOI: 10.1016/j.rser.2014.03.031
9. Liu L, Kong F, Liu X, Peng Y, Wang Q. A review on electric vehicles interacting with renewable energy in smart grids. Renewable and Sustainable Energy Reviews. 2015;51: 648-661. DOI:10.1016/j.rser.2015.06.036
10. European Commission. Community Research and Development and Development Information Service (CORDIS), eMobilita, Electromobility in urban transport: A multidimensional innovation (socio-economic and environmental effects). Available from: https://cordis.europa.eu/project/rcn/207049_en.html[Accessed: 2018-03-25] Electric Vehicles Integrated with Renewable Energy Sources for Sustainable Mobility http://dx.doi.org/10.5772/intechopen.76788
11. European Commission, Community Research and Development and Development Information Service (CORDIS), Ne Mo: Hyper-Network for electro Mobility. Available from: https://cordis.europa.eu/project/rcn/204973_en.html[Accessed: 2018-03-25]
12. European Commission, Community Research and Development and Development Information Service (CORDIS), OSEM-EV (Optimised and Systematic Energy Management in Electric Vehicles). Available from: http://cordis.europa.eu/project/rcn/194883_en.html [Accessed: 2018-03-25]
13. European Environment Agency. Electric vehicles and the energy sector—Impacts on Europe's future emissions. 2016. Available from: https://www.eea.europa.eu/themes/transport/electric-vehicles/electric-vehicles-and-energy [Accessed: 2018-02-20]
14. Tan KM, Ramachandara murthy VK, Yong JY. Integration of electric vehicles in smart grid: A review on vehicle-to-grid technologies and

optimisation techniques. Renewable and Sustainable Energy Reviews. 2016;53: 720-732. DOI: 10.1016/ j. rser. 2015.09.012

15. Dubarry M, Devie A, McKenzie K. Durability and reliability of electric vehicle batteries under electric utility grid operations: Bidirectional charging impact analysis. Journal of Power Sources. 2017; 358:39-49. DOI:10.1016/ j. jpowsour.2017.05.015
16. Shaukata N, Khan B, Ali SM, Mehmood CA, Khan J, Farid U, Majid M, Anwar SM, Jawad M, Ullah Z. A survey on electric vehicle transportation within smart grid system. Renewable and Sustainable Energy Reviews. 2018; 81:1329-1349. DOI: 10.1016/j.rser.2017.05.092
17. Pinson P, Christensen LEA, Madsen H, Sørensen PE, Donovan MH, Jensen LE. Regime-switching modelling of the fluctuations of offshore wind generation. Journal of Wind Engineering & Industrial Aerodynamics. 2008;96(2):2327-2347. DOI: 10.1016/j.jweia.2008.03.010
18. Jabr RA,Pal BC. Intermittent wind generation in optimal power flow dispatching. IET Generation, Transmission & Distribution. 2009; 3(1): 66-74. DOI:10.1049/iet-gtd:20080273
19. Bollen MHJ, Zhong J, Zavoda F, Meyer J, McEachern A, Corcoles Lopez F. Power quality aspects of smart grids. In: Proceedings of International Conference on Renewable Energies and Power Quality (ICREPQ'10); 23-25 March 2010; Granada, Spain. DOI: 10.24084/repqj08.58
20. Kempton W, Letendre SE. Electric vehicles as a new power source for electric utilities. Transportation Research Part D: Transport and Environment. 1997;2(3):157-175. DOI: 10. 1016/S1361-9209(97)00001-1
21. Clement-Nyns K, Haesen E, Driesen J. The impact of charging plug-in hybrid electric vehicles on a residential distribution grid. IEEE Transactions on Power Systems. 2010; 25(1):371-380. DOI: 10.1109/TPWRS.2009.2036481
22. Kempton W, Tomic J. Vehicle-to-grid power implementation: From stabilizing the grid to supporting large-scale renewable energy. Journal of Power Sources. 2005;144(1):280-294. DOI: 10.1016/j.jpowsour.2004.12.022, New Trends in Electrical Vehicle Powertrains
23. Kempton W, Udo V, Huber K, Komara K, Letendre S, Brunner SD, Pearre N. A test of vehicle-to-grid (V2G) for energy storage and frequency regulation in the PJM system. Technical Report, MAGIC Consortium, Jan. 2009. Available at:

http://www.magicconsortium.org/_Media/test-v2g-in-pjm-jan09.pdf [Accessed: 2018-02-20]
24. Pillai JR, Bak-Jensen B. Integration of vehicle-to-grid in the western Danish power system. IEEE Transactions on Sustainable Energy. 2011;2(1):12-19. DOI: 10.1109/TSTE.2010. 2072938
25. Galus MD, Koch S, Andersson G. Provision of load frequency control by PHEVs, controllable loads, and a cogeneration unit. IEEE Transactions on Industrial Electronics. 2011;58(10):4568-4582. DOI:10.1109/TIE.2011.2107715
26. Hay C, Togeby M, Bang NC, Søndergren C, Hansen LH. Introducing electric vehicles into the current electricity markets. Project Report, EDISON Deliverable D2.3. May, 2010
27. Energinet. dk. Energinet. DK's ancillary services strategy. Aug. 2011. Available from: www.energinet.dk [Accessed: 2018-02-20]
28. Lopes JAP, Almeida PMR, Soares FJ. Using vehicle-to-grid to maximise the integration of intermittent renewable energy resources in islanded electric grids. In: Proceedings of the International Conference on Clean Electrical Power (ICCEP '09); 9-11 June 2009; Capri, Italy. DOI: 10.1109/ICCEP.2009.5212041
29. Lopes JAP, Polenz SA, Moreira CL, Cherkaoui R. Identification of control and management strategies for LV unbalanced microgrids with plugged-in electric vehicles. Electric Power Systems Research. 2010;80(1):898-906. DOI: 10.1016/j.epsr.2009.12.013
30. Bessa RJ, Matos MA. Economic and technical management of an aggregation agent for electric vehicles: A literature survey. International Transactions on Electrical Energy Systems. 2011; 22:334-350. DOI:10.1002/etep.565
31. Birnie D. Solar-to-vehicle (S2V) systems for powering commuters of the future. Journal of Power Sources. 2009;186(2):539-542. DOI: 10.1016/j.jpowsour.2008.09.118
32. Bollen M, Hassan F. Integration of distributed generation in the power system. Chapter 5: Voltage Magnitude Variations. 5th ed. John Wiley & Sons Inc.; 2011. DOI: 10.1002/9781118029039
33. Corfee K, Korinek D, Cassel W, Hewicker C, Zillmer J, Pereira Morgado M, Ziegler H, Tong N, Hawkins D, Cernadas J. Distributed generation in Europe–Physical infrastructure and distributed generation connection, Memo. Available from: http://www.clean-coalition.org/site/wp-content/uploads/2012/11/Memo-1_Physical-Infrastructureand-DG-Interconnection.pdf[Accessed: 2018-02-20]

34. Etherden N, Bollen MHJ. Increasing the hosting capacity of distribution networks by curtailment of renewable energy sources. In: Proceedings of PowerTech;19-23June2011; Trondheim, Norway. DOI: 10.1109/PTC.2011.6019292 Electric Vehicles Integrated with Renewable Energy Sources for Sustainable Mobility http://dx.doi.org/10.5772/intechopen.76788
35. Carvalho P, Correia P, Ferreira L. Distributed reactive power generation control for voltage rise mitigation in distribution networks. EEE Transactions on Power Systems. 2008;23(2):766-772. DOI: 10.1109/TPWRS.2008.919203
36. Tonkoski R, Lopes LAC, El-Fowly T. Coordinated active power curtailment of grid-connected PV inverters for overvoltage prevention. IEEE Transactions on Sustainable Energy. 2011;2(2):139-147. DOI:10.1109/TSTE.2010.2098483
37. Fawzy T, Premm D, Bletterie B, Gorsek A. Active contribution of PV inverters to voltage control–From a smart grid vision to full-scale implementation. e & i Elektrotechnik und Informations technik. 2011;128(4):110-115. DOI: 10.1007/s00502-011-0820-z
38. Demirok E, Casado Gonzalez P, Frederiksen KHB, Sera D, Rodriguez P, Teodorescu R. Local reactive power control methods for overvoltage prevention of distributed solar inverters in low-voltage grids. IEEE Journal of Photovoltaics. 2011;1(2):174-182. DOI: 10.1109/JPHOTOV.2011.2174821
39. Marra F, Fawzy YT, Bülo T, BlažicBlazic B. Energy storage options for voltage support in low-voltage grids with high penetration of photovoltaic. In: Proceeding 2012 3rd IEEE PES International Conference and Exhibition on Innovative Smart Grid Technologies (ISGTEurope '13); 14-17October2012; Berlin. DOI:10.1109/ISGTEurope 2012.6465690
40. Liu X, Aichhorn A, Liu L, Li H. Coordinated control of distributed energy storage system with tap changer transformers for voltage rise mitigation under high photovoltaic penetration. IEEE Transactions on Smart Grid. 2012;3(2):897-906. DOI: 10.1109/TSG.2011. 2177501
41. Kempton W, Tomic J, Letendre SE, Brooks A, Lipman T. Vehicle to grid power: Battery, hybrid, and fuel cell vehicles as resources for distributed electric power in California. Working Paper Series ECD-ITS- RR-01-03, Jun. 2001. Available from: http://escholarship.org/uc/item/5cc9g0jp[Accessed: 2018-02-20]
42. Commission of the European Communitie. Regulation (EEC) No 4064/89 Merger Procedure. Mar. 1999. [Online]. Available from: www.ec.europa.eu [Accessed: 2018-02-20]

43. Galus MD, Zima M, Andersson G. On the integration of plug-in hybrid electric vehicles into existing power system structures. Energy Policy. 2010;38(11):6736-6745. DOI: 10.1016/j.enpol.2010.06.043
44. Fell K, Huber K, Zink B. Assessment of plug-in electric vehicle integration with ISO/RTO systems. Technical Report, Mar. 2010. Available from: http://www.isorto.org/atf/cf/%7B5B4E85C6-7EAC-40A0-8DC3003829518EBD%7D/IRC_Report_Assessment_of_Plugin_Electric_Vehicle_Integration_with_ISORTO_Systems_03232010.pdf [Accessed:2018-02-20]
45. Brooks A, Gage T. Integration of electric drive vehicles with the electric power grid—A new value stream. In: 18th Internat. Electric Vehicle Symposium and Exhibition, Oct. 2001, New Trends in Electrical Vehicle Powertrains
46. Bessa RJ, Matos MA, Soares FJ, Lopes JAO. Optimised bidding of an EV aggregation agent in the electricity market. IEEE Transactions on Smart Grid. 2012;3(1):443-452. DOI: 10.1109/TSG.2011.2159632
47. Lopes J A P, Soares F J, Almeida P M R. Integration of electric vehicles in the electric power system. In: Proceedings of the IEEE, 04 October 2010; pp.168-183.DOI: 10.1109/JPROC.2010.2066250
48. Galus, M D, Andersson, G. Demand management of grid-connected plug-in hybrid electric vehicles. In: Proceedings of the Energy 2030 Conference (ENERGY2008);17- 8 November 2008; Atlanta, GA, USA. DOI:10.1109/ENERGY.2008.4781014
49. Brenna M, Foiadelli F, Longo M. The exploitation of vehicle-to-grid function for power quality improvement in a smart grid. IEEE Transactions on Intelligent Transportation Systems. 2014;15(5):2169-2177. DOI: 10.1109/TITS.2014.2312206
50. Kanchev H, Di Lu F, Colas V, Lazarov B. Energy management and operational planning of a microgrid with a PV-based active generator for smart grid applications. IEEE Transactions on Industrial Electronics. 2011;58(10):4583-4592. DOI: 10.1109/TIE.2011.2119451
51. Liserre M, Sauter T, Hung JY. Future energy systems, integrating renewable energy sources into the smart power grid through industrial electronics. IEEE Industrial Electronics Magazine. 2010;4(1):18-37. DOI: 10.1109/MIE.2010.935861 Electric Vehicles Integrated with Renewable Energy Sources for Sustainable Mobility http://dx.doi.org/10.5772/intechopen.76788

Utilizing recycled materials by avoiding disposals in Eco-friendly Green Highways Construction

Robin L. Schroeder has been with the Federal Highway Administration for 15 years. For the past three years, he has worked in the Materials Branch of the Construction and Maintenance Division in Washington, D.C., where he is extensively involved with the use of recycled materials in highway construction. In that capacity, he was the primary coordinator of a symposium that was held last year in Denver, Colo., on the use of recycled materials in highway construction. He also helped to develop a report to Congress concerning recycled materials as required by Section 1038 of ISTEA. He earned a bachelor's degree in civil engineering from Oregon State University, and he is a registered professional engineer in the state of Washington.

Primary Source: TRBWeekly, https://trid.trb.org/view/413914
RecordURL:http://www.fhwa.dot.gov/publications/publicroads/94fall/p94au32.cfm
Availability: Find a library where documents are available. Order URL:http://worldcat.org/oclc/1586080
Corporate Authors: Federal Highway Administration
1200 New Jersey Avenue, S E, Washington, DC, United States 20590

1. R. J. Collins and S. K. Ciesielski. Recycling and Use of Waste Materials and Byproducts in Highway Construction, Volumes 1 & 2, 1993.
2. Imtiaz Ahmed. Use of Waste Materials in Highway Construction, Report No. FHWA/IN/JHRP-91/3,1991.
3. Cheryl Solomon. Slag--Iron and Steel, Annual Report, U.S. Department of the Interior, Bureau of Mines, Washington, D.C., 1992.
4. K. Sakai, et al." Properties of Granulated Blast-Furnace Slag Cement Concrete, "Proceedings of the Fourth International Conference on Fly Ash, Silica Fume, Slag and Natural Pozzolans in Concrete, Istanbul, Turkey, May 1992
5. S. Nagataki, et al. "Properties of Concrete Using Newly Developed Low-Heat Cements and Experiments with Mass Concrete Model, "Proceedings of the Fourth International Conference on Fly Ash, Silica Fume, Slag and Natural Pozzolans in Concrete, Istanbul, Turkey, May 1992.
6. K. Togawa, et al. "Study on the Effects of Blast-Furnace Slag on Properties of No-Slump Concrete Mixtures, "Proceedings of the Fourth International Conference on Fly Ash, Silica Fume, Slag and Natural Pozzolans in Concrete, Istanbul, Turkey, May 1992.
7. American Association of State Highway and Transportation Officials, Subcommittee on Construction, Quality Construction Task Force. "Use

of Waste Materials in Highway Construction," unpublished report, August 1993.
8. Availability of Mining Wastes and Their Potential for Use as a Highway Material--Executive Summary, Publication No. FHWA-RD-78-28, Federal Highway Administration, Washington, D.C., September 1977.
9. Tim Ramirez. 79-12 Steel Slag Aggregate in Bituminous Mixtures--Final Report, Pennsylvania Department of Transportation, 1992.
10. Characterization of Municipal Solid Waste in the United States: 1992 Update, Executive Summary. Report No. EPA/530-S-92-019, Environmental Protection Agency, Washington, D.C., 1992.
11. Engineering and Environmental Aspects of Recycled Materials for Highway Construction, Volume I: Final Report. FHWA Contract No. DTFH61-92-C-00060, Federal Highway Administration, Washington, D.C., 1993.
12. "Recycled plastic finds home in asphalt binder, "Roads & Bridges, March 1993, pp. 41-47.
13. Engineering News Record, February 22, 1993, p. 42.
14. Anton Horner. "Use of Scrap Tire Rubber in Carsonite Noise Barrier, "Symposium Proceedings - Recovery and Effective Reuse of Discarded Materials and Byproducts for Construction of Highway Facilities, October 1993.
15. Peter Eng. Alberta Transportation and Utilities, Research and Development Report No. ABTR/RD/TM-92/09, Project No. 91021, June 1993.
16. "'Gl asphalt' utilization dependent upon availability. "Roads & Bridges, February 1993, pp. 59-61.
17. Clean Washington Center. Using Recycled Glass as a Construction Aggregate: A Summary of the Glass Feedstock Evaluation Project. Seattle, 1993.
18. George Dewey, et al. "Municipal Waste Combustor Ash as an Aggregate Substitute in Bituminous Mixture, "Symposium Proceedings - Recovery and Effective Reuse of Discarded Materials and Byproducts for Construction of Highway Facilities, October 1993.
19. Three-Year Results on the Performance of Incinerator Residue in a Bituminous Base, Publication No. FHWA-RD-78- 144, Federal Highway Administration, Washington, D.C., 1978.
20. Macro Pasetto. "Cement Stabilization of Urban Solid Waste Slags and Ashes in Road Base and Sub-Base Layers, "Symposium Proceedings-Recovery and Effective Reuse of Discarded Materials and Byproducts for Construction of Highway Facilities, October 1993.

21. North Carolina Department of Transportation Materials and Tests Unit. "A Laboratory Evaluation on the Effects of Ground Tire Rubber on Strength Performance of Concrete, "Symposium Proceedings - Recovery and Effective Reuse of Discarded Materials and Byproducts for Construction of Highway Facilities, October 1993.
22. Dana N. Humphrey and Robert A. Eaton. "Tire Chips as a Subgrade Insulation—Field Trial, "Symposium Proceedings Recovery and Effective Reuse of Discarded Materials and Byproducts for Construction of Highway Facilities, October 1993.
23. Youjiang Wang, et al. "Fiber Reinforced Concrete Using Recycled Carpet Industrial Waste and Its Potential Use in Highway Construction," Symposium Proceedings - Recovery and Effective Reuse of Discarded Materials and Byproducts for Construction of Highway Facilities, October 1993.
24. Jeffery Groom, et al. "Use of Waste Nylon Fibers in Portland Cement Concrete to Reduce Plastic Shrinkage Cracking, "Symposium Proceedings - Recovery and Effective Reuse of Discarded Materials and Byproducts for Construction of Highway Facilities, October 1993.
25. Gregory Scot Gordon, et al. "Use of Waste Fibers in Asphalt Concrete, "Symposium Proceedings - Recovery and Effective Reuse of Discarded Materials and Byproducts for Construction of Highway Facilities, October 1993.
26. J.D. Brock."From Roofing Shingles to Roads,"TechnicalPaperT-20,1990.
27. Mary Stroup-Gardiner, et al." Permanent Deformation and Low Temperature Behaviour of Roofing-Modified HMA, "Symposium Proceedings - Recovery and Effective Reuse of Discarded Materials and Byproducts for Construction of Highway Facilities, October 1993.
28. "Shingle Scrap in Asphalt Concrete," unpublished report, Study No. 9PR1010, Minnesota Department of Transportation, 1991.
29. Donald Saylak, et al. "Applications for FGD Byproduct Gypsum, "Symposium Proceedings - Recovery and Effective Reuse of Discarded Materials and Byproducts for Construction of Highway Facilities, October 1993.
30. "Fly ash sets standard for recycled material use, "Roads&Bridges, November 1992, pp.50-56.
31. David Q. Hunsucker, et al. "Evaluating the Use of Ponded Fly Ash in Roadway Base Course, "Symposium Proceedings - Recovery and Effective Reuse of Discarded Materials and Byproducts for Construction of Highway Facilities, October 1993.

32. A.R. Dawson, et al. "Some British Experience of the Behavior of Furnace Bottom Ash and Slate Waste for Pavement Foundations, "Symposium Proceedings - Recovery and Effective Reuse of Discarded Materials and Byproducts for Construction of Highway Facilities, October 1993.
33. E. Eleni Vassiladou, et al. "Utilization of Fly and Bottom Ash as a Partial Fine Aggregate Replacement in Asphalt Concrete Mixtures, "Symposium Proceedings - Recovery and Effective Reuse of Discarded Materials and Byproducts for Construction of Highway Facilities, October 1993.
34. William Wolfe, et al. "Truck Ramp Construction from Clean Coal Technology Waste Products, "Symposium Proceedings - Recovery and Effective Reuse of Discarded Materials and Byproducts for Construction of Highway Facilities, October 1993.

Chapter 10 Maintenance of Safe Roads and Bridges

References:

1. Transportation Engineering & Planning, by Papacostas, P. D. Prevedouros, Prentice Hall, India
2. Manual for Road Investment Decision Mod Technical Papers, Consultants' Role for the Preparation of Project for the development of rural road infrastructure in India, by H. S. Bhatia, S. Goswami, Premal D. Mehta and Mrs. Rachna Chopra, for "Seminar on Integrated development Of Rural and Arterial Road Network for Socio-Economic Growth, Vol. I, 5-6 December 2003 by IRC".
3. Theory and Applications of Economics in Highway and Transport Planning, by Dr Vinay Maitri, Dr P. K. Sarkar, Standard Publishers Distributors
4. Traffic Engineering and Transport Planning, by Dr L. R. Kadiyali, Khanna Publishers
5. Road Development Plan Vision:2021, Ministry of Road Transport & Highways, Govt. Of India

Manual on Economic Evaluation of Highway Projects in India, IRC SP 30-1993, IRC

INDEX